BIOLOGICAL DIVERSITY: PROBLEMS AND CHALLENGES

The Pennsylvania Academy of Science Publications

Books and Journal

Editor: Shyamal K. Majumdar
Professor of Biology, Lafayette College
Easton, Pennsylvania 18042

1. *Energy, Environment, and the Economy,* 1981. ISBN: 0-9606670-0-8. Editor: S.K. Majumdar.
2. *Pennsylvania Coal: Resources, Technology and Utilization,* 1983. ISBN: 0-9606670-1-6. Editors: S.K. Majumdar and E.W. Miller.
3. *Hazardous and Toxic Wastes: Technology, Management and Health Effects,* 1984. ISBN: 0-9606670-2-4. Editors: S.K. Majumdar and E. W. Miller.
4. *Solid and Liquid Wastes: Management, Methods and Socioeconomic Considerations,* 1984. ISBN: 0-9606670-3-2. Editors: S.K. Majumdar and E.W. Miller.
5. *Management of Radioactive Materials and Wastes: Issues and Progress,* 1985. ISBN: 0-9606670-4-0. Editors: S.K. Majumdar and E.W. Miller.
6. *Endangered and Threatened Species Programs in Pennsylvania and Other States: Causes, Issues and Management,* 1986. ISBN: 0-9606670-5-9. Editors: S.K. Majumdar, F.J. Brenner, and A.F. Rhoads.
7. *Environmental Consequences of Energy Production: Problems and Prospects,* 1987. ISBN: 0-9606670-6-7. Editors: S.K. Majumdar, F.J. Brenner and E.W. Miller.
8. *Contaminant Problems and Management of Living Chesapeake Bay Resources,* 1987. ISBN: 0-9606670-7-5. Editors: S.K. Majumdar, L.W. Hall, Jr. and H.M. Austin.
9. *Ecology and Restoration of The Delaware River Basin,* 1988. ISBN: 0-9606670-8-3. Editors: S.K. Majumdar, E.W. Miller and L.E. Sage.
10. *Management of Hazardous Materials and Wastes: Treatment, Minimization and Environmental Impacts,* 1989. ISBN: 0-9606670-9-1. Editors: S.K. Majumdar, E.W. Miller and R.F. Schmalz.
11. *Wetlands Ecology and Conservation: Emphasis in Pennsylvania,* 1989. ISBN: 0-945809-01-8. Editors: S.K. Majumdar, R.P. Brooks, F.J. Brenner and R.W. Tiner, Jr.
12. *Water Resources in Pennsylvania: Availability, Quality and Management,* 1990. ISBN: 0-945809-02-6. Editors: S.K. Majumdar, E.W. Miller and R.R. Parizek.
13. *Environmental Radon: Occurrence, Control and Health Hazards*, 1990. ISBN: 0-945809-03-4. Editors: S.K. Majumdar, R.F. Schmalz and E.W. Miller.
14. *Science Education in the United States: Issues, Crises, and Priorities,* 1991. ISBN: 0945809-04-2. Editors: S.K. Majumdar, L.M. Rosenfeld, P.A. Rubba, E.W. Miller and R.F. Schmalz.
15. *Air Pollution: Environmental Issues and Health Effects,* 1991; ISBN: 0-945809-05-0. Editors: S.K. Majumdar, E.W. Miller, and J.J. Cahir.
16. *Natural and Technological Disasters: Causes, Effects and Preventive Measures*, 1992; ISBN: O-945809-06-9. Editors: S.K. Majumdar, G.S. Forbes, E.W. Miller, and R.F. Schmalz.
17. *Global Climate Change: Implications, Challenges and Mitigation Measures*, 1992; ISBN: 0-945809-07-7. Editors: S.K. Majumdar, L.S. Kalkstein, B, Yarnal, E.W. Miller, and L.M. Rosenfeld.
18. *Conservation and Resource Management,* 1993; ISBN: 0-945809-08-5. Editors: S.K. Majumdar, E.W. Miller, D.E. Baker, E.K. Brown, J.R. Pratt, and R.F. Schmalz.
19. *Biological Diversity: Problems and Challenges,* 1994; ISBN: 0945809-09-3. Editors: S.K. Majumdar, F.J. Brenner, J.E. Lovich, J.F. Schalles, and E.W. Miller.

BIOLOGICAL DIVERSITY: PROBLEMS AND CHALLENGES

EDITED BY

S.K. MAJUMDAR, Professor and Head
Department of Biology
Lafayette College
Easton, PA 18042

F.J. BRENNER, Professor
Department of Biology
Grove City College
Grove City, PA 16127

J.E. LOVICH, Wildlife Biologist
U.S. Department of the Interior
National Biological Survey
Palm Springs Field Station
North Palm Srings, CA 92258

J.F. SCHALLES, Associate Professor of Biology
Environmental Science Program Director
Creighton University
Omaha, Nebraska 68178

E.W. MILLER, Professor and Associate Dean (Emeritus)
The Pennsylvania State University
University Park, PA 16802

Founded on April 18, 1924

A Publication of
The Pennsylvania Academy of Science

Library of Congress Cataloging in Publication Data

Bibliography
Index
Majumdar, Shyamal K. 1938-, ed.

Library of Congress Catalog Card No.: 93-87463

ISBN-0-945809-09-3

A Publication of
The Pennsylvania Academy of Science, Easton, PA 18042

Printed in the United States of America by

Typehouse of Easton
Easton, PA 18042

PREFACE

There are few places on Earth that have not been modified by man. In the process of utilizing the Earth, the global environments have been greatly altered. As a consequence, many species of plants and animals no longer exist. Currently, we are losing plant and animal species at an accelerating rate, mainly due to human over-population and global economic activity. Uncontrolled growth, according to many experts, could threaten the survival of many species, including humans. The importance of these changes are only now becoming evident. Species that have evolved over millions of years must be preserved for the future welfare of the world's population. Biodiversity has become a topic of discussion by both the scientific community and the lay public. This book was developed by the Pennsylvania Academy of Science to address a variety of topics as they relate to biodiversity.

This volume, divided into four major parts, presents a comprehensive view of the worldwide problems facing biological diversity. It not only describes but analyzes a wide variety of ecosystems. Part I, consisting of five chapters, presents some of the recognized problems and developing policies that are being established to, if not stop, slow the decline and disappearance of plants and animals. More significant is the increase monitoring of ecosystems and the development of scientific procedures that hopefully will create a stable living environment.

Part II presents an evaluation of the changes and impacts of biodiversity. Statistical and theoretical aspects and diversity indices provide fundamental foundations. Specific aspects are then presented, such as pollutants in the aquatic system, deforestation, and green house warming. Part III is devoted to the role and implementation of programs. The problems of biological diversity are so complex that scientific consideration must be correlated with humanistic considerations. One of the major problems exists between scientific and management decisions and this topic is presented from a number of view points.

Part IV provides specific case studies of regional biodiversity. These examples are taken from a wide variety of environments. Of the ten chapters, two chapters each are devoted to vertebrates and insects, one to fish, and one to forests. Several of the studies concentrate on a region of the United States including such areas as Pennsylvania, Southeastern United States, Nebraska, Mississippi Delta, Great Lakes, and Pacific Northwest. Another study focuses on one region of Asia.

The co-editors of this volume wish to thank each of the authors for their excellent contribution to this volume. They have presented a perspective that is only possible from the scientists involved in research in a particular area. Because of its breadth of material, the volume will appeal not only to scientists, but to the individual who wants information on man's conflicts with the environment. The co-editors wish to thank their institutions for the space and facilities provided. Finally the co-editors express appreciation to their wives for their understanding of the time involved in such a project.

S.K. Majumdar
F.J. Brenner
J.E. Lovich
J.F. Schalles
E.W. Miller
Editors
February, 1994

ACKNOWLEDGEMENTS

The publication of this book was aided by contributions from The Pennsylvania Power and Light Company, Allentown, Pennsylvania and E. Willard and Ruby S. Miller book publication fund.

Biological Diversity: Problems and Challenges

Table of Contents

Preface V
Acknowledgements VI

Part One: Biodiversity: Definitions, Monitoring and Policies

Chapter 1: MONITORING AND PROTECTING BIOTIC DIVERSITY
C. Kenneth Dodd, Jr., National Biological Survey, Gainesville, FL 1

Chapter 2: BIODIVERSITY: A GLOBAL PERSPECTIVE
Gordon L. Kirkland, Jr., Director, Vertebrate Museum, Shippensburg University, Shippensburg, PA and Jay R. Stauffer, Jr., School of Forest Resources, The Pennsylvania State University, University Park, PA 12

Chapter 3: CONSERVATION OF BIODIVERSITY
Eugene P. Odum, Institute of Ecology, University of Georgia, Athens, GA 18

Chapter 4: U.S. POLICIES TO CONSERVE BIODIVERSITY
David E. Blockstein, Princeton University, Princeton Committee of the National Institute for the Environment, Washington, D.C. 26

Chapter 5: CONSERVATION OF BIODIVERSITY: THE INDIAN SCENARIO
Arun K. Sharma, University of Calcutta, Calcutta, India 41

Part Two: Biodiversity - Changes and Impacts

Chapter 6: STATISTICAL AND THEORETICAL ASPECTS OF DIVERSITY INDICES
Richard P. Steiner, Department of Mathematical Sciences, University of Ohio, Akron, OH 50

Chapter 7: DIVERSITY INDICES AND THEIR BIOLOGICAL IMPLICATIONS
Nancy McCreary Waters, Department of Biology, Lafayette College, Easton, PA 64

Chapter 8: ACID PRECIPITATION AND SURFACE WATER ACIDIFICATION: IMPLICATIONS FOR AQUATIC BIODIVERSITY
Patricia T. Bradt, Environmental Studies Center Laboratory, Lehigh University, Bethlehem, PA 76

Chapter 9: NATIVE PLANT BIODIVERSITY VS. THE INTRODUCED INVADERS: STATUS OF THE CONFLICT AND FUTURE MANAGEMENT OPTIONS
Curtis C. Daehler and Donald R. Strong, Center for Population Biology, University of California, Davis, CA 92

Chapter 10: IMPLICATIONS OF POLLUTANTS ON DIVERSITY OF AQUATIC BACTERIA AND FUNGI
Michelle A. Baker, Department of Biology, University of New Mexico, Albuquerque, NM; Jeffrey Merkle and S.K. Majumdar, Department of Biology, Lafayette College, Easton, PA 114

Chapter 11: TRIBUTYLTINS IN AQUATIC ECOSYSTEMS
Joseph J. Cooney, University of Massachusetts Harbor Campus, Boston, MA; James H. Weber, University of New Hampshire, Durham, NH; and Larry R. Sherman, University of Scranton, Scranton, PA 123

Chapter 12: GREENHOUSE WARMING: IMPACTS ON BIODIVERSITY
Miguel F. Acevedo, Department of Geography and Institute of Applied Sciences, University of North Texas, Denton, TX 136

Chapter 13: BIODIVERSITY AT RISK: DEFORESTATION
Mark S. Boyce, Department of Zoology & Physiology, University of Wyoming, Laramie, WY 151

Part Three: Rationale and Implementation Programs to Maintain and Increase Biodiversity

Chapter 14: MANAGING BIODIVERSITY: CONSERVATION VS. PRESERVATION
Robert L. Burgess, SUNY College of Environmental and Forestry, Department of Environmental and Forest Biology, Syracuse, NY 162

Chapter 15: PROTECTION OF BIOLOGICAL DIVERSITY: MISSING CONNECTION BETWEEN SCIENCE AND MANAGEMENT
Christine Schonewald-Cox, National Biological Survey, Division of Environmental Studies, University of California, Davis, CA 170

Chapter 16: THE ROLE OF SCIENCE IN PRESERVING BIODIVERSITY IN NATIONAL PARKS
John Lemons, Department of Life Sciences, University of New England, Biddeford, ME 184

Chapter 17: ETHICAL AND PHILOSOPHICAL ASPECTS OF BIODIVERSITY CONSERVATION
William H. Murdy, Emory University-Oxford College, Oxford, GA 199

Chapter 18: THE ROLE OF WILDLIFE MANAGEMENT IN THE MAINTENANCE OF BIODIVERSITY
William E. Magnusson, Department of Ecology, INPA, CP478, Brazil, South America 209

Chapter 19: POSSIBLE ROLE OF REPETITIVE DNA ELEMENTS IN PLANT DIVERSITY
Rajat K. Chaudhuri and Ila Chaudhuri, Molecular Biology Laboratory, Botany Department, University of Calcutta, Calcutta, India 221

Chapter 20: THE ROLE OF BOTANIC GARDENS
James P. Folsom, Huntington Botanical Gardens, San Marino, CA 241

Chapter 21: THE ROLE OF ZOOS AND AQUARIUMS IN THE CONSERVATION OF BIODIVERSITY
Donald E. Moore, Thompson Park Conservancy, Watertown, NY and Robert Wiese, AAZPA Conservation Center, Bethesda, MD 253

Chapter 22: BIODIVERSITY AND ENDANGERED SPECIES RECOVERY PROGRAMS
Craig Johnson, Department of the Interior, Washington, D.C. 265

Chapter 23: INCREASING BIOLOGICAL DIVERSITY THROUGH RESTORATION OF DISTURBED ECOSYSTEMS
Fred J. Brenner, Grove City College, Grove City, PA 281

Chapter 24: SINGLE SPECIES VS. COMMUNITY AND ECOSYSTEM MANAGEMENT: THEIR IMPACT ON BIOLOGICAL DIVERSITY
William C. Hunter, U.S. Fish and Wildlife Service, Atlanta, GA 302

Part Four: Case Studies of Biodiversity

Chapter 25: DIVERSITY OF TERRESTRIAL VERTEBRATES IN PENNSYLVANIA
C.J. McCoy, Kenneth C. Parkes and Janis C. Sacco, The Carnegie Museum of Natural History, Pittsburgh, PA 315

Chapter 26: DIVERSITY STATUS OF TERRESTRIAL INSECTS
Theodore Burk, Department of Biology, Creighton University, Omaha, NE 326

Chapter 27: DIVERSITY STATUS OF AQUATIC INSECTS
Mel Zimmerman, Biology Department, Lycoming College, Williamsport, PA 344

Chapter 28: CHANGES IN AMPHIBIAN BIODIVERSITY ASSOCIATED WITH 25 YEARS OF PINE FOREST REGENERATION: IMPLICATIONS FOR BIODIVERSITY MANAGEMENT
Bruce W. Grant, University of Georgia, Savannah River Ecology Laboratory, Aiken, SC; and Kent L. Brown and Gary W. Ferguson, Department of Biology, Texas, Christian University, Fort Worth, TX; and J. Whitfield Gibbons, University of Georgia, Savannah River Ecology Laboratory, Aiken, SC 354

Chapter 29: FACTORS INFLUENCING FISH SPECIES DIVERSITY

IN THE GREAT LAKES AND EFFECT OF RECENT INTRODUCTIONS ON LAKE ERIE
John M. Campbell, Department of Biology, Mercyhurst College, Erie, PA and Roger B. Kenyon, Pennsylvania Fish Commission, Fairview, PA 368

Chapter 30: BIODIVERSITY AND ZOOGEOGRAPHY OF NON-MARINE TURTLES IN SOUTHEAST ASIA
Jeffrey E. Lovich, National Biological Survey, North Palm Springs, CA 380

Chapter 31: VERTEBRATES IN THE PACIFIC NORTHWEST: SPECIES RICHNESS, ENDEMISM AND DEPENDENCY ON OLD-GROWTH FORESTS
R. Bruce Bury, National Biological Survey, Fort Collins, CO 392

Chapter 32: BIODIVERSITY OF PRESQUE ISLE
John M. Campbell, Department of Biology, Mercyhurst College, Erie, PA; and James K. Bissel, Department of Botany, Cleveland Museum of Natural History, Cleveland, OH; and Harry N. Cunningham and Edwin C. Masteller, Penn State - The Behrend College, Erie, PA; and Jean Stull, Waterford, PA 405

Chapter 33: THE PLATTE RIVER: A CONFLICT BETWEEN A MANAGED WATER RESOURCE AND REGIONAL BIODIVERSTIY
John F. Schalles, Department of Biology, Creighton University, Omaha, NE 423

Chapter 34: BIODIVERSITY OF THE MISSISSIPPI DELTA: PAST, PRESENT AND FUTURE
Leonard A. Brennan, Department of Wildlife and Fisheries, Mississippi State, Mississippi 444

Index 456

Biological Diversity: Problems and Challenges. Edited by S.K. Majumdar, F.J. Brenner, J.E. Lovich, J.F. Schalles, and E.W. Miller.

Chapter One

MONITORING AND PROTECTING BIOTIC DIVERSITY

C. KENNETH DODD, JR

National Biological Survey
U.S. Fish and Wildlife Service
412 N.E. 16th Avenue, Room 250
Gainesville, FL 32601

INTRODUCTION

Biological diversity is a topic of current interest in both the scholarly and popular press. Biologists, government agencies, and the general public debate definitions, sampling procedures, management and protection programs, and consequences of the loss of biodiversity. The destruction of centers of biodiversity, such as tropical moist forests, coral reefs, and wetlands, receives much attention. However, nearly all ecosystems are under environmental stress from human-caused phenomena, and some ecosystems almost have disappeared, such as vernal pools in California and the old growth coastal longleaf pine (*Pinus palustris*) forests of the southeastern United States.[1] With the loss of each unique ecosystem, community, species, or population, the earth's biotic diversity decreases.

The goal of biodiversity monitoring and protection should be the maintenance of biotic communities and ecosystems rather than the protection of individual animals, plants, or even species. In this paper, the term biodiversity encompasses both the number of species within an area (species richness) and the number of individuals per species (species diversity), and includes only species normally found within an ecosystem in relatively undisturbed habitats. Other factors, such as the presence of exotic species (e.g., starlings, English sparrows, common carp, walking catfish, kudzu, and Russian thistle in North America) and fragmented ecosystems that lead to an increase in ecotonal habitats, can artificially increase measures of diversity. Both species richness and diversity[2] are important. Other sections of this book treat different facets of biological diversity. In this chapter, I briefly review procedures for monitoring and protecting communities from human-caused changes and impacts that negatively affect natural diversity. Persons needing more extensive information should consult primary source books (e.g., 3, 4, 5, 6, 7, 8) and journals, particularly *Biological Conservation, Conservation Biology, Biodiversity and Conservation,* and *The Natural Areas Journal.*

MONITORING CHANGES AND IMPACTS

What Should Be Monitored?

Finances, personnel, and time constraints often dictate the extent of monitoring. Noss[9] suggested an integrated approach to monitoring biodiversity that incorporates biotic composition, structure, and function at the regional landscape, community-ecosystem, population-species, and genetic levels. Ideally, biologists at least should monitor whole communities[10] or ecosystems for changes in structure and composition that might signal stress. In practice, the resources or expertise to conduct sampling and monitoring often are not available, and researchers must limit their activities spatially or taxonomically. Biologists must decide goals and objectives at the start of a monitoring program to ensure that sampling protocols are valid and that underlying assumptions are met.

Ecosystems and Communities. Substantial research funding and technological support are necessary to monitor regional landscapes, ecosystems, and communities. Large-scale monitoring is undertaken both directly in the field and with sophisticated technology, such as remote sensing. However, direct ecosystem sampling often is not necessary to determine the gross effects of large-scale habitat changes. For example, tropical moist forests throughout the world are centers of biotic diversity, especially for invertebrates.[11,12] If these forests are destroyed, one may expect a large-scale loss of biotic diversity even without knowing all the species that occur within them. Likewise, many governmental jurisdictions require the development of land management plans. If plans allow the destruction of unique or diverse biotic communities, biotic diversity also will be lost.

Field surveys are necessary to inventory habitats and to correlate species' distributions with habitat type. These baseline data complement information from remote sensing and land use planning datasets. Follow-up field surveys, sometimes termed "ground truthing," document centers of endemism and diversity that might be overlooked by remote sensing techniques, and verify the accuracy of available data. Although the Amazon rain forest appears structurally uniform, for example, there are distinct centers of diversity and endemism reflecting faunal and floristic changes throughout the Quaternary.[13] Field surveys also document subtle changes that occur in the absence of gross physical habitat change. For example, amphibian populations have declined in many parts of the world, particularly in high mountains. Declines occurred in the absence of apparent environmental stress and even in remote regions. If amphibian communities were not monitored, biologists would overlook the declines and fail to investigate underlying causes.

Sampling protocols that are less labor intensive or expensive than complete faunal surveys can be used to provide baseline data on the overall community or ecosystem structure. One method is to sample indicator species. There are many problems underlying the assumptions of using indicator species as indices of population trends in other species and in habitat quality.[14] Indicator species should be used as measures of environmental stress on biodiversity only if more rigorous forms of monitoring cannot be used.[15] Examples of indicator species include keystone species, key common species, and species or groups with sensitive life history stages or habitat requirements.

Keystone Species. Keystone species are species that many other species rely upon for some critical aspect of their life history and survival. Because many species need the keystone species, one way to survey and protect biotic diversity would be to survey and protect the keystone species. For example, the gopher tortoise (*Gopherus polyphemus*) of the southeastern United States generally is confined to sandy upland soils of the coastal plain. Its range coincides to a large extent with the longleaf pine (*Pinus palustris*)—turkey oak (*Quercus laevis*)—wiregrass (*Aristida stricta*) forest community. There, the tortoise digs long burrows as retreats from cold, heat, and fire. The tortoise is a benign host, and more than 60 vertebrates and 300 invertebrates are either obligate or facultative burrow residents.[16] In addition, the floristic community within gopher tortoise habitat is quite diverse with hundreds of species recorded. If the survival of the tortoise is assured, then it is likely that the survival of the animal and plant communities that depend upon or co-occur with it will be assured. Other keystone species include non-insectivorous pollinating bats[17,18] and reef building corals.

Key Common Species. Wide-ranging common species with certain biological attributes (e.g., easily observed, easily studied, long-lived, occur in high population densities, and possess relatively small home ranges over a wide geographic region) may serve as indicators of biotic diversity. Field guides contain numerous examples of wide-ranging species that occur in different habitats. The extirpation of or serious population declines in key common species may not indicate the extent of impact on other species or identify the source of problems. However, the absence of key common species in appropriate habitats gives an indication of problems that need to be addressed. Key common species should be identified and censused throughout their range, and certain widely spaced populations should be targeted for long-term research.

For example, the eastern box turtle (*Terrapene carolina*) ranges from New England south to the Florida Keys and west to the Great Plains and Texas. By nearly all accounts, it is a common species. However, long-studied populations have declined dramatically even in protected areas[19,20,21]. Habitat destruction is severe throughout much of the range, the species is collected for pets in large numbers, and roads isolate populations leading to fragmentation, decreased population size, and direct mortality. Declines in box turtle populations may signal more serious environmental problems affecting local biotic diversity. Problems affecting them certainly affect other species in the same habitat.

Species with Sensitive Life History Stages or Habitat Requirements. Another method to monitor biotic diversity is to focus on a taxonomic group with a sensitive or a spatially limited life history stage. At such times, certain species or communities of species are concentrated and conspicuous. Information (e.g., numbers, age and size class structure, sex ratio) that might not be available even by long-term intensive sampling throughout the year can be gathered in a short period of time. For example, researchers could sample the species richness of tropical frog communities during the breeding season when adult males are conspicuous as they call for females. Larval frog or insect communities can be sampled in spatially limited breeding pools. Whole groups of insects are found in selected microhabitats, such as the clumps of dead leaves of tree branches suspended above the ground in Neotropical forests,[11] where they are censused easily. Aquatic invertebrates long have been used as indicators of human-caused environmental stress.

Taxonomic and Geographic Focus

Although biologists can appreciate the overall magnitude of biotic diversity within a region only through a broad-scale review, they may choose to focus attention on particular taxonomic groups, such as orchids, frogs, or beetles. A taxonomic focus allows intensive examination of animals or plants that require particular identification or sampling skills. Likewise, certain taxonomic groups or their life history stages might be particularly sensitive to stress and thus serve as indicators of the potential loss of diversity. For example, there is considerable interspecific variation in acid tolerance in adult amphibians, and eggs and larvae are more sensitive to acid conditions than breeding adults.[22,23,23] Ideally, specialists should sample taxonomic groups simultaneously. Examples of such an approach are contained in a recent volume comparing the biotas of four Neotropical rainforests[25] and in research conducted at the U.S. National Science Foundation's Long-Term Ecological Research Program sites.[26]

Although remote sensing techniques provide gross estimates of the loss of biotic communities, field biodiversity studies are needed for every taxonomic group in nearly every geographic region of the world. The region with the most urgent need for comprehensive inventory and monitoring stretches from southern Mexico to Brazil, especially in the tropical moist forests. However, the species richness and diversity of rainforests of southeast Asia, Africa, Australia, and islands such as Madagascar, New Guinea, and the Philippines also are poorly known.

Myers[27] delineated 10 global "hotspots" of plant biological diversity. These 10 areas comprise only 0.2% of the earth's surface yet contain an estimated 27% of all plant species in tropical forests. They include the eastern slopes of the Andes and western Ecuador that, like the other areas, are centers of invertebrate as well as vertebrate[28] diversity. However, in a practical sense, conservation rarely proceeds from a total ecosystem approach because of the juxtaposition of political boundaries across biological systems. As such, biodiversity studies usually focus on ecosystems and communities within political units, generally at a national level. Certain countries, such as Brazil, Colombia, Indonesia, Mexico, Madagascar, and Australia, are considered megadiversity countries[28] because they contain a disproportionately large amount of the world's biotic diversity. Worldwide programs designed to maximize research and protection of biotic diversity should concentrate efforts on diversity hotspots within the political framework of these nations.

Regional centers of biotic diversity within national or local jurisdictions are easily identified after systematic fieldwork. For example, the Florida Game and Fresh Water Fish Commission recently completed a review of Florida's 668 vertebrate species and subspecies. Species were rated on both biological and management-related scales, and variable scores were tabulated to pinpoint species and geographic regions for special survey, monitoring, research, and management.[29] For reptiles, a combined geographic point score identified three regions for special attention, the southeastern coast, the Central Florida Ridge, and the Apalachicola Basin. Biologists could use a similar approach to identify particularly diverse regions in almost any geographically defined area. Critical areas of concern could be pinpointed further by combining species distribution maps with data on environmental variables and land use planning using Geographic Information System (GIS) computer software.

Sampling Considerations

Biologists should design sampling procedures that obtain the maximum amount of information. The procedures adopted will, of course, depend on the available resources, taxonomic entities, and geographic area where monitoring is undertaken. Long-term intensive sampling provides the best database whereas short-term studies underestimate both species richness and diversity. For example, I sampled the herpetofauna of a 0.16 ha temporary pond in north Florida sandhills for 1,343 days. Although a few species dominated the community, the 42nd species was not collected until the last month of the 62.5 month study.[30] In more complex communities or in large geographic regions, the problems of determining species richness and diversity become more difficult. However, knowledge of every species within a community is not necessary to appreciate the extent of biodiversity.

Sampling techniques need to be standardized. If different sampling regimes are employed, it becomes very difficult to separate biological differences from procedural differences. If studies are to be comparable, they should cover similar amounts of territory using the same techniques over the same amount of time during the same seasons. It seems axiomatic that appropriate statistical techniques should be used. Likewise, indices of biotic diversity must be used carefully and only if assumptions are met.[2,9] (see also chapters X and Y).

A clear summary of standard sampling techniques that can be used across a range of habitats is not available for most taxonomic groups. To help researchers standardize data collection for biodiversity studies, the Smithsonian Institution initiated a project to develop a series of handbooks that will summarize and review sampling techniques and procedures. Specialists in biodiversity field studies will offer information on objectives, target organisms and habitats, background theory, experimental design, field methods, data handling, analysis and interpretation, special considerations, and recommendations for each technique. Technical sections will include a list of references where more detailed information is available. The amphibian handbook[31] is nearly finished, and the handbook on mammals is in preparation. Handbooks on birds, reptiles, and trees are planned.

In addition to sampling limitations imposed by resource and logistical considerations, the data obtained from sampling biological diversity must be assessed with caution. Mayr[32] stated that phenotypic, physiological, and behavioral variation is "largely ignored by ecologists, most of whom discuss the ecological requirements of a species in a strictly typological manner." Many ecologists seem to ignore or downplay population fluctuation and geographic variation in life history traits. Biologists need to guard against ecological typology thinking.

Biodiversity studies monitor a regional landscape, ecosystem, or community at a point in time. Often, many species are involved about which little is known. Obviously, large scale habitat reduction results in a loss of diversity, but even small scale modifications can affect diversity because species are not evenly distributed within a habitat. In addition, biological populations may fluctuate through natural processes of growth, dispersal, and extinction without any obvious source of stress. In some instances, natural factors may cause dramatic changes in community structure or composition. For example, temperate zone amphibians vary reproductive output considerably from one year to the next[33] and changes occur in even tropical frog communities because of cold weather.[34] Unless natural population fluctuation is separated from changes due to human-caused stress, population changes observed during biodiversity monitoring will be very difficult to interpret.[33,35]

Long-term Studies

The best way to separate human-caused from natural population fluctuations or changes is to monitor species and communities over a long time. The length of time a community should be monitored will depend on the interest and goals of the observer. It is unlikely that large-scale habitat monitoring should ever cease given present human population pressures. The logistic and financial difficulties associated with biodiversity fieldwork ensures that there probably will never be enough effort devoted to monitoring.

It is safe to suggest that communities should be monitored as long as possible. Unfortunately, I know of no easy formula to determine how long that might be because species have widely varying life spans and generation times. If researchers monitor a community through several generations of its dominant species, a long-term study might be of short duration if invertebrates were involved or centuries if the dominant species were large canopy forest trees. Even rather familiar small vertebrates such as amphibians routinely live 10 to 15 years. A long-term study could span several decades, yet comparable data from other studies on the same species might not be available with which to make comparisons and draw conclusions.

Long-term research is needed to provide sound baseline data, answer questions, and direct future research. It is particularly appropriate for studies that involve processes that are slow, rare or episodic, contain high variability, or are subtle or complex.[36,37] These are all factors that affect the biodiversity of communities. A system of long-term study sites needs to be developed internationally, nationally, and regionally. An example is the U.S. National Science Foundation's Long-Term Ecological Research Program,[26] but even this program is limited in its taxonomic and geographic scope. Funding agencies and employers needs to support long-term research instead of focusing on short-term studies with higher publication output.

PROTECTING AGAINST CHANGES AND IMPACTS

The best protection against human-caused changes and impacts that adversely affect biological diversity is constant vigilance of the environment coupled with basic research on species and communities. Biologists must be in the field observing and monitoring animals and plants to detect both gross and subtle changes occurring in ecosystem function and composition. Once changes are noted, a sound database will help identify viable management options. Basic research on systematics, life history, and community ecology forms the foundation of conservation efforts on behalf of biological diversity. In this regard, it may be necessary to change attitudes away from viewing research as science only if it involves high technology equipment and hypothesis testing with complex experimental designs. Both have their place in conservation, but alpha level fieldwork and systematics are needed desparately in most taxonomic groups.

Research and Monitoring

Background research is necessary on the distribution, life histories, effects of perturbation and stress, and conservation techniques involving species, communities, and ecosystems. It is hard to appreciate diversity if it is not identified. Systematic studies must be undertaken to identify species, especially in poorly understood but complex invertebrate groups. Genetic studies assist morphological studies by showing the realm of underlying variation within and among species. Genetic studies also identify cryptic species[38] and increase our level of knowledge concerning the extent of biological diversity.

While the macrodistribution of many species is generally known, the microdistribution often is known poorly. Biological surveys should be undertaken on a national[39] and local level to identify high diversity regions and to catalogue the distribution of resident species in various habitats. Once identified, conservation and government agencies should target areas of high diversity for acquisition and management. Research on distribution and community structure should be combined with long-term population monitoring and, eventually, modeling.

The life history and ecology of a few species are well-known but information is fragmented and poorly-known for the vast majority of species. In addition, specific aspects of a species' life history in one area may bear little resemblance to those in distant populations. Life-history variation, such as in some turtle species,[40] sometimes occurs even in nearby populations. Such variation has important ramifications for managing and protecting diversity because it will dictate which strategies are most appropriate.[41] Research on the life history of species used as diversity indicators is especially needed. Once baseline data are available on species and communities, the effects of stress and perturbation are separated more easily from natural variation.

Finally, there are many techniques that are used to sample, monitor, and protect species, communities, and their habitats. Examples include setting up biological reserves, connecting reserves with movement corridors, head-starting, surrogate parenting, repatriating and relocating species, and captive propagation. However, many techniques are advocated or adopted without adequate background research as to whether they work or are appropriate for the species, community, or habitat in question(e.g.,[42]). There is an urgent need to evaluate conservation and management techniques, especially on a long-term basis. Strategies and programs designed to protect biological diversity must operate within both biological and human-imposed constraints.[41] Failure to understand or adhere to these constraints ensures that conservation efforts likely will fail.

Habitat Protection

The only long-term solution to protecting biological diversity is to ensure habitat protection. However, setting aside a parcel of land or portion of an aquatic community, even a large parcel, may not be sufficient to ensure the integrity and functioning of an ecosystem. Global warming from hydrocarbon emissions, ozone depletion, increased atmospheric acidity, and oceanic and atmospheric pollution all transcend political boundaries and may have long-term impacts on a worldwide scale. In addition, the expanding human population and the increased materialistic expectations of people throughout the world threaten biological diversity even on protected lands. Unless these topics are addressed, reserve establishment by itself will not protect biological diversity.

Still, protecting habitats by setting restrictions on their use is the best way to protect biodiversity. Restrictions need not be exclusive, that is, some uses are compatible with the protection of biodiversity. Indeed, habitat protection in most economically underdeveloped regions will be accomplished only by ensuring the well-being of people inhabiting in or near the reserve.[6] In regions with strong economies or a long history of conservation management, habitat protection may allow some level of multiple use[43] or remain as near to wilderness as possible.[44] However, there is considerable disagreement concerning the levels of multiple use that can be allowed without jeopardizing biodiversity. If protected habitats are subject to exploitation for short-term gain in times of economic or other perceived crises, biological diversity can never be preserved.

Questions concerning reserve creation are discussed by Shafer.[7] Important additional topics to consider include a clear statement of the reserve's goals, conservation corridors (Will species use them? How wide should they be? In what types of habitats? Are there negative side-effects[45]), design (shape? large or small?), population size[46,47] and biology of the inhabitants, and long-term maintenance.

CONCLUSIONS

Monitoring changes and impacts on biotic diversity requires a substantial commitment of support and effort. Species, especially keystone species, key common species, and species with sensitive life history stages and microhabitats, communities, and ecosystems all need to be monitored. However, resource limitations may restrict monitoring spatially or taxonomically. Remote sensing technology may examine gross habitat changes, but fieldwork is needed to assess species richness and diversity, and to monitor changes in community composition and structure. Long-term studies using standardized sampling techniques are particularly required. The best defense against a loss of biotic diversity is a program that combines research on systematics, distribution, and life history with long-term community monitoring. Habitat protection will prevent the loss of biodiversity, but the concept of habitat needs to be expanded beyond a localized area to include global concerns. Unless habitat integrity can be ensured, efforts to conserve biodiversity will prove unsuccessful.

ACKNOWLEDGEMENTS

I thank R. Franz, L. Lefebvre, and J.D. Williams for comments on the manuscript.

REFERENCES

1. Noss, R.F. 1989. Longleaf pine and wiregrass: Keystone components of an endangered ecosystem. *Nat. Areas J.* 9:211-213.
2. Magurran, A.E. 1988. *Ecological Diversity and Its Measurement.* Princeton University Press, Princeton, NJ, pp. 179.
3. Soulé, M.E. and B.A. Wilcox (Eds.). 1980. *Conservation Biology.* Sinauer Associates, Inc., Sunderland, MA, pp. 584.
4. Frankel, O.H. and M.E. Soulé. 1981. *Conservation and Evolution.* Cambridge University Press, Cambridge, UK, pp. 327.
5. Soulé, M.E. (Ed.). 1986. *Conservation Biology: The Science of Scarcity and Diversity.* Sinauer Associates, Inc., Sunderland, MA, pp. 395.
6. Western, D. and M. Pearl (Eds.). 1989. *Conservation for the Twenty-first Century.* Oxford University Press, New York, NY, pp. 365.
7. Shafer, C.L. 1990. *Nature Reserves: Island Theory and Conservation Practice.* Smithsonian Institution Press, Washington, DC, pp. 189.
8. Wilson, E.O. (Ed.). 1988. *Biodiversity.* National Academy Press, Washington, DC, pp. 521.
9. Noss, R.F. 1990. Indicators for monitoring biodiversity: A hierarchical approach. *Conserv. Biol.* 4:355-364.
10. Begon, M., J.L. Harper and C.R. Townsend. 1986. *Ecology: Individuals, Populations and Communities.* Blackwell, Oxford, UK, pp. 876.

11. Erwin, T.L. 1991. How many species are there?: Revisited. *Conserv. Biol.* 5:330-333.
12. Gaston, K.J. 1991. The magnitude of global insect species richness. *Conserv. Biol.* 5:283-296.
13. Duellman, W.E. 1982. Quaternary climatic-ecological fluctuations in the lowland tropics: Frogs and forests, pp. 389-402. In: G.T. Prance (Ed.) *Biological Diversification in the Tropics,* Columbia University Press, New York, NY, pp. 714.
14. Landres, P.B., J. Verner and J.W. Thomas. 1988. Ecological uses of vertebrate indicator species: A critique. *Conserv. Biol.* 2:316-328.
15. Beanlands, G.E. and P.N. Duinker. 1983. *An Ecological Framework for Environmental Assessment in Canada.* Institute for Resource and Environmental Studies, Dalhousie University, Nova Scotia, Canada.
16. Jackson, D.R. and E.G. Milstrey. 1989. The fauna of gopher tortoise burrows, pp. 86-98. In: J.E. Diemer, D.R. Jackson, J.L. Landers, J.N. Layne and D.A. Wood (Ed.) *Gopher Tortoise Relocation Symposium Proceedings,* Florida Nongame Wildlife Program Technical Report No. 5, Tallahasee, FL, pp. 109.
17. Fleming, T.H. 1988. *The Short-tailed Fruit Bat.* University of Chicago Press, Chicago, IL, pp. 365.
18. Fujita, M.S. and M.D. Tuttle. 1991. Flying foxes (Chiroptera: Pteropodidae): Threatened animals of key ecological and economic importance. *Conserv. Biol.* 5:455-463.
19. Stickel, L.F. 1978. Changes in a box turtle population during three decades. *Copeia* 1978:221-225.
20. Schwartz, E.R., C.W. Schwartz, and A.R. Kiester. 1984. The three-toed box turtle in central Missouri, Part II: A nineteen-year study of home range, movements and population. *Publications of the Missouri Department of Conservation Terrestrial Series* 12:1-29.
21. Williams, E.C. and W.S. Parker. 1987. A long term study of a box turtle population at Allee Memorial Woods, Indiana, with emphasis on survivorship. *Herpetologica* 43:328-335.
22. Cook, R.P. 1983. Effects of acid precipitation on embryonic mortality of *Ambystoma* salamanders in the Connecticut Valley of Massachusetts. *Biol. Conserv.* 27:77-88.
23. Pierce, B.A. 1985. Acid tolerance in amphibians. *Bioscience* 35:239-243.
24. Warner, S.C., W.A. Dunson and J. Travis. 1991. Interaction of pH, density, and priority effects on the survivorship and growth of two species of hylid tadpoles. *Oecologia* 88: 331-339.
25. Gentry, A.H. (Ed.). 1990. *Four Neotropical Rainforests.* Yale University Press, New Haven, CT, pp. 627.
26. Swanson, F.J. and R.E. Sparks. 1990. Long-term ecological research and the invisible place. *Bioscience* 40:502-508.
27. Myers, N. 1988. Threatened biotas: "hotspots" in tropical forests. *Environmentalist* 8:187-208.
28. Mittermeier, R.A., J.L. Carr, I.R. Swingland, T.B. Werner and R.B. Mast. 1992. Conservation of amphibians and reptiles. In: K. Adler (Ed.) *Herpetology:*

Current Research on the Biology of Amphibians and Reptiles. SSAR Contributions to Herpetology, 9:59-80.

29. Millsap, B.A., J.A. Gore, D.E. Runde and S.I. Cerulean. 1990. Setting priorities for the conservation of fish and wildlife species in Florida. *Wildl. Monogr.* 111:1-57.
30. Dodd, C.K., Jr. 1992. Biological diversity of a temporary pond herpetofauna in north Florida sandhills. *Biodiver. and Conserv.* 1:125-142.
31. Heyer, W.R., M.A. Donnelly, R.W. McDiarmid, L.A. Hayek and M.S. Foster (Ed.). 1993. *Measuring and Monitoring Biological Diversity: Standard Methods for Amphibians.* Smithsonian Institution Press, Washington, DC, in press.
32. Mayr, E. 1970. *Populations, Species, and Evolution.* Belknap Press, Cambridge, MA, pp. 453.
33. Pechmann, J.H.K., D.E. Scott, R.D. Semlitsch, J.P. Caldwell, L.J. Vitt and J.W. Gibbons. 1991. Declining amphibian populations: the problem of separating human impacts from natural fluctuations. *Science* 235:892-895.
34. Heyer, W.R., A.S. Rand, C.A. Goncalvez da Cruz and O.L. Peixoto. 1988. Decimations, extinctions, and colonization of frog populations in southeast Brazil and their evolutionary implications. *Biotropica* 20:230-235.
35. Crump, M.L., F.R. Hensley and K.L. Clark. 1992. Apparent decline of the golden toad: underground or extinct? *Copeia* 1992:413-420.
36. Franklin, J.F. 1989. Importance and justification of long-term studies in ecology, pp. 3-19. In: G.E. Likens (Ed.) *Long-term Studies in Ecology,* Springer-Verlag, New York, NY, pp. 214.
37. Magnusson, J.J. 1990. Long-term research and the invisible present. *Bioscience* 40:495-501.
38. Daugherty, C.H., A. Cree, J.M. Hay and M.B. Thompson. 1990. Neglected taxonomy and continuing extinctions of tuatara (*Sphenodon*). *Nature* 347:177-179.
39. Kim, K.C. and L. Knutson (Ed.). 1986. *Foundations for a National Biological Survey.* Association of Systematics Collections, Lawrence, KS, pp. 215.
40. Gibbons, J.W. (Ed.). 1990. *The Slider Turtle.* Smithsonian Institution Press, Washington, DC, pp. 368.
41. Dodd, C.K., Jr. 1993. Strategies for snake conservation. In: R.A. Seigel and J.T. Collins (Ed.) *Snakes: Ecology and Behavior.* McGraw-Hill Book Co., New York, NY, in press.
42. Dodd, C.K., Jr. and R.A. Seigel. 1991. Relocation, repatriation, and translocation of amphibians and reptiles: Are they conservation strategies that work? *Herpetologica* 47:336-350.
43. Salwasser, H. 1991. New perspectives for sustaining diversity in U.S. National Forest ecosystems. *Conserv. Biol.* 5:567-569.
44. Noss, R.F. 1991. Sustainability and wilderness. *Conserv. Biol.* 5:120-122.
45. Simberloff, D. and J. Cox. 1987. Consequences and costs of conservation corridors. *Conserv. Biol.* 1:63-71.
46. Shaffer, M.L. 1981. Minimum population sizes for species conservation. *Bioscience* 31:131-134.
47. Shaffer, M.L. and F.B. Samson. 1985. Population size and extinction: A note on determining critical population sizes. *Am. Nat.* 125:144-152.

Biological Diversity: Problems and Challenges. Edited by S.K. Majumdar, F.J. Brenner, J.E. Lovich, J.F. Schalles, and E.W. Miller. © 1994, The Pennsylvania Academy of Science.

Chapter Two

BIODIVERSITY: A Global Perspective

GORDON L. KIRKLAND, JR.[1]
and
JAY R. STAUFFER, JR.[2]

[1]Vertebrate Museum
Shippensburg University
Shippensburg, PA 17257
and
[2]School of Forest Resources
The Pennsylvania State University
University Park, PA 16802

INTRODUCTION

Biological diversity, defined as the variability and variation among organisms as related to their ecology and habitats (Soule and Wilcox 1980, Norse et al. 1986, OTA-USC 1987), is being reduced at an alarming rate throughout the world. The loss of this diversity, which provides the basic resources for the life-support system on our planet, undermines the biota's capacity to respond to changing environmental conditions that in recent years are increasingly anthropogenic in origin. Economic development is being accelerated throughout the world to meet the needs of an ever-increasing human population, which already exceeds five billion individuals (McNeeley 1992). Such economic activity, coupled with rapid expansion of urban and suburban regions, has contributed to widespread habitat destruction and the potential for massive extinction of species, as well as a corresponding reduction of biodiversity, which is inclusive of genetic, population, species, and ecosystem diversities (Myers 1979, Lovejoy 1980, Wilson 1989).

EXTINCTIONS

Extinction has been an integral part of the evolution of life on earth since the appearance of the first life forms. In general, the probability of extinction of a given species follows an exponential decay function; thus, the percentage of species becoming extinct has remained relatively constant through time (Van Valen 1973, Wilson 1988). Nevertheless, over geologic time this relative constancy has been interrupted by periodic episodes of global extinctions, which appear to occur approximately every 26 million years (Raup and Sepkoski 1984). Despite these periodic episodes of global extinctions, the history of life on earth has been characterized by a trend of increasing biological diversity (Brown and Gibson 1983, Lovejoy 1986). Each period of global reduction in biodiversity has been followed by renewal of the trend of increasing biodiversity as groups of organisms have evolved and radiated to fill vacated and novel ecological niches (Brown and Gibson 1983). Although the causes of these massive species die-offs in the geological past may be debated, the fact that life on earth recovered from these periodic setbacks is clear. Today, the earth is experiencing another episode of accelerated extinction rates, but unlike previous extinction episodes, which apparently were caused by extrinsic factors (e.g., collisions with asteroids or the effects of continental drift), the current loss of species is directly attributed to a resident life form, *Homo sapiens*. Loss of biodiversity represents one of the major challenges facing humans at the close of the 20th century. This is because the global human population, which has achieved an exponential growth form, is having an increasingly massive impact on the earth's biodiversity, particularly in tropical latitudes. Unlike the past, when episodes of global extinction were followed by renewal of the trend of increasing biodiversity, the human-induced environmental changes that currently are accelerating global extinction rates are altering the environment in such a way as to prevent recolonization and to limit opportunities for speciation.

GLOBAL PATTERNS OF SPECIES DIVERSITY

Two geographic patterns of species richness or diversity are important in our current concern over the loss of biodiversity. The first is the familiar latitudinal gradient in species richness in which tropical habitats support the greatest number of species, and there is a gradient of decreasing species richness from the tropics to the poles. The second is the pattern of species diversity found on islands: small islands support fewer species than larger islands.

A number of hypotheses have been proposed to explain the greater species richness of tropical ecosystems compared to higher latitude systems (Baker 1970, Pianka 1988). Two of these deal with the greater and more stable primary productivity characteristic of many tropical ecosystems, which provides the requisite resource base to permit trophic specialization (Pianka 1988). An abundant and dependable food resource base permits many tropical species to become more specialized and to more finely divide available niche space than their higher latitude counterparts. The result is a greater number of species in the tropics; however, an important consequence is that in general each tropical species contains fewer individuals than species inhabiting higher latitudes. Because of their high degree of specialization, tropical species tend to be more sensitive to environmental perturbations than are their temperate zone counterparts.

In general, tropical communities are organized on the basis of interspecific interactions (e.g., competition, mutualism, predation, parasitism), whereas the physical environment is the principal force organizing temperate and higher latitude communities (Pianka 1988). This has resulted in fundamental differences between the abilities of species in tropical and temperate regions to withstand and adapt to anthropogenic disturbances. Species that have evolved in physically-controlled ecosystems of higher latitudes have broader limits of tolerance and niches in order to survive the vicissitudes of the physical environment to which they have been exposed during their evolutionary histories. These species are thus better adapted to survive human-induced environmental changes than their tropical counterparts, which have evolved in relatively benign physical environments. Because of their ecological specialization and narrow limits of tolerance, tropical species are more prone to extinction due to human-induced habitat alteration and destruction.

The trend of increasing species richness of communities with decreasing latitude is accompanied by a trend of decreasing size in the geographic ranges of species with declining latitude (Stevens 1989). In general, tropical species have smaller geographic distributions than temperate zone species. The smaller ranges of tropical species dictate that they also should have smaller total populations than temperate species. Once again, this has important implications for extinction rates of tropical species, because the probability of extinction increases with decreasing population size.

In general, species inhabiting the tropics are on the K-end of the r-K selection continuum. Due to the high species diversity in the tropics, interspecific interactions (competition, predation, herbivory, mutualism, and parasitism) play a greater role in the lives of tropical species than species inhabiting higher latitudes. In response, tropical species tend to channel their energy into mechanisms that permit them to maintain themselves and their offspring within the community in the face of these interspecific interactions. The reproductive strategies of more K-selected tropical species favor production of relatively few, well-provisioned young, which have high survivorship. Tropical species are thus less fecund than their more r-selected temperate zone counterparts. Because of their greater reproductive output, temperate zone species can recover from reduced population size more quickly than tropical species, and therefore can more quickly reduce the high probability of extinction associated with small populations. For example, it will require fewer females of a temperate fish species that produces several thousand eggs per female (e.g., common carp, *Cyprinus carpio*) to recolonize an area than a tropical species that produces 20-30 eggs per female (e.g., Lake Malawi zebra, *Pseudotropheus zebra*).

The second conspicuous geographic trend of species richness relates to islands: larger islands support more species than smaller islands. Islands also support fewer species than comparably-sized portions of continuous "mainland" habitat (Pianka

1988). In true islands there is a tendency for a 10-fold increase in island size to be accompanied by a doubling of the number of resident species (MacArthur and Wilson 1967, Pianka 1988). This has important implications as the biosphere is subdivided into increasingly smaller patches of natural habitat, interspersed with human-altered habitats. When a continuous habitat is fragmented, each fragment is initially supersaturated in terms of the number of species it can support. The biota of the island undergoes a process of relaxation (loss of species) until it reaches a sustainable level (Shafer 1990). A classic example of this is the avifauna of Barro Colorado Island, a land-bridge island which was formed when construction of the Panama Canal created Lake Gatun. Forty-five species of bird (22% of the avifauna present when the island was formed) had disappeared by 1970 (Willis 1972), and at least 50 species had disappeared by 1980 (Karr 1981), despite the fact that Barro Colorado Island has served as a nature reserve throughout its history.

FRAGMENTATION OF THE BIOSPHERE

The biosphere constitutes a mosaic of biotic communities. In prehistoric times, the nature of that mosaic was dictated by prevailing climatic and topographic conditions, along with natural disturbance events (e.g. fires, floods, storms, glaciations, etc.) Today, the environmental mosaic is increasingly dictated by human influences, the net effect of which has been to subdivide and dramatically reduce the size of natural systems, thus producing increasing fragmentation and insularization of the natural world. Overall, smaller patches harbor fewer species as predicted by the species-area curve (Simberloff 1984). Habitat fragments also have different ecological dynamics than they had when part of continuous biological communities. Such isolated habitat fragments are almost always subject to loss of species (Lovejoy 1986). Additionally, human-induced fragmentation of the landscape has eliminated dispersal corridors connecting similar ecosystems, thereby reducing gene flow between populations to such an extent that total genetic diversity has been reduced.

BIODIVERSITY AND THE TROPICS

The tropics represent a vast storehouse of global biological diversity. It is estimated that perhaps 50% of the extant species reside in only 6% of the earth's surface represented by tropical rain forests (Myers, 1988). The region is under severe human pressure to exploit its biotic resources; however, the biota of this region is not adapted to withstand human-induced destruction and fragmentation of communities. For example, most species of primary tropical rain forests do not occur in second growth stands, as is the case in forests of the temperate zone (Lovejoy 1986). Also, seed banks don't persist in the tropics. These characteristics help explain why New England's forests recovered from extensive cutting in the 19th century without major loss of diversity, but tropical forests likely will not recover from the destruction they are currently experiencing (Lovejoy 1986).

The processes that have contributed to the high species richness of tropical ecosystem render these systems particularly vulnerable to species loss due to human disturbance and destruction of natural habitats. This potential for loss of species is exacerbated in the tropics, because of the increased specialization of tropical species, which has resulted in a high degree of endemism, localization of species, and high local species richness.

The species-area relationship and the tendency for newly-created habitat islands to experience loss of species have serious implications for maintenance of biodiversity in an increasingly fragmented world. When these aspects of island biogeography are coupled with the sensitivity of tropical species to human disturbances, it is obvious that fragmentation of habitats in the species-rich tropics poses a much greater threat to the biodiversity of this region than similar disturbances in higher latitude regions.

LITERATURE CITED

Baker, H.G. 1970. Evolution in the tropics. Biotropica, 2:101-111.

Brown, J.H. and A.C. Gibson. 1983. Biogeography. C.V. Mosby Co., St. Louis, MO, 643 pp.

Karr, J.R. 1981. Avian extinction on Barro Colorado Island, Panama: a reassessment. Amer. Nat., 119:220-239.

Lovejoy, T.E. 1980. A projection of species extinctions. In the Global 2000 report to the President, V. 2, The Technical Report, Council on Environmental Quality and U.S. Department of State. U.S. Government Printing Office, Washington, D.C. 328-332 pp.

__________. 1986. Species leave the ark one by one. pp. 13-27 In The preservation of species. (B.G. Norton, ed.). Princeton Univ. Press, Princeton, N.J. 305 pp.

MacArthur, R.H. and E.O. Wilson. 1967. The theory of island biogeography. Monog. in pop. biol. #1, Princeton Univer. Press, Princeton, N.J. 203 pp.

McNeeley, J.A. 1992. The sinking ark: pollution and the worldwide loss of biodiversity. Biodiversity and Conservation, 1:2-18.

Myers, N. 1979. The sinking ark: a new look at the problem of disappearing species. Pergammon Press, Oxford, 307 pp.

__________. 1988. Tropical-forest species: going, going, going . . . Scientific American, 259:132.

Norse, E.A., K.L. Rosenbaum, D.S. Wilcove, B.A. Wilcox, W.H. Romme, D.W. Johnston and M.L. Stout. 1986. Conserving Biological Diversity in Our National Forests. The Wilderness Society, Washington, D.C., 116 pp.

OTA/USC, Office of Technology Assessment, U.S. Congress. 1987. Technologies to maintain biological diversity. OTA-F-330, Washington, D.C. 334 pp.

Pianka, E.R. 1988. Evolutionary ecology. 4th ed. Harper and Row, Publishers, New York. 468 pp.

Raup, D.M. and J.J. Sepkoski, Jr. 1984. Periodicity of extinctions in the geologic past. Proc. Natl. Acad. Si. USA 81:801-805.

Shafer, C.L. 1990. Nature reserves: island theory and conservation practice. Smithsonian Institution Press, Washington, D.C. 189 pp.

Simberloff, D.S. 1984. Mass extinction and the destruction of moist tropical forests. Zh. Obshch. Biol. 45:767-778.

Soule, M.E. and B.A. Wilcox, eds. 1980. Conservation biology: a ecological-evolutionary perspective. Sinauer Assoc. Sunderland, MA 395 pp.

Van Valen, L. 1973. A new evolutionary law. Evol. Ther. 1:1-30.

Willis, E.O. 1972. Populations and local extinctions of birds on Barro Colorado Island, Panama. Ecol. Monogr., 44:153-169.

Wilson, E.O. 1988. The current state of biological diversity. In: E.O. Wilson (ed.). Biodiversity. National Academy Press, Washington, D.C.

__________. 1989. Threats to biodiversity. Sci. Amer. 261(3):108-117.

Biological Diversity: Problems and Challenges. Edited by S.K. Majumdar, F.J. Brenner, J.E. Lovich, J.F. Schalles, and E.W. Miller. © 1994, The Pennsylvania Academy of Science.

Chapter Three

CONSERVATION OF BIODIVERSITY

EUGENE P. ODUM

Institute of Ecology
University of Georgia
Athens, GA 30602

INTRODUCTION

Much of the concern for biodiversity has focused on the species level even though it is self-evident that preservation of habitats is the key to conserving a diversity of species. Also, it is important to consider diversity below as well as above the species level; lack of genetic diversity, for example, is almost a certain road to extinction. Accordingly, a multilevel or hierarchical approach is required if we are serious about understanding and conserving biodiversity.

ECOLOGICAL FUNDAMENTALS OF DIVERSITY

First, let us review the basics which seem to be forgotten or not understood in much of the hoopla for preservation of biodiversity that we see in magazines, newspapers and even in professional journals. As a first order approach we need to recognize two equally important components, (1) the variety component and (2) the relative abundance component. The variety component is usually expressed as the number of kinds (species, genetic varieties, landscape or land use types) per unit of space or as a ratio of kinds to numbers. The relative abundance component involves the apportionment of individual units among the kinds. Various indices have been proposed to quantify this component which is perhaps best depicted graphically by dominance-diversity curves which plot relative importance (numbers, biomass, area) against rank order from common to rare. The frequently used Shannon Index gives weight to both variety and apportionment.

Preservation of biodiversity involves more that maintaining a large variety of life forms in the environment (like in a botanical garden or Zoo). Apportionment of species and habitats so that all the important ecosystem functions are maintained is more important. In general, relatively few kinds in each trophic group are common and many others rare. The rare component may play important roles in ecosystem function in addition to providing redundancy for the system. Too often the rare forms that provide the redundancy are the first to be affected by human alterations of the landscape.

DIVERSITY BELOW THE SPECIES LEVEL

Lifecycle stages of both plants and animals often contribute as much to the diversity of an ecosystem as several different species. Because they differ in feeding habits and habitat, a butterfly and its caterpillar, or a frog and its tadpole or a tree and its seedling provide more diversity than two closely related butterflies, frogs or trees. One can better assess organismic diversity by counting various life history stages as different units of diversity, than by counting species (so one need not be a skilled taxonomist to quantify diversity!).

I have already commented on the importance of genetic diversity. A lack of such diversity is of special concern in agriculture because so much of current world production of food crops involves only a handful of varieties which require massive water and chemical subsidies, and which are vulnerable to unexpected weather changes or new pests. As a hedge against the danger of putting all the eggs in too few baskets seed banks are being established. It has recently been noted that traditional (pre-industrial) agriculture as practiced in less developed countries of middle America and Asia provides the best of all gene banks because of the wide variety of food crops that are maintained in cultivation (Rhoades, 1990).

LANDSCAPE DIVERSITY

Scale is always an important consideration, especially when it comes to planning and management. Three decades ago, Whitaker (1960, 1965) suggested three scale levels, Alpha (within habitat), Beta (between habitat) and Gamma (regional) diversity. We can have very low Alpha diversity in the form of monoculture crop fields and forest plantations and yet maintain a high Beta and Gamma diversity on a large scale landscape if the monocultures are mixed in with a lot of other land-use and natural area components.

The landscape of the Department of Energy's National Environmental Research Park at the Savannah River Site (SRS) provides a good example of contrasting Alpha and Beta diversity. We have assessed habitat diversity of this 300 mi^2 (77,000 ha) reservation by gridding large scale aerial photographs (Odum and Kroodsma. 1977). Most of the former agricultural fields and old-fields, comprising about one-third of the area when the site was first set aside in 1951, have been converted to pine plantations. Despite the large area in monoculture, the overall habitat diversity remains high because plantations are intermixed with numerous hedgerows, patches of natural forest, stream valleys, impoundments and wetlands which have remained relatively undisturbed. Habitat diversity is similar to natural soil diversity (Figure 1). The Forest Service, which manages the timber on the SRS, has proposed extending pine plantations to increase pulp production even though the cost of preparing additional land, much of it unsuitable for pines, would be high. We projected that landscape diversity would be reduced sharply if pine monocultures were extended to 50 or 75% of the area (Figure 1). For the latter, the evenness index would decrease from 70 to 38 percent and the Shannon index from 2.3 to less than 1.0.

I believe that a landscape that is one-third "domesticated" (i.e., urban and intensively managed for crops, including short rotation forestry) and two-thirds "semi-natural" (i.e., supporting naturally developed ecosystems subject to non-destructive human use such as recreation, selective logging, research, etc.) is very desirable. The energy input and pollution output of the intensively developed part of the landscape is so high that a large solar-powered life-support area is required to buffer the impacts of the high energy sectors. This theme is developed in detail elsewhere (Odum 1989).

ECOTONES AND EDGES

It is important to distinguish between ecotones and edges. An ecotone is not

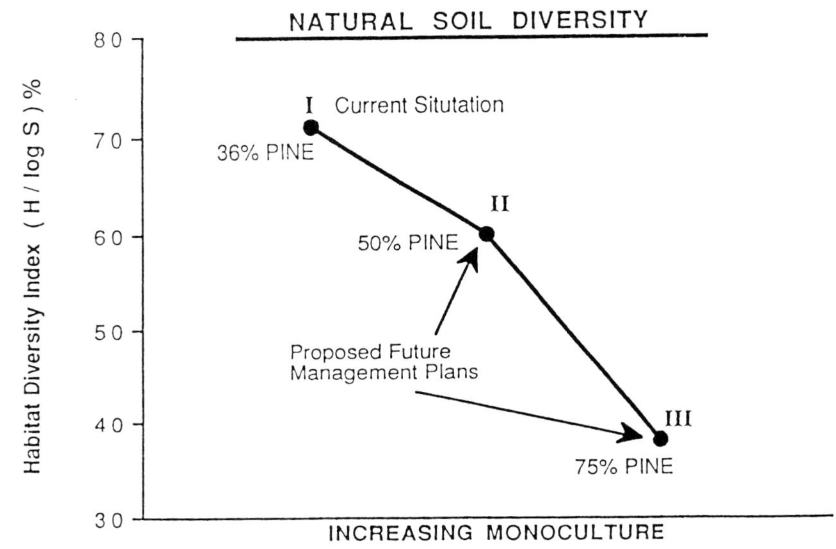

FIGURE 1. Habitat diversity on the 300 Mi^2 National Environmental Research Park at the Savannah River Site where one-third of the area is now in monoculture pine plantations. Because plantations are intermixed with a wide variety of more diverse habitats, landscape diversity is high and similar to natural soil diversity. Habitat diversity would decrease greatly if pine plantations were extended to 50 or 75% of the area. (After Odum, 1982; data from Odum and Kroodsma, 1977).

simply a boundry or edge; the concept assumes the existence of active interactions between two or more ecosystems (or patches) with properties, and frequently species that do not exist in either of the adjacent ecosystems (Decamps and Naiman, 1990). In addition to external processes causing discontinuities in natural gradients, internal processes such as sediment traps, root mats, special soil or water conditions, or animal activity (beaver, for example; Remillard, et al. 1987) may maintain an ecotone distinct from bordering communities (W.E. Odum, 1990).

Frequently, ecotones not only support unique species but have a greater variety and abundance of flora and fauna as compared to the adjoining more homogeneous communities. Accordingly, wide ecotones such as an inter-tidal zone or a savannah between forest and grassland enhance landscape diversity, and such areas should receive high priority for preservation. Game managers often recommend special planting between field and forest in order to increase numbers of game birds and mammals. In contrast, a sharp edge such as between a clearcut and an uncut forest may be a poor habitat, and the large amount of edge in a fragmented, human domesticated landscape usually reduces diversity (Saunders, et al. 1991). For example, forest interior songbirds, such as thrushes and many warblers, disappeared when the landscape was converted into small pieces of forest by intensive agricultural or suburban development in central Maryland (Whitcomb, et al. 1981). On the other hand, so-called forest-edge birds such as blackbirds, house wrens, house finches, cardinals and song sparrows increase with moderate suburban development that features a good deal of edge but also a lot of shrubbery and trees (and bird feeders!).

In fact, the landscape heterogeneity created by well planned residental development that includes set-aside natural areas (parks, greenways, etc.) and protected flood plains and watersheds can increase biodiversity. A Conservation Foundation study (Healy and Short 1981) reported that more and more rural land is being acquired by urbanites. No longer is land near large metropolitan areas owned by farmers, ranchers or commercial foresters. The urbanite often does not have the single minded desire or need to maximize yield. He or she may lease land for forestry or agriculture, but may use or lease for other purposes such as second homes, retirement homes, recreation, and so on. In many ways, this change in land ownership enhances diversity because old-growth forests are as highly valued as productive young ones, and small tracts formerly farmed are allowed to undergo natural succession. Such a desirable pattern is not likely to be sustained in the future unless states and counties become more active in mandating land-use planning than is the case at present.

Ecotones and sometimes edges not only contribute to landscape heterogeneity but they may also check the spread of disturbance across the landscape, as, for example, by providing natural fire breaks or blocking the spread of disease (Forman, 1987).

FOOD WEB DIVERSITY

Two major energy transfers from plants to consumers are commonly recognized—the grazing food chain (consumption of living plant tissues) and the detritus food chain (consumption and decomposition of non-living organic matter). These major flows may be subdivided into at least six distinct transfers that differ in the quality of the resource, the kinds of organisms that use the resource and the degree that they control nutrient cycling (Figure 2). The granivorous, nectarivorous, mycorrhizal and symbiotic nitrogen fixing pathways involve transfers of high quality photosynthates to specialized organisms which in turn provide important services to the plant. Without these services many plants would not survive and the ecosystem would be greatly degraded.

What I like to call "invisible diversity" or microbial diversity may be the most important aspect of Alpha diversity. Because microorganisms are so versatile, resilient and seemingly ubiquitous we usually assume that decomposition and recycling of nutrients will proceed no matter what we do to the plants and macrocumsumers. In mutualistic subsystems such as mycorrhizae and trees, it is becoming evident that their continued function can not be taken for granted. Vital partnershps do not develop spontaneously when trees are transplanted to new areas where colonizing spores of the right kind may be absent. Also, mycorrhizae themselves are easily killed or damaged by chemical additives (fertilizers, pesticides, herbicides). For example, fungicides applied to control a soil pathogen in Mexican coffee plantations killed the mycorrhizae, stopping mycorrhizal transport of phosphorus from soil to roots and causing phosphorus deficient coffee plants (Stephen Gliessman, personal communication).

Thus, toxic wastes, soil disturbance, eutrophication and other perturbations may have their effects at the ecosystem level by disrupting nutrient control subsystems, components that are rarely considered in impact assessment. Accordingly, great caution should be exercised in using broad-spectrum pesticides in diverse ecosystems.

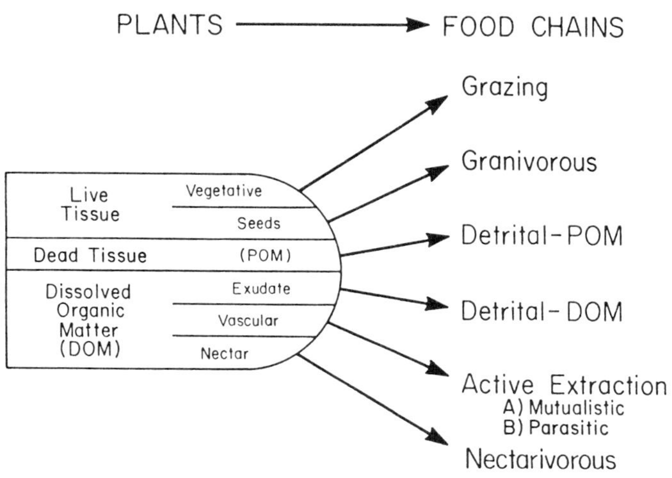

FIGURE 2. Diversity in the terrestrial food web resulting from a variety of autotroph-heterotroph interface energy flow pathways that differ in physical nature, resource quality and kinds of organisms involved. POM-particulate organic matter; DOM-dissolved organic matter. (Odum and Biever, 1984).

FUNCTIONAL DIVERSITY

There is increasing evidence that functional diversity may be more important than structural diversity. By functional diversity we mean variety in processes such as O_2 and CO_2 fluxes (productivity), microbial and chemical processes controlling nutrient concentration and recycling, nitrogen fixation, denitrification, mutualism, heterotrophic consumption and decomposition, food web complexity (Figure 2), and so on. The microcosm experiments of Van Vorris, et al. (1980) are especially interesting in this regard. These investigators assessed the response of old-field microcosms to high temperature or toxic (cadmium) stress by analyzing CO_2 flux and calcium leached from the system. An index of stability was devised that combined the extent of deviation from operating range (the resistence component) and the time required to return to normal after the acute stress (the resilience component). Results showed strong positive correlation between stability and functional diversity, but no correlation between stability and species diversity.

It is important to distinguish between resistence and resilience. An old-growth forest, for example, exhibits considerable resistence to weather perturbations (e.g. drought), but little resilience when severely disturbed by logging. In contrast, a young forest is less resistent but more resilient (i.e., bounces back faster). For example, after experimental logging, equilibrium in nutrient retention and recycling was restored rather quickly in the middle-aged forests on the well studied Hubbard Brook watersheds by what Marks (1974) has called "buried seed strategy". Seeds of the pin cherry, which remain viable in the soil for many years, germinated quickly and covered the watershed with a miniature forest that within 3-5 years reduced nutrient and soil loss caused by the logging (Bormann and Likens, 1979). Thus the diversity of recovery mechanisms may be greater in the earlier stages of succession.

Pimm (1992) applies resistence-resilience concepts to the population level in his efforts to establish a basis for determining how quickly species will recover from catastrophy, and which species are particularly vulnerable to extinction.

KEYSTONE GROUPS

In assessing the effects of removal or addition of species it is important to know if they might be what are known as "keystone species." Chapin, et al. (1992) defines a keystone species as "a functional group without redundancy." Loss of such species, or groups of species, can be expected to cause major changes in community structure and ecosystem function. In reviewing his long-term lake acidification studies, Schindler (1990) reports that there was a good deal of redundancy in the phytoplankton so that the primary production rate was not affected by a moderate lowering of pH. As acid-sensitive species dropped out, they were replaced by others in the same functional niche. There was much less redundancy in the benthos. Food chain structure and function of the whole lake was affected by the decline of certain benthic crustaceans because no other species appeared that could replace them as an important link in the food chain between primary production and fish. Accordingly, the crustaceans qualify as a keystone group. Preservation of keystone groups is especially important in the overall conservation of biodiversity.

A POSSIBLE BOTTOM LINE

It seems to me that the relationship between landscape heterogeneity and biodiversity can be modeled as a hump-backed subsidy-stress curve, that is, the convex curve obtained when "increasing perturbation" is plotted against "deviation from normal operating range" (Odum et al. 1979). In this case, we are plotting "landscape heterogeneity" against "diversity." Devoting a very large part of the landscape to production-oriented management (monocultures, etc.) results in low biodiversity. Increasing heterogeneity with land use planning that considers nonmarket life-support and estethic values as well as market economics will increase biodiversity (the "subsidy" component). Excessive fragmentation and edge which comes with unsustainable harvesting and haphazard, poorly planned economic development not only creates instability but also very quickly reduces biodiversity (the "stress" component).

Accordingly, what we hope for is that broad middle section of the convex performance curve where variety and redundancy sustain human quality of life. We see this kind of optimum diversity (neither too little nor too much) in many natural landscapes where succession and periodic disturbance (fire, storms, etc.) play major roles.

LITERATURE CITED

Bormann, F.H. and G.E. Likens. 1979. *Pattern and Process in a Forested Ecosystem.* Springer-Verlag, NY.

Chapin, F.S., E.D. Schulze and H.A. Moone. 1992. Biodiversity and ecosystem processes. Trends Ecol. Evol. 7(4):107-108.

Decamps, H. and R.J. Naiman. 1990. Towards an ecosystem perspective. Chap. 1 in: *The Ecology and Management of Aquatic-terrestrial Ecotones.* R.J. Naiman and H. Decamps, eds. Vol 4, Man and the Biosphere Series, Parthenon Publ. Group. Park Ridge, NJ.

Forman, R.T.T. 1987. The ethics of isolation, the spread of disturbance and landscape ecology, pp. 213-229 in: *Landscape Heterogeneity and Disturbance.* M.G. Turner, ed. Ecological Studies 64. Springer-Verlag, NY.

Healy, R.G. and J.L. Short. 1981. *The Market for Rural Land: Trends, Issues, Policies.* Conservation Foundation, Washington, DC.

Marks, P.L. 1980. The role of the pin cheery (*Prunus pennsylvania*) in the maintenance of stability in northern hardwood ecosystems. Ecol. Monogr. 44:73-88.

Odum, E.P. 1982. Diversity and the forest ecosystem. pp. 35-41 in: *Natural Diversity in Forest Ecosystems.* James and June Cooley, eds. USDA Forest Service Southeastern Experiment Station, Asheville, NC.

__________. 1989. *Ecology and Our Endangered Life-support systems.* Sinauer Associates, Sunderland, MA.

Odum, E.P. and R.L. Kroodsma. 1977. The power park concept. Amelorating man's disorder with nature's order. pp. 1-9 in: *Thermal Ecology II.* G.W. Esch and R.W. McFarlane, eds. Nat. Tech. Info. Service, Springfield, VA.

Odum, E.P., J.T. Finn and E.H. Franz. 1979. Perturbation theory and the subsidy-stress gradient. BioScience 29:347-352.

Odum, E.P. and L.J. Biever. 1984. Resource quality, mutualism, and energy partitioning in food chains. Am. Nat., 124:360-376.

Odum, W.E. 1990. Internal processes influencing the maintenance of ecotones: do they exist? Chap. 6 in: *The Ecology and Management of Aquatic-terrestrial Ecotones.* R.J. Naiman and H. Decamps, eds. Vol 4 *Man and Biosphere Series.* Parthenon Publ, Group, Park Ridge, NJ.

Pimm, S.L. 1992. *The Balance of Nature? Ecological Issues in the Conservation of Species and Communities.* Univ. Chicago Press, Chicago.

Remillard, M.M., G.K. Gruendling and D.J. Bogucki. 1987. Disturbance by beaver (*Castor canadensis* Kuhl) and increased landscape heterogeneity, pp. 103-122 in: *Landscape Heterogeneity and Disturbance.* M.G. Turner, ed. Ecol. Studies 64. Springer-Verlag, NY.

Rhoades, R.E. 1991. The world's food supply at risk. Nat. Geogr. 179(4):74-105.

Saunders, D.A., R.J. Hobbs and C.R. Margules. 1991. Biological consequences of ecosystem fragmentation: a review. Conserv. Biol. 5:1832.

Schindler, D.W. 1990. Experimental perturbations of whole lakes as tests of hypotheses concerning ecosystem structure and function. Oikos, 57 25-41.

Van Voris, P., P.V. O'Neill, W.R. Emanuel and H.H. Shugart. 1980. Functional complexity and ecosystem stability. Ecol. 61:1352-1360.

Whitcomb, R.F., J.F. Lynch, M.K. Klimikiewicz, C.S. Robbins, B.L. Whitcomb and D. Bystrak. 1981. Effects of forest fragmentation on avifauna of the eastern deciduous forest, pp. 125-205 in: *Forest Island Dynamics in Man-dominated Landscapes.* R.L. Burgess and D.M. Sharp, eds. Ecol. Studies 41. Springer-Verlag, NY.

Whittaker, R.H. 1960. Vegetation of the Siskiyou Mountains, Oregon and California. Ecol. Monogr. 30:279-338.

__________. 1965. Dominance and diversity in land plant communities. Science, 147:250-260.

Biological Diversity: Problems and Challenges. Edited by S.K. Majumdar, F.J. Brenner, J.E. Lovich, J.F. Schalles, and E.W. Miller.

Chapter Four

U.S. POLICIES TO CONSERVE BIODIVERSITY

DAVID E. BLOCKSTEIN
Princeton University
Committee for the National Institute for the Environment
730 11th St. NW
Washington, DC 20001-4521

INTRODUCTION

In the 1980's the term biological diversity burst out of the scientific journals where it had rested comfortably for decades of debate about the relationship between diversity and stability into the policy arena where the debate has been diverse and anything but stable. It became shortened as "biodiversity"[1] and, in 1992, made the headlines when the U.S. became the only industrialized nation to refuse to sign a global biodiversity treaty, which was signed by 158 countries.

This chapter explores the issue of biodiversity policy in the United States. It discusses the key legal tools available under the piecemeal approach that is presently being followed and presents a comprehensive approach as an alternative.

PARTIAL POLICIES, PIECEMEAL PRACTICES

The environmental policy of the United States is a collection of laws, Presidential executive orders, and regulations that have developed over the past century. The most comprehensive statement is the National Environmental Policy Act (NEPA), which lays out the right of Americans to a healthy environment. Although NEPA calls for "an environment which supports diversity," there is no U.S. national policy statement on biodiversity. Biodiversity had not been enunciated as a policy issue in 1969 when NEPA was approved and subsequent attempts to create a national biodiversity policy[2] have not yet been successful.

Despite this lack of an explicit statement, concerns about biodiversity underlie or are incorporated in a number of key environmental laws, including NEPA, the Endangered Species Act, the Marine Mammal Protection Act, the Clean Water Act, and the National Forest Management Act. The following is a review of some of the most significant of these laws and authorities. For a more detailed review that emphasizes the legal mandates of the Environmental Protection Agency, see Fischman.[3]

Endangered Species Act

Practices and attitudes towards biodiversity and, indeed, a sizeable share of natural resource management in the U.S. are greatly influenced by the powerful Endangered Species Act (ESA). Among other things, the ESA prohibits the "taking" (broadly interpreted as killing, harming, or harassing individuals directly or by degrading their habitat) of endangered species that have been listed under the Act. It also sets up procedures to ensure that federal actions will not jeopardize listed species or adversely modify their critical habitat. The definition of "species" includes distinct populations of vertebrates and subspecies of other plants and animals. Species are listed for biological reasons only, although economic criteria are incorporated in other aspects of the Act. As of 1992, some 1,300 taxa are listed as threatened or endangered. More than 750 of these occur in the United States.

There has been much controversy surrounding the ESA during its 20 year history. Advocates of unrestrained development point to conflicts where the resource base had been greatly destroyed and restrictions later imposed to protect endangered species (such as the threatened northern Spotted Owl) as a reason to weaken the Act.[4] However, the U.S. General Accounting Office[5] found that when the Fish and Wildlife Service and the National Marine Fisheries Service were asked to consider the impact of proposed development on an endangered species, in over 90% of the cases the projects were allowed to proceed. In nearly 90% of the remaining cases, the agencies offered alternative courses of action that would allow the project to continue and mitigate its impact on the species. Thus, claims that the ESA is preventing economic development are generally false.

Even among conservationists, the ESA is controversial. Some have argued that the species-centered approach of the ESA should be replaced by an ecosystem-centered approach.[6] They are concerned that having endangered species as the focal point means that tradeoffs will always favor the endangered species even at the expense of the ecosystem. This has happened in Nevada where a century-old irrigation boondoggle, the Newlands Reclamation Project has diverted so much water to agriculture that there was not enough to maintain the quality of both Pyramid Lake and Stillwater Wildlife Management Area.[7] Stillwater is much richer biologically as a refuge for migratory birds; Pyramid Lake is the home to the cui-ui, a large endangered fish that is of considerable importance to the local Indians. Although the ESA has not *required* agencies to trade off the refuge to maintain the endangered fish, that is the result of Department of the Interior decisions and court orders related to the ESA and Indian treaty rights. Conservation groups have been working to buy water rights and negotiate to reverse the desiccation of Stillwater,[7,8] but for now, the tradeoff favors the fish, to the detriment of the richer ecosystem.

In fact, the statutory goals of the ESA "to provide a means whereby the *ecosystems* (emphasis added) upon which endangered species and threatened species depend may be conserved . . ." are ecosystem focused, but centered around the location of endangered species. Two types of habitat protection are provided by ESA. First, actions by any party, public or private, (including habitat destruction) that *jeopardize* the recovery of the species are not allowed. Secondly, the law requires, with certain exceptions, designation of habitat critical to the survival of the listed species. Once designated as critical, no Federal agency or action can adversely alter the habitat, even if the modification does not jeopardize the species. However, the Fish and Wildlife Service has not clearly defined "adverse modification" in this context and it has generally been used synonomously with "jeopardy." These habitat conservation provisions give the ESA the potential to protect biodiversity beyond the individual species that has been listed. On the other hand, the law is reactive in that it only provides protection once a "species" is threatened with extinction and presumably has lost most of its genetic diversity. Additionally, ESA's habitat protection requirements only come into place once a species has been listed.

At least some of the apparent flaws in the ESA are due to its implementation and funding level, rather than the law itself.[9] The General Accounting Office[5,10] has generally been critical of the implementation of the ESA. Many duties of the responsible agencies have not been carried out. Critical habitat has been designated for less than 20% of the listed species.[5] Only 63% of the species under the jursidiction of the Fish and Wildlife Service, and only 29% of the species under the jurisdiction of the National Marine Fisheries Service had recovery plans approved as of 1991.[5] Proactive provisions of the law such as the requirement that the Secretaries of Interior and Agriculture develop and implement a program to conserve fish, wildlife and plants, including non-listed species, have never been implemented.

In addition, the listing process is slow, with a backlog of some 600 qualified species awaiting paperwork and some 3,000 other "candidate" endangered species awaiting technical review before they can be protected. A 1992 court decision (Fund for Animals and Defenders of Wildlife vs. Lujan) should help reduce this delay by requiring that groups of species be listed on an ecosystem basis whenever this is biologically appropriate. The process is sometimes highly politicized, especially in economically important cases such as the 1991 and 1992 listing of the northern Spotted Owl and certain populations of Pacific salmon as "threatened." Finally, funds are very limited. In 1992 the FWS received its largest appropriation ever to implement the Act - a paltry $42.3 million or about $10,000 per listed and candidate species.

Although many of these problems are due to inadequate implementation, rather than the legal language, the Endangered Species Act alone will not save biodiversity. Its reactive, save the species one-by-one approach is inadequate to protect a full range of diversity. The Nature Conservancy estimates that some 9,000 U.S. species may be at risk; every region of the U.S. is affected.[11] Although the Endangered Species Act is our biological safety net, it is sagging under the sheer weight of species needing protection.[12]

The Marine Mammal Protection Act

The Marine Mammal Protection Act (MMPA) is, based on its legal language, actually more stringent than the ESA. It prohibits the killing, taking, or harassing of marine mammals unless a permit is granted. Unlike the ESA, which only protects species once they have been listed, the MMPA assumes that all marine mammals should be absolutely protected, but allows exceptions if the potential permittee can prove that the species will not be harmed. Incidental taking in fisheries and other industries is also subject to a wide general permitting program. However, the MMPA is very limited in scope in that it only protects cetaceans, pinnipeds, and sea otters. Additionally, it only protects them from direct threats. Indirect threats, such as pollution, cannot be prevented under MMPA.

National Environmental Policy Act

The National Environmental Policy Act (NEPA) differs from ESA and MMPA in that it is largely procedural, and not prohibitive. It calls for full disclosure of environmental effects of a proposed major federal or federally-permitted project through an environmental impact statement (EIS). Alternative courses of action, including "no action" must be discussed. Projects that have a negative environmental impact are *not* automatically stopped by NEPA. However, under authority granted by the Clean Air Act, the Administrator of the Environmental Protection Agency (EPA) has the authority to provide substantive reviews of EIS's and delay projects with adverse impacts by referring them to the Council on Environmental Quality (CEQ) for consultation with the agency proposing the action.

The CEQ, which oversees the EIS process, has never issued explicit guidelines for consideration of impacts on biodiversity. However, in 1991-92, the CEQ held a series of workshops around the nation to discuss how ecological effects, including biodiversity, could be discussed in EIS's.[13] NEPA could be an important tool to protect diversity but more explicit guidelines are needed both for assessment and for determining what constitutes an "unsatisfactory" effect on the viability of a species or ecosystem. It is expected that such guidelines will be issued under the Clinton administration.

NEPA, like ESA, is reactive in that it only comes into effect after a project has been proposed. It has no provision for designating areas of high value where environmental values are the highest use.

Clean Water Act

The Federal Water Pollution Control Act or Clean Water Act (CWA) could be a powerful tool to protect biodiversity. Its objectives are to "restore and maintain the chemical, physical, and *biological integrity* of the Nation's waters" (emphasis added). These objectives certainly sounded like a biodiversity goal, but rarely have been interpreted this way.[3] Although the Act provides no definition of biological integrity, Karr and Dudley[14] define it as "the ability of an [aquatic] ecosystem to support and maintain a balanced, integrated, adaptive community of organisms having a species composition, diversity, and functional organization comparable to that of the natural habitats within a region."

In fact, aquatic diversity has rarely been raised as a policy issue in the national discussion of biodiversity. There were no speakers on freshwater aquatic biodiversity and only one on marine biodiversity at the 1986 national forum on biodiversity.[2] This is particularly disturbing when 78 (10%) of the remaining native freshwater fishes of North America are listed as endangered or threatened, and another 68 qualify for listing according to the American Fisheries Society.[15] They report a 45% increase in rare fishes in the 1980's.

The Clean Water Act prohibits the discharge of pollutants into U.S. waterways unless specifically permitted. States set standards for water quality, subject to Federal oversight and guidance. Water quality standards may be narrative or quantitative and often include protection of aquatic life among their objectives.[16] Although biological pollution (exotic species) is often a cause of loss of native aquatic diversity, no State has attempted to use CWA standards to control the introduction of exotics.

Provisions of the Clean Water Act that deal with wetlands are of particular importance for biodiversity. Only half of the 221 million acres of wetlands that covered the lower 48 states during colonial times still exist.[17] Wetlands are still being destroyed in the U.S. at rates of 35 acres/hour or 300,000 acres yer year.[18] Many imperiled components of biodiversity, including 43% of all listed endangered and threatened species, occur in wetlands during some part of their life cycle.[18]

Section 404 of the CWA provides the most comprehensive federal regulatory protection against actions that degrade wetlands.[16] It requires a federal permit for the discharge of dredged or fill material into wetlands and other waters. The Army Corps of Engineers has primary responsibility for the permit process. The Environmental Protection Agency (EPA), which reviews permit applications being considered by the Corps has established guidelines for both agencies to use in reviewing permit applications. To date, biodiversity has not been included in these guidelines.

From the mid-1980's through the early 1990's, the EPA several times revised their policies regarding how to "delineate" which wetlands are subject to Section 404. There has been intense controversy, focussing primarily on the value of small, seasonal, freshwater wetlands. Unfortunately, the value of these wetlands with respect to biodiversity has not been a contributing factor in this the debate. In 1992, Congress asked the National Academy of Sciences (NAS) to provide a scientific analysis of wetland delineation. The charge to the NAS includes investigation of methodologies to identify, measure and compare wetland values and functions. The NAS could discuss the value of wetlands for biodiversity and thus marry the issues of biodiversity and wetland conservation.

The Act gives the EPA authority for advanced identification (ADID) of valuable wetlands. Under this process EPA can identify in advance wetland sites that are suitable or unsuitable for discharge of dredged or fill material. Valuable sites could be put "off limits" and no permits would be granted for these sites. Through this process, Section 404 could be used for pro-active planning including for biodiversity. However, EPA has never used this "advance veto authority."[3]

In fact the EPA rarely has vetoed projects that the Corps of Engineers has decided to permit.[3] One notable exception took place in 1989 when EPA Administrator William Reilly stopped the proposed Two Forks dam because water diverted from the Platte River would harm the huge population of Sandhill Cranes that uses the Platte as a spring migratory stopover.

The World Wildlife Fund[16] has proposed that states develop comprehensive wetlands strategies. These strategies would be processes to bring together individuals and groups to help to identify wetland issues in the state and to develop solutions to achieve wetlands protection goals. Such strategies could enhance the protection of wetland-dependent elements of biodiversity, particularly if biodiversity conservation is identified as one of the goals. However, to be truly effective, the strategic process must also consider the role of non-wetland habitats and must deal with impacts (such as disruptions of river systems) that occur outside the state.

National Forest Management Act

The National Forest Management Act (NFMA) of 1976, which governs the actions of the Forest Service, contains the most direct biodiversity provision of any Federal law. Section 6(g)(3)(B) directs the Service, in its forest planning, to "provide for diversity of plant and animal communities based on the suitability and capability of the specific land area . . .". The regulations telling the Service how to implement this provision are even more specific, "fish and wildlife habitat shall be managed to maintain viable populations of native and desired nonnative vertebrates in the planning area. For planning purposes, a viable population shall be considered to have the estimated numbers and distribution of reproductive individuals to insure continued existence is well-distributed throughout the planning area" (36 Code of Federal Regulations 219.19).

The NFMA requires each national forest to develop a management plan. This process which began in the 1980's, has been marked with controversy. Most of the plans were appealed by conservationists, many under the biodiversity provision under NFMA. A group of botanists from the University of Wisconsin appealed the plans for the Chequamegon and Nicolet National Forests in Wisconsin, on the basis that the plans emphasized commodity (timber) production over diversity conservation. When these appeals were turned down by the regional office of the Forest Service, the botanists, joined by the local Sierra Club, sued the Forest Service. Regardless of the resolution of the case, it shows how the law and science can be used to attempt to conserve biological diversity.

The substantive power of the biodiversity requirements in NFMA plus the procedural requirements for full disclosure of probable environmental impacts under NEPA were used successfully by environmentalists in the early 1990s in court cases putting moratoriums on timber sales in the habitat of northern Spotted Owls.

Recently, the Forest Service has attempted to respond to its biodiversity mandate. In the late 1980's it established a committee on biodiversity, which then evolved into a program entitled "New Perspectives," which seeks to find ways to incorporate ecological values into forest management. However, the real test of how the Forest Service is acting to conserve biodiversity on the ground will take place during its next round of forest plans in the late 1990's and in the management actions that follow.

Other Laws

Many other laws, from the Migratory Bird Treaty Act to the National Plant Germplasm System Act, are directed towards the conservation of components of biodiversity. However, the Congressional Office of Technology Assessment[19] found that this piecemeal collection of laws is "too disjunct to address the full range of concerns over the loss of biological diversity." Without a recognition of the importance of diversity, this patchwork lacks a central focus. Different agencies have different mandates, management practices often conflict, and many so-called protected areas such as national parks and national wildlife refuges are too small or are not managed for the conservation of biodiversity.

As a result, the problem worsens exponentially, habitats continue to be fragmented and the list of candidates for the Endangered Species Act continues to grow. In the case of many systems from the southern California coastal sage scrub which basically consists of habitat islands in a simmering sea of development[20] to many western rivers, which have been buffeted by forces including channelization, water withdrawals and exotic species, we are facing ecosystem collapse.[21] Even an Endangered Species Act at full strength would not be enough to save them. We clearly need to bring together the various parts of our conservation policy repertoire.

A NEW WAY AHEAD

We need a comprehensive Federal approach to conserve biodiversity.[12,22] Piecemeal policy has led to a fragmented landscape with pockets of biodiversity and areas managed for favored species (largely game species), but with declining trends in most components of biodiversity.

Many of the ultimate causes of biodiversity loss - a human population whose numbers and consumption of resources are out of control - are unlikely to be addressed by a biodiversity policy. However, given this background of largely uncontrollable factors, it behooves us to take the biodiversity problem seriously and recognize that a planned approach to natural resource management is necessary to slow and eventually reverse the tide of devastation. We need a national policy,[23] implemented on a scale of bioregional ecosystems[11] - recognizing that the artificial boundaries of our designated public lands do not match ecological boundaries and thus complicate our ability to maintain biodiversity within them.[24]

We need to make biological diversity a primary goal of Federal land management. Agencies must adopt biodiversity as a goal and develop strategic approaches for its conservation. However, agencies must act in a united front. In late 1992, the Fish and Wildlife Service held a working meeting to develop an agency biodiversity strategy. However, it was immediately clear that many factors affecting the biota within the agency's charge were beyond the Service's jurisdiction to control. An agency strategy, while necessary, must be taken in concert with other agencies. From 1988 through 1992, Congressman James H. Scheuer introduced legislation making biodiversity conservation a national goal and mandating a Federal biodiversity strategy.[25] Because the private sector has a significant impact on the nation's biodiversity, private partnership must be attained in all aspects of a comprehensive biodiversity strategy.[26]

At least five fronts must be attacked simultaneously: research, protected areas, multiple use areas, ecological and species restoration, and education.[12,22]

Research

The first step in biodiversity protection should be to describe biodiversity and identify factors that are causing its decline. We know that the four horsemen of the biodiversity apocalypse are habitat destruction and fragmentation, exotic species, direct take, and pollution. A fifth horseman - climate change - may soon outrace all of them. Yet the interactions among these factors are not well understood.

The amount and direction of environmental research is wholly inadequate. The U.S. government spends about $2.4 billion on all environmental research annually.[27] This is about 3% of the total federal research and development effort. Less than $500 million goes to environmental biology.[28] With this in mind, it is no wonder that we don't know more about biodiversity and that the public has little understanding of the issue.

In addition, most of our environmental research is either narrowly prescriptive, trying to solve a particular environmental problem under a particular law or regulation, or is fundamental with little applied potential. Problems such as the interaction of habitat destruction and other causes of biodiversity loss fall in the chasm between the basic and narrowly applied research. For example, the Environmental Protection Agency in 1992 identified habitat and biodiversity as an "emerging ecological issue," but has not allocated any research money to the issue (EPA, Office of Research and Development, unpublished). No existing Federal agency can take the broad perspective that is needed.

This would change if an ambitious proposal to create a "National Institute for the Environment (NIE), is adopted. The NIE would be a lead agency for the science of understanding the environment.[29,30] It would support mission-oriented environmental research and training, like the National Institutes of Health supports biomedical research and training. The proposed agency would sponsor extramural panel-reviewed research on topics such as the relative importance of factors causing the decline of the Pacific salmon, the ecological impacts of ozone depletion, and bioregional studies such as understanding the interaction of people and the environment along the U.S. - Mexico border. It could look at big picture issues such as the relative importance of various stressors on biodiversity and could undertake retrospective analyses of land use patterns. The proposal for the NIE has been studied by the National Academy of Sciences and will be before Congress for consideration in 1993.

A related proposal is the Sustainable Biosphere Initiative (SBI) of the Ecological Society of America.[31] The SBI is an attempt by academic ecologists to set an agenda of research that is relevant to societal needs. It targets three broad areas of research - the ecological basis of biodiversity, the ecological basis of global change, and sustainable ecological systems. Although the Ecological Society has been lobbying hard for the adoption of the SBI by Federal agencies that fund environmental research,[32] these agencies are hampered by a narrow missions and limited budgets. The SBI will never be fully implemented unless the NIE is established.[33]

One hundred years ago, the U.S. had an active Biological Survey which cataloged the biotic wealth of the nation. This survey has essentially disappeared except for a vestigial program of the Fish and Wildlife Service. Increasing awareness of the biodiversity crisis brought calls for a national biological survey.[34] This led Congressman Scheuer to propose the creation of a National Center for Biological Diversity and Conservation Research. The purpose of this Center would be "to summarize and enhance the knowledge of the distribution, status, and characteristics of the biota in a manner that can be used in conservation and management."[25] Such an institution would allow us to know the true status of the various elements of biological diversity.

Although Congressman Scheuer retired in 1992 and Congress has not yet approved the National Biodiversity Center, there are still opportunities. A draft Presidential Executive Order creating a National Biodiversity Center at the Smithsonian Institution was prepared late in the Bush Administration, but was not signed by President Bush before he left office. President Clinton may sign this order as part of the U.S. obligations under the Global Biodiversity Treaty, which he signed in June 1993. The Center proposed in the Executive Order would be a clearinghouse, which will organize and provide access to information about the status of biodiversity. More recently, Secretary of Interior Bruce Babbitt proposed formation of a National Biological Survey (NBS) within the Department of the Interior.

Other needed research includes the development of a national map showing the distribution of areas with various types of legal protection as compared with the distribution of biological diversity. Such an approach, called "gap analysis" (because it identifies gaps in our legal protection system), has been initiated, using geographic information system (GIS) technology, in a project headed by Dr. Michael Scott of the U.S. Fish and Wildlife Service.[35] Thus far, the project has proceeded on a state-by-state basis and lacks national standardization. The National Biological Survey should adopt this project, provide national coordination and link it with finer scale approaches such as The Nature Conservancy's Natural Heritage System.

Protected Areas

The traditional approach to protection of biodiversity in terrestrial environments has been the creation of parks and reserves. The freshwater aquatic equivalent is designated of wild and scenic rivers (although biodiversity has never been a criteria for designation). However, only 3% of America's land surface are "protected" and only 0.2% of America's river miles are designated as wild and scenic.[36] Additionally, eight areas have been designated as national marine sanctuaries.

Biodiversity is not a criteria for selection of any of these protected areas. Many of these areas are not managed for the conservation of biodiversity. Additionally, only about half of the potential natural vegetation types in the U.S. are sufficiently represented in designated federal "reserves."[37] However, they could form the basis of a national biodiversity reserve system. To establish such a system, existing lands and waters must be managed with a common purpose of maintaining biodiversity (although not necessarily through a single management authority) and additional areas added in order that multiple examples of all ecosystem types be included. These areas must be large, in order that ecological processes, such as fire and other disturbances can continue. Protected areas must be physically linked by broad corridors to allow migration of animals and plants.[38]

"Protection" itself is somewhat of a misnomer because legal designation does not protect land and water from invasion by exotic organisms and airborne pollutants. Additionally, many of the protected areas are too small to maintain a full complement of species.[39] The concept of protected areas for biodiversity is incomplete for rivers and most streams which are open systems, although ponds and small lakes can be "protected" by inclusion in national parks and wilderness areas. Even though protection is not absolute, on a relative scale biodiversity is generally more likely to be maintained in areas that are specifically set aside for biodiversity conservation.

Multiple Use Areas

Most of the land and water in the U.S. can be considered as multiple use areas, where economic activities take place and have varying impacts on the biota. Aquatic habitats, wetlands, and riparian zones are particularly heavily impacted by human activities including pollution, dams, and withdrawal of water. Nationwide, only 23 million acres (20%) of riparian vegetation remains in natural or semi-natural condition.[40] Much terrestrial habitat has been altered and fragmented. Because the small fraction of the habitat that is being managed for conservation is insufficient to maintain the full complement of biodiversity, the battle to maintain biodiversity is likely to be won or lost on multiple use areas.

The agricultural landscape has been the most overlooked component of America with respect to biological diversity. Agriculture is America's largest industry and has impacts on biodiversity ranging from land conversion to contamination by pesticides and herbicides. Paradoxically, agriculture is absolutely dependent upon biodiversity for genetic improvement of crops to maintain yields under changing environmental conditions. Although a few provisions of the "Farm Bill" (Food, Agriculture, Conservation, and Trade Act of 1990) such as the conservation reserve program, "swampbuster" and "sodbuster" have been implemented which incorporate the use of subsidies as incentives for restoration or disincentives for cultivation of land that is marginal for agriculture but has particular environmental potential, much work remains to be done. We need to take a comprehensive look at the relationship between agriculture and biodiversity, including issues ranging from subsidies to germplasm. Perhaps, an enterprise so dependent on the genetic diversity of the few domesticated species can be transformed to help maintain the diversity of wild species as well.

Most of the federal laws mentioned previously affect multiple use areas. The environmental impact assessment process of NEPA affects public lands and major private activities that require federal permits. The permit process of Clean Water Act can have a more stringent effect controlling small scale private development of aquatic systems including wetlands and watersheds. Explicit directions for consideration of biodiversity in EIS's and development of quantitative biological criteria for water quality are necessary.

Under the CWA, water quality is determined by comparison of the river or stream in question with "unimpaired" waterways within the region that have the same designated use and similar physical characteristics. For more than a decade, an Index of Biotic Integrity (IBI) has been in use for this purpose.[41,42] By using unimpaired waterways as a reference, biointegrity is defined as the difference from a naturally-occurring baseline. Thus, no precise definition of biointegrity or biodiversity is necessary. Karr (pers. comm.) is now working to develop a similar reference criteria for biodiversity in terrestrial systems. If we have such criteria, we no longer will need to argue about the desired end state of biodiversity conservation, but will simply let nature be the determinant of success. An impoverished ecosystem would be one that is significantly different from the baseline system.

Ecological Recovery

Forty taxa of North American fishes have become extinct in the last century, 19 in the last 25 years.[43] Some 27% of extant North American fishes are endangered or near endangered and the number is increasing.[10] Similar results would likely be found for other taxa, were someone to make the compilation. We cannot afford to simply write off this substantial portion of our biodiversity. Although endangered species protection is expensive, it must be undertaken if we are to have any opportunity to continue to have native ecosystems. We should view recovery of endangered species as the short-term costs of correcting past mistakes. The long-term costs of extinction will surely exceed these short-term costs of restoration.

So many species are endangered (whether legally designated or not) that a species-by-species approach is not enough. Many endangered species share the same location and are affected by the same problems. For example, 10 (27%) of the 35 fish species in the Colorado River system are endangered, threatened or candidates for official listing.[43] Thus, we really are faced with failing ecosystems and the endangered species are just the tip of the iceberg.[21] This necessitates a causal rather than species-oriented approach to ecological recovery. We must identify the principal factors that have caused the decline in the biointegrity of the system and work to eliminate these factors and reverse their effects. Simply managing to artificially increase a species in a traumatized habitat will not be effective.

Most restoration must be considered to be experimental. Much research is needed to elucidate the workings of environmental systems, as well as how to try to put them back together once they have been altered.

Although restoration of degraded habitats is necessary, we must also be skeptical about the concept of mitigation compensation, which attempts to trade the destruction of one habitat (such as a wetland) for the "creation" or "restoration" of another. Although we can restore some ecosystems to a semblance of their natural state, it is foolish to destroy what now exists and pretend that we have the ability to recreate the system in its entirety elsewhere.

Education

None of the other steps will take place without education. As shown by the case of the snail darter, although laws can protect endangered species, most people simply do not care about the diversity of most taxa. Darters, louseworts, and beetles will never gain the public appeal of certain more "charismatic" birds and mammals. "Save the pupfish" only became a rallying cry, in part, because the name sounds "cute." Yet, pupfish are now being used as models in studies of hypertension due to their tolerance of hypersaline conditions. We need to educate the public about the economic, scientific, and ecological values of biodiversity.

Environmental education is essentially science education. It is education about the functioning of the natural world and its response to perturbation by people. Environmental education must occur on all levels. The consequences of ignorance make education far too important to be left to the designated educators. It is also unrealistic to only educate the young. If we wait until the "young people" are in charge, what biodiversity will remain for them to be in charge of? Not only must we educate the young, we must also educate the powerful.

ACHIEVING A COMPREHENSIVE APPROACH

A comprehensive approach can only be achieved by leadership from the highest levels - be it the President, the Congress or the governor of a State. The election of President Clinton and Vice-President Gore offers an opportunity for this leadership. The goal of biodiversity conservation must be articulated clearly as a statement of policy. For the U.S., a biodiversity strategy must involve all of the relevant federal natural resource management agencies along with scientific advice and public consent. It should incorporate national planning, but operate on the scale of designated regional ecosystems. Incentives should be provided for private sector participation. Action on the ground must be integrated with research and education, as described above.

The core of such a strategy is presently being attempted in California. In 1991, state and federal agencies signed a memorandum of agreement and created an executive council to set statewide goals for the protection of biodiversity and recommend statewide standards, and cooperate in policies, management, programs, research and monitoring, restoration and education. The State has been divided into regions, based on biomes and ecosystems. Bioregional councils will develop regional biodiversity strategies consistent with statewide goals and standards and will work with regional and local authorities to implement conservation strategies. The California strategy appears to contain all of the elements to conserve some of America's most diverse and threatened biota. The success of this strategy will depend upon political will, money, and an improvement of the science used in management.

Michigan, in 1992, became the first state to approve explicit biodiversity legislation, which was modeled after the Federal legislation. They will also develop a state-wide biodiversity strategy. Other states, including Florida and Kansas are considering similar legislative approaches.

These state-level ventures, if successful, could lead the way for a national biodiversity strategy. The vision of Congressman Scheuer for passage of a national biodiversity conservation law may finally be achieved.

ACKNOWLEDGEMENTS

This piece is dedicated to the memory of Joseph D. Pratt, whose infectious spirit and total dedication for biodiversity remains even after he has passed from this Earth. Thanks to Rob Fischman of Indiana University Law School, John Fitzgerald of Defenders of Wildlife, Debra Prybyla of the World Wildlife Fund and David Wilcove of the Environmental Defense Fund for their comments on versions of this manuscript.

LITERATURE CITED

1. Wilson, E.O., ed. 1988. *Biodiversity.* National Academy Press, Washington, DC. 251 pp.
2. Blockstein, D.E. 1988. U.S. legislative progress toward conserving biological diversity. *Conserv. Biol.* 2:311-313.
3. Fischman, R.L. 1992. Biological diversity and environmental protection. *Environmental Law.* 22:435-502.
4. Mann, C.C. and M.L. Plummer. 1992. The butterfly problem. *The Atlantic.* 269:47-70.
5. U.S. General Accounting Office. 1992. Endangered species act: types and number of implementing actions. GAO/RCED-92-131BR. Washington, DC. 40 pp.
6. Rohlf, D.J. 1991. Six biological reasons the endangered species act doesn't work and what to do about it. *Conserv. Biol.* 5:273-282.
7. Laycock, G. 1988. What water for Stillwater? *Audubon* 90(6):14-25.
8. Vetter, D. 1991. Desert Mirage? *Nature Conservancy* 41(5):22-27.
9. Tobin, R.J. 1991. *The Expendable Future: U.S. Politics and the Protection of Biological Diversity.* Duke Univ. Press, Durham, NC. 325 pp.
10. U.S. General Accounting Office. 1988. Endangered species management improvements could enhance recovery program. GAO/RCED-89-5. Washington, DC. 100 pp.
11. Council on Environmental Quality. 1990. Environmental Quality. 21st Annual Report. U.S. Government Printing Office, Washington, DC. 388 pp.
12. Blockstein, D.E. 1989. Toward a federal plan for biological diversity. *Issues in Science and Technology.* 5(4):63-67.
13. Council on Environmental Quality. 1993. Incorporating biological considerations into environmental impact analysis under the National Environmental Policy Act. U.S. Government Printing Office, Washington, DC. 79 pp.
14. Karr, J.R. and D.R. Dudley. 1981: Ecological perspective on water quality goals. *Environmental Manage.* 5:55-68.
15. Williams, J.E. and seven coauthors. 1989. Fishes of North America, endangered, threatened or of special concern: 1989. *Fisheries* (Bethesda) 14(6):2-20.
16. World Wildlife Fund. 1992. *Statewide wetlands strategies.* Island Press. Washington, DC. 268 pp.

17. U.S. Department of Interior Fish and Wildlife Service. 1990. *Wetlands: meeting the President's challenge.* U.S.F.W.S. Washington, DC.
18. Feierabend, J.S. 1992. *Endangered species, endangered wetlands: life on the edge.* National Wildlife Federation, Washington, DC. 49 pp.
19. U.S. Congress Office of Technology Assessment. 1987. *Technologies to maintian biological diversity.* OTA-F-330. U.S. Government Printing Office, Washington, DC.
20. Atwood, J.L. 1992. A maximum estimate of the California Gnatcatcher's population size in the United States. *Western Birds* 23:1-10.
21. Williams, J.E. and seven coauthors. 1985. Endangered aquatic ecosystems in North American deserts with a list of vanishing species in the region. *J. Ariz. - Nev. Acad. Sci.* 20:1-62.
22. Blockstein, D.E. 1992. An aquatic perspective on U.S. biodiversity policy. *Fisheries* (Bethesda) 17(3):26-30.
23. Titus, T.R. 1992. Biodiversity: the need for a national policy. *Fisheries* (Bethesda) 17(3):31-34.
24. Schonewald-Cox, C.E. 1988. Bringing boundaries to the protection of nature reserves. *BioScience* 38:480-486.
25. U.S. House of Representatives. 1990. *National biological diversity conservation and environmental research act.* U.S. House of Representatives Report 101-901. U.S. Government Printing Office, Washington, DC.
26. Keystone Center. 1991. *Final consensus report of the Keystone policy dialogue on biological diversity on federal lands.* The Keystone Center, Keystone, CO. 96 pp.
27. Mandula, B.B. and D.E. Blockstein. 1992. Federal funding for environmental research. *Environ. Sci. Technol.* 26:1496-1502.
28. National Science Foundation. 1992. Budget summary, fiscal year 1993. National Science Foundation, Washington, DC.
29. Howe, H.H., S.P. Hubbell and D.E. Blockstein. 1990. Rationale for the National Institutes for the Environment. The Environmental Professional 12:360-363.
30. Committee for the National Institute for the Environment. 1993. Executive Summary: A proposal for a National Institute for the Environment. *The Environmental Professional* 15:150-154.
31. Lubchenco, J. and 15 coauthors. 1991. The Sustainable Biosphere Initiative: an ecological research agenda. *Ecology* 72:371-412.
32. Janetos, A.C. 1992. The sustainable biosphere initiative and the federal agencies. *Bulletin of the Ecological Society of America* 73(1):25-28.
33. Carney, H.J. 1992. Implementation of the national and international sustainable biosphere initiatives. *Bulletin of the Ecological Society of America* 73:228-230.
34. Kim, K.C. and L. Knutson, eds. 1986. *Foundations for a national biological survey.* Association of Systematics Collections, Lawrence, KS. 215 pp.
35. Scott, J.M. 1990. Preserving life on Earth: we need a new approach. *J. Forestry.* 88(3):13-14.
36. Benke, A.C. 1990. A perspective on America's vanishing streams. *J.N. Am. Bethol. Soc.* 91:77-88.

37. Crumpacker, D.W., S.W. Hodge, D. Friendly and W.P. Gregg. 1988. A preliminary assessment of the status of major terrestrial and wetland ecosystems on Federal and Indian lands in the United States. *Conserv. Biol.* 2:103-115.
38. Hudson, W.E., ed. 1991. *Landscape Linkages and Biodiversity.* Island Press, Washington, DC. 194 pp.
39. Newmark, W.D. 1987. A land-bridge internal perspective on mammalian extinctions in western North American parks. *Nature* 325:430-432.
40. Hunt, C.E. 1988. *Down by the River.* The impact of federal water projects and policies on biological diversity. Island Press, Washington, DC. 266 pp.
41. Karr, J.R. 1981. Assessment of biotic integrity using fish communities. *Fisheries (Bethesda).* 6(6):21-27.
42. Miller, D.L. and 13 coauthors. 1988. Regional applications of an index of biotic integrity for use in water resource management. *Fisheries (Bethesda).* 13(5):12-20.
43. Williams, J.E. and R.R. Miller. 1990. Conservation status of the North American fish fauna in fresh water. *J. Fish Biol.* 37 (Supplement A):79-85.

Biological Diversity: Problems and Challenges. Edited by S.K. Majumdar, F.J. Brenner, J.E. Lovich, J.F. Schalles, and E.W. Miller.

Chapter Five

CONSERVATION OF BIODIVERSITY: The Indian Scenario

ARUN KUMAR SHARMA

Centre for Advanced Study in Cell & Chromosome Research
Department of Botany
University of Calcutta
35 Ballygunj Circular Road
Calcutta 700019, India

INTRODUCTION

The conservation of biological diversity is one of the most crucial issues facing the world in general, and developing world in particular, since most of the developing countries are located in the tropical and sub-tropical belts, rich in biodiversity. The establishment of gene bank is considered as an important strategy for the conservation of resources.

In India, among the eight centers of crop plant diversity in the world, one such gene center dealing with arid and semi arid plants is located. About 15,000 species of higher plants have so far been identified and enumerated in Indian flora and almost one third of this flora is endemic and unique. Out of 3000 economically important species, nearly 1000 are wild that provide sources of food; 500 are of medicinal use, more than 300 are related to cultivated plants and are useful for breeding programmes, and about 160 species are of minor or major cultivated types. To these numbers are added more than 1000 species which are used for various formulations.

CONSERVATION OF GENETIC RESOURCES

The importance of conservation of genetic resources is widely recognized. With regard to *in situ* conservation as far as the Indian scenario is concerned, one of their primary objectives is to identify specific areas designated to be *Biosphere Reserves* in this country based on the criteria of genetic diversity, natural growth, representation and inaccessibility as a conservation unit. A simultaneous programme has also been undertaken for identification of primary forests in this country, however small they may be, in view of the fact that they provide milieux for the natural selection to operate.

Several areas have so far been identified as potential biosphere reserves, these include: (1) Mysore Plateau - Wynad in the Nilgiris with the Silent Valley, (2) Khanchendzonga National Park on Sikkim, (3) the Nanda Devi Sanctuary, (4) Valley of Flowers and (5) Kedernath in U.P., (6) Simlipal and Jeypore Hill Forests stretch in Orissa, (7) Kanha National Park, Madhya Pradesh, (8) North Island of Andaman and Jarawa Tribal Reserves, (9) Sunderbans in West Bengal, (10) Namdapha, Lalichopra, Pakku, Tawang and Walong in Arunachal Pradesh, (11) Gulf of Mannar in Tamil Nadu and (12) Tura Ridge in Meghalaya. Several other areas to be identified as *Biosphere Reserves* are also under consideration.

The gene pools, including ecotypes, or cytotypes and the wild plants collected from forest or pasture and those which are related to the domesticated plants require to be preserved and perpetuated ensuring the genetic integrity of their natural state. The medium and long term conservation of all germplasm resources, the organization, exploration and collection of germplasm and exchange of resources, fall within the objectives of the gene bank.

The principal objectives are to organize cultivation of special groups of plants, establish field gene banks, herbal gardens, botanical research stations, identify and select natural habitats/forests and create genetic reserve by conservation of important wild species. Most such banks have a computerized data base.

The importance of data base management at both national and regional levels can hardly be overemphasized. The language of dissemination of information at the regional or global level needs to be standardized. The computerized information network is expected to help a coordinated effort in standardization of terminology and methodology for data recording, storage/retrieval, characterization and utilization. This would encourage international, regional and national coordination, would help in maintenance and distribution of stock and would also aid in establishing a management system. It will be a major input for national/regional economy through earning valuable foreign exchange.

Creation of herbarium facilities, tissue culture repository, glass-houses, nature of germplasms to be conserved and conditions for long and medium term storage are different facets of the activities of the gene bank. In addition, the field facilities, including botanical and herbal gardens, are necessary.

The genetic wealth of India is also known for its rich diversity of species used in medicine, including almost 1000 species to produce different formulations. Most of these still come from the wild plants occurring in the forests. Despite the fact that these natural resources are not infinite, though growing abundantly in the forests in the plains as well as mountains, a systematic attempt has been made to domesticate and bring these into cultivation. As a result, several such important species are on the verge of extinction. Examples may be cited of species of *Podophyllum, Coptis* and *Nardostachys.* At the global level, the importance of plants for the production of medicines is not new. At least 7000 medicinal compounds in the modern western pharmacopeia are plant based products. The industrialized world is conscious of the value of plant derived drugs. The pharmacists are focussing on synthetic production of natural compounds, but only a few have yet been synthesized and the rest are still extracted from the plants. Even then, the crude plant extract is often mere effective than the synthetic one.

Plant genetic resources are invaluable national and global assets in meeting the needs of increasing crop production, sustained productivity and human health. The rich genetic diversity of economic plants, located in different centers with potential value for current and future needs of humankind is increasingly being realized. Some of these resources, such as plant species of medicinal and aromatic value which occur in natural ecosystems, have been over-exploited and are either threatened, vulnerable or in danger of extinction. The genetic resources constitute a unique global heritage and their collection, utilization and conservation are of primary concern for human survival.

The strategy of conservation is guided by the nature of the plant species and the objectives and scope of conservation. The nature of the material is of major importance, including the duration of the life cycle, generation times, mode of reproduction, size of the individual or population and the ecological status names, whatever, weedy or domesticated. The objectives cover research, introduction, enrichment of diversity and may also determine the degree of integrity to be maintained. The strategy encompasses the entire methodology including a rational population size needed to be preserved.

The long term preservation of broad-based gene pool is generally desired in view of the fact that potentially useful species or wild relatives might be needed for immediate use in future with the advancement of our knowledge of the identification of precise sequences and their genetic transfer.

One of the important aspects of conservation is the safe and efficient collection and storage of information. It is important not merely to conserve areas, communities, populations or individual, but to seek information relating to their characteristics and quality for adequate safeguard and accessibility of information. This would be equally applicable to communities studied in *in situ* genetic reserves or biosphere reserves or to those maintained *ex situ* in gene banks, botanical gardens, herbal gardens or even plantations. The information should cover characteristics of site and status of the material. Chemical characteristics and clinical assays are essential inputs for documentation and utilization.

The gene pools (special ecotypes, morphotypes, chemotypes or cytotypes) of the wild species related to our crops, collected from forest, pastures or wastelands, need to be preserved in perpetuity ensuring the genetic integrity of their natural state. The gradual extinction of gene pools of many wild species used in forestry or pharmaceutical industry are to a great extent attributable to the destruction of indigenous forests; deforestation for agricultural or horticultural development; replacement by exotic species; damage by fire, accidental or deliberate (for example in shifting cultivation) or even introduction of new pests and pathogens or parasites. The long term conservation of the dynamics of primitive population is indeed a difficult task. It principally depends on protection of gene pools in areas under scientific control namely, in experimental stations or regions suitably dispersed to cover a range of communities and environments. All these strategies need to be utilized in gene banks. Conservation of genetic diversity is important not merely to improve present and future bioproductivity but also to ensure the very survival of the human race, which is entirely dependent on these components of the biosphere.

One of the causes of the rapid degradation of tropical forests is the simultaneous decline of tribal populations living in forests. Leaving aside their intrinsic value as one of the most important components of the biosphere, the tribal people have a vast knowledge of the properties of endemic medicinal and economic species. This knowledge is perpetuated through generations mainly by oral transmission. With the gradual disappearance of the tribes, the knowledge gained through long antecedent period of evolution is vanishing. The preservation of this ethnobotanical knowledge is essential for the welfare of the future generations.

Plant genetic resources are ultimately related to important programmes at national, regional and global levels. The two aspects of the programmes are the conservation of rapidly disappearing genetic stocks for present and future use and the judicious utilization of conserved materials in projects both for improvement of the stock and enrichment of genetic diversity. The success of these activities is dependent on the availability of information regarding the agronomic practices, both qualitative and quantitative, and to the genetic attributes of resistance to pests, diseases and stress in different polymorphic types.

TYPES OF SEEDS

The necessity of gene banks, as seed repository for the medium and long term conservation of seeds *vis-a-vis* germplasm of diverse crops, including medicinal and aromatic plants, is regarded as the most convenient method. It avoids the need of too frequent regeneration in the field. In such banks, the corp species are often categorized basically into two groups, based on their seed characteristics, as well as physiology and behaviour under storage.

Orthodox species cover the majority of agricultural and horticultural crop species. In this group, aging leads to seed death, influenced principally by the moisture/ content and temperature. The survival of seeds and the longevity can possibly be affected by controlling the environment during seed storage. In addition, moistening and drying before storage lower the temperature. The moisture content (up tp 5%) is inversely proportional to the longevity of the seeds.

Recalcitrant species include many species of tropical fruits and a number of forest plants of tropical and temperate origin. In contrast to the orthodox species, the rate of deterioration in recalcitrant seeds is difficult to control as they are not subject to desiccation and it is difficult to cool such seeds to sub-zero temperatures. The heavy freezing often leads to death due to the formation of ice. Moreover, in recalcitrant species of the tropics, if cooled to temperatures at the range of 10°C or less, there is the onset of chilling-injury. To secure short term storage, recalcitrant seeds are treated with a fungicide and kept moist under aerobic state.

TYPES OF GERMPLASM TO BE CONSERVED

The gene centres are normally responsible for: —

(i) Conservation of base collections for long terms;

(ii) Conservation of active collection for (a) medium term storage (b) regeneration (c) multiplication and distribution (d) evaluation and (e) documentation.

The duplicate collections of the basic materials are conserved for long term storage with adequate safeguards against loss or deterioration at different regions.

CONDITIONS FOR STORAGE

Longterm Storage: The viability of *orthodox* seeds can be significantly improved through control of storage environment. The increase in longevity with decrease in either the temperature or moisture content shows an exponential curve. The specified standards for storage at low temperature are -20°C or less in closed containers at a seed moisture content of 5 + 1% on wet basis. During storage, the content of seed moisture can be controlled by drying the seed to about 5% moisture level and subsequently maintaining the moisture content in hermetic sealed containers. It is desirable to provide a common environment for all orthodox species, so that a single gene bank (storage vault) can accommodate a large number of species. The longevity of seeds of different species may differ under different environmental conditions.

The standard temperature can be maintained in case it is desired to restrict the conservation only to a few species. However, stored accessions at -10°C require regular monitoring and regeneration to a much higher level than those conserved at -20°C. It is also possible to reduce the temperature much below -20°C. The extreme limit being investigated is storage in liquid nitrogen at -196°C. For orthodox species, cryopreservation is quite suitable.

Medium Storage: For medium term storage, the sample size must be larger than base collections since such collections are continually depleted because of their regular use in evaluation after distribution. Trials for regeneration too should be frequent to check again loss of viability and consequent loss of material. Thus, germplasm collections need to be kept at temperatures between 0°C and +10°C under medium term storage. The hermetic seal serves purpose of controlling the content of seed moisture.

Method of drying and determination of moisture content: The normally recommended moisture content is 4-6% for prescribed drying of seeds of orthodox species in gene banks. As the different accessions at the seed bank from different centres differ in their moisture content with great variation between species, equipments and standards set for commercial practice may not serve the purpose. The method should be easy of operation and applicable to a variety of species. Hot air drying method often varies with the design of the equipment. This method, unless combined with dehumidification, becomes unsuitable for humid tropical conditions.

The most convenient method for drying seeds to 5% moisture content is in a drying room maintained at about 15°C and 10-15% RH with adequate air circulation arrangement. This effect is achieved by an air dehumidifier with refrigeration arrangement for lowering the temperature and removal of dehumidifier-generated heat.

Herbarium Facilities: Plant Genetic Resources (PGR) programmes are always linked with collection, evaluation and conservation of authentic plant materials which can be verified at any period. The herbarium with good collection of authenticated specimens helps in the correct identification and taxonomic analysis of plants during the course of medium or long term preservation. The specimens of diverse morphotypes, cytotypes, ecotypes or chemotypes are essential for analysis. Samples collected from the field or from natural habitats of wild species are dried, pressed and labelled with all necessary information.

In order to have an overall conservation strategy, linked with the cultivation of endangered species, and breeding of new cultivars and their exploitation for medicinal and industrial purposes, the establishment of a chain of herbal gardens in different agro-climatic zones in different parts of the country is necessary. Repeated collection of wild plants can become detrimental to the very existence of such specialized plant communities, particularly in the developing regions. This practice needs to be halted in the interest of conservation. Important plant species, once identified and documented, should be established in herbal gardens — a method of *ex situ* preservation and multiplied to meet commercial demands. The efforts will go a long way in sustaining production. With suitable input, the increasing future requirements can also be met. The herbal gardens can serve as experimental farms for breeding and selection of new cultivars as well as demonstration of methodology for conservation and utilization.

Collection: It is undisputed fact that inexhaustible gene pools postulated by Vavilov are gradually drying up due to decimation of forest ecosystems and due to excessive collection of wild species, having medicinal, industrial or crop value. As countries in the tropical region of the globe have the natural resources, there has been a tendency to exploitation leading to the complete extinction of several such species. To a great extent, such destruction and erosion can also be attributed to the great success of the new varieties or strains selected by plant breeders which through large scale cultivation, have replaced the old diversity of primitive forms. The wild species of forest and scrub vegetation too are eliminated by degradation of forests. The magnitude of such loss can be visualized by the fact that almost one third of the forests cover may disappear by the end of the century, carrying with it the vast reservoir of tapped and untapped genetic resources. An adequate strategy at national and regional levels for their maintenance is imperative.

GENE BANK

India established its premier institution, the National Bureau of Plant Genetic Resources (NBPGR) at New Delhi in 1976 as a nodal organization to conduct, promote and coordinate all activities related to genetic resources of crops and their wild relatives with a view to ensure their continued availability in crop improvement. The Bureau has the national responsibility for export and import of plant germplasm for research purposes. Since 1976, exploration of plants, plant quarantine, exchange, evaluation, characterization, documentation and conservation of germplasms are the major activities at the Bureau. It also has the primary responsibility for maintaining basic collections in the National Gene Bank. The Bank includes a seed repository, a plant tissue repository, cryopreservation facilities and field gene banks and clonal repositories. National Computer Based Data Management System of Plant Genetic Resources looks after documentation and information retrieval, herbarium facilities and seed museum on wild and cultivated plants.

The regional centres of NBPGR have been located in the temperate regions at Simla; arid region at Jodhpur; semi-arid region at Akola and Amravati; humid tropical region at Trichur and humid sub-tropical regions at Shillong in addition to several other regional centres in Cuttack (Orissa), Ranchi (Bihar), Bhowali (U.P. Hills) and Hyderabad (A.P.). During the past decade, the Bureau has played a prominent role in introducing several promising species, forms and accessions of wheat, rice, millets, oilseeds, vegetables, fruits and commercial crops, amounting to cover hundred thousand samples from known sources abroad. In view of its charter, several expeditions in India and abroad have been made for the collection of diverse genetic resources from different regions. It has a *National Seed Repository* holding national collection compromising well over a hundred and twenty thousand samples (as repoted in G-15 document).

The first National Seed Repository or Gene Bank at NBPGR, New Delhi was established during 1980-85 with the mandate for long term conservation of germplasm of several agri-horticultural crops and their wild relatives. Infrastructural facilities for medium (+10°C) and long term conservation (-20°C) facilities have been generated with a capacity to hold about 250,000 samples. The Gene Bank is also provided with facilities for seed processing, drying and testing. The scientific programmes are directed towards the generation of basic information which may provide strong support for the safe conservation of resources, particularly of the orthodox seed species. The effective collection of the Gene Bank exceeds well over 150,000 samples. With the establishment of the more modern facilities for Gene Bank, the collections may ultimately exceed 600,000 samples under long term conservation of principally orthodox species.

The Indian Gene Bank has been designated as repository for global base collection for about thirteen crop species by the International Board for Plant Genetic Resources. The plant species to be investigated have recognized economic value as food, medicine or in industry. Priority is given to those species which have already proven medicinal or commercial potential, through recognized scientific or other parameters. The scope of conservation covers a wide spectrum including those species, which offer potential for development of transgenic lines through genetic manipulation. The selection of plants for Gene Bank is often based on overall considerations of morphological and chemical characteristics. It is because of the fact, that for certain species such as crop plants, the commerical product is visible seed, root, leaf or the whole plant whereas in medicinal and aromatic plants, the product more often is chemical which may be intermediate or a final chemically valuable compound. It is also necessary that identification of such races must be based on quick methods of chemical analysis of minute quantities of material. Materials identified on such criteria are treated as chemoraces, constituting a very important component of the Gene Bank. Such species need multiplication through *in vitro* technology.

A global interest has been generated in utilizing *in vitro* technique and biotechnology for the enrichment of genetic diversity, rapid multiplication, conservation, and gene transfer through protoplast or other means under strictly aseptic conditions.

The Gene Bank has the potential of conserving genetic diversity of orthodox crop species for periods ranging from 10 to 50 years approximately, without field regeneration and maintaining full genetic potential after storage. There are several species, however, which show recalcitrant seed behaviour in this respect. Morever, a number of species do not reproduce sexually and propagation is through vegetative means. Several species may be cross-pollinated and highly heterozygous or prove to have pest and pathogen problems in the field. For the conservation of such species, development of alternative strategies for genetic conservation is needed.

Lately, *in vitro* technologies have emerged as a powerful tool in the conservation of germplasm. A wide variety of tissues and cells, such as, protoplast cultures, callus, organs and meristems, shoot tip, axillary buds, pollen, anthers and even embryos including somatic embryo can be preserved for a long period of time. The production of artificial seed from somatic embryos, adopting alginate method, has proved to be very fruitful, in the conservation of even protoplasts. The cultures can be maintained in minimal media, growth retardants or with the aid of regulators. The reduction in temperature and light regime has proved to be of great use. The cultures can also be kept frozen in liquid nitrogen at -196°C.

The National Facility for Plant Tissue Culture Repository (NFPTCR) was established at the NPBGR, New Delhi in 1986. The facility is fully operative including programmes on cryopreservation of seeds, pollen, embryos, embryonic axes as well as *in vitro* cultures. It has been assigned the national mandate for the conservation of several asexually propagated crop plants covering bulbous, tuberous, and fruit plant species as well as plantation crops, medicinal and aromatic plants and endangered plant species. The conservation of wild and weedy species of future potential value fall within its mandate.

Different strategies are thus adopted in India for preventing the rapid decimation of dwindling resources. All possible methods for the conservation of such species together with sustainable development are imperative. The Gene Bank and germplasm centres in India, including the regional ones, are cooperating with the other regional centres in the tropical region of the globe.

RECOMMENDED BIBLIOGRAPHY

1. G-15 project on *establishment of gene banks for medicinal and aromatic plants and herbs in developing countries* (1991) presented for India in G-15 Summit, Caracas (Venezuela) and Senegal (1992).
2. *Impact of the development of Science and Technology on environment.* Presidential address at the 68th session of the Indian Science Congress (1981) delivered by A.K. Sharma at Varanasi, pp 1-196.
3. *Resources and Human well-being: Inputs from Science and Technology* (1987) Presidential Address at the 74th Session of the Indian Science Congress, delivered by Archana Sharma, Bangalore.
4. *Environmental Priorities In India And Sustainable Development* (1986) Presidential address at the 73rd Session of the Indian Science Congress delivered by T.N. Khoshoo at Delhi.
5. *World Resources 1990-1991,* published by World Resources Institute with UNEP and UNDP, 1992.

Biological Diversity: Problems and Challenges. Edited by S.K. Majumdar, F.J. Brenner, J.E. Lovich, J.F. Schalles, and E.W. Miller. © 1994, The Pennsylvania Academy of Science.

Chapter Six

STATISTICAL AND THEORETICAL ASPECTS OF DIVERSITY INDICES

RICHARD P. STEINER

Department of Mathematical Sciences
The University of Ohio
Akron, Ohio 44325-4002

INTRODUCTION

Diversity is perhaps one of the most important characteristics of biological communities. Much biological significance has been attached to diversity. Communities with high diversity are thought to be more stable, although possibly less productive, than communities with low diversity. While it is easy to possess a general "feel" for what constitutes biological diversity, precise conceptual definitions are a necessary foundation to the development of useful quantitative measures of diversity. Two general methods of analyzing diversity have been employed 1) analysis of the shapes of species-abundance curves and the fitting of mathematical distributions to these curves and 2) the calculation of summary measures called diversity indices.[1,2] In this chapter, the concept of diversity is explored and related to various diversity measures. Problems involved with measuring diversity are discussed and some suggestions are made.

THE CONCEPT OF DIVERSITY

What is diversity? Dictionary definitions include synonyms such as "difference", "distinct", and "variety". Variability is another word that may be used to describe diversity. Biological diversity is most often thought of in terms of species diversity, which can be thought of as the degree of taxonomic variability that is present from individual to individual (or area to area). Mathematically, information theory is related to the concept of variability and some diversity measures are based in information theory. To better understand how information is related to variability, consider the following illustration: Visualize a page of text containing row upon row of a single letter, "S". There is no variation from character to character and the page contains very little information. If there were 2000 such letters on the page, the information contained therein could easily be abbreviated as 2000 S with little or no loss in information. Now consider the page of text you are now reading. It contains many different letters with considerable variation from character to character and contains much more information than the page filled with the letter "S". Another important point is that it is also much more difficult to abbreviate or summarize the information on this page without substantial loss of information. So it is with biological diversity.

To formally define a concept of diversity, it must be recognized that diversity is a characteristic of a community, or a part of a community. While distinct physical boundaries for a community may not exist, for the purposes of quantification, the community must be specifically defined in time and space. Also, if the entire community is not being measured, the specific component must be defined. This well-defined entity may be referred to as the *statistical population.* Problems associated with the nature of the statistical population and the method by which data are collected from this population will be addressed later.

Once the population has been defined, what is meant by diversity must be examined. "When a population is classified into several classes, the *diversity* of the population is intended as a measure of the variability of the classes in it, in terms of the number of classes and their relative frequencies."[3] These "relative frequencies" may be the proportion of organisms in each class or it may be the proportion of biomass contributed by each class. In general, the setting is one in which some quantity (individuals, biomass, or some other measure of importance) is apportioned into **S** classes (species or other taxonomic or trophic groupings), with π_i representing the proportion in or contribution of class i (i = 1, ..., S), with the requirements that $\pi_i > 0$ and $\sum_{i=1}^{S} \pi_i = 1$[4]. For example, if the quantity is number of individuals, π_i is the number of individuals in class i divided by the total number of individuals in the population. This is referred to as *relative abundance.* If the quantity is biomass, π_i is the biomass of class i divided by the total biomass of the population. For the sake of generality, the term *relative importance* will be used to describe the π_i.

Biological diversity is not a single, objectively defined characteristic. It has long been viewed as having two components *richness* and *evenness* (eg.,[1] p. 149). Richness refers to the number of classes. Evenness is a measure of how close the distribution of importances among the classes is to a uniform distribution. A completely even population is one in which $\pi_i = 1/S$ for all i. Poole[2] asserts that the only truly objective measure of diversity is richness. Definitions which include relative importance as part of diversity (like the definition[3] quoted above), become subjective, depending on how the relationship between relative importance and diversity is defined.

Approaches to Diversity

There have been recent efforts to develop a unified approach to the concept of biological diversity and diversity indices. Notable efforts are those by Hill,[5] Patil and Tallie,[6] and C.R. Rao.[7] All of these authors view diversity as an average property of a community. Hill developed a family of diversity indices that are functions of weighted averages of powers of the relative importances.

Patil and Tallie[6] define a diversity measure of a community as its average rarity. Within this conceptual framework, they define a diversity index as

$\triangle(\pi) = \sum_{i=1}^{S} \pi_i R(i;\pi)$, where $R(i;\pi)$ is a measure of rarity for class i and is a function of $\pi = (\pi_1, \pi_2, \ldots, \pi_s)$, which is the relative importance vector. Thus the diversity index $\triangle(\pi)$ is a weighted average of the measures of rarity for each class. Widely used indices such as the Shannon and the Simpson can be expressed in this manner with appropriate choices of the rarity measure.

Patil and Tallie discuss two broad classes of rarity measures: *dichotomy,* in which the rarity of class (i) depends only on its measure of relative importance (π_i). For this type of rarity measure, $R(i;\pi)$ may be written as $R(\pi_i)$ emphasizing the dependence on π_i only. The term "dichotomy" is used to describe this type of rarity measure because the rarity of class i does not change if the other S-1 classes are grouped into a single class. A second type of rarity measure is referred to as *ranking*, in which the rarity of a class (i) depends only on the relative rank of its importance, and not on the actual value of π_i. For this type of rarity measure, $R(i;\pi)$ may be written as R(i), denoting the rarity of the class with rank i. (The class with highest importance receives a rank of 1.)

Rao[7] takes a different approach to diversity. Let **X** be a set of *m* measurements taken on each member of a population. The cumulative distribution function of **X** is $F(\mathbf{x})$. Now consider two individuals, one with $\mathbf{X} = \mathbf{x}_1$ and one with $\mathbf{X} = \mathbf{x}_2$, selected from the population. A nonnegative, symmetric difference function $d(\mathbf{x}_1, \mathbf{x}_2)$ is selected to measure the difference between the two individuals. A suitable difference measure may be chosen to fit the particular problem under study. A diversity index (or diversity coefficient, to use Rao's terminology) may be constructed as the average difference between pairs of randomly selected individuals in the population. Rao's diversity index may be written as

$$\begin{aligned}\triangle(\text{Rao}) &= E[d(\mathbf{x}_1, \mathbf{x}_2)] \\ &= \int d(\mathbf{x}_1, \mathbf{x}_2)\, dF(\mathbf{x}_1)\, dF(\mathbf{x}_2), \text{ when } \mathbf{X} \text{ is continuous.}\end{aligned}$$

The integration is replaced with summation when **X** is discrete.

This approach to diversity is particularly useful when the interest is in the diversity of continuous attributes of individuals, although it also admits discrete, qualitative attributes such as taxonomic or trophic groupings. In this sense Rao's approach is broader than that of Hill[5] or Patil and Tallie.[6] Yet the latter two approaches provide a more intuitive approach to diversity of qualitative attributes, such as species diversity. Still, Rao's approach appears to be well-suited to measuring genetic diversity (See p. 27 of [7] or p. 5 of [8]), supports a framework for decomposing the diversity of cross-classified populations and mixtures of populations,[8] and can produce a single diversity index based on multiple partitions of the population. This latter capability will be explored later in this chapter.

Properties of Diversity Indices

Certain useful properties of diversity indices have been proposed. Some of these are summarized below:

1. A population with one class ($S = 1$) should have a (minimum) diversity of zero (eg.[4,6,8]). Zero, then, should be the smallest value that $\triangle(\pi)$ can attain. In the average rarity framework, this implies that $R(1) = 0$ when $S = 1$ for both dichotomy and ranking type indices[6].
2. A diversity index should increase monotonically in value as the evenness of the population increases. This implies that, for a given S, $\triangle(\pi)$ should attain its greatest value for a completely even population. (The latter statement is Property 1 of Pielou[9]). Some indices (eg.[4,10]) are scaled to be on the interval(0, 1). For an argument against rescaling diversity indices see.[11]
3. Diversity should increase when a class is added to a population or importance is transferred from one class to another class with lower importance. (This is Criterion C2 of [6] and, if met, satisfies both Properties 1 and 2 of.[9] See Theorem 8.1, p. 556 of [6]).
4. A diversity index $\triangle(\pi)$ should be a concave function of π. This property is necessary for the analysis described in [6] for decomposing the diversity of a population that is a mixture of subpopulations. This property has been ascribed to Lewontin.[12]
5. The rarity measure $R(\pi_i)$ should be a *decreasing* function of π_i for dichotomy type indices, and R(i) should be an *increasing* function of *i* for ranking type indices, both on the interval (0, 1). (This is Criterion C1 and Theorem 6.1 of [6]).

Properties 3-5 are equivalent—that is, if any one of them are met, they all are met—for the ranking type indices.[6] Properties 3 and 5 are both satisfied for dichotomy type indices if the function $V(\pi_i)$ are defined as $V(\pi_i) = \pi_i R(\pi_i)$ for $0 < \pi_i \leq 1$ with $V(0) = 0$, is a concave function on the interval $0 \leq \pi_i \leq 1$[6].

In this section the concept of diversity has been discussed and some general frameworks for quantifying diversity have been presented. The next section provides an overview of several specific diversity indices.

QUANTIFICATION OF DIVERSITY

Several diversity indices are briefly discussed, including the general logic behind each measure, the aspect of diversity it best captures, and its underlying assumptions. Where available, information on estimating the diversity measures from sample data is included also.

Simpson Index

Perhaps more appropriately called the Gini-Simpson index, it was developed by Gini,[13] and first applied in biology by Simpson.[14] Three common formulations exist:

$$\triangle_1(\text{Simpson}) = \sum_{i=1}^{S} \pi_i^2$$

$$\triangle_2(\text{Simpson}) = 1 - \sum_{i=1}^{S} \pi_i^2$$

$$\triangle_3(\text{Simpson}) = \left[\sum_{i=1}^{S} \pi_i^2\right]^{-1}$$

The version $\triangle_1$ is sometimes referred to as a coefficient of concentration of dominance. When the π_1 are relative abundances, i.e., the proportion of the population belonging to class i, $\triangle_1$ is interpretable as the probability of two randomly selected individuals from the population belonging to the same class. This measure has its minimum for a completely even population (with $S>1$) and its maximum when $S=1$. This situation is exactly opposite of Properties 1 and 2, above. For this reason, Simpson's index is often formulated as $\triangle_2$ or $\triangle_3$, which satisfy these properties.

Simpson's index measures the relative dominance of a few classes, rather than overall evenness.[2] It is heavily weighted toward the classes with highest relative importance and is less sensitive to changes in richness,[15] because the addition of classes with low relative importance have a small effect on the index (Classes with small π_i have even smaller π_i^2.)

When the π_i are relative abundances and the index is to be interpreted as the probability of two randomly selected individuals belonging to the same class ($\triangle_1$) or different classes ($\triangle_2$), the assumptions underlying the Simpson index are that a random sample of individuals is taken from a large (technically infinite) population. In ecological studies a random sample of individuals may be very difficult, if not impossible, to obtain because of the "patchiness" of natural populations and the indistinctness of individuals in some plant species. Some of the difficulties of sampling are discussed in [9] and.[15] In the case when diversity is calculated for a finite, completely enumerated population, Simpson's index should be calculated as follows:

$$\triangle'_1(\text{Simpson}) = \sum_{i=1}^{S} \frac{N_i}{N}\left(\frac{N_i - 1}{N-1}\right)$$

$$\triangle'_2(\text{Simpson}) = \sum_{i=1}^{S} \frac{N_i}{N}\left(\frac{N - N_i}{N-1}\right)$$

where N_i is the number of individuals in class i and N is the total number of individuals in all classes. The measure $\triangle'_2$ is recommended by Hurlbert.[16]

Shannon-Wiener Index

This widely used index was derived by Shannon and Wiener, working independently of each other.[15] The index derives from information theory and can be thought of as a measure of the uncertainty involved with predicting the class to which a randomly selected individual belongs.[2] This definition assumes that the relative importances are relative abundances, or at least that the probability of selecting an individual belonging to a particular class is proportional to the relative importance of that class (regardless of the measure of importance used). Earlier in this chapter, information was related to the concept of variability. It can be seen that in a population with high variability from individual to individual (such as a completely even population with several classes), it would be difficult to predict class membership of a randomly selected individual. The Shannon-Wiener index is formulated as follows:

$$\triangle(\text{Shannon}) = -\sum_{i=1}^{S} \pi_i \log_D(\pi_i).$$

The units of measure of the index depend upon the base of the logarithm, D; Poole[2] suggests the base e (natural logarithms) for most applications. As with Simpson's index, a random sample from a large (infinite) population is assumed. Shannon's index is biased, but usually the bias is small.[2,15] Poole[2] and Magurran[5] give expressions for $\triangle$(Shannon) and its variance that adjust for the bias.

In the case when diversity is calculated for a finite, completely enumerated population, a different information theoretic measure, Brillouin's index is used. This diversity index is defined as follows, with D, N_i, and N defined as above:

$$\triangle(\text{Brillouin}) = \frac{1}{N} \log_D(N!) - \sum_{i=1}^{S} \log_D(N_i!).$$

Entropy Measures

Another class of information theoretic measures called entropies have received considerable attention over the last two decades. The seminal work appears to be that of Rényi.[17] Several diversity indices that are modifications of Rényi's entropy of order α have been proposed. Three of these are described below:

1. Rényi.

$$\triangle_\alpha(\text{Rényi}) = \frac{1}{1-\alpha} \log_D \sum_{i=1}^{S} \pi_i^\alpha, \text{ with } \alpha > 0, \text{ but } \alpha \neq 1.$$

2. *Havrda and Charvát.*[18,19]

$$\triangle_\alpha(\mathrm{HC}) = \frac{1}{\alpha - 1}\left[1 - \sum_{i=1}^{S} \pi_i^\alpha\right], \text{ with } \alpha > 0, \text{ but } \alpha \neq 1.$$

3. *Saud, Khan, and Khan.*[4]

$$\triangle_\alpha(\mathrm{SKK}) = 1 + \frac{1}{1-\alpha}\left[\log_D \sum_{i=1}^{S} \pi_i \, D^{[\alpha-1]\pi_i}\right], \text{ with } \alpha > 0, \text{ but } \alpha \neq 1.$$

These entropy-based indices are related to the more familiar measures previously discussed in this section. For $\alpha = 1$, $\triangle_1$(Rényi) and $\triangle_1$(HC) are defined in a limiting sense[19] and become Shannon's index, $\triangle$(Shannon). Also, notice that $\triangle_2$(HC) $= \triangle_2$(Simpson).

The "diversity of order α" developed by Saud, Khan, and Khan is not equivalent to Shannon's index for any value of α, but for a completely even population $\triangle_\alpha$(SKK) $= \triangle_2$ (Simpson). For smaller values of α the $\triangle_\alpha$(SKK) is more sensitive to classes with lower relative importance, but as α increases the index is more heavily influenced by classes with high relative importance.

Diversity Measures of Hill

Hill[5] developed one of the first unified approaches to biological diversity indices. As mention earlier, this approach is based on functions of weighted averages of powers of the relative abundances. Specifically, the index may be written as:

$$\frac{\left[\sum_{i=1}^{S} w_i \pi_i^{\alpha-1}\right]^{-\frac{1}{\alpha-1}}}{\sum_{i=1}^{S} w_i}$$

where the w_i are the weights. The weights chosen by Hill are the relative importances themselves, i.e. $w_i = \pi_i$. So Hill's index is

$$\triangle_\alpha(\mathrm{Hill}) = \left[\sum_{i=1}^{S} w_i \pi_i^\alpha\right]^{-\frac{1}{\alpha-1}}$$

Thus Hill's index is a family of diversity measures indexed by α. Like the index of Saud, Khan, and Khan,[4] Hill's index is more sensitive to classes with lower relative importance when α is small, but is more heavily influenced by classes with high relative importance as α increases. Several choices of α yield interpretable and familiar measures of diversity. Listing these in ascending order of α can give some insight as to the aspect of diversity these measures emphasize, as well as how they compare to each other:

$\triangle_{-\infty}$(Hill) = $1/\pi_{[s]}$ = reciprocal of the relative abundance of the **least** important class

$\triangle_0$(Hill) = S = total number of classes

$\triangle_1$(Hill) = $D^{\triangle(\text{Shannon})}$, where D is the logarithmic base used in calculating $\triangle$(Shannon)

$\triangle_2$(Hill) = $\triangle_3$(Simpson)

$\triangle_{\infty}$(Hill) = $1/\pi_{[1]}$ = reciprocal of the relative abundance of the **most** important class

As in the case of $\triangle_1$(Rényi) and $\triangle_1$(HC), $\triangle_1$(Hill) is defined in the limit as α approaches 1, yielding the result given above. Hill's approach is closely related to Rényi's[17] entropy since $\triangle_\alpha$(Rényi) = $\log_D \triangle_\alpha$(Hill).

Diversity Measures of Patil and Tallie

A more recently developed approach to a unified concept of diversity is that of Patil and Tallie.[6] As discussed earlier, they define a diversity index as average rarity:

$$\triangle(\pi) = \sum_{i=1}^{S} \pi_i R(i;\pi).$$

Different choices of the rarity measure $R(i;\pi)$ yield different diversity indices. Rarity measures of the dichotomy type, which depend on π_i($R(i;\pi) = R(\pi_i)$), yield some of the familiar diversity indices:

$$R(\pi_i) = \begin{cases} \pi_{[s]}^{-2} & \text{if } i = S \\ 0 & \text{otherwise} \end{cases} \quad \text{yields } \triangle_{-\infty}(\text{Hill}) \tag{1}$$

$$R(\pi_i) = 1/\pi_i \quad \text{yields } \triangle_0(\text{Hill}) = S \tag{2}$$

$$R(\pi_i) = (1 - \pi_i)/\pi_i \quad \text{yields } \triangle_{-1}(\text{PT}) = S - 1 \text{ (See below.)} \tag{3}$$

$$R(\pi_i) = \log_D(\pi_i) \quad \text{yields } \triangle(\text{Shannon}) \tag{4}$$

$$R(\pi_i) = \pi_i \quad \text{yields } \triangle_1(\text{Simpson}) \tag{5}$$

$$R(\pi_i) = 1 - \pi_i \quad \text{yields } \triangle_2(\text{Simpson}) \tag{6}$$

$$R(\pi_i) = \begin{cases} \pi_{[1]}^{-2} & \text{if } i = 1 \\ 0 & \text{otherwise} \end{cases} \quad \text{yields } \triangle_{\infty}(\text{Hill}) \tag{7}$$

Patil and Tallie[6] also propose an index for which the rarity measure is defined as

$$R(\pi_i) = \begin{cases} (1 - \pi_i^{\beta}) / \beta \text{ if } \beta \neq 0. \\ -\log_D(\pi_i) \quad \text{if } \beta = 0 \end{cases}$$

The resulting diversity index can be written as

$$\triangle_{\beta}(PT) = \frac{1}{\beta}\left[1 - \sum_{i=1}^{S} \pi_i^{\beta+1}\right]$$

which Patil and Tallie call the "diversity index of degree β". When $\beta = 0$, $\triangle_0(PT) = \triangle(\text{Shannon})$ and when $\beta = 1$, $\triangle_1(PT) = \triangle_2(\text{Simpson})$. The index $\triangle_{-1}(PT) = S-1$ has a minimum value of zero, as suggested in Property 1 above. In fact, $\triangle_{\beta}(PT)$ satisfies Property 1 for all values of β; Properties 3 and 4 are satisfied if and only if $\beta \geq -1$[6]. This diversity index of degree β is identical to the entropy measure of Havrda and Charvát[18,19]: $\triangle_{\beta}(PT) = \triangle_{\alpha}(HC)$ with $\alpha = \beta + 1$.

Diversity indices based on ranking type rarity measures are attractive in that they do not depend on knowing (or estimating) the *values* of the relative importances of the classes, but depends only on *ranking* the classes by relative importance. Define a ranking type diversity index as

$$\triangle_{\rho}^{(rank)}(PT) = \sum_{i=1}^{S} \pi_{[i]} R(i),$$

where $\pi_{[i]}$ = the relative importance of the class with rank i; $\pi_{[1]} \geq \pi_{[2]} \geq \ldots \geq \pi_{[S]}$. The rarity measure for this index is

$$R(i) = \begin{cases} (1 - \rho^{i-1}) / (1 - \rho) & \text{if } \rho > 0 \text{ but } \rho \neq 1 \\ i - 1 & \text{if } \rho = 1 \end{cases}$$

When $\rho = 1$, The rarity measure R(i) represents the number of classes with lower relative importance than class i.

Both $\triangle_{\beta}(PT)$ and $\triangle_{\rho}^{(rank)}(PT)$ are families of diversity indices with varying sensitivities to classes with low relative importance. For smaller values of β, classes with low relative importance have a greater effect on $\triangle_{\beta}(PT)$, but as β increases, the index becomes less sensitive to classes with low relative importance and more influenced by classes with high relative importance. For $\triangle_{\rho}^{(rank)}(PT)$ the effect of varying ρ is the opposite. The index $\triangle_{\rho}^{(rank)}(PT)$ is *less* sensitive to classes with low relative importance when ρ is small and more heavily influenced by classes with high relative importance. As ρ increases the influence of classes with high importance lessens and the index becomes more sensitive to classes with low importance.

Diversity Measures of Rao

The approach of Rao[7,8] is based on the average difference of attributes between two randomly selected individuals from the population, and is thus conceptually somewhat different from other approaches to the quantification of diversity. Rao's index is $\triangle(\text{Rao}) = E[d(\mathbf{x}_1,\mathbf{x}_2)]$, where $d(\mathbf{x}_1,\mathbf{x}_2)$ is the nonnegative, symmetric distance measure for individuals with m x 1 measurement vectors $\mathbf{X} = \mathbf{x}_1$ and $\mathbf{X} = \mathbf{x}_2$. When $\mathbf{X}$ is continuous with variance-covariance matrix Σ, a reasonable choice of difference function is

$$d(\mathbf{x}_1,\mathbf{x}_2) = (\mathbf{x}_1 - \mathbf{x}_2)'\,\mathbf{A}\,(\mathbf{x}_1 - \mathbf{x}_2),$$

where $\mathbf{A}$ is a m x m scaling matrix. If $\mathbf{A}$ is an identity matrix, the resulting difference function is the squared Euclidean distance between the individuals. With this choice of distance function, Rao's diversity index is

$$\triangle(\text{Rao}) = 2\ \text{trace}\ (A\Sigma).$$

When $\mathbf{A} = \mathbf{I}$,

$$\triangle_1(\text{Rao}) = 2\ \text{trace}\ (\Sigma) = 2\sum_{r=1}^{m} \delta_r^2,$$

where δ_r^2 is the variance of the r-th component of the $\mathbf{X}$ measurements.

When the measurements represent m different discrete classifications of the individuals in the population, an alternative difference function is required. Let X_r be the r-th classification, with S_r classes and let π_{ri} be the probability an individual is in the i-th class of the r-th classification. (Rao's development is based on probabilities of class membership, but the index may still be used when the π_{ri} are other measures of relative importance.) In this setting a possible difference function[7] is

$$d(\mathbf{x}_1,\mathbf{x}_2) = m - \sum_{r=1}^{m} \delta_r,$$

where $\delta_r = 1$ if the two individuals belong to the same class in the r-th classification and zero otherwise. Notice that two individuals that belong to the same class in all r classifications will have $d(\mathbf{x}_1,\mathbf{x}_2) = 0$ (no difference). The maximum difference, $d(\mathbf{x}_1,\mathbf{x}_2) = m$, occurs for two individuals that are in different classes in all r classifications. The diversity index for this setting is

$$\triangle_2(\text{Rao}) = \sum_{r=1}^{m}\left(1 - \sum_{i=1}^{S_r} \pi_{ri}^2\right) = \sum_{r=1}^{m} \triangle_2(\text{Simpson})_r,$$

where $\triangle_2(\text{Simpson})_r$ is Simpson's index computed for the r-th classification.

For a single classification (m = 1), the subscript r may be suppressed and $\triangle_2$(Rao) = $\triangle_2$(Simpson). When there are multiple classifications, Rao's diversity is the sum of the separate Simpson diversities. For example, suppose that each individual in a bird community is classified as to species and feeding guild (m = 2). Then the diversity of the community as measured by $\triangle_2$(Rao) is species diversity plus guild diversity.

Richness and Evenness

Two components of diversity, richness and evenness, are confounded in many diversity indices. To overcome this difficulty, techniques have been developed to extract separate measures of richness and evenness from diversity indices.

A richness measure credited[6,16] to MacArthur[20] is the number of classes a hypothetical completely even population must have in order to have the same diversity as the population under study, as measured by a specified diversity index. This measure may be referred to as the *numbers equivalent* of a diversity index $\triangle$ and may be denoted as S($\triangle$). Numbers equivalents for several diversity indices are given below:

1. *Simpson.*

$$S(\triangle_1(\text{Simpson})) = \triangle_3(\text{Simpson})$$
$$S(\triangle_2(\text{Simpson})) = \triangle_3(\text{Simpson})$$

2. *Shannon-Wiener.*

$$S(\triangle(\text{Shannon}))\ D^{\triangle(\text{Shannon})}$$

3. *Havrda and Charvát.*

$$S(\triangle_\alpha(\text{HC})) = [\ 1- (\alpha -1)\triangle_\alpha(\text{HC})\ ]^{-1/(\alpha-1)}$$
$$\log_D S(\triangle_\alpha(\text{HC})) = \triangle_\alpha(\text{Rényi})$$

4. *Patil and Tallie.*

$$S(\triangle_\beta(\text{PT}))\ ]\ 1 - \beta\triangle^\beta(\text{PT})\ ]^{-1/\beta}$$
$$= \triangle_{\beta+1}(\text{Hill})$$

$$S(\triangle_1^{(\text{rank})}(\text{PT})) = 2\left(\sum_{i=1}^{S} i\pi_{[1]}\right) - 1$$

The numbers equivalents for the index of Havrda and Charvát and the index of Patil and Tallie are, of course, equal: $S(\triangle\alpha(\text{PT})) = S(\triangle^\alpha(\text{HC}))$ with $\alpha = \beta + 1$. It can also be seen that Hill's index is the numbers equivalent of $\triangle^\alpha$(HC) and $\triangle^\beta$(PT), and that Rényi's entropy is the log numbers equivalent. Numbers equivalents can be convenient measures of richness and lend insight into relationships among the various diversity indices. However, the simplest richness measure is the number of classes, S, which is its own numbers equivalent.

Evenness measures are constructed from diversity indices by comparing the observed diversity value to the value for a completely even population with the same number of classes. Property 2, above, states that for given S, a diversity index should attain its maximum for a completely even population. Indices for which this is true include $\triangle_2$(Simpson), $\triangle$(Shannon), $\triangle_\beta$(PT) for $\beta \geq -1$, and $\triangle_\rho^{(rank)}$(PT) for $\rho \geq 0$.

One approach[9] is to construct an evenness measure, J, as simply the ratio of the observed value of the diversity index, $\triangle$, to the maximum value, $\triangle_{max}$, that the index can attain for the number of classes (S) in the population. This evenness measure thus takes the form

$$J = \frac{\triangle}{\triangle_{max}}$$

A second evenness measure[16] which removes dependence on sample size and number of classes[9] is

$$V = \frac{\triangle - \triangle_{min}}{\triangle_{max} - \triangle_{min}}$$

where $\triangle_{min}$ is the minimum value of the diversity index $\triangle$ can attain for a given. S. Pielou[9] gives the value of $\triangle_{max}$ and $\triangle_{min}$ for Brillouin's index (H_{max} on p. 15 and H_{min} on p. 16). Hurlbert[16] gives $\triangle_{max}$ and $\triangle_{min}$ for his recommended indices, including Simpson's index. Nayak[19] gives expressions for $\triangle_{max}$ for the Shannon-Wiener index, Havrda and Charvát entropy, and Rényi's entropy.

A third approach, discussed by Pielou[9] and based on the work of Lloyd and Ghelardi,[21] expresses evenness (ϵ) as the ratio of the number of classes required for a population following MacArthur's broken stick law to have the same (Shannon) diversity as the observed population (S') to the actual number of classes (S). Thus, $\epsilon = S'/S$. The quantity S' is actually a numbers equivalent type of measure. Therefore, the evenness measure ϵ can be generalized to any diversity index $\triangle$, using the completely even population as the standard (rather than the broken stick model), by replacing S' with the numbers equivalent $S(\triangle)$. The resulting evenness measure is given by $\epsilon' = S(\triangle)/S$. Hill[5] proposed a family of evenness measures $E_{a,b} = \triangle_a(\text{Hill})/\triangle_b(\text{Hill})$ of which $\epsilon' = E_{a,0}$ is a special case.

A problem with J, V, ϵ, and ϵ' is that the number of classes in the population, S, is needed to compute $\triangle_{max}$ and $\triangle_{min}$. The value of S is rarely known and usually must be estimated from a sample. It is well known that the number of classes in a sample tends to underestimate S, and that the number of classes observed in a sample is dependent on the sample size. For this reason, Peet[11] argues against scaled diversity indices as measures of evenness, stating that such scaling increases the dependence of the measure on sample size. The problem is more pervasive than this, however, since all diversity indices to a degree depend on all classes being represented.

CONCLUSIONS

Fisher, Corbet, and Williams[22] are credited with the first use of a diversity index in biology. Since that time there has been an explosion of diversity indices in the literature and considerable debate as to what constitutes the "diversity" of a biological community. In this chapter the concept of diversity and several approaches to quantifying diversity have been discussed. What follows here are some observations and recommendations.

First of all, one cannot refer to *the* diversity of a community, for there are various interwoven attributes of the organisms that make up the community. Although species diversity is perhaps the most common measure, there are other important classifications. These might include trophic level, feeding guild, heigh class (of plants in a forest), for example. More than one type of diversity should be examined.

Regardless of the type of diversity being examined, diversity is an average characteristic of a community. It is a weighted average and the weights are the relative importances of the classes. Again, one cannot speak of *the* relative importance, for several different measures of importance might be used — each reflecting a different aspect of the classes. Numbers of individuals are most commonly used, but, depending on the nature of the classes, biomass, productivity, number of breeding pairs, or some other measure may be used. Hill[5] proposed an interesting measure of importance based on productivity. For each type of classification one should, where possible look at diversity based on multiple importance measures.

Which diversity index should be used? Despite the plethora of indices available several of the families of indices are transformations of each other, taking a different viewpoint of the same general concept. It seems rather remarkable that some form of Simpson's measure occurs as a special case in almost every family of indices. An exception is the index of Saud, Khan, and Khan,[4] which is an extension of Simpson's index. Rao's[7] approach to diversity produces a Simpson-based measure for multiple classifications. This does not necessarily mean Simpson's index is the "best", but it is familiar, intuitive, and fits into a variety of diversity frameworks.

Simpson's index is more sensitive to the classes with high relative importance and less sensitive to classes with low relative importance. The families of indices discussed in this chapter are parameterized according to varying sensitivity to high and low relative importance. A reasonable course of action may be to choose two indices, one near each end of the sensitivity spectrum. Also, richness and evenness should be examined separately as they are both of interest and are confounded in most diversity indices.

How does one summarize or compare a multitude of diversity measures? Multivariate analysis of variance may be used to compare vector-valued diversity measures among different communities. Several diversity measures may be summarized through principal component analysis.

Problems still face users of diversity indices. Sampling issues including inability to select individuals at random are major concerns when using species diversity indices. Future research should be aimed at developing diversity measures that circumvent these problems or developing techniques for adequately sampling biological communities.

REFERENCES

1. Odum, E.P. 1971. *Fundamentals of Ecology.* W.B. Saunders, Philadelphia, PA, 574 pp.
2. Poole, R.W. 1974. *An Introduction to Quantitative Ecology.* McGraw-Hill, New York, NY, 532 pp.
3. Caso, C. and Gil, M.A. 1988. The Gini-Simpson index of diversity: estimation in the stratified sampling. *Communications in Statistics - Theory and Methods.* 17(9):2981-2995.
4. Saud, I., Khan, M.Z., and A.A. Khan. 1984. Diversity of order α. *Journal of Theoretical Biology.* 107:521-530.
5. Hill, M.O. 1973. Diversity and evenness: a unifying notation and its consequences. *Ecology.* 54:427-432.
6. Patil, G.P. and C. Tallie. 1982. Diversity as a concept and its measurement. *Journal of the American Statistical Association.* 77:548-561.
7. Rao, C.R. 1982. Diversity and dissimilarity coefficients: a unified approach. *Theoretical Population Biology.* 21:24-43.
8. Rao, C.R. 1982. Diversity: its measurement, decomposition, apportionment and analysis. *Sankhua, Seris A.* 44:1-22.
9. Pielou, E.C. 1975. *Ecological Diversity.* John Wiley & Sons, New York, NY, 165 pp.
10. Fager, E.W. 1972. Diversity—a sampling study. *American Naturalist.* 106:293-310.
11. Peet, R.K. 1975. Relative diversity indices. *Ecology.* 56:496-498.
12. Lewontin, R.C. 1972. The apportionment of human diversity. *Evolutionary Biology.* 6:381-398.
13. Gini, C. 1912. Variabilitá e Mutabilitá. *Studi Economicoaguridici della facotta di Giurispurdenza dell Universite di Cagliari,* a III, Parte II.
14. Simpson, E.H. 1949. Measurement of Diversity. *Nature.* 163:688.
15. Magurran, A.E. 1988. *Ecological Diversity and its Measurement.* Princeton University Press, Princeton, NJ, 179 pp.
16. Hurlbert, S.H. 1971. The nonconcept of species diversity: a critique and alternative parameters. *Ecology.* 52:577-586.
17. Rényi, A. 1961. On measures of entropy and information. In: J. Neyman (Ed.) *Fourth Berkley Symposium on Mathematical Statistics and Probability.* University of California Press, Berkley, CA, pp. 547-561.
18. Havrda, M.E. and F. Charvát. 1967. Quantification method of classification processes: concept of structural α-entropy. *Kybernatika.* 3:30-35.
19. Nayak, T.K. 1985. On diversity measures based on entropy functions. *Communications in Statistics - Theory and Methods.* 14(1):203-215.
20. MacArthur, R.H. 1965. Patterns of species diversity. *Biological Reviews of the Cambridge Philosophical Society,* 40:510-533.
21. Lloyd, M. and R.J. Gehlardi. 1964. A table for calculating the 'equitability' component of species diversity. *Journal of Animal Ecology.* 33:217-225.
22. Fisher, R.A., Corbet, A.S., and C.B. Williams. 1943. The relation between the number of individuals and the number of species in a random sample of an animal population. *Journal of Animal Ecology,* 12:42-58.

Biological Diversity: Problems and Challenges. Edited by S.K. Majumdar, F.J. Brenner, J.E. Lovich, J.F. Schalles, and E.W. Miller.

Chapter Seven

DIVERSITY INDICES AND THEIR BIOLOGICAL IMPLICATIONS

NANCY McCREARY WATERS

Department of Biology
Lafayette College
Easton, PA 18042

INTRODUCTION

If community ecology represents one of the earliest areas of self-conscious ecology (McIntosh 1985), then diversity is probably one of the earliest abstract constructs of ecology to be quantified. Wallace (1878) observed that tropical animal life was generally more abundant and varied on a per-unit-area basis than animal life in temperate communities. This intuitive concept of diversity seems simply and easily defined, making it a popular subject for investigations of quantitative ecology. The presupposition that diversity is actually a measurable, quantifiable characteristic of a community has generated a plethora of indices, both of diversity and of its complement, community similarity (Green 1980). However, Peet (1974) suggested that diversity is in fact defined by the indices used to measure it, while Hurlburt (1971) declared diversity to be a non-concept, despairing of its usefulness to modern ecology. To some extent, the development and complexity of ideas and indices of diversity parallels advances in technology and our ability to acquire, organize and process information (Green 1980).

An informal, highly non-random examination of a trio of recent ecology texts indicates that diversity simply is a count of the number of species (Ricklefs 1990, Smith 1992, Stiling 1992). Although Charles Elton in 1927 commented that the total number of animal species reflected some important ecological principle, exactly what the nature of the principle was remained unclear (McIntosh 1985). By the late 1960's, it was not clear what was meant by diversity also, since elements of species richness and evenness had become part of most diversity indices, which in turn were variously related to theories of stability, succession, and disturbance. Further, since calculations of diversity depend upon concrete measurements of communities, any factors which contribute to bias in obtaining those measurements likewise generate bias in estimates of diversity.

To demonstrate, sampling issues present just one component of bias-controlling efforts. A cursory examination of a manual on vegetation science enumerated no fewer than ten approaches to plant ecology involving quantitative methods including, among others, the crown-diameter method, the point-intercept method, the quadrat-charting method, the line-intercept method, the count-plot method, the Bitterlich variable radius method, and the point-centered quarter method (Mueller-Dombois and Ellenberg 1974). In addition to choice of sampling method, bias can be introduced by design, frequency, seasonal timing, degree of replication, and spatial scale (Green 1979, Clarke and Green 1988). Coupled with the increased complexity of univariate and multivariate statistical approaches to the analysis of community data (Gauch 1982), it is clear that efforts to control the introduction of bias in quantitative ecology are ongoing, and this scenario excludes the problems uniquely associated with studies of animal or non-terrestrial communities!

This deceptively simple construct exerts influence most recently in questions of conservation of habitats and preservation of species. The term **biodiversity** has been introduced in the vernacular to refer to the variety of global life forms, the ecological functions they perform, and the genetic variation they contain (Smith 1992). Yet until we come to some consensus about the biological implications of ecological diversity, any attempts to preserve or maintain biodiversity on a practical level are unlikely to succeed.

The ensuing chapter, while not an exhaustive treatment of diversity indices, nonetheless presents ecological principles encompassed in estimates of diversity. It discusses two commonly used indices of diversity, placing them in the context of quantitative ecology. It then examines biological implications of diversity to community and ecosystem theories like stability and disturbance. Finally, it concludes with examples of and recommendations for environmental applications employing diversity estimates.

DIVERSITY AS AN ECOLOGICAL PRINCIPLE

Antecedents to Diversity Estimates:

To some extent, diversity as a concept was rooted in the recognition of pattern in ecology (Hutchinson 1953), or the physical arrangement of individuals with respect to one another. Temporal changes in pattern were thought to reflect changes in individual organisms. The observed pattern reflected some causative process or mechanism, such as predation or competition. Interestingly, there appeared some over-arching uniformity to the quantification of kinds of pattern, such that many examined communities yielded a log-normal distribution of species abundance

(Preston 1948), and all had approximately the same numerical constant by which they were quantified (Hutchinson 1953). However, the implications of this consistency in ecology were not well-understood. As a result, Hutchinson (1953) concluded that,

> "It is likely that something very important is involved here, but for the present what it may be is a mystery, a very good thing with which to end a discourse."

The discourse, of course, was only beginning. In the classic "Homage to Sant Rosalia," Hutchinson (1959) formally linked diversity to niche diversification as a result of trophic characteristics, and communities with higher diversity were thought capable of "...seiz(ing) new evolutionary opportunities" (Hutchinson 1959). Diversity (here, number of species) was linked with evolutionary flexibility and stability. Hutchinson's question, "Why are there so many kinds of living things?," generated a search for some unifying theory on diversity for the next quarter century (Margalef 1963, Brown 1981, Ghilarov 1984).

Hutchinson (1961) further suggested that spatial-temporal changes on a short time scale allow aquatic plankton to coexist with greater diversity than would otherwise be likely, given rates of competitive exclusion and characteristics of the physical environment. Heterogeneity in space could derive from heterogeneity in a temporally changing environment; thus, the seeds were sown for future studies of non-equilibrium theories of diversity and the role played by disturbance, both of which are important to present concerns for biodiversity.

These earlier studies described diversity strictly in terms of the number of species present, or **species richness,** for which some indices had already been developed (among them, Gleason 1922, Margalef 1957, Menhinick 1964). Meanwhile, MacArthur (1960) explored theoretical and empirical impacts of changes in relative species abundance, comparing impacts among different communities. Although concerned primarily with equilibrium conditions, MacArthur (1960) allowed that relative abundance in non-equilbrium conditions likewise followed log-normal distributions characteristic of species richness. However, of greater import was the following assertion—when relative abundance of a species increased in an equilibrium community, a concomitant decrease occurred in the relative abundance of another species.

This concern about relative abundance exemplifies the developing need for quantifying another aspect of diversity, namely **evenness,** or equitability. Pielou (1975) uses evenness to describe the apportionment of the total individuals in the community among the different species present; she also discusses a number of indices which measure variously species richness and evenness. Since individual indices are discussed elsewhere in this volume (see R. Steiner), the discussion herein will restrict itself to two commonly used indices with direct application to conservation practices.

Assumptions and Use of Diversity Indices:

Both mathematical and biological correlations exist among a number of diversity indices (DeBenedictis 1973, Hill 1973). For example, generic-level diversity in a Pennsylvania stream invertebrate community demonstrated the same trends across five locations, four sample periods, two seasons, and two years whether diversity was calculated by Simpson's (1949) Index or by Shannon's (1948) Index (Opyoke and McCreary Waters, in review). The widespread use of Simpson's Index reflects its biological intuitive nature, since it has been said to describe the probability of an interspecific encounter, or PIE (Hurlburt 1971).

However, Simpson's Index is sensitive to the abundance of more plentiful species in a sample, having been characterized as an expression of "dominance concentration" (Whittaker 1964). Further, it assumes that data are obtained randomly for a community (Brower et al. 1990); thus, the index may be partially dependent on sample size (Washington 1984). Many variations of Simpson's Index have included attempts to overcome such limitations and to relate diversity to heterogeneity (McIntosh 1967, Peet 1974, Washington 1984).

Shannon's Index (variously called Shannon-Wiener and Shannon-Weaver, see Brower et al. 1990) is perhaps the most widely used index based upon information theory (Peet 1974). In general, information theoretical indices described the degree of uncertainty that a randomly selected individual in a community might be of a given species—for less diverse communities, one could be more certain of the identity of a randomly selected individual, while in more diverse communities, uncertainty would increase (Brower et al. 1990). Shannon (1949) equated the uncertainty with "entropy," which N. Abramson validated in 1963 as the equivalent of entropy in thermodynamics (Washington 1984).

Shannon's Index assumes use of a random sample from relatively large aggregations (*e.g.,* a randomly obtained sample of species from a community with virtually every species represented), tends to be biased toward rarer species (Brower et al. 1990), and is less sensitive to the increased information gained in larger-sized samples (Peet 1974). Further, it has been criticized due to its lack of biological intuitiveness (Hurlburt 1971; Washington 1984). It is, however, an attractive concept, reducing the information obtained from the many species and individuals in an area to a single comparable figure (Washington 1984). Its widespread adoption also has contributed to bringing ecology closer to the realm of 'hard' physical science; use of such terms as entropy, information bits, and mathematical communication, goes far to satisfying what Frank Egler (1986) termed "physics envy" in the discipline of ecology.

RELATIONSHIPS TO COMMUNITY AND ECOSYSTEM THEORY

Diversity and Stability:

Early diversity studies suggested that high complexity, exemplified by diversity, increased stability of ecosystems; McNaughton (1977) reviewed the importance of this relationship to the First International Congress of Ecology. However, as McNaughton pointed out in that review, there is no consensus among ecologists as to exactly which properties of ecosystems diversity is supposed to stabilize! For example, Margalef (1963) stated that an ecosystem with unstable species composition may exhibit stable energy flow as a result of high complexity.

Loucks (1970) summarized the literature in a graphical display; he found general acceptance that increasing diversity was directly related to increases in succession, stability, and annual primary production. However, all of these described relationships have been debated by at least some authors (Loucks 1970). Pimm (1984) explained some discrepancies among diversity-stability relationships by pointing out that early studies demonstrated less oscillation and fluctuation in growth of individual populations the more complex the theoretical model. For communities with numerous pathways of energy transfer (trophic structure), the loss of any single pathway would have less impact than in a community with limited trophic complexity. Since this reflected the conventional wisdom of "not putting your eggs in one basket" (Pimm 1984), formal mathematical arguments were not pursued.

Diversity's relationship to stability is complicated by the definition of stability. For example, Pimm (1984) tabulated four different variables associated with the general term stability; coupled with the ongoing debate on a definition of diversity, it is no wonder that Pimm likewise listed four components of diversity!

Specifically, Pimm (1984) distinguished stability, resilience, persistence, and resistance from one another: **stability** is attributed to a system only if all variables return to the initial equilibrium following perturbation. Stability may be local (only true for small perturbations) or global (true for all possible perturbations) in nature. Regardless, stability is not a continuous variable; systems are either stable or unstable. Since stability by this definition requires equilibrium, continuously cycling ecosystems are inherently unstable, as are any systems during non-equilibrium conditions.

Of greater interest in questions of maintaining biodiversity, perhaps because they are more practically useful, are the remaining variables. **Resilience** describes the degree of rapidity by which equilibrium is restored following a perturbation, and is related to **persistence,** or how long a variable remains unchanged in the face of perturbation. Both resilience and persistence are often expressed in units of time, *e.g.,* the years needed to restore a fire-ravaged pine forest, or the years during which a scrub-shrub community is expected to remain intact as a successional sere.

Finally, **resistance** describes the degree to which a variable is changed following perturbation. Perturbations themselves may impact these components of stability and, ultimately community diversity, since changes occur on various scales of time and space, as well as in varying degrees of intensity. The impact of perturbations and other forms of disturbance on diversity will be discussed later. For now, it is clear that these characteristics are interrelated in many complicated ways, all of which impact predictions regarding diversity.

This relationship is further complicated by differences among theoretical arguments and related evidence from field studies; in general, field studies targeting global stability have compared existing systems to one another, rather than comparing them to existing hypothetical systems, which often have targeted local stability of species abundance (Pimm 1984). The result is that field studies have provided general views of ecosystems, rather than impacts of changes in individual species' abundances. McNaughton (1977) compared the plethora of theoretical and mathematical examinations of diversity-stability to the paucity of empirical examinations of the relationship, suggesting that, "...(T)he marked stability of attitudes regarding diversity-stability relationships in ecosystems arises primarily from a low diversity of empirical tests of the hypothesis."

Pimm (1984) drew a litany of conclusions regarding diversity and stability, particularly as they describe connectedness of ecosystem components. His assertions are twofold:

> With more species present in the community: (1) the less connected they should be, if the community is to be stable, (2) the less resilient the constituent populations, (3) the greater the resultant change from a single population removal, and (4) the longer the community will persist in the absence of a species removal.
>
> With greater connectedness in a community: (1) the fewer species it must have to be stable, (2) the more likely other species will be lost in the event of one species removal, but the more resilient they will be, (3) its composition will be more persistent, and (4) its biomass will be more resistant in the event of a population removal.

These conclusions are valuable because they explain in detail the controversies regarding diversity-stability relationships. In addition, they suggest important, practical questions for future examination: How quickly do species composition and biomass return to equilibrium, how long does composition remain unchanged, how much change occurs following addition (or removal) of a species? These questions must be answered for ecosystems currently prone to anthropomorphic perturbations, if biodiversity is to be maintained in such ecosystems.

Diversity and Disturbance:

Since Hutchinson's (1961) suggestion that heterogenous conditions over time might contribute to heterogeneity (*i.e.,* diversity) in space, ecologists have examined and questioned the long-held assumption of the need for equilibrium conditions in ecological models. Disturbance, also referred to as perturbation (although see Pickett and White 1985), interferes with the establishment of equilibrium in a community. Ecologists have more recently argued that disturbance effectively prevents most, if not all, communities in nature from achieving competitive equilibrium (Drury and Nisbet 1973, Connell 1978, Miller 1982). Such non-equilibrium communities defy explanation based on equilibrium theories of ecology. More importantly, biodiversity in non-equilibrium communities is likely to respond differently than expected if predictions are based on models requiring equilibrium conditions.

One of the most comprehensive examinations of the way in which disturbance impacts diversity is Michael Huston's seminal paper (1979); Huston explains patterns of species diversity by examining differences in how quickly community components approach, but do not achieve, competitive equilibrium. In these non-equilibrium communities, the frequency by which populations are reduced due to disturbance is balanced by the rate at which populations are competitively displaced. Therefore, diversity should be highest at intermediate levels of disturbance. When disturbance is frequent, the approach to competitive equilibrium and exclusion is slowed, individual population levels are repeatedly reduced or locally extinct, and diversity is low. With infrequent disturbance, populations interact and proceed to competitive exclusion, generating low diversity. The cause of disturbance can be a variety of mechanism(s) — environmental fluctuation, interactions among newly invading populations, human intervention — the prediction of higher diversity at intermediate levels of disturbance remains consistent.

Quantifying the impact of disturbance means defining its characteristics. Initial studies focussed on the influence of frequency of disturbance, examining the importance of interval length between each disturbance event (Huston 1979, Sousa 1979, Denslow 1980, Carpenter and McCreary 1985). These studies often used an expression of disturbance rate employing total area disturbed per unit time, or a probability of disturbance per unit time. For example, Shannon diversity increased with interval of disturbance in a mixed phytoplankton community (Gaedeke and Sommer 1986). Related to frequency, the timing or phasing (Abugov 1982) of disturbance can affect diversity, depending on how coincident disturbance is with pulses of growth and colonization. Repeated disturbance of stream substrates delayed recovery of stream macroinvertebrate diversity (Reice 1985).

Miller (1982) drew attention to the importance of considering the size of the area disturbed, and how size interacts with frequency to magnify or mitigate the impact of disturbance on diversity (for a review, see Pickett and White 1985). For example, nesting behavior by centrarchid sunfish in a Wisconsin lake created open, colonizable patches of substrate; these patches were of the same size scale as observed patterns of plant species, and are probably responsible for maintaining diversity and preventing competitive exclusion in this sytem (Carpenter and McCreary 1985, McCreary 1991). At a much larger spatial scale, Krummel *et al.* (1987) used fractal analyses to find that landscape patterns in patches of human-disturbed forest differed according to the size of the patches.

Finally, Malanson (1984) suggested that intensity of the disturbance regime should be an added dimension in the study of how disturbance alters diversity in communities. Particularly useful in disturbance due to fire, flooding and herbivory, intensity has potential for use in quantifying impacts of artificial disturbances, such as radiation and pollution disturbance (Malanson 1984).

Although some components of the diversity-disturbance relationship remain unknown, it is increasingly apparent that diversity is adversely affected at either very low or very high disturbance regimes, whether related to size, frequency or intensity. For geographical areas where concerns over biodiversity abound, the emerging recommendation is one of prudence and restraint by humans, with minimal interference in biological functions—neither increasing or creating undue "artificial" disturbance, nor inhibiting the occurrence of natural disturbances, such as fire, which are important to maintaining diversity.

ENVIRONMENTAL APPLICATION FOR USE OF DIVERSITY INDICES

Given that all diversity estimates suffer from some inherent limitations, scientists can nonetheless employ these quantitative measures to examine empirical changes in populations and communities. Anthropomorphic change presents one of the more compelling reasons for continued empirical measurement of diversity; pollution, deforestation, desertification, and climatic change have substantially altered many ecosystems, and are likely to continue adversely affecting global biodiversity.

Diversity estimates have had widespread application in aquatic systems, so much so that a number of indices have been developed strictly for use in lakes and rivers (for a review, see Washington 1984). This may be due in part to the degree of governmental regulation and control of waterways, as compared to terrestrial ecosystems. It may also reflect dissatisfaction by scientists and engineers with difficult-to-quantify characteristics such as "water quality" (Washington 1984).

Regardless of the cause, diversity of life in aquatic systems has been the basis of many water-related applications. For example, the New Jersey Department of Environmental Protection (NJDEP), in compliance with Environmental Protection Agency (EPA) guidelines, requires a biological survey and ecological analysis of any receiving stream in order to grant any Discharge Allocation Certificate; that survey and analysis is interpreted and enforced by the NJDEP to mean a quantitative assessment of benthic invertebrate community diversity (Evrard, pers. comm.).

Washington (1984) found that only a few diversity indices, including Simpson's Index, were suitable for discerning pollutant effects in aquatic ecosystems, particularly since the EPA is mainly concerned about sensitivity to changes in rarer species. Interestingly, in the same study diversity was more useful than indicator organisms (biotic indices) in describing impacts of water pollutants. However, a study by Kremen (1992) integrated diversity estimates of forests with species assemblages as indicators, thereby incorporating diversity estimates into conservation management practices in terrestrial systems. There is, therefore, evidence that empirical estimates of diversity can ultimately aid efforts to conserve biodiversity both in aquatic and in terrestrial ecosystems.

RECOMMENDATIONS AND CONCLUSIONS

Concerns over protection of biodiversity on the planet occur at many levels. Efforts of many conservation groups have focussed on protecting individual species recognized as endangered, or habitat areas supporting unique communities (Scott, *et al.* 1987). While monitoring rare species can address immediate needs and can be scientifically defensible (Murphy and Noon 1992), it provides little or no background for analysis of long-term biodiversity trends.

With estimates of extinction rates from 1000 to 5000 species per year (Myers 1979), determining risk of presently non-endangered species is a formidable task. Furthermore, there is widespread disparity in endangeredness among life forms—Mlot (1989) cites global data that over 19,000 plant species are known to be endangered, rare, or otherwise threatened, a number 15 times that for invertebrate species, 19 times that for bird species, and 43 times that for mammal species, yet maintaining plant biodiversity often appears to be far less politically desirable that doing so for higher animal biodiversity. Developing a plan which reflects the varied population-level constituents in community dynamics is complicated. The genetic diversity presently at risk seems astronomical (Hedrick and Miller 1992); estimating vulnerability of species at risk in the future seems impossible. Yet individual species concerns must be part of any attempt to maintain community-level biodiversity.

To some extent, use of biologically intuitive diversity indices in a regional program of inventory may provide baseline community data to assess their potential vulnerability to declines in biodiversity. Such a program might mirror the International Biological Program from the 1960's (McIntosh 1985), where ecology moved out of an exclusively academic domain into "big biology" with large scale programs originating from diverse institutions. In contrast to programs like NAPAP (the National Acid Precipitation Assessment Program), an inventory for conserving biodiversity needs to be a *science-driven* effort, not a *policy-driven* one (Russell 1992). Along these lines, recent legislation from California (Natural Communities Conservation Planning) to Pennsylvania (National Heritage Corridor; Lehigh River Foundation) has generated cooperative public-private partnerships with strong scientific participation to inventory and ultimately protect expanses of natural habitats from further change.

Such an inventory for conservation of biodiversity might now include more integrated, multidisciplinary management approaches with greater technological sophistication. An example is the geographic information systems approach presented by Scott *et al.* (1987), which employs estimates of diversity and species richness of vegetation in models incorporation characteristics of hydrology, geology, soils, and topography of areas. Likewise, initiatives such as the IUBS-SCOPE-UNESCO Harvard Forest Workshop are beginning efforts coordinated by our scientific community (Solbrig 1992). This approach can enable conservation of natural communities with more common plant and animal assemblages, ultimately to protect the entire specturm of biodiversity.

ACKNOWLEDGEMENTS

The author wishes to extend special thanks to Ms. Lisa Galloway Evrard for comments on initial versions of the manuscript and supporting information. Additionally, thanks are extended to three anonymous reviewers, for their valuable comments, and to Dr. John S. Waters and Dr. Robert P. MacIntosh for their special guidance.

LITERATURE CITED

Abugov, R. 1982. Species diversity and phasing of disturbance. *Ecol.* 63:289-293.

Brower, J., J. Zar, and C. Von Ende. 1990. *Field and Laboratory Methods for General Ecology.* Wm. C. Brown, Pubs., Dubuque, IA. 237 pp.

Brown, J.H. 1981. Two decades of Homage to Santa Rosalia: Toward a general theory of diversity. *Amer. Zool.* 21:877-888.

Carpenter, S.R. and N.J. McCreary. 1985. Effects of fish nests on pattern and zonation of submersed macrophytes in a softwater lake. *Aq. Bot.* 22:21-32.

Clarke, K.R. and R.H. Green. 1988. Statistical design and analysis for a 'biological effects' study. *Marine Ecology Progress Series* 46:213-226.

Connell, J.H. 1978. Diversity in tropical rain forests and coral reefs. *Science* 199: 1302-1310.

DeBenedictis, P.A. 1973. On the correlations between certain diversity indices. *The Amer. Nat.* 107:295-302.

Denslow, J.S. 1980. Patterns of plant species diversity during succession under different disturbance regimes. *Oecol. (Berl.)* 46:18-21.

Drury, W.H. and J.C.T. Nisbet. 1973. Succession. *J. Arnold Arb.* 54:331-368.

Egler, F.E. 1986. 'Physics envy' in ecology. *Bull. Ecol. Soc. Amer.* 67:233-235.

Evrard, Lisa Galloway. Omni Environmental Corporation, Inc., Princeton, NJ. Personal Communication.

Gaedeke, A. and U. Sommer. 1986. The influence of the frequency of periodic disturbance on the maintenance of phytoplankton diversity. *Oecol. (Berl.)* 71:25-28.

Gauch, H.G., Jr. 1982. *Multivariate Analysis In Community Ecology.* Cambridge University Press, New York, NY. 298 pp.

Ghilarov, A.M. 1984. The paradox of the plankton reconsidered; or why do species coexist? *Oikos* 43:46-52.

Gleason, H.A. 1922. On the relation between species and area. *Ecol.* 3:158-162.

Green, R.H. 1979. *Sampling Design and Statistical Methods for Environmental Biology.* Wiley and Sons, Inc., New York, NY. 257 pp.

Green, R.H. 1980. Multivariate approaches in ecology: the assessment of ecological similarity. *Ann. Rev. Ecol. Syst.* 11:1-14.

Hedrick, P.W. and P.S. Miller. 1992. Conservation genetics: techniques and fundamentals. *Ecol Appl.* 2:30-46.

Hurlbert, S.H. 1971. The nonconcept of species diversity: a critique and alternative parameters. *Ecol.* 52:577-586.

Huston, M. 1979. A general hypothesis of species diversity. *The Amer. Nat.* 113:81-101.

Hutchinson, G.E. 1953. The concept of pattern in ecology. *Proc. Acad. Nat. Sci.* Phil. 105:1-12.

Hutchinson, G.E. 1959. Homage to Santa Rosalia, or Why are there so many kinds of animals? *The Amer. Nat.* 93:145-159.

Hutchinson, G.E. 1961. The paradox of the plankton. *The Amer. Nat.* 95:137-145.

Kremen, C. 1992. Assessing the indicator properties of species assemblages for natural areas monitoring. *Ecol. Appl.* 2:203-217.

Krummel, J.R., R.H. Gardner, G. Sugihara, R.V. O'Neill, and P.R. Coleman. 1987. Landscape patterns in a disturbed environment. *Oikos* 48:321-324.

Loucks, O.L. 1970. Evolution of diversity, efficiency and community stability. *Amer. Zool.* 10:17-25.

MacArthur, R.H. 1960. On the relative abundance of species. *The Amer. Nat.* 94:25-36.

Malanson, G.P. 1984. Intensity as a third factor of disturbance regime and its effect on species diversity. *Oikos* 43:411-413.

Margalef, R. 1957. La teoria de la informacion en ecologia. *Mem. Real Acad. Cienc. Art. Barcelona* 32:373-449.

Margalef, R. 1963. On certain unifying principles in ecology. *The Amer. Nat.* 97:357-374.

McCreary, N.J. 1991. Competition as a mechanism of submersed macrophyte community structure. *Aq. Bot.* 41:177-193.

McIntosh, R.P. 1985. *The Background of Ecology: Concept and Theory.* Cambridge University Press, New York, NY. 383 pp.

McNaughton, S.J. 1977. Diversity and stability of ecological communities: a comment on the role of empiricism in ecology. *The Amer. Nat.* 111:515-525.

Menhinick, E.F. 1964. A comparison of some species-individuals diversity indices applied to samples of field insects. *Ecol.* 45:859-861.

Miller, T.E. 1982. Community diversity and interaction between the size and frequency of disturbance. *The Amer. Nat.* 120:533-536.

Mlot, C. 1989. Blueprint for conserving plant diversity. *BioScience* 39:364-368.

Mueller-Dombois, D. and H. Ellenberg. 1974. *Aims and Methods of Vegetation Ecology.* Wiley and Sons, Inc. New York, NY. 547 pp.

Murphy, D.D. and B.R. Noon. 1992. Integrating scientific methods with habitat conservation planning: Reserve design for Northern Spotted Owls. *Ecol. Appl.* 2:3-17.

Myers, N. 1979. *The Sinking Ark.* Pergamon Press, Oxford, UK.

Opyoke, J.R. and N.J. McCreary Waters. Patterns of seasonal movement in the benthic invertebrate community of Oughoughton Creek. *Fresh Biol., in review.*

Peet, R.K. 1974. The measurement of species diversity. *Ann. Rev. Ecol. Syst.* 5:285-307.

Pickett, S.T.A. and P.S. White, eds. 1985. *The Ecology of Natural Disturbance and Patch Dynamics.* Academic Press Inc., New York, NY. 472 pp.

Pielou, E.C. 1975. *Ecological Diversity.* John Wiley and Sons, New York, NY. 165 pp.

Pimm, S.L. 1984. The complexity and stability of ecosystems. *Nature* 307:321-326.

Preston, F.W. 1948. The commonness and rarity of species. *Ecol.* 29:254-283.

Reice, S.R. 1985. Experimental disturbance and the maintenance of species diversity in a stream community. *Oecol. (Berl.)* 67:90-97.

Ricklefs, R.E. 1990. *Ecology.* W.H. Freeman and Co., New York, N.Y. 896 pp.

Russell, M. 1992. Lessons from NAPAP. *Ecol. Appl.* 2:107-110.

Scott, J.M., B. Csuti, J.D. Jacobi and E. Estes. 1987. Species richness. *BioScience* 37:782-788.

Shannon, C.E. 1948. A mathematical theory of communication. *Bell System Tech. J.* 27:379-423, 623-656.

Simpson, E.H. 1979. Measurement of diversity. *Nature* 163:688.

Smith, R.L. 1992. *Elements of Ecology.* Harper Collins Pub. Inc., New York, NY. 617 pp.

Solbrig, O.T. 1992. The IUBS-SCOPE-UNESCO program of research in biodiversity. *Ecol. Appl.* 2:131-138.

Sousa, W.P. 1979. Disturbance in marine intertidal boulder fields: the non-equilibrium maintenance of species diversity. *Ecol.* 60:1225-1239.
Stiling, P.D. 1992. *Introductory Ecology.* Prentice Hall, Englewood Cliffs, NJ. 597 pp.
Wallace, A.R. 1878. *Tropical Nature and Other Essays.* MacMillan Pubs., New York, NY.
Washington, H.G. 1984. Diversity, biotic and similarity indices: A review with special relevance to aquatic ecosystems. *Water Res.* 18:653-694.
Whittaker, R.H. 1965. Dominance and diversity in land plant communities. *Science* 147:250-260.

Biological Diversity: Problems and Challenges. Edited by S.K. Majumdar, F.J. Brenner, J.E. Lovich, J.F. Schalles, and E.W. Miller. © 1994, The Pennsylvania Academy of Science.

Chapter Eight

ACID PRECIPITATION AND SURFACE WATER ACIDIFICATION: Implications for Aquatic Biodiversity

PATRICIA T. BRADT

Principal Research Scientist
Environmental Studies Center
Chandler-Ullmann Building
Lehigh University
17 Memorial Drive East
Bethlehem, PA 18015

INTRODUCTION

Biological diversity is receiving national attention as U.S. Congress considers legislation to preserve the remaining species on earth. Much attention has focused on losses of species in the tropics and rainforests. However, species are disappearing in the temperate zone also, from habitat destruction, from introduction of exotics and from pollution, for example. This paper discusses the effect of acid precipitation on sensitive aquatic ecosystems in northeastern North America.

Biological diversity in time and space has been a major component in the development of the science of ecology.[1] Biological diversity has at least three meanings; genetic diversity, species diversity and ecological diversity.[2] Steele[3] proposes a fourth meaning for biodiversity - functional diversity, the variety of response to environmental change. Change has always been a determining factor in ecosystems, but scientists and the informed public are currently very concerned not only about the rate of changes, but also about the large scale over which these changes are occurring.[3] Human activities are now inducing environmental changes at greater rates and over much larger areas than have ever been recorded in geological time.

Acid precipitation is defined as any precipitation falling with a pH less than 5.6, the expected pH of pure water in equilibrium with atmospheric carbon dioxide.[4] Acid precipitation is falling over eastern North America[5] and northern and central Europe.[6] Acid precipitation continued to fall over the U.S. in 1990.[7] Acid falls not only in precipitation, but also in acid dusts, e.g. dry deposition, that is more difficult to measure than wet deposition (rain, snow, mist). On a regional basis, dry deposition is estimated to be roughly equal to wet deposition.[5]

Acid precipitation and dry deposition are caused by the release of the air pollutants sulfur dioxide (SO_2) and oxides of nitrogen (NO_X) into the atmosphere. These pollutants react with atmospheric moisture to form sulfuric and nitric acid. These strong acids are then carried by the prevailing winds, west to east, and fall to earth as precipitation, dilute solutions of sulfuric and nitric acids. Therefore, air pollutants released in the midwest may come to rest as acid precipitation in the northeastern U.S. or eastern Canada.

Waters sensitive to acidification (e.g. decrease in pH) are "soft" waters, that have minimal acid neutralizing minerals dissolving from underlying bedrock and soils. "Soft" waters are limited in the chemical ability to neutralize incoming acid from precipitation. Regions with poorly buffered waters include those areas underlain by insoluble bedrock such as quartz, granite, gneisses and insoluble sandstones and shales.[8] Such waters are often located in mountainous areas, but low altitude waters on sandy (e.g. Cape Cod) or other poorly buffered soils may also be at risk.

In the U.S., poorly buffered waters are located in New England, in northern Florida, in the northern Midwest and in mountainous area of New York, Pennsylvania, New Jersey, West Virginia and Virginia.[9,10,11] In Canada, sensitive waters are located in Ontario, Quebec, New Brunswick, Newfoundland and Nova Scotia.[12] In Europe, sensitive waters are found in southern Scandinavia and in sections of central Europe and the British Isles.[8,13]

When acid precipitation falls on sensitive waters, the pH of the water may fall to a level stressful, or even lethal, to acid sensitive organisms. Reproducing organisms, their eggs and young, are usually more vulnerable than older, stronger adults to chemical stresses. The first sign of acid stress may be reproductive failure in sensitive groups. The fish population may be composed of a few older fish, with few young, due to reproductive failure.

FACTORS CONTROLLING EXTENT AND RATE OF ACIDIFICATION

Lakes at high risk from acidification are lakes high in altitude and small in size (Figure 1). For the Pocono region of northeastern Pennsylvania these lakes are higher than 500 M and smaller than 20 ha.[14] High, small lakes are usually headwater lakes, isolated from excess human activities.[15] Such lakes are the pristine, oligotrophic, wilderness lakes prized by humans for fishing and other recreation. Likewise, headwater streams are often first acidified. The small drainage basins, thin mountain soils and high dependence on precipitation causes such headwaters to be vulnerable because they are limited in acid neutralizing ability. Such sensitive waters hundreds of miles from pollution sources may thus be affected by acid precipitation.

CHEMICAL EFFECTS OF ACIDIFICATION

Acidification of surface waters may be measured not only by decreases in pH, but also by decreases in acid neutralizing capacity (ANC). ANC (analysis similar to total alkalinity) is measured by calculating the amount of dilute acid required to depress the pH of the water being tested. If a large amount of acid is required the water is high in ANC, but if only a small amount of acid depresses pH, the water is low in ANC, expressed as ueq L^{-1} of calcium carbonate ($CaCO_3$). As acid precipitation continues, ANC decreases and may reach zero or negative values. Decreases in ANC over time signal acidification.[16] ANC has no direct effects on the biota, but changes in ANC are accompanied by pH, calcium and aluminum changes that may affect the biota directly or indirectly.[17] During acidification specific conductance, a measure of ionic concentration may also decrease and this decrease may precede a decrease in ANC during acidification.[18,19,20]

Levels of ANC associated with different sensitivities of surface waters are shown in Table 1. Once ANC measures zero or negative, the water body is "acidified" and pH may vary greatly. If pH depressions coincide with sensitive life stages, reproduction is impaired or young cannot survive; populations decrease.

Acidification rates may be measured by the annual loss of ANC. In 11 lakes in the Pocono mountains, with 5 or more ANC readings over 5 or more years, mean ANC loss was 5.1 ueq per year[21], a very rapid rate of acidification. In a survey of

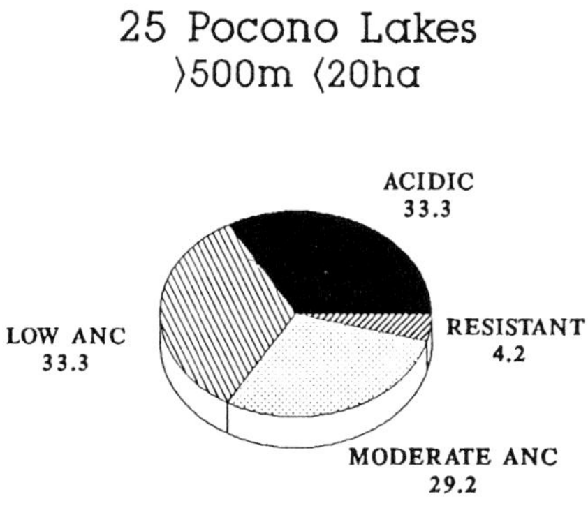

FIGURE 1. Sensitivity (based on acid neutralizing capacity, ANC) to acidification of 25 Pocono Mountain lakes at altitudes above 500m and with surface areas less than 20 hectares.

TABLE 1

Categories of Sensitivity of Surface Waters to Acidification Acid Neutralizing Capacity (ueq l^{-1})
(Linthurst et al. 1986)

Resistant	>201
Moderate ANC	51 to 200
Low ANC	1 to 50
Acidic	≤0.0

274 lakes in the Adirondack mountains of New York, 80% of the lakes had declining ANC, with a median loss of 50 ueq L^{-1} over the last 40 to 50 years.[15] As surface waters acidify, metals such as aluminum, manganese and zinc become more soluble,[22] and consequently are leached from soils and bedrock. Levels of these metals in surrounding waters may rise. Inorganic, monomeric aluminum levels may increase to levels toxic to the biota.[17] Aluminum may be toxic to fish in concentrations of 0.1 to 0.2 mg L^{-1}, levels found in some acidified waters.[23] Aluminum is most soluble at pH 5.2 to 5.4 so it is at these pH levels that toxicity is greatest.[24]

Another chemical effect of acidification is a decrease in the cations calcium, magnesium[20] and sodium. These three elements are usually low in mountain soils and are leached out of the drainage basin soils by continued acid precipitation. Calcium and ANC are highly correlated ($p < 0.001$) indicating the close relationship of these two factors. Levels of these elements in water may rise temporarily during acidification ([16,25]), but eventually levels decrease as the elements are leached into water and exported downstream. We have documented decreases in calcium and magnesium ($p < 0.02$) over a 5 year period in an acidifying Pocono lake.[20] Acidic lakes, therefore, may have lower calcium and magnesium levels than less acidic lakes.[19]

Another effect of surface water acidification is increasing transparency of the water,[26] measured by the depth of visibility of a black and white disk (Secchi disk). Acid lakes, therefore, are beautiful and clear. We have documented a statistically significant ($p < 0.01$) decrease in Secchi depth over 5 years in an acidifying Pocono lake.[20]

BIOLOGICAL EFFECTS

When waters are chemically stressed, aquatic species sensitive to the pollutants are severely stressed physiologically and their numbers may decrease or they may disappear completely. Number of aquatic taxa may decrease as pollution sensitive species are lost at a greater rate than pollution tolerant species become established.[27] A chemical stress such as acidification may cause decreases in numbers of plant and animal species, and biological diversity, e.g. biodiversity, is lowered.

The reduction in numbers of species makes the aquatic ecosystem more vulnerable to change,[8] less resilient and less able to return to a pre-stress condition. Aquatic ecosystems in midwestern and northeastern U.S. are becoming impoverished and simplified as aquatic taxa disappear in conjunction with increasing acidification.[28]

A problem that arises with determining acid sensitive and acid tolerant taxa is that these taxa may differ depending on geographical location. For example, among phytoplankton, Dinophyceae (Pyrrhophyta) dominate in some acidic Canadian and Scandinavian lakes[29] while Chrysophyceae often dominate in southern Canadian and Pocono lakes.[18,19,30] Among benthic invertebrates, European species tend to be more acid sensitive than North American members of the same groups. Okland and Okland[31] propose the North American groups appear less acid sensitive because they have evolved more effective mechanisms for dealing with extreme environmental changes over longer periods and over larger areas than the European fauna.

Such differences are apparent among the Isopoda, Amphipoda and Mollusca.

We have collected *Caecidotea communis* (Isopoda, formerly *Asellus communis*) from an acidic Pocono lake (pH <5.0, ANC <0.0 ueq l^{-1}) and more than 50% of these organisms survived 14 days at pH 3.65 in the laboratory. The European *Asellus aquaticus* is more acid sensitive and has not been collected from waters with pH below 4.8.[32] Such differences in sensitivities complicates extrapolation of data from one region to another. Collins et al.[33] propose that changes in North American benthic fauna have not been as obvious as those in Scandinavia because aluminum levels are lower in the U.S.

Schindler et al.[28] used data for northeast and midwestern U.S. from Linthurst et al.[10] to model changes in aquatic taxa with acidification. They conclude that 1506 lakes in Pocono/Catskill region have experienced a median decrease of 55 ueq L^{-1} ANC, a median decrease of 1.6 pH units and a median increase of 92 ueq L^{-1} SO_4. ANC loss and sulfate gain for the region are greater than for other 8 other U.S. subregions. Losses of aquatic taxa accompany these chemical changes. Based on known acid tolerances, Schindler et al.[28] predict losses in the Pocono/Catskills region of 31% of algal taxa. 4% of rotifer taxa, 18% of crustacea taxa, 46% of insect taxa, 69% of leech taxa, 52% of mollusc taxa and 14% of fish taxa. Percentages of taxa lost are highest in the Pocono/Catskill region, and second highest in the Adirondacks.

We have been studying the biology and chemistry of Pocono acidifying lakes since 1981. Analyses of our data reinforce these concerns about losses of aquatic diversity from surface water acidification.

There are many ways to measure biological diversity. Species richness is a very useful measure of diversity.[34] Species density, number of species per unit area, is the most commonly used measure of species richness. Other diversity measurements include: diversity and evenness indices for individual collections and similarity indices for comparison of 2 communities.[35]

In this paper taxa richness and frequency abundance are used as measures of biological diversity. The commonly used diversity indices are not sensitive enough to discern subtle changes in the benthic invertebrate community during acidification.[36] Community comparison indices have been more successful in comparing communities in treated (limestoned) and reference lakes.

ECOSYSTEM BIOTIC COMPONENTS

All major trophic levels of aquatic ecosystems may be affected by acidification.[8] How are the various biotic components of aquatic ecosystems affected? We will now consider these biotic components.

PRODUCERS: PHYTOPLANKTON AND MACROPHYTES

Phytoplankton

Acidic waters do have fewer species of algae[37,38] and algae dominating in acidic waters vary depending on geographic region (see above). The greatest change in species numbers occurs between pH 5.0 and pH 6.0.[37] Significant reductions in species richness occurred below pH 5.6 in an artificially acidified Canadian lake.[39] Dominant algae at low pH levels are usually Chrysophyceae or Dinophyceae,[30,40] depending on the region.[41]

When summer data for 11 Pocono lakes at various stages of acidification, over 1 to 10 summers, are compiled there are fewer Cyanobacteria species at lower ANC levels ($p < 0.03$) and fewer total phytoplankton species at lower calcium levels (Figure 2). Calcium is highly correlated with ANC ($r \pm 0.95$, $p < 0.001$), therefore most low ANC lakes are also low in calcium. These analyses support the observation of fewer phytoplankton species at lower ANC and calcium levels.

Macrophytes

In Adirondack lakes ANC is correlated with macrophyte species richness.[42] In the Poconos there were fewer macrophyte species in an acidic lake.[19] During a 4 year study, there was a decrease ($p < 0.03$) in macrophyte taxa richness in an acidifying lake.[20]

In many acidic lakes, mats of *Sphagnum* moss may cover the lake bottom,[43] overgrowing other benthic plants, decreasing the number of competing species and causing community changes, and even water chemistry changes.

CONSUMERS: MICROORGANISMS, ZOOPLANKTON, BENTHIC INVERTEBRATES, FISH AND AMPHIBIANS

Microorganisms (Microconsumers)

The decomposition of organic matter is usually reduced at pH levels below 5.5 and organic material may accumulate.[8,44] Acidic lakes have lower densities of bacteria than less acidic lakes.[45] Nitrification may be inhibited or cease below pH 5.5 and the nitrogen cycle is disrupted.[46] Mineralization is reduced and leaves and dead wood accumulate on the lake bottom.[8]

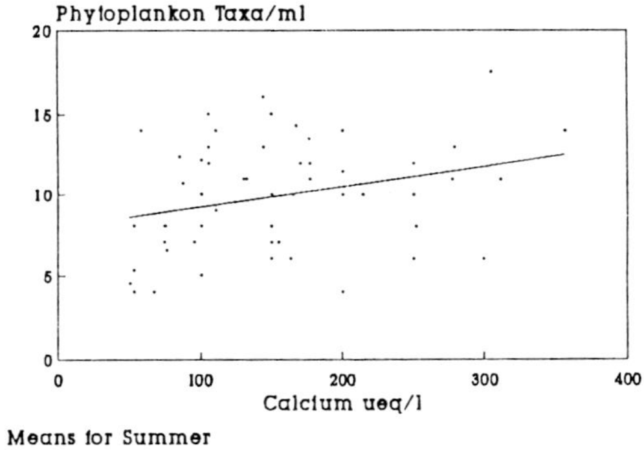

FIGURE 2. Regression of number of total phytoplankton taxa per milliliter versus calcium levels. Values for phytoplankton are from summer samples only. $r = .29$, $y = 0.013x + 7.94$, $df = 53$, $p < 0.03$.

Zooplankton

As with other biota, there are fewer species of zooplankton in acidic and acidifying lakes.[13,37,47] Numbers of rotifers and cladocerans, especially, are reduced.[48,49] There were fewer numbers and taxa of rotifers in an acidic Pocono lake than in 2 less acidic lakes.[19] In the acidic lake only one rotifer species, *Keratella taurocephala*, was collected and it is acid tolerant.[50] In a 4 year study of an acidifying Pocono lake rotifer numbers decreased ($p < 0.01$) while increases ($p < 0.01$) were documented in the acid tolerant cladoceran, *Bosmina longirostris* and the acid tolerant copepod, *Diaptomus minutus*.[20]

When summer data for zooplankton populations in 11 Pocono lakes over 1 to 10 summers are compared with mean summer surface ANC, rotifer species richness ($p < 0.01$, $r = .41$) and total zooplankton species richness (Figure 3) are both correlated with ANC. The relationship between ANC and taxa richness of zooplankton are, thereby, substantiated.

Benthic Macroinvertebrates

Acidic lakes and streams have fewer species of benthic invertebrates.[31,37,51] Community changes also occur as surface waters acidify; there are fewer Ephemeroptera, Gastropoda, Pelecypoda,[31] Amphipoda[52,53] and Hirudinea.[28] Not only are numbers of sensitive organisms reduced, but taxa richness of these groups, e.g. Ephemeroptera, may also decrease.[54] As acidification proceeds sensitive species are gradually replaced by acid tolerant ones.

As stated previously, acid sensitivity among reportedly sensitive groups may vary among species. *Amnicola limnosa* is apparently the most acid tolerant Gastropoda and has been collected at pH 4.0.[55] Likewise, sediment burrowing fingernail clams (Sphaeriidae) are acid tolerant pelecypods[56] and have been collected at pH levels down to 5.4.[57] Among larger Pelecypoda, the Unionidae are the most acid sensitive.[49]

Over a three year study of 3 Pocono lakes with different ANC levels, there were

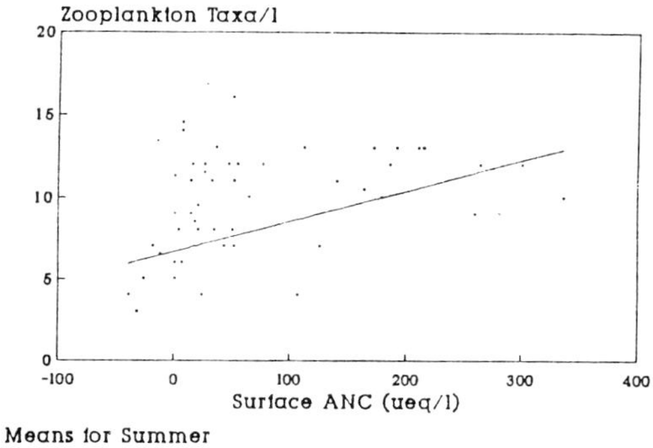

FIGURE 3. Regression of number of total zooplankton taxa per liter versus acid neutralizing capacity (ANC). Values are from summer samples only. $r = .45$, $y = 0.0187x + 6.69$, $df = 53$, $p < 0.01$.

no significant differences ($p > 0.05$) among annual mean taxa richness among the lakes. However, there were significant differences ($p < 0.05$) among the communities, e.g. more Ephemeroptera, Gastropoda and Pelecypoda in the circumneutral lake and more Chironomidae, Certatopogonidae and invertebrate predators in the acidic lake.[36]

In a limestoned Pocono lake, taxa richness increased ($p < 0.01$) (Bradt, P.T., unpublished data). Three years postlimestoning taxa richness in the limestoned lake was significantly higher ($p < 0.01$) than at the acidifying reference lake.[20] There were also more Ephemeroptera and Amphipoda in the limestoned lake and more Isopoda in the acidifying lake.[20]

As with other trophic levels, the structure of the benthic invertebrate community may vary with latitude and longitude. For example, in the Poconos the isopod *Caecidotea communis* is frequently collected in acidic lakes, but in Adirondack lakes it is seldom collected (Figure 4).[58] A few acid tolerant Crangonyctidae (Amphipoda) have been collected in acidic Adirondack lakes,[58] but only the acid sensitive amphipod, *Hyalella azteca*,[53] is found in Pocono lakes. Some leeches are acid tolerant,[59] while others are acid sensitive.[13,28] Among crayfish, there are differences in acid tolerance among species.[60] *Cambarus robustus* and *C. bartoni* are acid tolerant[61] and are found in acidic Adirondack lakes and occasionally in Pocono lakes. However, the acid sensitive *Orconectes virilis* disappeared when pH reached 5.0 in an experimentally acidified Canadian lake.[62] These variations in acid sensitivity among the same orders and families make generalizations about sensitivity above the genus level very difficult.

Regression analyses of annual means of ANC, calcium and number of invertebrates (spring, summer, fall - 12 to 30 Ekman grab samples) from 8 Pocono lakes over 1 to 7 years are shown in Figures 5 and 6. Numbers of Ephemeroptera are correlated with ANC (Figure 5) and numbers of Amphipoda with calcium (Figure 6). Analyses of these data indicated the following additional relationships; percentage of Mollusca correlated with calcium ($p < 0.01$), $r = .56$), percentage of *Caecidotea communis* negatively correlated with ANC ($p < 0.03$, $r = -.45$) and total biomass (wet weight) correlated with calcium ($p < 0.01$, $r = .70$).

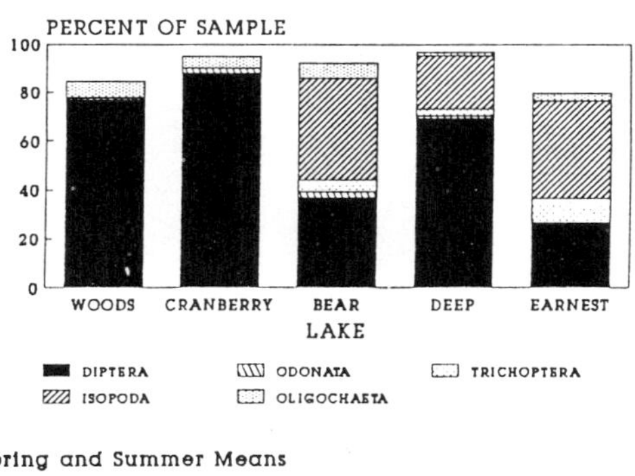

FIGURE 4. Percent of macroinvertebrates in acidic lakes in the Adirondack Mountains (New York) and the Pocono Mountains (Pennsylvania). Adirondacks = Cranberry Pond and Woods Lake. Poconos = Bear Lake, Deep Lake, Lake Earnest.

Fish

There is a large amount of literature concerning the effects of surface water acidification on fish.[63] Acidifying surface waters usually have more older, larger fish and fewer young fish.[24] Cyprinidae appear to be the most sensitive.[49] Many surveys have demonstrated that acidifying surface waters have fewer species of fish.[29,37,49,64,65,66] Geographical differences are present, as expected, and acid tolerant fish vary in different regions. In the Poconos, for example, yellow perch and pumpkinseed sunfish appear to be the most acid tolerant. In one acidic lake (pH <5.0) a population of stunted pumpkinseed sunfish has evidently undergone physiological adaptation and occassionally reproduces successfully. In the Adirondacks brook trout are found in acidifying waters, but suffer reproductive failure at pH <5.0.[16]

Amphibians, Birds and Mammals

Amphibians breed in water and acidification, especially of temporary breeding

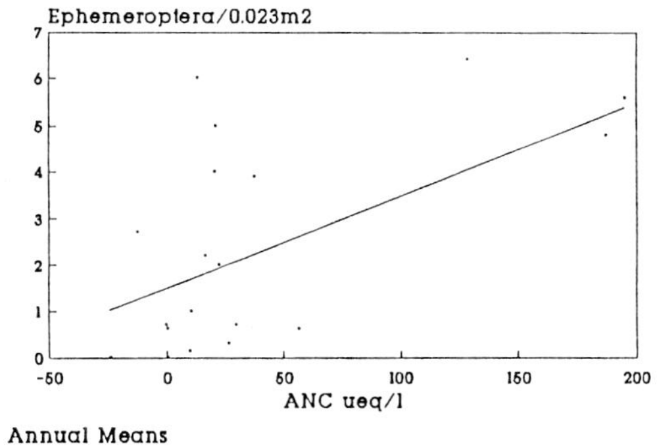

FIGURE 5. Regression of number of Ephemeroptera versus acid neutralizing capacity (ANC). Annual means. r = .62, y = 0.02x + 1.5, df = 18, p<0.01.

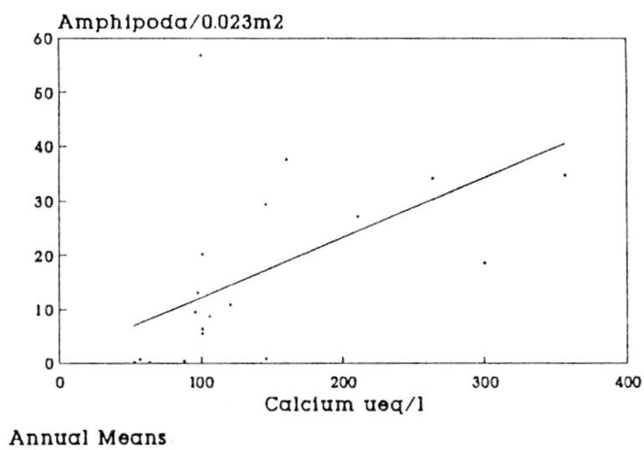

FIGURE 6. Regression of number of Amphipoda versus calcium levels. Annual Means. r = .54, y = 0.11x + 1.25, df = 18, p<0.01.

ponds, may cause stress to eggs and larvae.[67] Sadinski and Dunson[68] report decreases in Jefferson's salamanders and pH levels low enough to endanger both spotted salamanders and wood frogs in central Pennsylvania. There are major concerns about the decline in amphibians worldwide and acid rain is suspected as a factor.[69]

As numbers and taxa of fish decline, the birds and mammals that feed on them may also decline. For example, loons, diving ducks, herons and kingfishers may move elsewhere when fish are scarce. Mammals such as otters and mink will also suffer when fish numbers decrease. Raccoons eat clams and mussels and must find alternate prey when acidification reduces these important food sources.

CONCLUSIONS

Reduced taxa richness of phytoplankton, macrophytes, microorganisms, zooplankton, macrophytes and fish in acidifying surface waters represents an important decrease in aquatic biological diversity. Given the large number of sensitive waters in North America (Table 2), these losses could be substantial. Concurrent with losses of valuable aquatic species are community changes as tough acid tolerant species become established and sensitive species are lost. The form of community change depends on many factors, such as geographical location, competition and colonization abilities.

Surface water acidification has not ceased with the passage of the Clean Air Act in 1990. Precipitation continues to be acidic[7] and will probably continue to be so

TABLE 2

Numbers of Lakes (>4 ha) Sensitive to Acidification in North America

	UNITED STATES (Linthurst et al. 1986)		
	Moderate ANC (51-200 ueq l^{-1})	Low ANC (1-50 ueq l^{-1})	Acidic ($\leq$0.00 ueq l^{-1})
Northeast			
Adirondacks	450	321	138
Pocono/Catskills	378	116	78
New England	2066	602	109
Upper Midwest	2207	767	147
Southeast	498	283	463
	CANADA (Kelso et al. 1986)		
Ontario		40,662	2,020
Quebec		229,711	6,185
Newfoundland/Labrador		75,181	2,497
New Brunswick		1,718	554
Nova Scotia		6,144	3,115

for many more years. As controls are implemented on sulfur dioxide and oxides of nitrogen, there should be a gradual increase in precipitation pH. It will probably be many years before these pH changes are reflected in surface water chemistry. Meanwhile lakes and streams are still acidifying in sensitive regions of North America and Europe; the extent and rate of surface water acidification is not definitively known.

As acidification continues and ANC decreases, there will undoubtedly be more losses of important aquatic species. Given the regression lines here shown for sensitive taxa and taxa richness, we can expect continued reductions in acid sensitive species until the pH of precipitation starts to increase. Once pH starts to rise, recovery of acidifying waters may be impeded by low levels of calcium in acidifying waters.[70]

In the Poconos, there is a mean annual ANC loss of 5.1 ueql^{-1}. There are at least 40 lakes with ANC levels less than 50 ueq l^{-1}.[21] So at the current rate of acid deposition these 40 lakes could become acidic in the next 10 years. Concurrent with this rapid acidification rate would be the losses of valued taxa when ANC, calcium and pH decrease. Such taxa losses would include fewer phytoplankton and zooplankton taxa (including rotifers), fewer mayflies and amphipods, lower percentages of molluscs and lower total biomass. Percentages of Isopoda would increase. Such changes in the lower trophic levels may have far reaching and cascading impacts on the ecosystem. As ecosystems change under stress, species are lost, processes are altered and may be irreversibly damaged. Leopold[71] observed "If the biota, in the course of eons, has built something we like but do not understand, then who but a fool would discard seemingly useless parts? To keep every cog and wheel is the first precaution of intelligent tinkering."

There is great concern over the loss of aquatic biodiversity, not only in the tropics, but also in temperate regions.[71,73] Biologists are also very concerned about the loss of invertebrate species, especially.[74] Invertebrates are the most diverse group on earth and account for 90% of the animal biomass and for 95% of all animal species.[75] Schindler et al.'s[28] projections concerning loss of invertebrate and vertebrate aquatic species may come true, in fact, they may already be true. Protecting aquatic diversity in the temperate zone is a very difficult task.[72] Unfortunately, most scientists do not have the resources to collect regularly the essential data to determine what is happening to surface waters in acid impacted regions. Collecting and compiling such data on aquatic biodiversity should have a high national priority. Such databases would provide invaluable references against which changes from other anthropogenic effects, such as global climate change, could be measured. We may lose species before we even know they are there. Aquatic biologists know this is happening, the scientific evidence is definitive and overwhelming. Who is listening?

LITERATURE CITED

1. Ray, G.C. and J.F. Grassle. 1991. Marine biological diversity. BioScience 41(7):453-457.

2. Office of Technology Assessment (OTA). 1987. Technologies to maintain biological diversity. U.S. Congress. OTA-F-330, Washington, D.C. 334 pp.
3. Steele, J.H. 1991. Marine functional diversity. BioScience 41(7):470-474.
4. Haines, T.A. 1982. Introduction to Symposium. In R.E. Johnson, (Ed.), Acid Rain/Fisheries. Amer. Fish. Soc., Bethesda, MD. pp. 3-4.
5. Stensland, G.J., D.M. Whelpdale and G. Oehlert. 1986. Precipitation chemistry. In Acid deposition, Long-term trends. National Academy Press, Washington, D.C. pp. 128-199.
6. Semb, A. and H. Dovland. 1986. Atmospheric deposition in Fenno-Scandia: Characteristics and Trends. Water, Air and Soil Pollut. 30:5-16.
7. National Atmospheric Deposition Program (NADP). 1991. NADP/NTN Annual data summary. Precipitation chemistry in the United States. 1990. Natural Resource Ecology Laboratory. Colorado State University. Fort Collins, Co. 475 pp.
8. Overrein, L.N., H.M. Seip and A. Tollan. 1980. Acid precipitation - Effects on forest and fish. Final Report, SNSF Project. 1972-1980. Oslo, Norway. 175 pp.
9. Office of Technology Assessment (OTA). 1984. Acid rain and transported air pollutants: Implications for public policy. U.S. Congress, OTA-0-204, Washington, D.C. 323 pp.
10. Linthurst, R.A., D.H. Landers, J.M. Eilers, D.F. Brakke, W.D. Overton, E.P. Meier and R.E. Crowe. 1986. Characteristics of lakes in the eastern United States. Vol. I. Population descriptions and physico-chemical relationships. EPA/600/4-86/007a. U.S. Environmental Protection Agency, Washington, D.C.. 136 pp.
11. Kaufmann, P.R., A.T. Herlihy, J.W. Elwood, M.E. Mitch, W.S. Overton, M.J. Sale, J.J. Messer, K.A. Cougan, D.V. Peck, K.H. Reckhow, A.J. Kinney, S.J. Christie, D.D. Brown, C.A. Hagley and H.I. Jager. 1988. Chemical characteristics of streams in the Mid-Atlantic and Southeastern United States. Vol. I. Population descriptions and physico-chemical relationships. EPA/600/3-88/021a. ERL-COR-482. U.S. Environmental Protection Agency. Office of Research and Development, Washington, D.C. 397 pp.
12. National Research Council Canada (NRCC). 1981. Acidification in the Canadian aquatic environment: Scientific criteria for assessing the effects of acidic deposition on aquatic ecosystems. NRCC No. 18475. Environmental Secretariat, NRCC/CRCC, Ottawa. 369 pp.
13. Bowman, J.J. 1987. The chemical and biological features of poorly buffered Irish lakes. In H. Barth, (Ed.), Reversibility of acidification. Elsevier Applied Science, London. pp. 46-63.
14. Bradt, P.T. and L. Wimpfheimer. 1987. Survey of Pocono lakes to evaluate sensitivity to acidification. Tech. Rpt., Environmental Studies Ctr., Lehigh University, Bethlehem, PA. 24 pp.
15. Asbury, C.E., F.A. Vertucci, M.D. Mattson and G.E. Likens. 1989. Acidification of Adirondack Lakes. Environ. Sci. Technol. 23(3):362-365.
16. Schofield, C.L. 1982. Historical fisheries changed in the United States related to decreases in surface water pH. In R.E. Johnson, (Ed.), Acid Rain/Fisheries. Amer. Fish. Soc., Bethesda, MD. pp. 57-67.

17. Baker, J.P. 1990a. Chemical and physical characteristics that influence biological response. In Baker, J.P., D.P. Bernard, S.W. Christensen, M.J. Sale, J. Freda, K. Heltcher, D. Marmorek, L. Rowe, P. Scanlon, G. Suter, W. Warren-Hicks and P. Welbourn. 1990. Biological effects of changes in surface water acid-base chemistry. NAPAP report 13. National Acid Precipitation Assessment Program, Acidic deposition: State of science and technology. Vol. II. pp. 47-66.
18. Kelso, J., R.J. Love, J.H. Lipsit and R. Dermott. 1982. Chemical and biological status of headwater lakes in the Sault Ste. Marie District, Ontario. In F.M. D'itri, (Ed.), Acid precipitation: Effects on ecological systems. Ann Arbor Sci., Ann Arbor, MI. pp. 165-207.
19. Bradt, P.T., J.L. Dudley, M.B. Berg and D.S. Barrasso. 1986. Biology and chemistry of three Pennsylvania lakes: Responses to acid precipitation. Water, Air and Soil Pollut. 30:505-513.
20. Bradt, P.T., J.L. Dudley and J.A. Remenar. 1989. Biology and chemistry of two acid-stressed lakes in the Pocono Mountains: Changes in limestone treated and reference lakes. Tech. Complt. Rpt., PA Power and Light Co., Allentown, PA. 183 pp.
21. Bradt, P.T. and J.A. Remenar. 1990. Extent and rate of acidification in the Pocono Mountains: Relationship to biodiversity. Paper presented at National Acid Precipitation Assessment Program (NAPAP), Acidic deposition: State of science and technology. Hilton Head, S.C. February 1990.
22. Schofield, C.L. 1976. Acidification of Adirondack lakes by atmospheric precipitation: Extent and magnitude of the problem. Final Rpt. D.J. Project F-28-R, New York Dept. Environ. Conserv., Albany, NY.
23. Schofield, C.L. and J.R. Trojnar. 1980. Aluminum toxicity to brook trout (*Salvelinus fontinalis*) in acidified waters. In T.Y. Toribara, M.W. Miller and R.E. Morrow (Eds.), Polluted Rain., Plenum Press, New York. pp. 341-366.
24. Baker, J.P. 1982. Effects on fish of metals associated with acidification. In R.E. Johnson, (Ed.), Acid Rain/Fisheries. Amer. Fish. Soc., Bethesda, MD. pp. 165-176.
25. Bradt, P.T., M.B. Berg, D.S. Barrasso and J.L. Dudley. 1984. The biological and chemical impact of acid precipitation on Pocono Mountain lakes. Tech. Complt. Rpt., PA Power and Light Co., Allentown, PA. 215 pp.
26. Yan, N.D. 1983. Effects of changes in pH on the transparency and thermal regimes of Lohi Lake, near Sudbury, Ontario. Can. J. Fish. Aquat. Sci. 40:621-626.
27. Ford, J. 1989. The effects of chemical stress on aquatic species composition and community structure. In S.A. Levin, M.A. Harwell, J.R. Kelly and K.D. Kimball (Eds.), Ecotoxicology: Problems and approaches. Springer-Verlag, New York. pp. 99-144.
28. Schindler, D.W., S.E.M. Kaslan and R.H. Hesslein. 1989. Biological impoverishment in lakes of the midwestern and northeastern United States from acid rain. Environ. Sci. Technol. 23:573-580.
29. Conway, H.L. and G.R. Hendrey. 1982. Ecological effects of acid precipitation on primary producers. In F.M. D'itri, (Ed.), Acid precipitation: Effects on ecological systems. Ann Arbor Sci., Ann Arbor, MI. pp. 277-295.

30. Hendrey, G.R. 1982. Effects of acidification on aquatic primary producers and decomposers. In R.E. Johnson, (Ed.), Acid Rain/Fisheries. Amer. Fish. Soc., Bethesda, MD. pp. 125-135.
31. Okland, J. and K.A. Okland. 1986. The effects of acid deposition on benthic animals in lakes and streams. Experientia 42:471-486.
32. Okland, K.A. 1980. Ecology and distribution of *Asellus aquaticus* (L.) in Norway, including relation to acidification in lakes. SNSF Project IR 52/80. Oslo, As, Norway. 70 pp.
33. Collins, N.C., A.P. Zimmerman and R. Knoechel. 1981. Comparisons of benthic fauna and epifauna in acidified and nonacidified Ontario lakes. In R. Singer, Ed., Effects of acidic precipitation on benthos. North American Benthological Soc., Springfield, IL. pp. 35-48.
34. Magurran, A.E. 1988. Ecological diversity and its measurement. Princeton University Press., Princeton, NJ. 179 pp.
35. Wolda, H. 1981. Similarity indices, sample size and diversity. Oecologia 50:296-302.
36. Bradt, P.T. and M.B. Berg. 1987. Macrozoobenthos of three Pennsylvania lakes: Responses to acidification. Hydrobiologia 150:63-74.
37. Kelso, J.R.M., C.K. Minns, J.E. Gray and M.L. Jones. 1986. Acidification of surface waters in eastern Canada and its relationship to aquatic biota. Can. Spec. Publ. Fish. Aquat. Sci. No. 87. Dept. Fish. Oceans. Scient. Inform. Public. Ottawa, Canada. 42 pp.
38. Seigfried, C.A., J.A. Bloomfield and J.W. Sutherland. 1989. Acidity status and phytoplankton species richness, standing crop and community composition in Adirondack, New York, U.S.A. lakes. Hydrobiologia 175:13-32.
39. Findlay, D.L. and S.E. Kasian. 1986. Phytoplankton community response to acidification of Lake 223, experimental lakes areas, Northwestern Ontario, Water, Air and Soil Pollut. 30:719-726.
40. Yan, N.D. and P. Stokes. 1978. Phytoplankton of an acidic lakes and its response to experimental alterations of pH. Environ. Conserv. 5:93-100.
41. Dillon, P.J., N.D. Yan and H.H. Harvey. 1984. Acidic deposition: Effects on aquatic ecosystems. CRC Critical Reviews in Environ. Control. 13:167-194.
42. Roberts, D.A. 1984. Macrophytes. In R. Singer and C.W. Boylen, (Eds.), Biological field survey of northeastern acidified lakes. Final Rept. U.S. Environmental Protection Agency/North Carolina State University Acid Deposition Program Contract No, APP-0313-1983. Troy, NY.
43. Hendrey, G.R. and F.A. Vertucci. 1980. Benthic plant communities in acidic Lake Colden, New York: Sphagnum and the algal mat. In D. Drablos and A. Tollan, (Eds.), Ecological impacts of acid precipitation. SNSF Project, Oslo, Norway. pp. 314-315.
44. Bernard, D.P. 1990. Effects of acidification on microbial communities. In Baker, J.P., D.P. Bernard, S.W. Christensen, M.J. Sale, J. Freda, K. Heltcher, D. Marmorek, L. Rowe, P. Scanlon, G. Suter, W. Warren-Hicks and P. Welbourn. Biological effects of changes in surface water acid-base chemistry. NAPAP report 13. National Acid Precipitation Assessment Prog., Acidic deposition: State of science and technology. Vol. II. pp. 66-74.

45. Rao, S.S. and B.J. Dutka. 1983. Influence of acid precipitation on bacterial populations in lakes. Hydrobiologia 98:153-157.
46. Rudd, J.W.M., C.A. Kelly, D.W. Schindler and M.A. Turner. 1988. Disruption of the nitrogen cycle in acidified lakes. Science 240:1515-1517.
47. Malley, D.F., D.L. Findlay and P.S.S. Chang. 1982. Ecological effects of acid precipitation on zooplankton. In F.M. D'itri, (Ed.), Acid precipitation: Effects on ecological systems. Ann Arbor Sci., Ann Arbor, MI. pp. 297-327.
48. Brezonik, P.L., T.L. Crisman and R.L. Schulze. 1984. Planktonic communities in Florida softwater lakes of varying pH. Can. J. Fish. Aquat. Sci. 41:46-56.
49. Eilers, J.M., G.L. Lien and R.G. Berg. 1984. Aquatic organisms in acidic environments: A literature review. Wisc. Dept. Nat. Resource Tech. Bull. 150. 18 pp.
50. Schaffner, W.R. 1989. Effects of neutralization and addition of brook trout (*Salvelinus fontinalis*) on the limnetic zooplankton communities of two acidic lakes. Can. J. Fish. Aquat. Sci. 46:295-305.
51. Kimmel, W.G., D.J. Murphy, W.E. Sharpe and D.R. DeWalle. 1985. Macroinvertebrate community structure and detritus processing rates in two southwest Pennsylvania streams acidified by atmospheric deposition. Hydrobiologia 124:97-102.
52. Raddum, G.G. and A. Fjellheim. 1984. Acidification and early warning organisms in freshwater in western Norway. Verh. Int. Ver. Limnol. 22:1973-1980.
53. Stephenson, M. and G.L. Mackie. 1986. Lake acidification as a limiting factor in the distribution of the freshwater amphipod *Hyalella azteca*. Can. J. Fish. Aquat. Sci. 43:288-292.
54. Mackay, R.J. and K.E. Kersey. 1985. A preliminary study of aquatic insect communities and lea decomposition in acid streams near Dorset, Ontario. Hydrobiologia 122:3-11.
55. Servos, M.R. and G.L. Mackie. 1986. The effect of short-term acidification during snowmelt on selected Mollusca in south-central Ontario. Can. J. Zool. 64:1690-1695.
56. Rooke, J.B. and G.L. Mackie. 1984. Growth and production of three species of molluscs in six low alkalinity lakes in Ontario. Can. J. Zool. 62:1474-1478.
57. Dermott, R., J.R.M. Kelso and A. Douglas. 1986. The benthic fauna of 41 acid sensitive headwaters lakes in north central Ontario. Water, Air and Soil Pollut. 28:283-292.
58. Bradt, P.T. 1991. Benthic macroinvertebrates of Adirondack lakes: Impact of acidification and mitigation. Final Rpt. to Lake Acidification Mitigation Program (LAMP). Dept. Natural Resources, Cornell Univ., Ithaca, NY. 87 pp.
59. Zischke, J.A., J.W. Arthur, K.J. Nordlie, R.O. Hermanutz, D.A. Standen and T.P. Henry. 1983. Acidification effects on macroinvertebrates and fathead minnows (*Pimephales promelas*) in outdoor experimental channels. Water Res. 17:47-63.
60. Rowe, L. 1990. Effects of acidification of benthic macroinvertebrates. In Baker, J.P., D.P. Bernard, S.W. Christensen, M.J. Sale, J. Freda, K. Heltcher, D. Marmorek, L. Rowe, P. Scanlon, G. Suter, W. Warren-Hicks and P. Welbourn. Biological effects of changes in surface water acid-base chemistry. NAPAP report 13. National Acid Precipitation Assessment Program. Acidic deposition: State of science and technology. Vol. II. pp. 105-113.

61. Electric Power Research Institute (EPRI). 1985. Lake acidification mitigation project (LAMP). Second Annual Tech. Rpt. Nov. 1984-Sept. 1985. Contract RP2337-1.
62. Davies, I.J. 1989. Population collapse of the crayfish *Orconectes virilis* in response to experimental whole-lake acidification. Can. J. Fish. Aquat. Sci. 46:910-922.
63. Baker, J.P. 1990b. Effects of acidification on fish. In Baker, J.P., D.P. Bernard, S.W. Christensen, M.J. Sale, J. Freda, K. Heltcher, D. Marmorek, L. Rowe, P. Scanlon, G. Suter, W. Warren-Hicks and P. Welbourn. 1990. Biological effects of changes in surface water acid-base chemistry. NAPAP report 13. National Acid Precipitation Assessment Program. Acidic deposition: State of science and technology. Vol. II. pp. 113-135.
64. Arnold, D.E., R.W. Light and V.J. Dymond. 1980. Probable effects of acid precipitation on Pennsylvania waters., EPA-600-80-012. Corvallis Environmental Res. Lab., U.S. Environmental Protection Agency, Corvallis, OR. 20 pp.
65. Harvey, H.H. and C. Lee. 1982. Historical fisheries changes related to surface water pH changes in Canada. In R.E. Johnson, (Ed.), Acid Rain/Fisheries. Amer. Fish. Soc., Bethesda, MD. pp. 45-55.
66. Bergquist, B.C. 1991. Extinction and natural recolonization of fish in acidified and limed lakes. Nordic. J. Freshw. Res. 66:50-62.
67. Freda, J. 1990. Effects of acidification on amphibians. In Baker, J.P., D.P. Bernard, S.W. Christensen, M.J. Sale, J. Freda, K. Heltcher, D. Marmorek, L. Rowe, P. Scanlon, G. Suter, W. Warren-Hicks and P. Welbourn. Biological effects of changes in surface water acid-base chemistry. NAPAP report 13. National Acid Precipitation Assessment Prog., Acidic deposition: State of science and technology. Vol. II. pp. 135-151.
68. Sadinski, W.J. and W.A. Dunson. 1990. The effects of low pH on the animal communities of temporary ponds in central Pennsylvania. In J.A. Lynch, E.S. Corbett and J.W. Grimm, (Eds.), Atmospheric deposition in Pennsylvania: A critical assessment. Environmental Resources Research Instit., The Pennsylvania State Univ., University Park, PA. p. 111.
69. Barinaga, M. 1990. Where have all the froggies gone? Science 247:1033-1035.
70. Howells, G. 1987. Acidity mitigation in a small upland lake. In H. Barth, (Ed.), Reversibility of acidification. Elsevier Applied Science, London. pp. 104-111.
71. Leopold. A. 1950. A sand county almanac, and sketches here and there. Oxford University Press, New York.
72. Franklin, J.F. 1988. Structural and functional diversity in temperate forests. In E.O. Wilson (Ed.), Biodiversity. National Academy Press, Washington, D.C. pp. 166-175.
73. Mlot, C. 1991. Giving freshwater its due. BioScience 41(11):756-757.
74. Wilson, E.O. 1988. The current state of biological diversity. In E.O. Wilson (Ed.), Biodiversity. National Academy Press, Washington, D.C. pp. 3-18.
75. National Science Board. 1989. Loss of biological diversity: A global crisis requiring international solutions. Comm. on International Science's Task Force on Global Biodiversity. National Science Foundation, Washington, D.C. 19 pp.

Biological Diversity: Problems and Challenges. Edited by S.K. Majumdar, F.J. Brenner, J.E. Lovich, J.F. Schalles, and E.W. Miller. © 1994, The Pennsylvania Academy of Science.

Chapter Nine

NATIVE PLANT BIODIVERSITY VS. THE INTRODUCED INVADERS: Status of the Conflict and Future Management Options

CURTIS C. DAEHLER* and **DONALD R. STRONG**

Center for Population Biology
University of California, Davis
Davis, CA 95616

> *[In Argentina] . . . the cardoon has a far wider range . . . very many (probably several hundred) square miles are covered by one mass of these prickly plants, and are impenetrable by man or beast. Over the undulating plains, where these great beds occur, nothing else can live. (Darwin 1839:138-139)*[1]

> *In Ceylon . . . a verbenaceous plant introduced from the West Indies less than 50 years ago . . . now covers thousands of acres with its dense masses of foliage, taking complete possession of land where cultivation has been neglected or abandoned, preventing growth of any other plant, and even destroying small trees. (Wallace 1889:29-30)*[2]

INTRODUCTION

Naturalists over the past 200 years have witnessed spectacular biological invasions. The cardon thistle described by Darwin (*Cyanara cardunculus*) remains a nuisance today in the Pampas of South America, although mechanized land tilling prevents establishment of the massive monocultures dominant 150 years ago.[3]

*Corresponding author

The plant described by Wallace, *Lantana camara* var. mixta, similarly remains a well established pest in Ceylon.[4] Both invasions were the result of transoceanic plant introductions that would have been far less likely in the absence of human assistance. This paper addresses the threats of human-induced biological invasions to native plant biodiversity and discusses options for lessening these threats now and in the future.

Biological invasions are certainly natural phenomena that long antedate humankind. Plants and animals all have natural dispersal mechanisms allowing them the potential to colonize new areas. Following colonization, vicariant diversification and radiation of the species may take place. As a result, the region of highest current species diversity of a taxon is not necessarily its center of origin.[5] For example, marsupials did not origniate in Australia, yet they are currently most diverse there. Historically, certain events such as climatic change following an iceage, or formation of connections between continents or oceans have spawned massive invasions. For example, following the completion of the Central American land bridge during the Pliocene, invasions from North to South America were legion. The majority of montane plants currently found in South America are of North American origin, while many lowland plant species presently found in Central America are thought to have invaded from South America.[6] Despite these past exchanges, the present rate of intercontinental and transoceanic biotic exchange is unprecedented (Vermeij 1991), and is a direct consequence of human inventions that promote global dispersion of biota. Alteration of remaining nature reserves by biological invasions and the resulting homogenization and decline in diversity of the Earth's biota is a problem which must be considered in any plan for maintaining biodiversity.

What is a Biological Invasion?

Most arrivals and colonizations do not result in biological invasions. An invasion occurs only when an introduced species reproduces *and* expands its range beyond the site of introduction. Rates of introduced species establishment undoubtedly vary from habitat to habitat but are surprisingly low. For example, of 348 alien plant species germinating in association with a wool processing factory in Britain seventy years ago, none remain today.[7] Historical evidence suggests that biological invasions as defined above are realized by only a select few species.

How are Species Introduced?

Alien species have long been introduced both deliberately and unintentionally by humans. Table 1 lists some possible modes of accidental introduction. Overall species introduction rates in most areas of the world have increased at least linearly and perhaps exponentially over the past 100 years.[9] While some modes of introduction, like wool processing and ship ballast contamination, have become less important in recent years due to technological changes, other modes of chance introduction, like those resulting from human travel, have likely increased.

Figure 1 illustrates a positive relationship between annual human visitation of nature reserves and the number of introduced species catalogued in those reserves. The number of alien species in a reserve was not found to be correlated with the size of the reserve (at least for South Africa).[10] While it is possible that reserves with

TABLE 1

Sources of Accidental Species Introductions [Crawley (7)[1], Heywood (8)[2]]

shoes, clothing[1,2]
automobiles and machinery - tires, mud[1,2]
railroads/trains[1]
bird seed[1,2]
packing material and crates[2]
imported fruits and vegetables[1]
soil in potted plants[2]
escapes from botanical gardens[1]
escapes from cultivated gardens[1]
escapes/discards from aquaria[1]
grain, animal fodder[1]
wool and hides[1,2]
ship ballast[1]

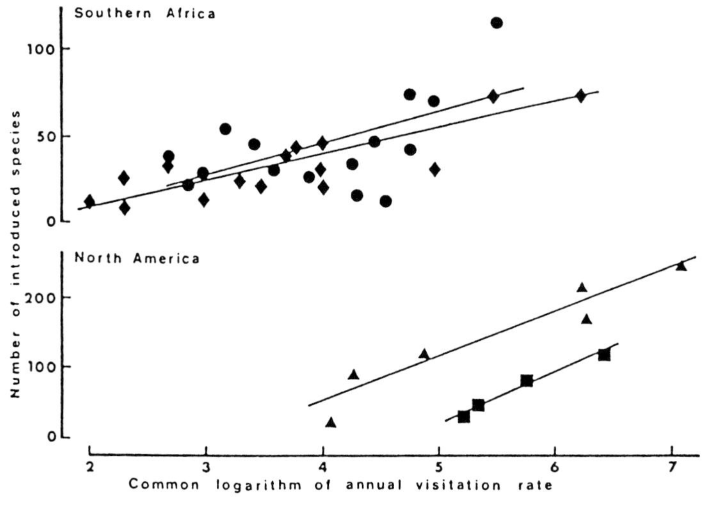

FIGURE 1. Number of introduced species found in nature reserves vs. log of annual human visitation rate.

● tropical dry forest/woodland (S. Africa)
♦ evergreen sclerophyllous forest/shrub/woodland (S. Africa)
▲ broadleaf forest/woodland (N. America)
■ desert/semi-desert (N. America)

Slope of the regression varies among continents and biomes, however the relationship is always positive (reprinted with permission, from Usher[11]).

greater annual visitation also have larger surrounding human populations, the data nevertheless suggest that future maintenance of natural reserves may require regulation of human visitation.[11]

Deliberate introductions by humans are surprisingly common and pose an interesting problem. Many people are intrigued by "exotic" plants or animals, and exotics are perpetuated in lawns and gardens, or as pets. In the horticultural trade, alien plant species are among the best sellers. Promotion of native plant sales through educating people of the value and uses of their native plants,[12] or levying a moderate tax for sale of unessential alien species may help to alleviate our infatuation with exotics. Intentional introductions also occur when people relocate to a new geographic region. For nostalgic reasons, people often desire to emulate the landscape of their former homes at their new place of residence. Many introductions can be attributed to this phenomenon. For example, during the European colonization of Australia, hundreds of horticultural plants were introduced from Europe by "acclimatization societies"[13] and many bird invasions of North America and Hawaii were spawned through intentional introductions by "civic" societies.[14,15]

Why are Biological Invasions so Difficult to Predict?

The search for characteristics of plants and animals that confer an *a priori* propensity for invasion has met limited success. Table 2 lists some of the attributes of invasive species and weedy (colonizing) species that have been proposed over the past half-century. Authors of such lists acknowledge that a given invader need not (and will not) have all of the characteristics advocated. Nevertheless, such lists have poor predictive value. For example, virtually all plant taxa are endowed with at least one of the characteristics from this compiled list, therefore, this list cannot be used to unambigously identify which species will be an invader and which species will not. A second problem is that some of the characteristics, such as genetic diversity and phenotypic plasticity, are multifaceted and difficult to quantify; while "ability to rapidly alter genome in response to selection" is impossible to predict *a priori*. A more predictive approach for identification may be to examine the native range of the potential invader. In several plant genera, positive correlation has been observed between the geographical or environmental amplitude of a species in its native range and its ability to become established into new ranges on different continents.[20,21] For example, Roy et al.[21] analyzed the global range 28 species of Brome grass and found a positive correlation between the number of environmental zones occupied in the native range (mostly within Europe) and the number of zones occupied outside the native range (mostly on other continents). Distance from the native range was not found to be an important factor in establishment. Using this correlation, the authors were able to identify three species which were outliers and had not yet invaded as many environmental zones as would be predicted from the regression. The authors suggest that these three species, *Bromus squarrosus, B. arvenis,* and *B. secalinus* are serious potential invaders. This type of analysis is especially promising because it makes concrete predictions which can be tested either experimentally or with time.

Another strategy for predicting invasions is to focus on characterization of invasion-prone environments. Many invasions are associated with severe disturbances

TABLE 2

Attributes of invading or colonizing species as proposed by Baker (16),[1] Safriel and Ritte (17),[2] Ehrlich (18),[3] DiCastri (19)[4]

Attribute
Plants only
no special requirements for germination[1]
discontinuous (self controlled) germination[1]
seeds capable of long dormancy period[1]
rapid seedling growth[1]
early flowering[1]
pollination by wind or non-specialized pollinator[1]
high seed output, in wide range of environmental conditions[1]
high seed predation in native environment[1]
different phenology from natives[1]
ability to create seed bank[4]
edible fruits or seeds transported by animals[4]
polyploidy[4]
seed dispersal by wind or animals[4]
vigorous vegetative production if perennial[1]
brittle rhizome or rootstock base if perennial[1]
ability to regenerate from rootstock fragments if perrenial[1]
Plants or Animals
good disperser[1,2]
wide potential niche[4]
subject to r-selection[4]
able to shift between r and K strategy[2,3]
short generation time[3]
special competitive adaptation - choking growth, allelopathy[1]
rapid growth in early stage of colonization[2]
ability to maintain large population in invaded area[2]
ability to undergo rapid genetic change in response to selection[2]
high ratio of birth rate:death rate[2]
high genetic variability in source and founder populations[2,3]
high phenotypic plasticity[2]
large native range[3]
abundant in native range[3]
associated with *Homo sapiens*[3]
longer growing/active season than natives
self-compatible, not obligatory selfer (excludes mammals)[1]
larger than most relatives[3]
short, simple life cycle[4]
Animals only
high vagility[3,4]
high mobility[4]
broad diet[3]
gregarious[3]
small body size[4]

generated by humans. For example, common invasion sites have included agricultural fields, roadsides, towns and wasteland. This observation has led to an advocation that successful biological invasions require disturbance in order to make some resource available to an invader.[22,23] Unfortunately, a free resource is similar to an "open niche" in that it is difficult to identify *a priori*. An additional challenge with the "disturbance hypothesis" is to phrase it in a falsifiable way. The introduction of an alien propagule, in and of itself, can constitute a disturbance of the ecosystem. Furthermore, all ecosystems are subject to frequent natural disturbance at some level, therefore an hypothesis suggesting general disturbance is required for invasion has little predictive value. If the "disturbance hypothesis" is phrased more specifically by defining disturbance as "human-induced alterations in the ecosystem," then the hypothesis can be falsified by numerous examples of alien species invasions into natural areas in the apparent absence of human disturbance ([10,25] and references therein, as well as several examples from Table 3). In summary, while many biological invasions are certainly associated with obvious disturbances, various formulations of the "disturbance hypothesis" have yet to demonstrate much predictive value.

Another observation is that biological invasions have been less common in extreme environments such as deserts and arctic climates. Presumably, fewer organisms are adapted to survive and reproduce in the harshest climates, making the pool of potential invaders smaller. A less fully explored hypothesis is that biomass of a habitat might be used to predict the habitat's degree of invasibility.[7] Figure 2 shows invasion of the alien grass *Anthoxantham* into various vegetation ensembalages in California. The degree of invasion suggests an exponentially decreasing function of existing biomass.[26] The relationship seems to apply well even when the existing biomass is alien. Further experimental studies are needed to address

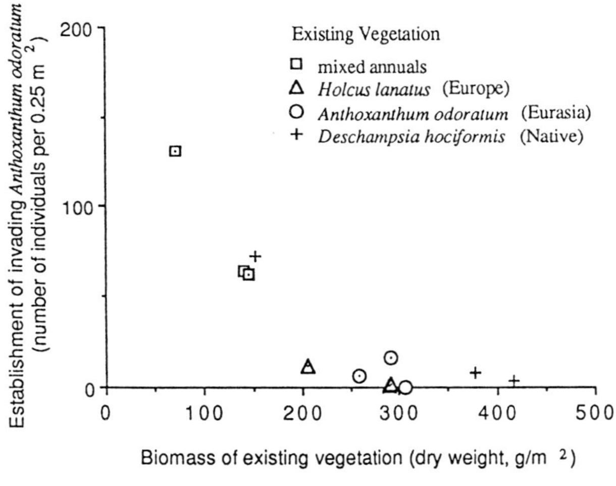

FIGURE 2. Establishment of invading *Anthoxanthum* vs. biomass of existing vegetation. Communities of high biomass are less vulnerable to invasion (redrawn from Peart[26]).

this relationship. Perhaps supplementing native biomass near reserve edges may be used as a future strategy to prevent invasion.

One final enigma involved in predicting biological invasions is the role of timing and chance in establishment and persistence of an introduced species.[27] An introduced seed swept into a stream during a rainy season may end up at the bottom of a lake, while a seed entering the same stream at a time of low flow may become stranded along the bank, then germinate and survive to reproduce a thousand seeds. Demographic and environmental stochasticity becomes an especially important factor in establishment when initial propagule size is small.[27,28]

Threats of Introduced Species to Native Plant Populations

A decrease in local plant diversity is one of the most common phenomena associated with the invasion of an alien.[29] For example, in areas of the South African fynbos biome invaded by *Hakea sericea*, the number of different plant species encountered per square meter is halved as compared to ajoining areas not invaded by *Hakea*.[29] These reductions in biodiversity are not necessarily accompanied by native species extinctions, since natives may continue to thrive in refuges free of the alien, or even coexist (usually at lower densities) with the alien. But aliens can, usually indirectly, threaten native plant species with extinction. Alien plants, animals and pathogens can all reduce indigenous plant population sizes dramatically. When the population size of a species is reduced, the likelihood of extinction due to demographic stochasticity increases. An accompanying decrease in within-species genetic diversity may also increase the probability of extinction by limiting the range of disease resistance or environmental tolerance within the species. Some plants which have accumulated a seed bank are buffered against extinction, however, in the long term, even these plants become threatened as the seed bank is depleted or habitat conducive of germination and growth is permanently lost. Table 3 identifies a variety of introduced species that threaten native plants. This list differs substantially from a list of economically important agricultural weeds.[70] A few examples from this table are discussed in detail below to illustrate the diversity of alien threats, ranging from crowding or shading natives to ecosystem level effects, such as altering the fire regime of the invaded area.

A Note on Competitive Superiority of Introduced Plant Species

The importance of competition in structuring natural ecosystems has been the subject of much debate.[71] It is virtually impossible to demonstrate direct competition through non-manipulative observations, and experimental designs demonstrating natural competition for specific resources are especially difficult to formulate and interpret.[72] There are, however, a few studies which persuasively demonstrate differential competitive abilities for specific resources. For example, Caldwell et al.[31] used a dual isotope technique to show superior competitive uptake of phosphorus by the alien *Agropyron desertorum* as compared to a native *Agropyron spicatum*. Commonly, introduced species are inferred to be superior competitors to natives

TABLE 3

A Survey of Some Introduced Species Which Threaten Native Plant Diversity in Natural Areas

Introduced Species	taxon	native to	introduced to	reported effect	threatened	ref
Plants						
Acacia spp.	shrub	Aust.	S. Afr.	N	g,s	30
Agropyron desertorum	grass	Eurasia	N. Amer.	F,D	g,s,t	31,32
Ammophila arenaria	grass	Europe	Calif.	G,C	g,s	33,34
Bidens frondosa	herbac.	N. Amer.	Europe	C	g	35
Bromus japonicus	grass	Europe	N. Amer.	L	g	36
Bromus tectorum	grass	Europe	N. Amer.	F,C	*	37
Carpobrotus edulis	herbac.	S. Afr.	N. Amer.	C,D	g,s	38
Cenchrus ciliaris	grass	Africa	Aust.	F,D	*	39
Chromolaena odoratum	vine	N. Amer.	Africa	F,D,S	g,t	40
Chrysanthemoides monilifera	shrub	S. Afr.	Aust.	G,C	g,s	41
Datura inoxia	herbac.	S. Amer.	S. Afr.	H,D	r	42
Elaeagnus angustifolia	tree	Eurasia	N. Amer.	D	r	43
Hakea sericea	shrub	Aust.	S. Afr.	C,S,F	g,s,t	29,44
Hyparrenia rufa	grass	Africa	S. Amer.	F,D	*	45,46
Ligustrum robustum	shrub	Ceylon	Island	D	g,s	47
Lonicera japonica	vine	Europe	Island	S	t	48
Melaleuca quinquenervia	tree	Aust.	N. Amer.	S	t	49
Melia azedarach	tree	Asia	S. Afr.	C	t	40
Melinis minutiflora	grass	Africa	Island	F,D	*	50
Mesembryanthemum crystallinum	herbac.	Africa	Calif.	D,L	g	51
Mimosa pigra	shrub	S. Amer.	Aust.	S,D,R	g,s	52
Myrica faya	tree	Island	Island	N,L	*	53
Nicotiana glauca	shrub	S. Amer.	S. Afr.	H,D	r	42
Pennisetum polystachyon	grass	Africa	Aust.	F,C	g,*	54
Pinus radiata	tree	N. Amer.	S. Afr.	S,L,H	g,s,t	29,44
Psidium cattleianum	shrub	S. Amer.	Island	D	g,s	47
Pueraria lobata	vine	Asia	N. Amer.	D	s,t	55
Rhododendron ponticum	shrub	Asia	Ireland	S,L,R	g,s,t	56
Robina pseudacacia	tree	N. Amer.	Europe	N	g,s,t	35
Rubus moluccanus	shrub	S. Asia	Island	D	g,s	35,47
Sesbania punicea	shrub	S. Amer.	S. Afr.	D,G	r	57
Solidago gigantea	herbac.	N. Amer.	Europe	D	r	58
Tamarix spp.	shrub	Eurasia	N. Amer.	D,H,G	r	42
Trifolium subterraneum	herbac.	Europe	Aust.	N	g	3
Animals						
Bubalis bubalis	swamp buffalo	Asia	Aust.	B,T	g,s	59
Capra hircus	goat	Eurasia	Island	B,T	g,s,t	60
Cervis elaphas	deer	N. Amer.	S. Amer.	B	t	61
Hemitragus jemlahicus	tahr	S. Asia	S. Afr.	B,T	g,s,t	24
Iridomyrmex humilis	ant	S. Amer.	S. Afr.	R	s,t	62
Mus musculus	mouse	Europe	Island	R	t	63
Myocaster coypus	coypu	S. Amer.	England	B	r	56
Oryctolagus cuniculus	rabbit	Europe	S. Amer.	B	g,t,*	64
Sus scrofa	pig	Eurasia	N. Amer.	B,T	g,s,t	65,66

Table 3 *continued*

Introduced Species	taxon	native to	introduced to	reported effect	threatened	ref
Pathogens						
Ceratocystis ulmi	fungus	Asia	N. Amer.	K	t (elm)	67
Cronartium ribicola	fungus	Eurasia	N. Amer.	K	t (pine)	67
Endothia parasitica	fungus	Asia	N. Amer.	K	t (chestnut)	68
Phytophthora cinnamoni	fungus	E. Aust.	W. Aust.	K	t,s	69

C = competition for nutrients, water or light probable - increase in alien is correlated with a decrease in native(s) - mechanism uncertain
D = dense growth displaces native(s)
F = alters fire regime - fires more frequent, more intense, or more widespread, preventing establishment of natives
G = alters geomorphology - causes accretion of substrate or alters drainage
H = alters hydrology - lowers water table, high transpiration during dry season
L = litter accumulation inhibits natives
N = nitrogen fixation - increases N content of soil, promoting subsequent invasions
S = shades natives - eventually killing
B = browsing extensively
R = inhibits recruitment of natives though seed predation
T = trampling or digging - kills vegetation, causes soil compaction and erosion of top soil
K = kills many infected plants, epidemic-like spread

Threatened - key

t = native trees
r = native riverine plants or inhabitants of seasonal waterways
g = native grasses or other herbaceous plants
s = native shrubs
* = may alter ecosystem function, preventing or altering natural succession.

for resources such as nutrients, water, light, or space. While some studies of introduced plants cited in Table 3 were experimental, others simply report a correlation between increased prominence of the alien and decreased native species population size, diversity, or vigor.

Tamarix spp. - A Deep-rooted Problem

Several species of *Tamarix* from Europe have been introduced to parks and reserves in the Southwestern United States. These willow-like shrubs up to 8 m tall form dense masses along seasonally flooded waterways and ponds.[73] In addition to crowding out native plants, the *Tamarix* possess deep penetrating roots which, unlike most natives, tap into the water table throughout much of the year.[42] The resulting incessant transpiration can lower the water table, affecting water availability for the entire ecosystem.[42] In contrast to natives that are commonly washed from banks of waterways during flooding, the deep-rooted *Tamarix* are not cleared by heavy waterflow. Tenacious *Tamarix* stabilizes banks, accreting sediment, and causing reduced river channel width and increased over-bank flooding.[74] Since *Tamarix* seeds are dispersed long distances through waterways, their spread has been difficult to control.[42] Large protions of several park budgets are spent on physical removal of *Tamarix*.[42] There is a possibility that a recently discovered plant pathogen may be used to control *Tamarix* infestations.[68]

Iceplant - A Halodistributor

Mesembryanthemum crystallinum (Aizoaceae) is a succulent, low-growing annual from Africa which has invaded coastal grasslands of California. Seedlings become established in open areas and form densely vegetated patches.[51] The plant is an halodistributor, concentrating salt from lower soil layers within its leaves. When the plant dies and decomposes, the resulting high osmotic content of the surface soil inhibits recruitment of many native grasses.[51] By this mechanism, the plant can spread and maintain dominance in areas previously occupied by natives. Perhaps because of their attractive flowers and brilliant red late-summer leaves, iceplants (especially the perennial species such as *Carpobrotus edulis*) have been popularly used for landscaping and freeway margins in the San Francisco Bay area.

Chromolaena odoratum *- A Fired-up Vine*

This vine from South America infests several reserves in Africa. The climbing plant rapidly smothers all types of native vegetation, including large trees.[40] It can be especially prominent at ecotones between forest and grassland, where it poses a serious fire threat. The vines burn readily even when green, and can allow fires to penetrate native forest canopies which otherwise would not burn.[40] In addition to this vine, numerous alien grasses also affect natural fire cycles in a variety of ecosystems around the world, typically making fires more frequent, more intense and intolerable to native species.[3,39]

Argentine Ant - A Negligent Forest Manager

Iridomyrmex humilis (Argentine ant) has recently invaded undisturbed reserves in South Africa, where it often excludes native ant species.[62] Native ants disperse and transport underground the seeds of several native trees of the family Proteaceae. The unharmed seeds can then germinate from abandoned native ant nests. The invading Argentine ant, on the other hand, carries seeds only short distances and rarely delivers them underground.[62] As a result, most seeds which are left exposed are subject to rapid predation, and recruitment of Proteaceae is extremely low or non-existent in areas invaded by the Argentine ant.[62] Future plant community structure is expected to be altered and the endemic Proteaceae are threatened by this invasion-induced "collapse of an ant-plant mutualism".[62]

Asian Swamp Buffalo - Voracious Eater and Plant Invasion Catalyst?

The Asian swamp buffalo, *Bubalis bubalis*, was introduced to Australia in the 1800's as a work animal. A feral population of as many as 250,000 became established from liberated animals in the late 1980's.[52] These buffalo live in wetland reserve areas where their grazing has made several native herbaceous plants rare.[59] In addition, their trampling has killed adult trees by altering hydrology, while water consumption by the buffalo has affected water availability during the dry season.[52] Buffalo overgrazing of native plants has been implicated as a cause of invasions of unpalatable *Mimosa pigra* (giant senstive plant) from South America.[52] This plant grows as imposing monocultures in many areas of North Australia. Many other introduced grazing mammals, especially feral goats and feral pigs threaten native vegetation in natural areas around the world.[60]

Phytophthora cinnamoni - *An Introduced Pathogen*

Phytophthora cinnamoni is a fungus that causes an often fatal root disease. More than 1000 plant species are thought to be susceptible to the fungus.[75] The disease was presumably introduced from eastern Australia to western Australia on infected nursery material and has infected more than 300,000 hectares of jarrah forest.[75] The fungus can cause death within weeks in 50% of species infected.[69] Most native trees and shrubs of Australia are susceptible to the disease, which spreads rapidly through waterlogged soils.[69] Several of the dominant native forest species such as *Eucalyptus marginata* and *Banksia grandis* have completely disappeared from large areas as a result of this epidemic.[68,69] Control of the disease will probably depend on the establishment of resistant strains of plants.[69] The situation is similar for the American chestnut, a tree species formerly dominant in many deciduous hardwood forests of the eastern United States. In a matter of decades, the American chestnut has been nearly wiped out by an invading fungal epidemic from Asia.[68]

Which Plant Species are Most Threatened by Aliens?

Oceanic islands have long been known for their great susceptibility to biological invasions.[76,77] Presumably this phenomenon is due to the evolution of isolated flora and fauna in the absence of certain mainland taxa, such as mammalian grazers or large predators.[77] The small size and high perimeter-to-interior ratio of islands may also increase their susceptibility to invasion, providing more low-biomass edge habitat which may be more hospitable to initial colonization by aliens. High rates of endemism, in association with limited or non-existent areas of refuge from aliens make island populations especially susceptible to extinction.[78]

Mainland species with narrow native ranges also are especially endangered by alien invasions. In South Africa, 44 out of 70 rare native plant taxa studied are directly threatened by invasive aliens.[79] The many rare plant species which appear to require extremely specialized environments within nature reserves will be at great risk as reserves contract and become altered by increasing alien biomass.

DISCUSSION

Preventing Introductions

Preventing the introduction of aliens species, especially plants and insects, seems a daunting task given the many possible modes of import and the escalating frequency of intercontinental travel. Clearly it is not feasible to search every international package, shipment, or traveler for stowaway propagules of alien plant seeds or insects. In many countries, government agencies (such as the United States Department of Agriculture (USDA) and state agencies in the United States) inspect some foreign shipments for certain known agricultural pests. In deciding which aliens will be screened, agencies usually consider both the potential economic impact (primarily agricultural) of the alien and some crude estimate of the probability of establishment of the alien.[80,81] The true effectiveness of these programs is difficult to evaluate in the absence of a control, however it is safe to say that at least some of alien propagules that could have become established populations have been intercepted. Any significant increase in the effectiveness of these interception programs would probably require substantial economic investment, since detection of small seed propagules and insects is presently labor intensive. Given that the probability of establishment of most introduced species is known to be very low, in many areas it may be more cost-effective to allow small, difficult to detect propagules to be extinguished by nature. Further support for not intensifying stowaway interception programs comes from a study by Groves,[82] which concludes that most problematic plant invasions in nature areas stem from deliberate horticultural introductions. Efforts could instead be concentrated on traveler education and curtailing public demand for unessential exotic plants.

Preventing Establishment

Somewhat successful programs have been enacted to detect growing propagules of recently introduced agricultural weeds.[9,80] The USDA and state agencies assemble lists of known agricultural pests and Holm et al.[70] provides an excellent reference describing the world's worst agricultural weeds and their sites of infestation. Government agents attempt to identify and eradicate any establishing small patches of those species on a priority weed list. A similar comprehensive program should be established to monitor aliens which threaten nature reserves. Many of the species in Table 3 are not agricultural weeds and are therefore not monitored by the USDA. Conversely, many agricultural weeds do not seem to threaten nature reserves. A comprehensive list of alien plant and animal species that have become established and have caused problems in nature reserves around the world would be an extremely useful tool. Reserve employees (and volunteers) could be trained to identify eliminate selected plants from this list. In some reserves, easily identified aliens which are expected to become problematic could be described in a vistor's brochure, and token rewards could be provided to visitors who collect or identify a noxious alien. Such a program should have a high potential for preventing many new invasions, however, since reserves are not always isolated units, regional management practices independent of reserve boundaries may become essential. For example, problems may arise when aliens have already become well established outside the reserve and are regularly dispersed in through rivers or neighboring cultivated fields.[42]

Eradication or Control of Invading and Established Aliens

A. Large Grazing Mammals

The devastating effect of introduced grazing mammals on native vegetation of both islands and continents has been well documented.[60,61,64,65] Large grazers like goats, pigs, and cattle were first introduced to many uninhabited islands by European explorers to ensure a ready supply of meat on future voyages.[60] On continents, grazers have often escaped from domestication to establish feral populations in nearby natural areas. Large feral animals are relatively simple to eradicate by shooting, trapping or poisoning, however there are often some obstacles to their eradication. For example, animal rights groups frequently oppose killing the animals. In some cases, this conflict has been successfully circumvented by live-capturing animals and relocating them.[42] In other areas, sport hunters have opposed the eradication of introduced grazers like deer.[61] Since recovery of the native community has sometimes been reported following eradication of the grazers,[83] removal of alien grazers from at least large portions of reserves should be seriously considered in most cases, especially when experimental areas fenced to exclude alien grazers show a re-establishment of native plants.

B. Alien Plants - Physical and Mechanical Control

If strong early detection programs are initiated as advocated above, physical removal of aliens could be a simple and effective method of alien control or eradication. Mechanical control is similar to physical removal and involves destroying plants through tilling land, crushing, or mowing. This method alone is often not successful with robust plants, however success may be improved when mechanical distruption is used in combination with chemical control. Mechanical control can sometimes backfire, spreading vegetative propagules beyond the original site of infestation. For example, when areas of the Tijuana estuary were disked to alleviate an infestation by *Chysanthemum* sp., the plant rapidly rebounded with increased dominance.[12] Another problem is that a different invader may simply fill in open space rapidly generated by mechanical control.[82]

C. Alien Plants and Insects - Chemical Control

Local chemical control of many agricultural pests has been successful from the agriculturalist's point of view. The obvious disadvantage of chemical control in nature areas is that agents are not species specific. Spraying a pesticide in any natural area will inevitably affect nearby natives, although this effect can be minimized with spot applications. Another disadvantage of chemical control is the health hazards associated with many agents, especially when used near water supplies.[82] These undesirable effects must be carefully weighed against the benefit (see [84]). One advantage of a chemical agent may be immediate control of large infestations at generally lower cost than labor-intensive physical removal. Eradication is usually only realistic for small, well-confined populations. When chemical control is enlisted, repeated periodic applications are almost always required and the alien may return in full force if a treatment period is missed.[81] As with mechanical control, any large open areas that result following chemical control may simply become colonized by other aliens.[82]

D. Alien Plants and Insects - Biological Control

Biological control is most used after an alien has become well established. A problem with biological control is uncertainty of success. Past experience shows that 40-48% of biocontrol agents have achieved some degree of success as measured by a reduction in the alien population density.[85] Releasing multiple control agents may increase the likelihood of success.[86] If a biological control agent for the alien of interest has not yet been identified, then there is the disadvantage of additional time-consuming screenings required prior to any release. Advantages to biocontrol are that released agents are often species specific, and do not constitute a human health hazard. In addition, biological control may require only limited investments if the control agent becomes firmly established. The often slower, "natural" action of biological control as compared to mechanical and chemical control may be more likely to lead to regrowth of native vegetation as the alien recedes, rather than simple replacement with another alien.

One important aspect of biocontrol is that the alien is generally not eradicated. Rather, a successful biocontrol effort leads to the alien population falling to low levels that are not threatening to other species. Reduction in alien plant populations by biocontrol can be spectacular. Figure 3 shows a local example of the successful biocontrol of *Sesbania punicea* by the weevil *Neodiplogrammus quadrivittatus.* The plant was introduced to South Africa from South America and grew vigorously to make impenetrable thickets, especially near river banks.[57] The weevil *N. quadrivittatus,* which has a stem-boring larva, was imported from South America in 1984 along with a second seed eating weevil (*Ryssomatus marginatus*). Within four years, and with relatively little expense, satisfactory control has been reported in many areas.[57] Biocontrol will become an increasingly attractive option as biocontrol agent "libraries" grow. A concern with biological control is the possibility that introduced control agents may alter natural ecosystems in the long term by directly or indirectly affecting native species abundance through competition or predation.[87] For example, a biotype of the European weevil *Rhinocyllus conicus* introduced to western North America for control of alien thistles is also found among the insects in native California thistle heads.[88] Long term effects of *R. conicus* on the native thistles are unknown.[88]

E. Integrated Approaches

Some combination of the above control methods tailored to the alien species of concern may be most effective. An understanding and consideration of population biology and ecology of the invader is important in deciding which control approaches might be most successful.[89] Control options that are sometimes applicable

FIGURE 3. A dense thicket of introduced *Sesbania punicea,*thriving at Grahamstown, South Africa (December 1984) (top). The same site four years later following successful biological control (December 1988) (middle). The control agent, *Neodiplogrammus quadrivittatus*, whose stem-boring larvae can kill even the largest shrubs (bottom). (Photos provided by J.H. Hoffman and V.C. Moran).

include utilization of the alien for food or forage, burning, or cultural control (for example, encouraging farmers to practice sound grazing management).[89] Difficulties with integrated control arise because the different groups involved such as government agencies, landowners and reserve managers must all be committed to implementing the control plan.[89]

F. Importance of Follow-Through

Lack of follow-through and poor long-range planning have been implicated in the failure of many control programs. Eradication of a well established alien is rarely possible, therefore, most control programs must have long range funding with a goal of keeping alien populations at acceptable levels.[82] Careful monitoring of success (or failure) throughout a control program is also essential. Different areas may be found to respond better with certain methods or treatments, and by tailoring the control program to the alien's response, effectiveness of the comprehensive program can be maximized.

CONCLUSIONS

As habitat destruction increases worldwide and natural areas diminish, it will become increasingly necessary to defend nature reserves from invasion by alien species in the interest of protecting native plant diversity. Traditional methods for identifying invaders *a priori* are unreliable or impractical. Many of the species in Table 3 are known to be invaders in nature areas of multiple islands or continents; therefore, one starting point for preventing future invasions would be to assemble a more comprehensive list of plants and animals which have proven to be problematic in natural areas, including any control methods attempted. This list could be sorted according to climatic region to give reserve managers some indication of which species might be expected to appear as problematic invaders of their reserves in the future. Early detection could then increase the likelihood of eradication, or at least provide adequate time to plan a comprehensive control program. As for aliens which are already well established, biological control methods, accompanied by efforts to maintain the environment most conductive to native species (for example, by maintaining the natural fire regime) appear to offer the best long range solutions to limiting the extent of alien domination. Issues that will require much additional study and consideration are the effect of global climate change on future invasion patterns, and the potential for future invasions of reserves by genetically engineered organisms. Fresh approaches are required for characterizing and modeling biological invasions, and it appears that development of models with predictive power will require inclusion of empirically measured ecosystem-specific and invader-specific parameters at the sacrifice of generality.

ACKNOWLEDGEMENTS

We are grateful to J.H. Hoffmann and V.C. Moran for providing photographs from their biocontrol work with *Sesbania*. We thank Carla D'Antonio for comments which improved the manuscript. This work was supported by National Geographic Society grant #4703-92 and National Science Foundation grant #BSR 9007429 to DRS.

REFERENCES

1. Darwin C. 1839. *Journal of Researches into the Geology and Natural History of the various countries visited by H.M.S. Beagle under the command of Captain Fitzroy.* Henry Colburn. London.
2. Wallace, A.R. 1889. *Darwinism, an exposition of the theory of natural selection with some applications,* Macmillan and Co. London, 494 p.
3. Mack, R.N. 1989. Temperate Grasslands Vulnerable to Plant Invasions: Characteristics and Consequences, pp. 155-180. In: J.A. Drake et al. (Eds.) *Biological Invasions, A Global Perspective.* SCOPE 37, John Wiley and Sons, New York, 525 p.
4. Moldenke, H.N. and A.L. Moldenke. 1983. Verbenaceae, pp. 196-488. In: M.D. Dassanayake and F.R. Fosberg (Eds.) *A Revised Handbook to the Flora of Ceylon, Vol. IV.* Amerind Publishing, New Delhi.
5. Croizat, L. 1962. *Space, time form: the biological synthesis.* L. Croizat, Caracas, Venezuela. 881 p.
6. Vermeij, G.J. 1991. When biotas meet: understanding biotic interchange. *Science.* 253:1099-1104.
7. Crawley, M.J. 1986. What makes a community invasible? pp. 429-453, In: A.J. Gray, M.J. Crawley and P.J. Edwards (Eds.) *Colonization, Succession and Stability.*
8. Heywood, V.H. 1989. Patterns, Extents and Modes of Invasions by Terrestrial Plants, pp. 31-60. In: J.A. Drake et al. (Eds.) *Biological Invasions, A Global Perspective.* SCOPE 37, John Wiley and Sons, New York, 525 p.
9. Rejmanek, M., C.D. Thomsen and I.D. Peters. 1991. Invasive vascular plants of California, pp. 81-102. In: R.H. Groves and F. DiCastri (Eds.) *Biogeography of Mediterranean Invasions.* Cambridge Univ. Press. Cambridge, England. 485 p.
10. MacDonald, I.A.W., F.J. Powrie and W.R. Siegfried. 1986. The differential invasion of southern Africa's biomes and ecosystems by alien plants and animals, pp. 209-228. In: I.A.W. MacDonald, F.J. Kruger and A.A. Ferrar (Eds.) *The Ecology and Management of Biological Invasions in Southern Africa.* Oxford Univ. Press, Cape Town, 324 p.
11. Usher, M.B. 1988. Biological invasions of nature reserves: a search for generalizations. *Biol. Conserv.* 44:119-135.
12. Zedler, J.B. 1992. Invasive exotic plants: threats to coastal ecosystems. *Proc. S. Calif. Acad. Sci.* In Press.
13. Goves, R.H. 1991. A short history of biological invasions of Australia, pp. 59-64. In: R.H. Groves and F. DiCastri (Eds.) *Biogeography of Mediterranean Invasions.* Cambridge Univ. Press. Cambridge, Enlgand. 485 p.

14. Moulton, M.P. and Pimm, S.L. 1986. Species introductions to Hawaii, pp. 231-249. In: H.A. Mooney and J.A. Drake (Eds.) *Ecology of Biological Invasions of North America and Hawaii.* Springer-Verlag. 321 p.
15. Vuilleumier, F. Invasions in the mediterranean avifaunas of California and Chile, pp. 327-358. In: R.H. Groves and F. DiCastri (Eds.) *Biogeography of Mediterranean Invasions.* Cambridge Univ. Press. Cambridge, England. 485 p.
16. Baker, H.G. 1965. Characteristics and modes of origin of weeds, pp. 147-172. In: H.G. Baker and G.L. Stebbins (Eds.) *The Genetics of Colonizing Species.* Academic Press, New York, 588 p.
17. Safriel, U.N. and U. Ritte. 1980. Criteria for the identification of potential colonizers. *Biol. J. Linn. Soc.* 13:287-297.
18. Ehrlich, P.R. 1986. Which animal will invade? pp. 79-95. In: H.A. Mooney and J.A. Drake (Eds.) *Ecology of Biological Invasions of North America and Hawaii.* Springer-Verlag, New York. 321 p.
19. DiCastri, F. 1990. On invading species and invaded ecosystems: the interplay of historical chance and biological necessity. In: F. DiCastri, A.J. Hansen and M. Debussche (Eds.) *Biological Invasions of Europe and the Mediterranean Basin.* Kluwer Publishers, Boston. 463 p.
20. Forcella, F., J.T. Wood and S.P. Dillon. 1986. Characteristics distinguishing invasive weeds within *Echium* (Bugloss). *Weed Research.* 26:351-364.
21. Roy, J., ML. Navas and L. Sonie. 1991. Invasion by annual brome grasses: a case study challenging the homocline approach to invasions, pp. 207-224. In: R.H. Groves and F. DiCastri (Eds.) *Biogeography of Mediterranean Invasions.* Cambridge Univ. Press. Cambridge, England. 485 p.
22. Fox, M.D. and B.J. Fox. 1986. The susceptibility of natural communities to invasion, pp. 57-66. In: R.H. Groves and J.J. Burden (Eds.) *Ecology of Biological Invasions, An Australian Perspective.* Cambridge Univ. Press. Canberra. 159 p.
23. DiCastri, F. 1991. An ecological overview of the five regions of the world with a mediterranean climate, pp. 3-16. In: R.H. Groves and F. DiCastri (Eds.) *Biogeography of Mediterranean Invasions.* Cambridge Univ. Press. Cambridge, England. 485 p.
24. MacDonald, I.A.W., L.L. Loope, M.B. Usher and O. Hamann. 1989. Wildlife conservation and the invasion of nature reserves by introduced species: A global perspective, pp. 215-256. In: J.A. Drake et al. (Eds.) *Biological Invasions, A Global Perspective.* SCOPE 37, John Wiley and Sons, New York, 525 p.
25. Rejmanek, M. 1989. Invasibility of plant communities, pp. 369-388. In: J.A. Drake et al. (Eds.) *Biological Invasions, A Global Perspective.* SCOPE 37, John Wiley and Sons, New York, 525 p.
26. Peart, D.R. 1985. Experimental analysis of succession in a grassland at Sea Ranch, California. Ph.D. Thesis. University of California, Davis. 92 p.
27. Crawley, M.J. 1989. Chance and timing in biological invasions, pp. 407-424. In: J.A. Drake et al. (Eds.) *Biological Invasions, A Global Perspective.* SCOPE 37, John Wiley and Sons, New York, 525 p.
28. Hengeveld, R. 1989. *Dynamics of Biological Invasions.* Chapman and Hall, New York. 160 p.

29. Breytenbach, G.J. 1986. Impacts of alien organisms on terrestrial communities with emphasis on communities of the south-western Cape, pp. 229-238, In: I.A.W. MacDonald, F.J. Kruger and A.A. Ferrar (Eds.) *The Ecology and Management of Biological Invasions in Southern Africa.* Oxford Univ. Press, Cape Town, 324 p.
30. Witkoswski, E.T.F. 1991. Effects of invasive alien acacias on nutrient cycling in the coastal lowlands of the Cape fynbos. *J. Appl. Ecol.* 28:1-15.
31. Caldwell, M.M., D.M. Eissenstat, J.H. Richards M.F. Allen. 1985. Competition for phosphorus: Differential uptake from dual-isotope-labeled soil interspaces between shrub and grass. *Science.* 229:384-386.
32. Elloitt, K.J. and A.S. White. 1989. Competitive effects of various grasses and forbs on ponderosa pine seedlings. *For. Sci.* 33(2):356-366.
33. Barbour, M.G. and Johnson, A.F. 1977. Beach and dune. In: *Terrestrial Vegetation of California.* M.G. Barbour and J. Major (Eds.) John Wiley and Sons, New York, pp. 223-261.
34. Slobodchikoff, C.F. and J.T. Doyen. 1977. Effects of *Ammophila arenaria* on sand dune arthropod communities. *Ecology.* 58:1171-175.
35. Sykora, K.V. 1990. History of the impact of man on the distribution of plant species. F. di Castri, A.J. Hansen and M. Debussche (Eds.). In: *Biological Invasions of Europe and the Mediterranean Basin.* Kluwer Publishers, Boston. pp. 37-50.
36. Whisenant, S. 1990. Postfire population dynamics of *Bromus japnicus. Am. Midl. Nat.* 123:301-308.
37. Melgoza, G., R.S. Nowak and R.J. Tausch. 1990. Soil water exploitation after fire: competition between *Bromus tectorum* (cheatgrass) and two native species. *Oecologia.* 83:7-13.
38. D'Antonio, C.M. and B.E. Mahall. 1991. Root profiles and competition between the invasive exotic perennial, *Carpobrotus edulis*, and two native shrub species in California coastal scrub. *Am. J. Bot.* 78(7):885-894.
39. D'Antonio, C.M. and P.M. Vitousek. 1992. Grass invasions, the grass/fire cycle, and global change. *Ann. Rev. Ecol. Systemat.* 23:63-88.
40. Macdonald, I.A.W 1983. Alien trees, shrubs, and creepers invading indigenous vegetation in the Hluhluwe-Umfolozi Game Reserve Complex in Natal. *Bothalia.* 14:949-959.
41. Weiss, P.W. and I.R. Noble. 1984. Status of coastal dune communities invaded by *Chrysanthenmoides monilifera. Aust. J. Ecol.* 9:93-98.
42. Loope, L.L., P.G. Sanchez, P.W. Tarr, W.L. Loope and R.L. Anderson. 1988. Biological invasions of arid land nature reserves. *Biol. Conserv.* 44:95-118.
43. Knopf, F.L. and T.E. Olson. 1984. Naturalization of Russian-olive: Implications to Rocky Mountain wildlife. *Wild. Soc. Bull.* 12:289-298.
44. Versfeld, D.B. and B.W. van Wilgen. 1986. Impact of woody aliens on ecosystem properties. I.A.W. Macdonald, F.J. Kruger, and A.A. Ferrar (Eds.). In: *The Ecology and Management of Biological Invasions in Southern Africa.* Oxford University Press, Cape Town. pp. 239-246.
45. Janzen, D. 1988. Management of habitat fragments in a tropical dry forest. *Ann. Miss. Bot. Gar.* 75:105-116.

46. Janzen, D. 1988. Tropical dry forests: The most endangered major tropical ecosystem. E.O. Wilson (Ed.). In: *Biodiversity.* National Academy Press. Washington. pp. 130-137.
47. Lorence, D.H. and R.W. Sussman. 1986. Exotic species invasions into Mauritius wet forest remnants. *J. Trop. Ecol.* 2:147-162.
48. Thomas, L.K. 1980. The impact of three exotic plant species on a Potomac Island. Natural Park Service Scientific Monograph Series. 13:1-179.
49. Myers, R.L. 1983. The susceptibility to invasion by the exotic tree *Melaleuca quinquenervia* in southern Florida. *J. Appl. Ecol.* 20:645-658.
50. Hughes, R.F., P.M. Vitousek and T. Tunison. 1991. Alien grass invasion and fire in the seasonal submontane zone of Hawaii. *Ecology* 72(2):743-746.
51. Vivrette, N.J. and C.H. Miller. 1977. Mechanism of invasion and dominance of coastal grassland by *Mesembryanthemum crystallinum. Ecol. Monogr.* 47:301-318.
52. Lonsdale, M. and R. Braithwaite. 1988. The shrub that conquered the bush. *New Sci.* 120(1634):52-55.
53. Vitousek, P.M. and L.R. Walker. 1989. Biological invasion by *Myrica faya* in Hawaii: Plant demography, nitrogen fixation, ecosystem effects. *Ecol. Monogr.* 59:247-265.
54. MacDonald, I.A.W. and G.W. Frame. 1988. The invasion of introduced species into nature reserves in tropical savannas and dry woodlands. *Biol. Conserv.* 44:67-93.
55. Cullota, W. 1991. Biological immigrants under fire. *Science.* 254:1444-1447.
56. Usher, M.B. 1986. Invasibility and wildlife conservation: invasive species on nature reserves. *Phil. Trans. R. Soc. Lond. B.* 314:695-710.
57. Hoffmann, J.H. and V.C. Moran. 1991. Biological control of *Sesbania punicea* (Fabaceae) in South Africa. *Agricult., Ecosys. and Envir.* 37:157-173.
58. Kornas, J. 1990. Plant invasions in central Europe: historical and ecological aspects. F. di Castri, A.J. Hansen, and M. Debussche (Eds.). In: *Biological Invasions of Europe and the Mediterranean Basin.* Kluwer Publishers, Boston. pp. 19-36.
59. Williams, C.K. and M.G. Ridpath. 1982. Rates of herbage ingestion and turnover of water and sodium in feral swamp bufallo, *Bubalis bubalis,* in relation to primary production in a cyperaceous swamp in monsoonal Northern Australia. *Aust. Wild. Res.* 9:397-408.
60. Coblentz, B.E. 1978. The effects of feral goats (*Capra hircus*) on island ecosystems. *Biol. Conserv.* 13:279-286.
61. Veblin, T.T., M. Mermoz, C. Martin and T. Kitzberger. 1992. Ecological impacts of introduced animals in Nahuel National Park, Argentina. *Conserv. Biol.* 6(1):71-83.
62. Bond, W.J. and P. Slingsby. 1984. Collapse of an ant-plant mutualism: The Argentine ant, *Iridomyrmex humilis* and myrmecochorous Proteaceae. *Ecology.* 65: 1031-1037.
63. Breytenbach, G.J. 1986. Dispersal: the case of the missing ant and the alien mouse. *S. Afr. J. Bot.* 52:463-466.

64. Fuentes, E.R., F.M. Jaksic and J. Simonetti. 1983. European rabbits versus native mammals: Effects on shrub seedlings. *Oecologia.* 58:411-414.
65. Singer, F.J., W.T. Swank and E.E.C. Clebsch. 1984. Effets of wild pig rooting in a deciduous forest. *J. Wildl. Manage.* 48:464-473.
66. Bowman, D.M. and W.J. Panton. 1991. Sign and habitat impact of banteng (*Bos javanicus*) and pig (*Sus scrofa*), Cobourg Peninsula, northern Australia. *Aust. J. Ecol.* 16:15-17.
67. Bingham, R.T., R.J. Hoff and G.I. McDonald. 1971. Disease resistance in forest trees. *Ann. Rev. Phytopathol.* 2:433-452.
68. Quimby, P.C., Jr. 1982. Impact of diseases on plant populations. In: R. Charudattan and H.L. Walker (Eds.) *Biological Control of Weeds with Plant Pathogens.* John Wiley and Sons, New York. pp. 47-60.
69. Weste, G. and P. Taylor. 1971. The invasion of native forests by *Phytophthora cinnamomi.* I. Brisbane Ranges, Victoria. *Aust. J. Bot.* 19:281-294.
70. Holm, L.G., D.L. Plucknett, J.V. Pancho and J.P. Herberger. 1977. *The World's Worst Weeds, Distribution and Biology.* East-West Center, Univ. Press of Hawaii, Honolulu. 609 p.
71. Strong, D.R., D. Simberloff, L.G. Abele and A.B. Thistle. 1984. *Ecological Communities: Conceptual Issues and the Evidence.* Princeton Univ. Press. 613 p.
72. Peters, R.H. 191. *A Critique for Ecology.* Cambridge Univ. Press., New York. 366 p.
73. Whitson, T.D. (Ed.) *Weeds of the West.* Western Society of Weed Science, University of Wyoming, Jackson WY, 630 p.
74. Graf, W.L. 1978. Fluvial adjustments to the spread of tamarisk in the Colorado Plateau region. *Geol. Soc. Amer. Bull.* 89:1491-1501.
75. von Broembson. 1989. In: J.A. Drake et al. (Eds.) *Biological Invasions, A Global Perspective.* SCOPE 37, John Wiley and Sons, New York, 525 p.
76. Wallace, A.R. 1880. *Island Life.* Macmillan and Co., London.
77. Elton, C.S. 1958. *The Ecology of Invasions by Animals and Plants.* Methuen, London, England. 181 p.
78. Loope, L.L. and Mueller-Dombois, D. 1989. Characteristics of invaded islands with special reference to Hawaii. In: J.A. Drake et al. (Eds.) *Biological Invasions, A Global Perspective.* SCOPE 37, John Wiley and Sons, New York, 525 p.
79. Hall, A.V., B. deWinter, S.P. Fourie and T.H. Arnold. 1984. Threatened plants in Southern Africa. *Biol. Conserv.* 28:5-20.
80. Navaratnam, S.J. and A. Catley. 1986. Quarantine measures to exclude plant pests, pp. 106-112. In: R.H. Groves and J.J. Burden (Eds.) *Ecology of Biological Invasions, An Australian Perspective.* Cambridge Univ. Press. Canberra. 159 p.
81. Kingsolver, C.H., J.S. Melching and K.R. Bromfield. 1983. The threat of exotic plant pathogens to agriculture in the United States. *Plant Diseases.* 67:595-600.
82. Groves, R.H. 1989. Ecological control of invasive terrestrial plants, pp. 437-462. In: J.A. Drake et al. (Eds.) *Biological Invasions, A Global Perspective.* SCOPE 37, John Wiley and Sons, New York, 525 p.
83. Brockie, R.E., L.L. Loope, M.B. Usher and O. Hamann. 1988. Biological invasions of island nature reserves. *Biol. Conserv.* 44:9-36.

84. Levin, S. A. 1989. Analysis of risk for invasions and control programs, pp. 425-436. In: J.A. Drake et al. (Eds.) *Biologial Invasions, A Global Perspective.* SCOPE 37, John Wiley and Sons, New York, 525 p.
85. DeBach, P. and D. Rosen. 1991. *Biological Control by Natural Enemies.* Cambridge Univ. Press, New York. 440 p.
86. Moran, V.C., S. Neser and J.H. Hoffmann. 1986. The potential of insect herbivores for the biological control of invasive plants in South Africa, pp. 261-268. In: I.A.W. MacDonald, F.J. Kruger, and A.A. Ferrar (Eds.) *The Ecology and Management of Biological Invasions in Southern Africa.* Oxford Univ. Press, Cape Town, 324 p.
87. Ehler, L.E. 1990. Environmental impact of introduced biological-control agents: implications for agricultural biotechnology, pp. 85-96. In: J.J. Marois and G. Bruening (Eds.) In: *Risk Assessment in Agricultural Biotechnology: proceedings of the international conference,* Nat. Res. Pub. #1928, University of California.
88. Turner, C.E., R.W. Pemberton and S.S. Rosenthal. 1987. Host utilization of native *Cirsium* thistles (Asteraceae) by the introduced weevil *Rhinocyllus conicus* (Coleoptera: Curculionidae) in California. *Envir. Entomol.* 16(1):111-115.
89. Kluge, R.L., H.G. Zimmermann, C.J. Cilliers and G.B. Harding. Integrated control for invasive alien weeds, pp. 295-304. In: I.A.W. MacDonald, F.J. Kruger, and A.A. Ferrar (Eds.) *The Ecology and Management of Biological Invasions in Southern Africa.* Oxford Univ. Press, Cape Town, 324 p.

Biological Diversity: Problems and Challenges. Edited by S.K. Majumdar, F.J. Brenner, J.E. Lovich, J.F. Schalles, and E.W. Miller.

Chapter Ten

IMPLICATIONS OF POLLUTANTS ON DIVERSITY OF AQUATIC BACTERIA AND FUNGI

MICHELLE A. BAKER,[1] JEFFREY MERKLE and SHYAMAL K. MAJUMDAR

Department of Biology
Lafayette College
Easton, PA 18042

INTRODUCTION

Carbon and nutrient cycling are fundamental processes carried out by microorganisms in aquatic ecosystems (Perry, et al. 1989). Bacteria and fungi perform a crucial function by accumulating these nutrients and making them available to higher trophic levels as well as serving as primary producers (Francis, et al. 1984).

In recent years, pollution of aquatic ecosystems by human activity has become widespread (Henderson-Sellers and Markland, 1987). Biological diversity measurements provide important information about the functional status of ecosystems (Atlas, 1984). Stability of an ecosystem usually requires some degree of biological diversity that may be disrupted by pollution (Atlas, 1984). In polluted conditions, the effects of stress on the environment are often assessed using diversity indices. Current research has monitored diversity of microorganisms in order to determine the effects of stress on the microbial community itself, and to interpolate stress effects on higher trophic levels of ecosystems (Atlas, 1984).

Microbial diversity and abundance in aquatic ecosystems are influenced by

---

[1]Current address: Department of Biology, University of New Mexico, Albuquerque, NM 87131.

anthropogenic stresses on the environment such as nutrient loading, acidification, mine drainage, heavy metal contamination, and industrial pollution. Effects of pollution on microbial community diversity and function may have ecosystem level impacts. This chapter reviews current research on the effects on anthropogenic pollutants on bacterial and fungal diversity and activity in aquatic environments.

EFFECTS OF NUTRIENT LOADING FROM HUMAN WASTEWATER AND AGRICULTURAL RUNOFF

One of the most common sources of pollution in freshwater systems is human or animal wastes (Henderson-Sellers and Markland, 1987). Human waste can enter waterways via septic tank leakage and as effluent from water treatment plants. Animal wastes often enter aquatic systems by runoff from agricultural plots. Agricultural runoff also includes high concentrations of phosphorus and nitrogen from fertilizers and may contain pesticide residues. Most studies have focused on microbial changes due to the nutrient loading of agricultural runoff. The high nutrient content found in human wastewater and agricultural runoff can lead to rapid eutrophication of aquatic systems (Henderson-Sellers and Markland, 1987). In general, eutrophication leads to lower plant and animal species diversity, and a change from aerobic to anaerobic processing of nutrients by the microflora (Henderson-Sellers and Markland, 1987).

Effects of human wastewater and agricultural runoff on aquatic microorganisms have been explored by several authors. Studies have reported a decrease in bacterial and/or fungal diversity due to contamination by human or animal wastes. For example, fewer numbers of bacterial genera associated with decomposition activities were found on leaves submerged in coves of Lake Wallenpaupack that were suspected of pollution by septic tank leakage and agricultural runoff compared to leaves in an unpolluted cove (Baker, 1992). Similarly, fewer numbers of fungal genera were identified from leaves at the polluted sites (Baker, 1992). Gray (1951) found differences in the composition of the bacterial community at sites of a stream that received agricultural runoff containing animal and human waste.

In a study of the effects of sewage on fungi in a stream, Cooke (1961) observed fewer fungal species at the site of discharge from a wastewater treatment plant. Further downstream from the outfall, the number of fungal species increased. Cooke also concluded that soil fungi from terrestrial runoff can become easily adapted to polluted water, as most of the fungi identified were of terrestrial origin (Cooke, 1961). Conway (1970) observed a limited aquatic fungal flora in a stream polluted by organic material, including sewage, compared to the flora found in unpolluted streams. On the other hand, no difference in the community structure of aquatic Hyphomycetes was identified in a river that received sewage treatment plant effluents, indicating that aquatic hyphomycetes may not be an appropriate indicator of sewage pollution (Suberkropp, et al. 1988).

Other studies have observed pollution associated changes in microbial function or activity. For example, Dierberg and Brezonik (1981) reported a decrease in the rate of leaf litter associated nitrogen fixation in a sewage enriched cypress dome, compared to a natural cypress dome. Baker (1992) observed fewer numbers of starch and protein hydrolysing bacteria on leaves submerged in a cove at Lake Wallenpaupack suspected of pollution by septic tank leakage compared to an unpolluted cove. Daehler and Majumdar (1992) reported that decomposing leaf litter from Lake Wallenpaupack coves suspected of pollution contained high concentrations of iron and manganese, suggesting presence of iron and manganese oxidizing bacteria at these sites. In general, contamination from nutrient loading by human and animal wastes appears to limit the diversity and activity of aquatic bacteria and fungi.

EFFECTS OF ACIDIFICATION

Microorganisms are extremely sensitive to environmental factors such as temperature, pressure and especially pH (Atlas, 1984). Bacteria may be important in regulating the pH of water receiving acidic deposition from the atmosphere (Kelly, et al. 1982). Acid rain contributes high amounts of nitrate and sulfate ions which are used by bacteria as electron acceptors in anaerobic environments (Kelly, et al. 1982); however, acidification of waterways due to acid rain may adversely affect the structure of pH sensitive microbial communities (Francis, et al. 1984), especially the decomposers (Majumdar, et a!. 1987).

Several studies have reported inhibition of microbial decomposition activity as a result of acidification. Kok and Van der Velde (1991) observed reduced activity of pectinases as a result of acidification in *in vitro* experiments. Francis, et al. (1984) reported reduced decomposition rates of leaf litter submerged in acidified Adirondack lakes. The same study also found lower numbers of bacteria in the water of acidified lakes (Francis, et al. 1984). While decomposition activities appear to be inhibited by acidification in some studies, Hoeniger (1985) found no significant difference in the rate of cellulose decomposition in acidified Canadian lakes.

Perry, et al. (1989) reported lower abundance of bacteria on leaves submerged in an acid-stressed lake. Chemical differences in the leaf litter were also observed (Perry, et al. 1989). Lower concentrations of phosphorus, chlorine, potassium, magnesium and calcium were observed in leaves from acidified lakes. Since microorganisms are the primary mediators of nutrient and element cycling in lakes (Henderson-Sellers and Markland, 1987; Perry, et al. 1989; Kok and Van der Velde, 1991), the chemical changes may reflect differences in the microbial community as a result of acidification.

Limestone treatment is often used to temporarily curb acid effects in lakes and streams. In a Pennsylvania study, increased bacterial numbers were found in a limestone treated lake compared to an acid stressed lake; numbers of heterotrophic and nitrifying bacteria were also significantly higher in the limed lake (Majumdar, et al. 1987).

EFFECTS OF MINE DRAINAGE

Acidified conditions in aquatic systems can also result in contamination from mine drainage. Acid mine drainage, like acidification by atmospheric deposition, affects microbial processes, especially decomposition (Carpenter, et al. 1983; Wassel and Mills, 1983). Carpenter, et al. (1983) concluded that reduced leaf decomposition rates resulting from acid mine drainage were caused by changes in the microbial community. The bacterial community at a polluted site was less diverse than at an unpolluted site.

The number of bacteria in aquatic systems may also be reduced by mine drainage. Direct counts of bacteria at a site in a lake receiving acid mine drainage were lower than at unimpacted sites (Wassel and Mills, 1983). Fewer heterotrophic bacteria and reduced bacterial diversity were also observed (Wassel and Mills, 1983).

Effects of acid mine drainage on fungi in aquatic environments are less well documented. Maltby and Booth (1991) found that aquatic Hyphomycetes were the group of fungi most affected by coal mine effluents. Leaf decomposition rate and fungal diversity were reduced at sites below input from the mine (Maltby and Booth, 1991).

EFFECTS OF HEAVY METALS

Heavy metals can enter aquatic systems from various sources, especially from industrial pollution and mining. Toxic effects of heavy metals are often associated with acidification, since acid conditions increase dissociation of the metals into the aqueous environment (Desjardins, et al. 1988). Several studies report that high concentrations of heavy metals may result in selection for metal-resistant strains of bacteria and/or decreases in microbial diversity (Dean-Ross, 1990; Desjardins et al. 1988; Houba and Remacle, 1980).

In a study of the effects of zinc on attached bacteria in artificial streams, Dean-Ross (1990) found that increased concentrations of zinc resulted in an increased number of zinc tolerant strains and an overall decrease in bacterial diversity. The bacterial biomass in the zinc contaminated artificial streams was lower than uncontaminated streams (Dean-Ross, 1990).

Similarly, Desjardins, et al. (1988) found a high incidence of metal resistant bacterial strains in a marsh contaminated with heavy metals from mine tailings. In a study of the effects of cadmium on aquatic bacteria, Houba and Remacle (1980) also reported that the density of cadmium resistant bacterial strains was greater at heavily contaminated sites. Although heavy metals have been reported to reduce bacterial diversity, this was not observed in a river contaminated with heavy metals (Dean-Ross and Mills, 1989).

Fewer studies have reported the effect of heavy metals on aquatic fungi. Abel and Barlocher (1984) reported that high concentrations of cadmium inhibited sporulation in aquatic Hyphomycetes. The toxicity of cadmium appeared to be species related, which may suggest resistance (Abel and Barlocher, 1984). Babich and Stotzky (1977) also report on the species-related toxic effect of cadmium. Since there appears to be a species-related toxicity, cadmium and other metals may select for resistant fungal strains, resulting in a change in fungal community structure.

EFFECTS OF PETROLEUM HYDROCARBONS

Petroleum hydrocarbons include crude oil, refined petroleum products like gasoline, and other toxic components. Anthropogenic sources and natural inputs of petroleum hydrocarbons are responsible for this pollution (Bartha and Atlas, 1977). The ecological impacts of chronic, low-level inputs are "quite different than those associated with small routine or large accidental spillage" (Bartha and Atlas, 1977). Large, infrequent inputs of oil cause stress to the environment, whereas chronic inputs allow for the adaptation of the microbial community (Bartha and Atlas, 1977; Griffiths, et al., 1981; Alexander and Schwarz, 1980; McKinley et al. 1982). The effects of petroleum on the environment generally relate to the location, amount, and type of oil spilled (Bartha and Atlas, 1977). Several studies have been conducted showing the differential effects of petroleum pollution on microbial diversity and population size.

Following input of petroleum hydrocarbons into certain ecosystems, microbial populations are sometimes found to increase. This can be attributed to an increase in hydrocarbon-utilizers and in microorganisms that can utilize the metabolites produced by them, with an overall increase in the number of microorganisms in the community (Bartha and Atlas, 1977; Atlas, et al. 1976). In an investigation into the effects of Prudhoe crude oil on several different microbial communities, bacterial increases were observed due, in large part, to increased numbers of oil degrading *Pseudomonads* (Atlas, et al. 1976); however, species diversity seemed to be unaffected. In the same study, fungal populations were found to be high in some areas of Cape Simpson "devoid of bacteria" since these fungi were thought to be tolerant to both oil and low pH.

Westlake, et al. (1974, cited in Bartha and Atlas, 1977) found that both incubation temperature and oil type were important in causing population increases. Different crude oils enabled different microorganisms to prosper. In addition, the same crude oil at different temperatures selectively increased different microbial populations (Westlake, et al., 1974).

Alexander and Schwarz (1980) examined the effects of South Louisiana and Kuwait Crude Oils on bacterial glucose utilization and found no significant short-term effect on glucose uptake or mineralization. McKinley, et al. (1982) noted no overall effect of petroleum hydrocarbons on heterotrophic activity. Lignin or cellulose mineralization were inhibited by refined petroleum products (gasoline, diesel fuel) though neither were significantly affected by Prudhoe Bay crude oil (McKinley, et al. 1982).

Griffiths, et al. (1981) found that after extended exposure to two different crude oils, pelagic heterotrophic communities were not adversely affected. In addition, in the presence of crude oil, glucose and glutamate uptake rates were generally reduced, although a wide range of effects were seen.

Crude and refined oil products have also been shown to have a negative impact on aquatic ecosystems. Crow, et al. (1975) found that in salt marsh estuaries, the number of cellulolytic microorganisms decreased after exposure to petroleum hydrocarbons. Walker and Colwell (1976, cited in Bartha and Atlas, 1977), using a mixture of aromatic and paraffinic compounds, found that these hydrocarbons can limit the growth of sediment and water bacteria.

Toluene, benzene, and ethylbenzene, components of petroleum products, have been reported to be toxic to microorganisms (Liu, 1973); however, longer aliphatic side-chain aromatics were shown to stimulate metabolic activity of the same microorganisms. Atlas (1975, cited from Bartha and Atlas, 1977) found that some crude oils contain toxic components that delay hydrocarbon biodegradation and increase populations of hydrocarbon-utilizing bacteria until evaporation.

Different hydrocarbon components in varying concentrations have been shown to be responsible for the different effects of petroleum pollution on aquatic microorganisms (Alexander and Schwarz, 1980; Atlas et al., 1976, Bartha and Atlas, 1977). The changes brought about in microbial communities following petroleum hydrocarbon input is also a function of the toxicity of the metabolic products of hydrocarbon utilizers (Atlas, et al. 1976), dissolution and emulsification in water, evaporation rate, as well as temperature, wind velocity, salinity, and turbulence (Bartha and Atlas, 1977; referenced in McKinley, et al. 1982).

CONCLUSION

While the effects of pollution depend on the initial state of the environment, contamination of aquatic systems by anthropogenic pollution in general, may limit the growth, diversity and activity of bacteria and fungi. Low level, chronic inputs of these contaminants, while not always affecting species diversity, tend to select for those microorganisms that are better able to utilize these inputs.

It should also be noted that, at any one time, several of the pollutants discussed in this review may be contaminating a single system. For example, acidity increases the ability of metal ions to dissociate into the aqueous environment, thereby increasing probability of toxic effects (Desjardins, et al 1988). Thus, a combination of pollutants may cause more excessive damage than the studies on any one pollutant would indicate.

Measurement of the diversity of microorganisms under environmental stress provides information on the functional status of the ecosystem (Atlas, 1984). Studies have shown that in polluted conditions, the original members of the microbial community are often replaced or become dominated by microorganisms that have the ability to survive under the stressed condition, resulting in a change in community structure and function. Since microorganisms are involved in fundamental nutrient and energy cycling processes, perturbations to the microbial community may have ecosystem-level effects (Atlas, 1984).

REFERENCES

Abel, T.H. and F. Barlocher. 1984. Effects of cadmium on aquatic hyphomycetes. *Applied and Environmental Microbiology.* 48(2):245-251.

Alexander, S.K. and J.R. Schwarz. 1980. Short-term effects of South Louisiana and Kuwait crude oils on glucose utilization by marine bacterial populations. *Applied and Environmental Microbiology.* 40:341-345.

Atlas, Ronald M. 1984. Use of microbial diversity measurements to assess environmental stress. Pages 540-545 In (International Symposium on Microbial Ecology, eds.) *Current Perspectives in Microbial Ecology: Proceedings of the Third International Symposium on Microbial Ecology at Michigan State University 7-12 August, 1983.* American Society of Microbiology, Washington, D.C.

Atlas, R.M. 1975. *Applied Microbiology.* 30, 396-403. Cited from Bartha and Atlas, 1977.

Atlas, Ronald M., Edmund A. Schofield, Frank A. Morelli and Roy E. Cameron. 1976. Effects of petroleum pollutants on arctic microbial populations. *Environmental Pollution.* 10:35-43.

Babich, H. and G. Stotzky. 1977. Sensitivity of various bacteria, including Actinomycetes, and fungi to cadmium and the influence of pH on sensitivity. *Applied and Environmental Microbiology.* 33:681-695.

Baker, Michelle A. 1992. Bacteria and fungi associated with decomposing leaves submerged in Lake Wallenpaupack. *Senior Honors Thesis.* Department of Biology, Lafayette College, Easton, PA.

Bartha, R. and R.M. Atlas. 1977. The microbiology of aquatic oil spills. Pages 226-266. In: *Advances in Applied Microbiology.* Academic Press, New York.

Carpenter, John, William E. Odum and Aaron Mills. 1983. Leaf litter decomposition in a reservoir affected by acid mine drainage. *Oikos.* 41:165-172.

Conway, Kenneth E. 1970. The aquatic hyphomycetes of central New York. *Mycologia.* 62:516-530.

Cooke, William Bridge. 1961. Pollution effects on the fungus population of a stream. *Ecology.* 42(1):1-17.

Crow, S.A., M.A. Hood and S.P. Meyers. 1975. In *Impact of the Use of Microorganisms on the Aquatic Environment* (A.W. Borquin, D.G. Ahearn and S.P. Meyers, eds.), EPA 660-3-75-001. pp. 221-227. EPA Res. Cent., Corvallis, Oregon.

Daehler, Curtis C. and Shyamal K. Majumdar. 1992. Elemental analyses of leaf litter and sediments at three coves sites varying in their water quality inputs in Lake Wallenpaupack, Pennsylvania, USA. *Environmental Technology.* 13:1-13.

Dean-Ross, Deborah. 1990. Response of attached bacteria to zinc in artificial streams. *Canadian Journal of Microbiology.* 36:561-566.

Dean-Ross, Deborah and Aaron Mills. 1989. Bacterial community structure and function along a heavy metal gradient. *Applied and Environmental Microbiology.* 55(8):2002-2009.

Desjardins, Robert M., Wayne C. Bradbury and Patricia L. Seyfried. 1988. Effects of metals from mine tailings on the microflora of a marsh treatment system. Pages 491-501 In: *Aquatic Toxicology and Hazard Assessment.* American Society for Testing and Materials, Philadelphia. 579 pp.

Dierberg, Forrest E. and Patrick 1. Brezonik. 1981. Nitrogen fixation (acetylene reduction) associated with decaying leaves of pond cypress (*Taxodium distichum* var *nutans*) in a natural and a sewage enriched cypress dome. *Applied and Environmental Microbiology*. 41(6):1413-1418.

Francis, Arokiasmy J., Helen L. Quinby and George R. Hendrey. 1984. Effect of lake pH on microbial decomposition of allochthonous litter. Pages 1-21 In: G.R. Hendrey (Ed.) *Early Biotic Responses to Advancing Lake Acidification*. Butterworth Publishers, Boston.

Gray, E. 1951. The ecology of the bacteria of Hobson's Brook, a Cambridgeshire chalk stream. *Journal of General Microbiology*. 5:840-859.

Griffiths, R.P., T.M. McNamara, B.A. Caldwell and R.Y. Morita. 1981. Field observations on the acute effects of crude oil on glucose and glutamate uptake in samples collected from arctic and subarctic waters. *Applied and Environmental Microbiology*. 41:1400-1406.

Henderson-Sellers, B. and H.R. Markland. 1987. *Decaying Lakes: The Origin and Control of Cultural Eutrophication*. John Wiley and Sons, NY.

Hoeniger, Judith F.M. 1985. Microbial decomposition of cellulose in acidifying lakes of south central Ontario. *Applied and Environmental Microbiology*. 50(2):315-322.

Kelly, C.A., John W.M. Rudd, R.B. Cook, D.W. Schindler. 1982. The potential importance of bacterial processes in regulating rate of lake acidification. *Limnology and Oceanography*. 27(5):868-882.

Kok, C.J. and G. Van der Velde. 1991. The influence of selected water quality parameters on the decay rate and exoenzymatic activity of detritus of *Nymphaea alba* L. floating leaf blades in laboratory experiments. *Oceologia*. 88:311-316.

Liu, D.L.S. 1973. Micro. Degradation Oil Pollut., Workshop, 1972 La State Univ. Publ. No. LSU-Sg-73-01, pp. 95-104.

Houba, C. and J. Remacle. 1980. Composition of the saprophytic bacterial communities in freshwater systems contaminated by heavy metals. *Microbial Ecology*. 6:55-69.

Majumdar, S.K., O. DeLucia, R.W. Holt, C.C. Derivaux, R.J. Marcus, T.J. Irwin, L.C. Mineo, P.M. Steed and R.L. Morris. 1987. Effects of limestone on microbial communities and sediment chemistry in acid-stressed Pennsylvania lakes, USA. Pages 465-472 In: (R. Perry, R.M. Harrison, J.N.B. Bell and J.N. Lester, Eds.) *Acid Rain: Scientific and Technical Advances*. Selper Limited.

Maltby, Lorraine and Rosemary Booth. 1991. The effect of coal mine effluent on fungal assemblages and leaf breakdown. *Water Research*. 25(3):247-250.

McKinley, Vicky L., Thomas W. Federle and J. Robie Vestal. 1982. Effects of petroleum hydrocarbons on plant litter microbiota in an arctic lake. *Applied and Environmental Microbiology*. 43:129-135.

Perry, J.A., R. Zeyen, M. Newsom and G. Ahlstrand. 1989. X-ray microanalysis of leaf litter decomposition in lakes. *Bioscience*. 39(4):260-263.

Suberkropp, K., A. Michelis, H.J. Lorch and J.C.G. Ottow. 1988. Effect of sewage treatment plant effluents on the distribution of aquatic hyphomycetes in the River Erms, Schwabiche Alb, F.R.G. *Aquatic Botany*. 32:141-153.

Walker, J.D. and R.R. Colwell. 1976. *Canadian Journal of Microbiology*. 22:886-891. Cited in Bartha and Atlas, 1977.

Wassel, Raymond A. and Aaron Mills. 1983. Changes in water and sediment bacterial community structure in a lake receiving acid mine drainage. *Microbial Ecology*. 9:155-169.

Westlake, D.W.S., A. Rubson, R. Phillippe and F.D. Cook. 1974. *Canadian Journal of Microbiology*. 20:915-918. Cited in Bartha and Atlas, 1977.

Biological Diversity: Problems and Challenges. Edited by S.K. Majumdar, F.J. Brenner, J.E. Lovich, J.F. Schalles, and E.W. Miller. © 1994, The Pennsylvania Academy of Science.

Chapter Eleven

TRIBUTYLTINS IN AQUATIC ECOSYSTEMS

JOSEPH J. COONEY,[1] JAMES H. WEBER[2] and LARRY R. SHERMAN[3]

[1]University of Massachusetts
Boston, MA 02125-3393
[2]University of New Hampshire, Durham, NH 03824-3598
[3]University of Scranton, Scranton, PA 18510-4626

INTRODUCTION

Organotins are compounds with one or more carbon-tin bonds. Addition of an organic group to tin increases tin's volatility, its lipid solubility and its toxicity and also alters its adsorptive properties. Tin can have a valence of 2+ or 4+, but Sn^{4+} predominates in the aquatic environment, thus tin can form up to four Sn-C bonds. About 5% of the tin produced is used for production of organotins, and some 50,000 tons of organotin chemicals are produced worldwide each year.[1]

The first application of an organotin was in 1925 as a mothproofing agent. In the 1930's organotins began to be used as stabilizers for organic chemicals used in transformers and capacitors and to stabilize polyvinylchlorides. In the 1950's the biocidal properties of organotins were explored by G.J.M. van der Kirk and his colleagues at the Institute of Organic Chemistry at Utrecht, The Netherlands.[2] Since then organotins have found ever-wider uses as fungicides, bactericides and insecticides, including uses as wood preservatives and as slimicides in cooling water systems. In the 1960's tributyltin compounds (TBT's) were used to kill snails which are intermediate hosts in the transmission of schistosomiasis.[3,4] They were also incorporated in paints used as antifouling agents on the nets of salmon farm pens and on boat hulls, where they are far more effective than the copper compounds which are the alternative active agent in antifouling paints. Indeed, tin has more uses than any other metal in the number of its organic applications.[1,5,6,7]

In 1954, a disastrous attempt was made to use organotin compounds for treatment of staphylococcal infections and the subsequent death of 102 people severely hampered the growth of use of organotin compounds.[7] In the 1970's tributyltin compounds were gradually recognized as a toxicant for non-target organisms. TBTs cause morphological aberrations in oysters and mussels[8] and they accumulate in the flesh of scallops, oysters[9] and salmon[10] held in pens treated with TBT-containing paint, thus entering the human food chain. Populations of the dogwhelk, *Nucella lapillus,* in waters with relatively low levels of TBT show a high incidence of imposex, a condition in which female animals grow male sex organs and the population cannot reproduce. Imposex can occur at concentrations of TBT as low as 1 ng/liter (1 part per trillion).[11] TBT has been called, "perhaps the most acutely toxic chemical to aquatic organisms ever deliberately introduced to water."[1]

TBT thus attracted the attention of environmental protection agencies in a number of countries. In 1984 TBT's use in antifouling paints was regulated in France and The United Kingdom; Germany and Switzerland followed suit in the next three years.[12,13] Several states in the United States also regulated use and in 1988 the U.S. regulated TBT in antifouling paints.[14] All the regulations are similar: paints with TBT may not be used on boats shorter than 25 meters in length. The reasoning is that large vessels spend most of their time at sea so TBT leaching from their paint is diluted in a great volume of ocean water. Smaller vessels, however, - particularly recreational boats - spend most of their time near shore or in harbor where TBT leaching from paint will be concentrated in sediments. TBT may also be used on aluminum-hulled boats where copper in paints can contribute to hull electrolysis. Thus, use of TBT has been regulated, not banned. Nevertheless, before the regulations were passed the organotins caused considerable environmental damage and it is expected that it will be many years before some aquatic systems recover.

ANALYTICAL METHODS

General Aspects

Attention has focused on butyltins because of their toxicity via antifouling paints and on methyltins which can be formed in the environment by biotic and abiotic means.[15,16,17,18] Determination of methyl- and butyltin compounds typically include the three general steps of extraction, concentration-derivatization, and detection.[19] The extractant may be simply 1 M HCl at ambient temperature followed by extraction into an organic solvent in the case of fresh or marine water samples. The extraction step is more difficult for biotic samples of fish, shellfish or plant material, and is most difficult for sediments. For biological materials and sediments, a strong, concentrated aqueous acid such as 6 M HCl is usually used with or without an organic solvent. Such harsh extraction steps are typically several hours long and include heating and/or sonication.

Both concentration- and derivatization-steps are usually necessary to concentrate the organotin because the analytes typically exist at ng/L concentrations in water and ng/g concentrations in biological materials and sediments. Derivatization increases the volatility of the analytes making it easier to concentrate them. There are two common methods for derivatization. In the first, the Grignard reaction converts organotin cations into neutral, relatively volatile compounds. This method usually requires extraction into an organic solvent, drying the solvent, running the Grignard reaction and evaporating the solvent to concentrate the analyte. Then *part* of the resulting solution is chromatographed using gas chromatography (GLC) to separate the analytes for detection. The second method involves hydride formation and is simpler. The extracted analyte solution is adjusted to about pH 5 and sodium borohydride is added. The resulting volatile, neutral organotin hydrides are easily concentrated and separated on a chromatographic column. The hydride method is shorter and uses *all* of the extract containing the analytes.

The most commonly used detectors are atomic absorption spectrometry (AAS) with a quartz furnace and mass spectrometry (MS). AAS has an absolute limit of detection for organotin compounds of approximately 20 pg in solutions of standards and approximately 1 ng in environmental extracts, while the limit of detection for MS is a few nanograms. MS is superior for identification of derivatized organotins because the parent ion can often be observed. AAS relies on the assumption that an analyte with the identical (GLC) *retention time* is the same compound as the standard. Thus, AAS is superior for low-concentration determinations, but MS is most useful for confirmation of compound identity.

Validation of the Analytical Procedure

Validation of the analytical procedure is required to demonstrate that no organotin compound has gained or lost an organogroup, and that a reasonable percent of analytes are detected.[20] Validation is done by adding known amounts of organotin analytes to fresh environmental samples (spiking), and carrying the sample through one of the procedures. In practice, this validation is very difficult for two reasons. First, one wishes to spike samples so that resulting concentrations are near environmental concentrations. However, the amount of the recovered spike must be corrected for analyte naturally present before calculating recovery, and accurate calculations require that the spiked amount be several times larger than the natural amount. If the spike is considerably more than the natural amount, percent recovery is unrealistically high. This occurs because environmental samples usually contain analytes at sites that range from weakly-bound to strongly-bound, i.e. easily- to difficulty-extractable. If too much spike is added, unextracted analyte from strongly-bound sites will not appreciably decrease the percent recovery, which may be unrealistically high. The second reason for difficulty in validation is that in many environmental samples, particularly in heterogeneous sediments, it is impossible to assure that the spike is in the same chemical sites as the natural analytes. Again, the percent recovery might be unrealistically high.

CONTAMINATION IN THE ENVIRONMENT

Methyl- and Butyltins in the Aquatic Environment

Concentrations of methyl- and butyltins in biotic and abiotic samples are illustrated in Table 1. Concentrations in the same sample type (e.g. Ontario sediments) can vary from non-detectable to thousands of nanograms per gram, reflecting polluted and pristine sampling areas. Butyltins arise from antifouling boat paints and methyltins form by methylation of inorganic tin in the environment. Furthermore, methyltin compounds are widely used in industry and high concentrations in some sediments probably reflect pollution from point sources. Methyl- and butyltins have 1,000- to 10,000-fold higher concentrations in sediments and biological materials than in water. For example, oysters from the Great Bay Estuary, New Hampshire bioaccumulate $MeSn^{3+}$ about 5,000-fold relative to the surrounding water.

A comparison of oysters from the non-polluted Great Bay Estuary and the polluted Loire River, France illustrates the effects of using antifouling paints on boats. Oyster samples from the Great Bay Estuary contained only one detectable organotin, about 110 ng/g $MeSn^{3+}$, while oyster samples from the Loire River had 2200 ng/g Bu_3Sn^+ and no methyltins. Bu_2Sn^{2+} and $BuSn^{3+}$ in Loire River oysters probably originate from degradation of Bu_3Sn^+ (Table 1).

Antifoulants

The most serious environmental consequences from use of organotins has come from the use of TBT's as antifoulants. Structures exposed to an aquatic environment are subject to both calcareous organisms such as barnacles, oysters, corals

TABLE 1

Typical Concentrations (ng Sn/g for Solids and ng Sn/mL in Water) of Methyl- and Butyltin Compounds in the Aquatic Environment[a]

Sample	$MeSn^{3+}$	Me_2Sn^{2+}	Me_3Sn^+	$BuSn^{3+}$	Bu_2Sn^{2+}	Bu_3Sn^+
Ontario Freshwater[21,22]	nd-1.1	nd-0.32	nd-0.18	nd-5.7	nd-1.4	nd-2.3
Great Bay Estuary Water[23]	0.02	0.01	0.01	nd-0.2	nd	nd-0.3
San Diego Bay sea-water[24,25]	nd-0.008	nd-0.045	nd	nd-0.04	nd-0.15	nd-0.20
Great Bay Estuary Oysters[20]	95-120	nd	nd	nd	nd	nd
Loire River Oysters[20]	nd	nd	nd	580	840	2200
Great Bay Estuary Aquatic Plants[23]	50-2000	nd-50	nd-400	nd-100	5-200	nd-50
Ontario Sediments[21,22,26]	nd-17,000	nd-170	nd-750	nd-4,700	nd-8,500	nd-11,000
San Diego Bay Sediments[21,24]	2-3	2	nd-0.2	3-19	3-22	nd-3

[a]nd = not detected

and tubeworms and to non-calcareous organisms including algae, bryozoa and tunicates. Fouling targets include ships, buoys, pilings, sea walls, oil rigs, cables and water intake pipes. Fouling adds weight to the structure, destroys streamlining (increases drag), decreases the lumen in pipes and can penetrate protective coatings which enhances corrosion. Paints containing TBT are the only effective means for controlling the zebra mussel, *Dreissena polymorpha,* in fresh waters, including the Great Lakes.[27] The economic loss is high.[4] The use of TBT-containing paints on US vessels instead of less-effective copper-based antifoulants is estimated to save $400 to 500 million per year in fuel costs, calculated at $16/barrel. This represents a saving of two billion gallons of precious fossil fuel each year.[28]

In the early 1960s it was discovered that addition of TBT compounds such as TBTO, TBTF and TBTS to elastomer (rubber) based paints yielded slow-release antifouling material which could prevent growth of fouling organisms for several years. Because of their hydrocarbon chains, trialkyltins are soluble in the elastomers. The emission and longevity of the antifoulant coating can be controlled by its geometry, crosslink density and specific additives.[4] The TBT was often loaded in excess of matrix solubility. It diffused from the coated surface rapidly, causing high concentrations of TBT in the water nearby. With some coatings an oily film appeared when a newly-painted article was first placed in the water. The paint continued to release TBT through its lifetime by a diffusion-dissolution mechanism. A large number of antifouling paints was put on the market; some suffered from poor adhesion, possibly from cross-linking effects of the TBT, but all had good antifouling properties.[4]

Most of the antifoulant coatings were marketed without adequate environmental studies. TBTs were known to be excellent molluscicides, primarily because they are adsorbed on organic materials found in aquatic systems. They were considered innocuous for most aquatic life because of their low solubility (ng/L). However, early studies at normal antifouling emission rates proved them to be detrimental to fish fry[29] and since then they have been shown to have adverse effects on almost all marine life.[21]

More recent formulations contain considerably less organotin and many antifouling paints are co-polymers, which hydrolyze slowly, leaching TBT onto the surface of the vessel. These formulations are just as effective as those with high concentrations but cause less difficulty in the environment. Nevertheless, butyltin compounds have become ubiquitous. They have been found in all areas of the world, including New Zealand,[31] Australia,[31] European coastal waters,[32,33,34] Canada[22,35] and the USA.[36,37] They also occur in fresh water[38,39] and in sewage sludge.[40]

INTERACTIONS WITH BIOTA

Toxicity

In general, triorganotins are more toxic than di- or monosubstituted organotins. Tetrasubstituted compounds are considered toxic only if they are metabolized to trisubstituted compounds. Among trisubstituted organotins, propyl-, butyl-, pentyl-, phenyl- and cyclohexyltins are generally most toxic; and a high octanol:water partition coefficient, *Kow*, and a high total molecular surface area predict greater toxicity.[41]

Understanding the role of organotins in aquatic systems is complicated by the fact that they can undergo disproportionation reactions, such as:

$$2R_3SnX \longleftrightarrow R_2SnX_2 + R_4Sn.$$

The extent of disproportionation reactions in the aquatic environment is poorly understood.[42,43]

TBT is used extensively in agriculture, especially by fruit growers, for it is effective against algae, fungi, insects and mites, but it is toxic to a variety of non-target organisms, too, at levels which have been found in polluted environments. These include macro- and microorganisms.[3,41,43,44] In general, eggs, larvae and juvenile forms are more sensitive than adults.[45,46] Some aquatic organisms can bioaccumulate tin compounds, including plants, shellfish and some finfish and both predators and filter feeders. Some microorganisms, which are at the base of the food web, are also sensitive to organotins.[41,47,48] Two international symposia have focused on the measurement of tin compounds in the marine environment and on their toxicity to marine animals.[49,50]

Considerable attention has been given to effects on the dogwhelk, *N. lapillus*, and other shellfish, which appear at levels as low as 1-10 parts per billion.[51] In the dogwhelk exposure to TBT led to a significant increase in testosterone.[52] Molluscs may be unusually sensitive because they have relatively low activities of cytochrome *P*-450 and mixed function oxidases which leads to TBT accumulation in tissues because TBT is not metabolized rapidly.[52a] The incidence of imposex in *N. lapillus* has been suggested as an indicator of TBT pollution as an alternative to more expensive chemical monitoring.[53]

Organotins penetrate biological membranes more readily than inorganic tins and they are more likely to accumulate in lipid-rich tissues or organelles. TBT is a membrane-active compound which acts as an ionophore facilitating halide-hydroxyl exchanges or by interfering with energy transduction processes in bacteria, chloroplasts and mitochondria.[41,54,55] TBT also inhibits energy-linked reactions in *Escherichia coli*,[56] and it interacts with red blood cell membranes causing the cells to lyse.[57] Although trisubstituted organotins are considered more toxic than their mono- or disubstituted homologues, this is not always the case for monobutyltin is more toxic to some yeasts,[58] to methanogenic and sulfate-reducing bacteria[59,60] and to bacteriophage T4.[61] Moreover, the chemical and physical environment can influence the toxicity of tin compounds to microorganisms.[62]

TBT inhibits hepatic microsomal cytochrome *P*-450 in fish.[63] Di- and trisubstituted organotins inhibit a number of enzymes and disubstituted tins are active against dithiol-containing enzymes.[41,64]

Effects on humans

Inorganic tin compounds are relatively non-toxic and few clinical or epidemiological examples exist for their intoxication. Organotins are considerably more toxic, but their toxicity depends upon the carbon chain length and the number of Sn-C bonds. The compounds most toxic to humans are the di- and trisubstituted methyl and ethyl compounds. As the chain length of the organic substituents increases beyond two, the compounds become more toxic to aquatic life but less toxic to humans. Except for cyclic compounds, organotin derivatives with five or more carbons in each organic group have little toxicity.[65] Nevertheless, longer chain compounds as well as short-chain organotins can produce acute skin burns that appear 1-8 hours later if the exposed area is not washed immediately.[65] Exposure to organotin vapors or direct eye contact causes lachrymation and sever suffusion of the conjunctiva within minutes which can persist up to five days (Sherman unpub.) Although no epidemiological data exist for humans, most organotins, especially dibutyl- and dioctyltins, have severe effects on the immune system and cause premature atrophy of the thymus gland in rats.[66]

In 1954 a proprietary formulation, "Stalinon," was marketed in France for oral administration for boils. The formulation was primarily linoleic acid, but was contaminated with diethyltin diiodide (LD_{50} in rats: 0.7 mg/Kg) and may have contained mono-, tri- and tetraethyltin.[7] The formulation led to the death of more than 100 people and the intoxication of 200 or more. Clinical data indicated cerebral edema of the white matter. Since the "Stalinon Affair," a great deal of control has been exerted in the manufacture and marketing of organotins and no further human deaths have been attributed to the compounds.[65]

Interactions of Microorganisms with Tin Compounds

Degradation of TBT occurs by sequential removal of butyl groups, and in some cases the butyl moieties can be hydroxylated.[67] Degradation has been reported in seagrass beds,[68] a bacterium,[69] fungi,[70] invertebrates and fish,[71] and mammals.[72]

In the marine water column TBT can be degraded to DBT and MBT with a half-life of several days;[73] degradation by photolysis alone has a half-life of >89 days,[74] but it can be stimulated in the presence of light when microalgae are present[75] and if nutrients are added.[76]

Most TBT is found in sediments and the surface microlayer.[74,77,78] Biological and chemical degradation in marine and freshwater sediments appears to be slow with a half-life of months,[74] years,[79] or absent.[80] Abiotic degradation can be significant at the sediment-water interface. Attempts to isolate from sediments microorganisms which can degrade TBT have been largely unsuccessful. Thus, while TBT degradation occurs in the aquatic ecosystem, the relative contributions of chemical, physical and biological mechanisms are uncertain, particularly in sediments.

Among the microorganisms which survive the toxic effects of tin compounds are some which can methylate one or more tin compounds. Methyltin compounds have been found in water, sediments and biota of aquatic systems (Table 1[36,43]). They may arise from anthropogenic sources or from biotic or abiotic methylation, and the relative contribution of each method is uncertain. Methylation can occur in pure cultures[81] and in natural populations[16,82,83] and anaerobically[15] as well as aerobically. Butylmethyltins have been found in harbor sediments.[26] Since TBT from ships is the principle source of tin in the harbors, this suggests that butyltins can be partially degraded and then methylated in situ. Difficulties in isolating pure cultures which can methylate tin compounds, methylation in natural samples which contain mixed cultures, and methylation by a co-culture containing two *Bacillus* spp.[18] suggest that microbial consortia can be involved. There is presently no firm evidence for the biosynthesis of organotins having organic groups larger than one carbon.

RECOMMENDATIONS AND CONCLUSIONS

In the last decade methods have been developed which will detect and quantify organotins in the complex matrices associated with animal tissues and environmental samples. We have realized that it is not the total tin burden of a system which is important but the amounts of individual tin and organotin species and the methods have been expanded to include separation of these species. Additional methods development is needed, for the methods available are time consuming.

Considerable effort has been expended in determining the toxicity of tributyltin to a variety of aquatic species, sometimes without examining mono- and dibutyltins to see if they are also toxic. There is evidence that they can be as toxic as TBT to certain microorganisms. This work should be expanded to include other organotins, including ethyl-, cyclohexyl- and phenyltins which have extensive commercial use. Furthermore, little is known of the bioaccumulation, or lack of it, in aquatic and terrestrial ecosystems.

Factors which control the degradation of organotins, particularly in sediments, are poorly understood. Such factors have potential for treatment of waste streams and for cleaning contaminated environs. Attention should be given to photosynthetic organisms and to consortia of microorganisms.

The physiology and biochemistry of organisms which metabolize tin compounds are also poorly understood, as are the physiological and genetic systems which regulate such metabolism.

A number of countries have regulated the use of TBT in antifouling paints. The regulations restrict the use of paints containing TBT to vessels longer than 25 meters. This will reduce the number of "hot spots" in marinas, harbors and shallow coastal waters where the greatest effect has occurred on living organisms. However, these restrictions put the greatest financial burden on some of those who can least afford it, small boat users and independent fishermen. Undoubtedly, research is needed in preparing antifouling materials which do not cause serious damage to the environment. Organotins are superior to copper-based antifoulants, especially in areas of high fouling, such as Alexandria, Egypt.[84] They are also very effective as insecticides. Since more than 70,000 organotin compounds exist, it is likely that some of them have more favorable environmental properties than the TBT compounds used to date.

The most serious effect on biodiversity in the aquatic environment has been imposex and resulting sterility in gastropods. Regulations enacted by several countries have helped, but anecdotes abound wherein boats from a country with regulations are taken to a country without such regulations to be painted with paint containing TBT. The continued use of TBT paints has seriously depleted some wildlife forms and may put others on the endangered list.

The Directorate General of the Commission of the European Community has established a task group to prioritize areas where research is most needed.[12] In addition to aspects listed above, the group has asked for more data on TBT distribution in coastal and lake waters, on contamination in sewage sludge and river sediments, on movement of organotins in the aquasphere and on bioconcentration and its impact on non-target organisms.

LITERATURE CITED

1. Maguire, R.J. 1987. Environmental aspects of tributyltin. *Appl. Organomet. Chem.* 1:475-498.
2. Luijten, J.G.A. 1972. Applications and biologial effects of organotin compounds, pp. 931-974. In: A.K. Sawyer (Ed.) *Organotin Compounds,* vol. 3, Marcel Dekker, Inc., New York.
3. Thayer, J.S. 1984. *Organometallic compounds and living organisms.* Academic Press, New York.
4. Cardarelli, N.F. 1985. Controlled release of organotins. *Rev. Si, Ge, Sn, Pb Cmpds.* 8:169-189.
5. Champ, M.A. 1986. Organotin symposium: introduction and overview. *Oceans 86 Proceedings.* vol. 4, not paginated, IEEE Service Center, Piscataway, NJ.
6. Davies, A.G. and P.J. Smith. 1982. Tin, pp. 519-627. In: F.G.A. Stone and W.W. Abel (Ed.) *Comprehensive Organometallic Chemistry: The Synthesis, Reactions and Structures of Organometellic Compounds,* vol. 2, Pergamon Press, New York.
7. Zuckerman, J.J., R.P. Reisdorf, H.V. Ellis III and R.R. Wilkinson. 1978. Organotins in Biology and the Environment, pp. 388-424. In: F.E. Brinckman and J.M. Bellama (Ed.) *Organometals and Organometelloids, Occurrence and Fate in the Environment.* Amer. Chem. Soc. Symp. Ser., vol. 82, Amer. Chem. Soc., Washington, DC.
8. Waldock, M.J. and J.E. Thain. 1983. Shell thickening in *Crassostrea gigas:* organotin antifouling or sediment induced? *Mar. Pollut. Bull.* 14:411-415.
9. Davies, I.M., J.C. McKie and J.D. Paul. 1986. Accumulation of tin and tributyltin from antifouling paint by cultivated scallops (*Pectin maximus*) and Pacific oysters (*Crassostrea gigas*). *Aquaculture* 55:103-114.
10. Short, J.W. and F.P. Thrower. 1986. Accumulation of butyltins in muscle tissue of Chinook salmon reared in sea pens treated with tri-*n*-butyltin, *Oceans 86 Proceedings,* vol. 4, pp. 1177-1181, IEEE Service Center, Piscataway, NJ.
11. Gibbs, P.E. and G.W. Bryan. 1986. Reproductive failure in populations of the dog-whelk, *Nucella lapilus,* caused by imposex induced by tributyltin from antifouling paints. *J. Mar. Biol. Assoc. U.K.* 66:767-777.
12. Müller, M.D., L. Renberg, G. Rippen. 1989. Tri-butyltin in the environment-sources, fate and determination, an assessment of present status and research needs. *Chemosphere* 18, 2015-2042.
13. Huggett, R.J., M.A. Unger. P.F. Seligman and A.O. Valkirs. 1992. The marine biocide tributyltin. *Environ. Sci. Technol.* 26:232-237.

14. U.S. Public Law 100-330 102 Stat. 605. Organotin antifouling paint control act of 1988, June 16, 1988.
15. Gilmour, C.C., J.H. Tuttle and J.C. Means. 1985. Tin methylation in sulfide bearing sediments, pp. 239-258, In: A.C. Sigleo and A. Hattori (Ed.) *Marine and Estuarine Geochemistry,* Lewis Publishers, Inc., Chelsea, MI.
16. Guard, H.E., A.B. Cobet and W.M. Coleman, III. 1981. Methylation of trimethyltin compounds by estuarine sediments. *Science* 213:770-771.
17. Hallas, L.E., J.C. Means and J.J. Cooney. 1982. Methylation of tin by estuarine microorganisms. *Science* 215:1505-1507.
18. Makkar, N.S. and J.J. Cooney. 1990. Methylation of monomethyltin by a bacterial coculture. *Geomicrobiol.* J. 8:101-107.
19. Donard, O.F.X. and F.M. Martin. 1992. Hyphenated techniques applied to environmental speciation studies. *Trends Anal. Chem.* 11:17-26.
20. Han, J.S. and J.H. Weber. 1988. Speciation of methyl- and butyltin compounds and inorgnic tin in oysters by hydride generation atomic absorption spectrometry. *Anal. Chem.* 60:316-319.
21. Maguire, R.J. 1991. Aquatic environmental aspects of non-pesticidal organotin compounds. *Water Pollut. Res. J. Can.* 26:243-360.
22. Maguire, R.J., R.J. Tacz, Y.K. Chau, G.A. Bengert and P.T.S. Wong. 1986. Occurrence of organotin compounds in water and sediment in Canada. *Chemosphere* 15:253-274.
23. Weber, J.H., J.S. Han and R. Francois. 1988. Speciation of methyl- and butyltin compounds in compartments of the Great Bay Estuary. pp. 395-400, In: M. Astruc and J.N. Lester (Ed.) *Heavy metals in the Hydrological Cycle,* Selper, Ltd., London.
24. Tugrul, S., T.I. Balkas and E.D. Goldberg. 1983. Methyltins in the marine environment. *Mar. Pollut. Bull.* 14:297-303.
25. Valkirs, A.O., P.F. Seligman, P.M. Stang, V. Homer, S.H. Lieberman, G. Vafa and C.A. Dooley. 1986. Measurement of butyltin compounds in San Diego Bay. *Mar. Pollut. Bull.* 7:319-324.
26. Maguire, R.J. 1984. Butyltin compounds and inorganic tin in sediments in Ontario. *Environ. Sci. Technol.* 18:291-294.
27. Mackie, G.L., W.N. Gibbons, B.W. Muncaster and I.M. Gray. 1989. The zebra mussel, *Dreissean polymorpha:* a synthesis of European experiences and a preview for North America. Environment Ontario, Queen's Printer for Ontario, 75 pp.
28. Champ, M.A. and W.L. Pugh. 1987. Tributyltin antifouling paints: introduction and overview. *Oceans 87 Proceedings,* vol. 4, pp. 1296-1308, IEEE Service Center, Piscataway, NJ.
29. Good, M.L. and D.S. Dundee and G. Swindler. 1979. Bioassay and environmental effects of OTC antifoulants. *Proc. 6th Int. Symp. Controlled Release Bioactive Mat.* II-1, II-2 New Orleans, LA, Controlled Release Society, Akron, Ohio.
30. King, N., M. Miller and S. deMora. 1989. TBT levels for sea water, sediment and selected marine species in coastal Northland and Auckland, New Zealand. *New Zealand J. Mar. Freshwater Res.* 23:287-294.
31. Butley, G.E. and M.S. Scammell. Research on TBT in Australian estuaries. *Appl. Organomet. Chem.* 5:99-105.

32. Alzieu, C., P. Michel, J. Sanjuan and B. Averty. 1990. Tributyltin levels in French Mediterranean coastal waters. *Appl. Organomet. Chem.* 4:55-61.
33. Cleary, J.J. and A.R.D. Stebbing. 1985. Organotin and total tin in coastal waters of southwest England. *Mar. Pollut. Bull.* 16:350-355.
34. Quevauviller, P.H. and O.F.X. Donard. 1990. Variability of butyltin determination in water and sediment samples from European coastal environments. *Appl. Organomet. Chem.* 4:353-367.
35. Cullen, W.R., G.E. Eigendorf, B.U. Nwata and A. Takatsu. 1990. The quantitation of butyltin and cyclohexyltin compounds in the marine environment of British Columbia. *Appl. Organomet. Chem.* 4:581-590.
36. Byrd, J.T. and M. Andreae. 1982. Tin and methyltin species in seawater: concentrations and fluxes. *Science* 218:565-569.
37. Seligman, P.F., J.G. Grovhoug, A.O. Valkirs, P.M. Stang, R. Francham, M.O. Stallard, B. Davidson and R.F. Lee. 1989. Distribution and fate of tributyltins in the U.S. marine environment. *Appl. Organomet. Chem.* 3:31-47.
38. Schebek, L., M. Andreae and H.J. Tobschall. 1991. Methyl- and butyltin compounds in water and sediments of the Rhine River. *Environ. Sci. Technol.* 25:871-878.
39. Fent, K. and J. Hunn. 1991. Phenyltins in water, sediment and biota of freshwater marinas. *Environ. Sci. Technol.* 25:956-963.
40. Fent, K. and M.D. Muller. 1991. Occurrence of organotins in municipal wastewater and sewage sludge and behavior in a treatment plant. *Environ. Sci. Technol.* 25:489-493.
41. Cooney, J.J. and S. Wuertz. 1989. Toxic effects of tin compounds on microorganisms. *J. Industrial Microbiol.* 4:375-402.
42. Blunden, S.J., P.Harston, R. Hill and J. Wardell. 1990. Disproportionation of triorganotin species, R_3SnX. *Appl. Organomet. Chem.* 4:383-385.
43. Thompson, J.A.J., M.G. Sheffer, R.C. Pierce, Y.K. Chau, J.J. Cooney, W.R. Cullen and J.R. Maguire. 1985. *Organotin compounds in the aquatic environment.* NRCC Assoc. Comm. Scientific Criteria Environ. Qual., Pub. No. NRCC 22494, National Research Council of Canada, Ottawa, Ontario, 284 pp.
44. Hall, L.W., Jr. and A.E. Pinkney. 1985. Acute and sublethal effects of organotin compounds on aquatic biota: an interpretative literature evaluation. *CRC Crit. Rev. Toxicol.* 14:159-209.
45. Beaumont, A.R. and M.D. Budd. 1984. High mortality of the larvae of the common mussel at low concentrations of tributyltin. *Mar. Pollut. Bull.* 15:402-405.
46. Fent, K. 1991. Bioconcentration and elimination of tributyltin chloride by embryos and larvae of minnows *Phoxinus phoxinus. Aquat. Tox.* 20:147-158.
47. Beaumont, A.R. and M.D. Newman. 1986. Low levels of tributyltin reduce growth of marine microalgae. *Mar. Pollut. Bull.* 17:457-461.
48. U'ren, S.C. 1983. Acute toxicity of bis(tributyltin) oxide to a marine copepod. *Mar. Pollut. Bull.* 14:303-306.
49. Anonymous. 1986. *Oceans 86 Proceedings,* vol. 4, IEEE Service Center, Piscataway, NJ.
50. Anonymous. 1987. *Oceans 87 Proceedings,* vol. 4, IEEE Service Center, Piscataway, NJ.

51. Oehlmann, J., E. Stroben and P. Fiorini. 1991. The morphological expression of imposex in *Nucella lapilus* (Linnaeus) (Gastropoda:Muricidae). *J. Moll. Stud.* 57:375-390.
52 Spooner, N., P.E. Gibbs, G.W. Bryan and L.J. Good. 1991. The effect of tributyltin upon steroid titres in the female dogwhelk, *Nucella lapillus,* and the development of imposex. *Mar. Environ. Res.* 32:37-49.
52a. Lee, R.F. 1991. Metabolism of tributyltin by marine animals and possible linkages to effects. *Mar. Environ. Res.* 32:29-35.
53. Ellis, D.V. 1991. New dangerous chemicals in the environment: lessons from TBT. *Mar. Pollut. Bull.* 22:8-10.
54. Aldridge, W.N. 1986. The toxicology and biological properties of organotin compounds. pp. 245-261, In: N.F. Cardarelli (Ed.) *Tin as a vital nutrient,* FL.
55. Selwyn, M.J. 1976. Triorganotin compounds as ionophores and inhibitors of ion translocating ATPases. *Adv. Chem. Ser.* 157:204-226.
56. Singh, A.P. and K. Singh. 1985. Differential sensitivity to TBT of cytochrome-containing and cytochrome-deficient cells of *E. coli* SASX76. *Experientia* 41:764-767.
57. Gray, B.H., M. Porvaznik, C. Flemming and L.H. Lee. 1987. Tri-*n*-butyltin: a membrane toxicant. *Toxicology* 47:35-54.
58. Laurence, O.S., J.J. Cooney and G.M. Gadd. 1989. Toxicity of organotins towards the marine yeast *Debaryomyces hansenii. Microb. Ecol.* 17:275-285.
59. Belay, N., B.S. Rajagopal and L. Daniels. 1990. Effects of alkyltin compounds on hydrogen-oxidizing anaerobic bacteria. *Curr. Microbiol.* 20:329-334.
60. Boopathy, R. and L. Daniels. 1991. Pattern of organotin inhibition of methanogenic bacteria. *Appl. Environ. Microbiol.* 57:1189-1193.
61. Doolittle, M.M. and J.J. Cooney. 1992. Inactivation of bacteriophage T4 by organic and inorganic tin compounds. *J. Indust. Microbiol.* 10:221-228.
62. Hallas, L.E., J.S. Thayer and J.J. Cooney. 1982. Factors affecting the toxic effect of tin on estuarine microorganisms. *Appl. Environ. Microbiol.* 44:193-197.
63. Fent, K. and J.J. Stegeman. 1991. Effects of tributyltin chloride in vitro on the hepatic microsomal monooxygenase system in the fish *Stenotomus chrysops. Aquat. Tox.* 20:159-168.
64. Davies, A. and P.J. Smith. 1980. Recent advances in organotin chemistry. *Adv. Inorg. Chem. Radiochem.* 23:1-77.
65. Blunden, S.J. and C.J. Evans. 1985. *Organotin compounds.* Publication 711, Internat. Tin Res. Inst. Oxbridge, Middlesex, UK. 44 pp.
66. Seinen, W. and M.I. Willems. 1976. Toxicity of organotin compounds. I. Atrophy of thymus and thymus-dependent lymphoid tissue in rats red $DOTC_{12}$. *Toxicol. Appl. Pharmacol.* 35:63-75.
67. Casida, J.E., E.C. Kimmel, B. Holm and G. Widmark. 1971. Oxidative dealkylation of tetra-, tri- and dialkyltins and tetra- and trialkylleads by liver microsomes. *Acta Chem. Scand.* 25:1497-1499.
68. Levine, S.N., D.T. Rudnick, J.R. Kelly, R.D. Morton, L.A. Buttel and K.A. Carr. 1990. Pollutant dynamics as influenced by seagrass beds: experiments with tributyltin in *Thalassia* microcosms. *Mar. Environ. Res.* 30:297-322.

69. Barug, D. 1981. Microbial degradation of bis(tributyltin) oxide. *Chemosphere* 10:1145-1154.
70. Orsler, R.J. and G.E. Holland. 1982. Degradation of tributyltin oxide by fungal cultures. *Int. Biodeterior. Bull.* 18:95-98.
71. Lee, R.F. 1985. Metabolism of tributyltin oxide by crabs, oysters and fish. *Mar. Environ. Res.* 17:145-148.
72. Kimmel, E.C., R.H. Fish and J.E. Casida. 1977. Bioorganotin chemistry. Metabolism of organotin compounds in microsomal monooxygenase systems and in mammals. *J. Agric. Food Chem.* 25:1-9.
73. Seligman, P.F., R.J. Maguire, R.F. Lee, K.R. Hinga, A.O. Valkirs and P.M. Stang. 1991. Persistence and fate of tributyltin in aquatic ecosystems. In: M. Champ and P.F. Seligman (Ed.) *Organotins: environmental fates and effects.* Elsevier Science Publishing, Inc., London.
74. Maguire, R.J. and R.K. Tacz. 1985. Degradation of tri-*n*-butyltin species in water and sediment from Toronto Harbor. *J. Agric. Food Chem.* 33:947-953.
75. Lee, R.F., A.O. Valkirs and P.F. Seligman. 1989. Importance of microalgae in the biodegradation of TBT in estuarine waters. *Environ. Sci. Technol.* 23:1515-1518.
76. Meyers-Schulte, K.J. and C.A. Dooley. 1990. ^{124}Sn as a tracer of tributyltin degradation in seawater. *Mar. Chem.* 39:339-354.
77. Maguire, R.J. and R.K. Tacz. 1987. Concentration of tributyltin in the surface microlayer of natural waters. *Water Pollut. Res.* J. 22:227-233.
78. Hall, L.W., M.J. Lenkevich, W.S. Hall, A.E. Pinkney and S.J. Bushong. 1987. Evaluation of butyltin compounds in Maryland waters of Chesapeake Bay. *Mar. Pollut. Bull.* 18:78-83.
79. deMora, S.J., N.G. King and M.C. Miller. 1989. Tributyltin and total tin in sediments: profiles and the apparent rate of TBT degradation. *Environ. Technol. Lett.* 10:901-908.
80. Adelman, D., K.R. Hinga and M.E.Q. Pilson. 1990. Biogeochemistry of butyltins in an enclosed marine ecosystem. *Environ. Sci. Technol.* 24:1027-1032.
81. Brinckman, F.E. and W.P. Iverson. 1975. Chemical and bacterial cycling of heavy metals in the estuarine system. pp. 314-342 In: T.M. Church (Ed.) *Marine chemistry in the coastal environment,* Amer. Chem. Soc. Symp. Ser. vol. 18 Amer. Chem. Soc., Washington, DC.
82. Chau, Y.K., P.T.S. Wong, O. Kramer and G.A. Bengert. 1981. Methylation of tin in the aquatic environment. *Proc. Third Int. Conf. Heavy Metals in the Environ.* pp. 641-644, CEP Consultants, Ltd., Edinburgh.
83. Cooney, J.J., L.E. Hallas and J.C. Means. 1981. Tin and microbes in the Chesapeake Bay, USA. *Proc. Third Int. Conf. Heavy Metals in the Environ.* pp. 413-482, CEP Consultants, Ltd. Edinburgh.
84. Ghanem, N.A., M.M. Ald El-Maleh and M.H. Ayeub. 1980. New effective combined-mechanism antifouling coating for fouling rich regions. *Seventh Int. Symp. Controlled Release Bioactive Mat.,* Fort Lauderdale, FL, Controlled Release Soc., Akron, OH.

Biological Diversity: Problems and Challenges. Edited by S.K. Majumdar, F.J. Brenner, J.E. Lovich, J.F. Schalles, and E.W. Miller.

Chapter Twelve

GREENHOUSE WARMING: Impacts on Biodiversity

MIGUEL F. ACEVEDO
Department of Geography and Institute of Applied Sciences
University of North Texas
Denton, TX 76203

INTRODUCTION

Understanding the effects of climate on biological diversity has a long and varied scientific research history. Studies on the latitudinal gradients in species richness, the relationships of climate and vegetation distribution, and changes in species diversity during past climate cycles, such as species extinctions and expansion and contraction of geographical ranges, are examples of the recognition given to the role of climate in shaping biological diversity.

In this chapter we will explore the potential impacts of future climate changes on biological diversity, specifically those induced by the greenhouse effect. Potential impacts range from changing patterns of community, landscape and regional diversity, to extinction of endangered species. The analysis focuses on the potential impacts on the diversity of terrestrial vegetation and fauna. Similar concepts could be applied to freshwater and marine ecosystems, but these systems are not discussed here (for example, see Ray et al., 1992 and Alexander, 1992). Indirect effects of climate change are likely for these systems; for example, alteration of coastal habitat from sea level rise and water quality changes induced by climate change. (Smith and Tirpak, 1989; Harte et al., 1992).

For the purposes of estimating impacts, it is useful to separate the direct effects of the build-up of greenhouse gases and of the resultant thermal, precipitation and radiation changes from those indirectly induced by climate change, such as habitat alterations due to a rise in sea level. This chapter will concentrate on direct effects of changes in temperature, precipitation and radiation. Assessing the indirect effects requires translation of the climate changes into environmental stresses. Direct effects of atmospheric build-up of greenhouse gases will only be discussed in regard to potential direct impacts of increased CO_2 concentration. An overview of these effects is found in the "Overview of Effects of Climate Change on Biodiversity" section.

Current concerns about loss of biodiversity due to global changes induced by anthropogenic activities deal mostly with the impacts of rapid deforestation, habitat destruction and fragmentation occurring in high diversity areas, such as the tropical rain forest (e.g. Wilson, 1988; Terborgh, 1992). Effects of climate change would combine with the effects of habitat fragmentation and alteration. Many assessments suggest that climate change will induce further loss of diversity (Smith and Tirpak, 1989; Myers, 1992). Migration barriers introduced by development would retard or impede migration responses to climate change. Forest reserves have been established assuming that current climate patterns will persist and therefore further climatic stress on diversity is likely (Peters, 1988, 1992).

To discuss the consequences of future climate change on diversity, expected scenarios of temperature, precipitation and radiation changes have to be made explicit. The use of General Circulation Models (GCM) for specification of scenarios is discussed in the "Climate Change Scenarios" section, along with a description of the IPCC scenarios (Houghton et al 1990, 1992). Making explicit the scale and measure to examine diversity can help to reduce ambiguity in the assessment of climate change effects. Diversity at regional, landscape and local scales may respond differently to climate change. Species richness is selected in this chapter as the measure of diversity. Details are given in the "Scales and Metric" section.

Once the scenarios of climate change are made explicit, the sensitivity of species richness to climate at different scales is needed to extrapolate diversity changes due to climate change. The existing empirical relations between climate and diversity are examined in the "Climate-Diversity Relationships" section. Finally, "Assessing Potential Impacts on Diversity" combines the scenarios of "Climate Change Scenarios" and the relations given in section "Climate-Diversity Relationships" to discuss the potential consequences of climate change on biodiversity. It is not the purpose of this chapter to make a comprehensive analysis, rather it reviews the main concepts involved and proposes possible avenues for performing the assessments.

CLIMATE CHANGE SCENARIOS

Many studies show that increasing atmospheric concentrations of greenhouse gases— carbon dioxide, methane, chlorofluorocarbons, nitrous oxide and water vapor—will lead, on average, to increasing temperatures at the Earth's surface (for example see Houghton et al., 1990). These studies also suggest that precipitation patterns will change and that sea levels will rise. Some greenhouse gases are more effective than others at changing climate. Carbon dioxide accounts for more than half of the enhanced greenhouse effect.

Due to uncertainties about the future course of emissions, global climate change is usually estimated by scenario-consequence analysis; that is, by quantitative examination of the differential consequences of a set of explicit plausible assumptions. The quantitative tools most used are General Circulation Models (GCM). Uncertainties about the response of climate to the increase in atmospheric concentrations of greenhouse gases are accounted for by sensitivity analysis on the relevant parameters of the GCM's (Schneider, 1992; Schneider et al., 1992). Several sets of scenarios have been proposed and examined using GCM output (for some examples see Smith and Tirpak, 1989; Schneider, 1992 and Schneider et al., 1992).

The report of Group I of the Intergovernmental Panel on Climate Change (IPCC) (Houghton et al. 1990) proposes a set of four scenarios for reductions in emissions. The "no reduction" (business as usual) scenario estimates a rate of temperature increase of 0.3°C per decade during the next century, or approximately 3°C above current values by the end of next century. The uncertainty range of this prediction is 0.2 to 0.5°C per decade. Under this scenario, an average rate of global mean sea level rise of about 6 cm/decade during the next century is expected, due to thermal expansion of the oceans and the melting of some land ice. The other three IPCC scenarios include progressive levels of emissions control and estimate rates as low as 0.1°C/decade.

Six new IPCC emissions scenarios (Houghton et al. 1992) take into account new policies and recent political changes, and introduce these factors into the "business as usual" scenario. Incorporating radiative effects of sulfate aerosols, effects of CO_2 fertilization and feedback from stratospheric ozone depletion, reduces the estimate of global average temperature increase to 20-30% below the IPCC "business as usual" estimate (3.3°C) for the time period 1990-2100. These new estimates are more uncertain and still above the limits of natural variability. The warming rate is estimated to be five times the rate observed during the last century (Wigley and Raper, 1992).

Regional climate change scenarios are different from the global mean, and the uncertainty is larger (e.g., Schneider et al., 1992). For Central North America and Southern Europe, some models predict higher than the global mean temperature and reduced precipitation and soil moisture (Mitchell et al. 1990). Climate sensitivity to CO_2 increase is especially uncertain in the tropics (Rind et al. 1990). Generalizations, however, are often made with respect to latitude and it is commonly stated that climate change will be above the global mean in the higher latitudes, and below the mean in the intertropical regions. It is likely that geographic sensitivity to climatic change is more complex than a simple dependence on latitude (Moran, 1992).

GCM's simulate the globe with fairly coarse resolution (typical grid sizes vary from 8° x 10° x 4° x 5° and the smallest is approximately 1°), making it difficult to derive estimations on the regional and local scales, which are needed to understand the impacts on biological systems (Schneider et al., 1992). Despite their limitations, GCM's are the best existing tools for impact studies (Goodess and Palutikof, 1992). The combination of GCM output and synoptic climatology has been proposed as a tool to generate regional scenarios and has been tested in PAN EARTH case studies for China, Sub-Saharan Africa and Venezuela (Robock et al. 1992a).

For example, a combination of GCM output and local climatology suggest that doubled CO_2 scenarios for Venezuela include different temperature increments for the dry and rainy season. These scenarios also prescribe concomitant precipitation increase and radiation decrease, during the rainy season, for medium to high sensitivity of climate to greenhouse gases concentrations (Robock et al. 1992b). For the medium sensitivity scenario, the estimated changes for the wet season are: 2°C in the maximum temperature, 3°C in the minimum temperature, and a 20% increase in precipitation. The temperature increment for the dry season is 1°C higher than the rainy season temperature increment. Examples of generation of regional scenarios by combining GCM output and observational records are discussed by Schneider et al. (1992). In a recent application, Robinson (1992) estimated the number of days per month exceeding a given threshold for North Carolina, by using regression equations relating threshold crossing numbers to monthly average temperature, in conjunction with GCM estimates of future average temperature.

OVERVIEW OF EFFECTS OF CLIMATE CHANGE ON BIODIVERSITY

As mentioned in the introduction to this chapter, a complete assessment of greenhouse warming effects on biodiversity of vegetation and fauna should consider three important types of effects: (1) direct effects of increased CO_2 concentrations and other greenhouse gases, (2) direct effects of changes in temperature, precipitation and radiation induced by increasing atmospheric concentrations of greenhouse gases, and (3) indirect effects of other environmental changes caused by greenhouse warming, such as those induced by a rise in sea level. The relative importance of each one of these types of effects depends on multiple factors. It is difficult to determine the sensitivity of the particular vegetation and fauna examined to each one of the effects. The indirect effects can be more important in many circumstances (Smith and Tirpak, 1989; Franklin et al., 1992 and Harte et al., 1992).

The direct impacts of increased CO_2 concentration are due to ecophysiological responses. For example, vegetation displays a differential photosynthetic response, C3 plants being more sensitive than C4 plants to increasing CO_2 concentrations, therefore determining relative densities of these plant groups. Another example is provided by Norby et al. (1992) who found that yellow-poplar trees (*Liriodendron tulipifera* L.) increase the leaf-level photosynthesis rate in enhanced concentrations of CO_2, but did not increase whole-plant carbon storage. Instead, carbon allocation patterns are changed, producing fewer leaves and more fine roots in an enriched CO_2 atmosphere. These changes represent a potential increase in water-use efficiency, which suggests further interactions with the direct responses to climate changes.

The direct effects of climate change are due to physiological responses to thermal, radiative and precipitation changes. For example, transition times among phenological phases will depend on accumulated degree-days, completion of sexual cycle depends on water stress and accumulated degree days, productivity decreases

decreases with water stress, etc. Responses of animals to elevated temperature and its potential consequences for diversity are examined by Dawson (1992), Tracy (1992) and Rubenstein (1992), and a particular example of developmental response to rainfall patterns is analyzed by Murphy and Weiss (1992).

Alternative species responses to climate change are adaptation, dispersal and migration. It is often emphasized that changes in the variance of climate parameters, rather than the mean, will be more important in affecting species distributions (e.g. Graham, 1992; Peters, 1992). Extrapolation of species level responses to ecosystem effects is difficult (Smith and Tirpak, 1989). Incomplete knowledge exists for the relationship of species patterns to ecosystem patterns. The number of species is greater than the number of roles at each trophic level, which seems to imply redundancy (Solbrig, 1992).

The indirect effects of climate change on biodiversity constitute responses to changes in environmental parameters influenced by climate. Changes in water level, salinity, pH, habitat, food availability, trophic relations and disturbance regimes are examples (Smith and Tirpak, 1989; Franklin et al., 1992; Harte et al., 1992). As already mentioned in the introduction, this chapter is concerned only with the direct effects of climate change. Issues related to the direct effects of greenhouse gases and indirect effects of climate change are not examined further.

SCALES AND METRICS

Defining spatial scale is important to understanding patterns of biodiversity. Local, patch, landscape, region, island and continental scales are important (Solbrig, 1992). For example, the concept of *alpha or community or within-habitat* diversity applies to the local scale, that of *gamma or between communities* to the landscape scale and that of *epsilon* to a broader geographic area or regional scale. *Beta* and *delta* diversities are used as rates of change of alpha and gamma diversity, respectively, along gradients (Whittaker, 1977). Most papers only deal with alpha, beta and gamma diversities.

Making explicit the scale of discussion helps in understanding the effects of climate change on biodiversity. For example, alpha or community diversity changes as the presence and proportion of each species is affected by differential responses to the same climate variables when subject to the same environmental constraints of soil, slope, etc. Landscape or gamma diversity will be affected by changes in the presence of some species in each of the habitats considered within the landscape. Variability of climate and environmental constraints within the landscape will differentially affect the unit or habitats of the landscape. At a regional scale, spatial changes in boundaries of biomes are important. Beta diversity patterns change as alpha diversity is differentially affected along the gradient, changing rapidly when it approaches ecotones. Climate change can affect beta diversity by changing boundaries and by changing communities within boundaries (Neilson, 1990).

A considerable number of diversity metrics or measures exist (see for example Magurran, 1988). In this chapter, we will concentrate on species richness. The use of more complex or complete measures to assess the potential effects of climate change on diversity are beyond the current understanding of the climate-diversity relationships.

CLIMATE-DIVERSITY RELATIONSHIPS

Single-factor explanations for patterns or gradients in species diversity have been inadequate (Solbrig 1992, Colwell 1992). The combination of climate with biogeography and contingencies, disturbance, immigration, and competition is likely involved in determining these patterns (Colwell, 1992). Factors have differential effects according to scale and consequently the underlying relationships to climate may be different for each scale.

Regional Scale

Currie (1991) has reported empirical relationships of regional scale species richness to climate variables in North America. At grid cells of 2.5° x 2.5° (south of 50°N), climate variables representing the common hypothesis of environmental determinants of diversity are correlated with species richness. Mean annual temperature, precipitation, potential evapotranspiration (PET) and actual evapotranspiration (AET), total solar radiation, elevation, seasonality of temperature and precipitation, historical factors (represented by the grid cell being glaciated or not in the past) and annual net primary productivity (NPP) were used as environmental variables. Species richness was derived from the overlap of species geographical distributions maps on each one of the grid cells.

These empirical relations (Currie, 1991) show mid latitude peaks for birds, mammals and amphibians, but monotonic decrease with latitude for trees and reptiles. Tree species richness is correlated with most of the factors, but especially with AET. Vertebrate species richness shows correlations with all climate variables, but especially with PET, solar radiation and mean annual temperature. All vertebrate groups show a saturating relationship with PET. The results indicate that more than 90% of regional species richness can be explained by environmental factors alone, excluding biotic interactions. Keeping PET constant, elevation and AET explain some differences. For example, bird, mammal and tree richness increase with elevation, but for amphibian and reptile richness decrease. Again holding PET constant, richness in birds and mammals decreases for higher AET, but increases for reptiles and amphibians. When more resolution is used for the taxa, by examining orders and families, the relationships with PET are not monotonic but show peaks. The historical effects are not very important, which suggests that richness could have reached the limits imposed by climate since the last glacial period.

The results obtained by Currie (1991) are consistent with the hypothesis that energy availability determines diversity, as originated by Hutchinson (1959) and studied by several researchers during the sixties and seventies (see references in Brown, 1981), and during the last decade by Wright (1983), Abramsky and Rosenzweig (1984), Turner et al. (1987) and Currie and Paquin (1987), among others.

Primary productivity would set maximum values for plant diversity and the best correlate to NPP on the large scale is AET. Richerson and Lum (1980) found plant richness in California to be related to temperature and precipitation (see also Westman and Malanson, 1992), the two main factors in AET. The results given in Currie (1991) seem to indicate that animal richness is related to PET as a measure of ambient-energy levels.

At the regional scale, the relation between species number and productivity can be seen as the relation between the saturation level of species-area curves and NPP (Lieth, 1975). Several methods for NPP determination can be used; for example Lieth's methods, which are based on temperature and precipitation as limiting factors, and the Chikugo model (Uchijima and Seino, 1988), which makes NPP linearly dependent on net radiation and the relative dryness index. A common way of relating vegetation and climate at large scales is to construct classification schemes; for example, the ratio PET/precipitation is used in the Holdridge scheme, assuming PET is proportional to temperature.

The mean latitudinal geographical range of many species shows a positive correlation (Stevens, 1989; France, 1992) to latitude. Stevens (1989) calls this relation the Rapoport rule and suggests that exposure to a larger annual range of climatic conditions in higher latitudes has favored evolution of broader climatic tolerances as latitude increases. This rule is used by Stevens (1989) to suggest that a narrow range of climatic tolerances at low latitudes will make "mass effects" (Shmida and Wilson, 1985) important in determining diversity and therefore help to explain latitudinal gradients in species diversity.

Environmental factors such as temperature and precipitation are important in shaping biogeographic ranges of plant species (e.g., Webb, 1992) and plant distribution according to physiognomy and structure (Woodward, 1992). On a broad scale, the ranges are limited by the maximum allowable ratio of sustained metabolic rate and basal metabolic rate (Parsons, 1991).

Other empirical data show positive correlations of vegetation diversity (measured by family richness) with increasing minimum temperature in North and Central America, as well as with increasing precipitation (mediated by the leaf area index, LAI) in Africa (Woodward, 1992).

Landscape and Local Scales

The energy hypothesis does not seem to completely explain landscape and local patterns. Superimposed on the regional patterns is the variability on the landscape and local scale (among and within habitats), which suggests that species richness is controlled by different factors according to scale (Currie, 1991): climate for the regional scale, habitat heterogeneity for the landscape scale, and biotic interactions for the local scale. Currie's results can be seen as "capacity rules," but landscape and local patterns would require "allocation rules" (Brown, 1981), based on other factors such as body size (Brown and Nicoletto, 1991) and biotic interactions.

Assuming similar climate conditions for the whole landscape, topographic and edaphic heterogeneity could determine differences in microclimate, soil moisture and other habitat variables among patches or facets, which will represent differences in PET and AET at this scale. Therefore, spatial variation in energy could determine diversity patterns at the landscape scale. For plant richness AET will integrate water availability and temperature. Faunal richness would be dependent on habitat variability and dynamics (Seagle and Shugart, 1985), and since disturbance would play a role in driving the mosaic dynamics, richness may depend on the disturbance patterns.

At a local scale, topographic and edaphic factors operate together with biotic interactions to determine diversity patterns. Mass effect and ecological equivalency factors (Schmida and Wilson, 1985) would be important at this scale. Climate control of demographics in some populations (Cawker 1980) indicates that climate also plays an important role at this scale. Community diversity can be related to large scale environmental variables, such as climate, by correlating the local species richness from sites at geographical locations with different climatic conditions (for example see Gentry, 1982, 1988). Comparison at landscape and local scales requires estimation of the relation of species to area, since saturation cannot be assumed a *priori* for these scales. An example of the importance of including area is the comparison presented by Gentry (1982) for plant species richness in tropical forests.

ASSESSING POTENTIAL IMPACTS ON DIVERSITY

Changes in Diversity Patterns

The potential use of diversity-PET relationships (Currie, 1991) to infer potential effects of climate change on diversity at a large regional scale requires assessments of the changes in PET and AET as a function of the greenhouse effect scenarios. However, current GCM soil hydrology is not adequate to rely on as a guide to changes in PET and AET. Rind et al. (1990) found a large increase in PET for the U.S. during the next century using transient runs and doubled CO_2 runs with the Goddard Institute of Space Studies (GISS) GCM. Global precipitation cannot keep up with increased water demand over land. Lower than expected values for AET/PET are obtained in GCMs for the control run, due to the lack of a vegetated surface; these low values for AET/PET keep AET reasonable given the high values for PET obtained with GISS. Rind et al. (1990) suggest that the supply-demand index of precipitation minus PET will decrease for the U.S. during the next century, due to this large increase in PET. The GCM results imply that either soil moisture or vegetation biomass will decrease, suggesting that a decrement of plant diversity is likely in areas where PET increase cannot be compensated by increasing precipitation. Even in areas at high latitudes where large precipitation increases are expected, forests may be very sensitive to PET changes (Bonan et al. 1990).

Shifts in lifezones using the Holdridge scheme and four GCM's show that regions at mid-latitudes in the Northern Hemisphere have a tendency for drying and therefore forested areas may shift to shrublands and grasslands. However, results for tropical areas vary among GCMs. The United Kingdom Meteorological Office (UKMO) GCM shows extensive drying in the tropical forested areas, suggesting transition from mesic (sum of moist, wet and rain lifezones) forests to woodland. But the Oregon State University (OSU) GCM shows expansion of tropical forests. Most of the expansion in tropical forests is accounted for by dry tropical forests (Smith et al., 1992). Northern shifts of Holdridge life-zone boundaries and transition to subtropical moist forest are reported by Harris and Cropper (1992) for Florida.

Shifts to dryer life-zones are consistent with a suggested decrease in diversity due to decreasing AET, which in turn will be produced by higher PET and lower water availability. In these arguments, the possible feedback effects of transpiration reduction in a CO_2 enriched atmosphere are not included (see the "Overview of Effects of Climate Change on Biodiversity" section). For a consideration of these effects, see Woodward (1992).

Changes in vegetation distribution according to scenarios of climate change are estimated by Woodward (1992), using the relationships of plant physiognomy and structure to temperature and precipitation.

At the landscape and local scale, assessment of the impacts is more uncertain due to the lack of resolution of the GCM's at those scales (see the "Climate Change Scenarios" section) and the lack of single relationships linking diversity to climate factors. Inferences about local impacts are possible by using community level diversity changes on large scale environmental and geographical gradients. An example will be discussed in the *Effects on High Diversity Tropical Forests* section. Evaluation of forest composition changes and of habitat alterations by means of forest simulators have been shown to be a viable approach (Shugart and Smith, 1992; Botkin and Nisbet, 1992). Changes in species richness can also be inferred from area reduction and species-area curves, given that this relationship is robust (Murphy and Weiss, 1992).

Adaptation and Migration Responses

Adaptation responses will depend on the rate of climate change (Smith and Tirpak, 1989). Relatively low values of stress will induce adaptation responses if energy is available, but higher levels of stress will likely induce migration responses. The effects of climate change on diversity could be intensified at the boundaries of species distributions and at ecotones. Individuals at distribution range boundaries typically have environmental stress levels exceeding those of more central areas (Parsons, 1991), and consequently they should be more sensitive to an increase of stress.

Higher evolutionary change is possible at an intermediate distance from the center of a geographical range, where climatic stress is present, but sufficient metabolic energy can accommodate for change. At species boundaries, high environmental stress leads to phenotypic variability, but a low metabolic energy requirement adaptation precludes range expansion into more stressful habitats (Parsons, 1991).

The fast rate at which future climate change is expected to occur suggests that adaptation responses may not compensate species loss (Peters, 1992; Webb, 1992; Dawson, 1992). Rapid evolutionary change is possible in slowly varying environmental regimes with narrow intensity but stasis is shown in rapidly varying regimes (Sheldon, 1990). Species with narrow niche requirements could either remain static or go extinct as the environment shifts, but more generalized species could persist and respond to selection (Riclefs and Latham, 1992).

A common prediction for species response is the shifting of geographical ranges northward and to higher elevations (Peters, 1992; Davis and Zabinsky, 1992). According to the current scenarios for a 3°C/ century rate of temperature increase (see "Climate Change Scenarios" section), several Km/yr of northward migration or several m/yr in elevation are needed to keep pace with shifting isotherms (Peters, 1988, 1992; Adams and Woodward, 1992; Smith and Tirpak, 1989). However, paleo-records indicate rates of 0.2-0.4 Km/yr for most tree species (Smith and Tirpak, 1989; Davis and Zabinsky, 1992). There is debate on whether migration lagged behind climate in rapid warming phases of the early Holocene.

Important differences also occur between species (Adams and Woodward, 1992). Differential rates of shifting will introduce new biological interactions due to different plant and animal association in a given area (Ford, 1982; Peters, 1988, 1992). Biological factors as determinants of diversity will then play an important role in determining further changes in diversity. Communities and ecosystems are not expected to shift as units, but will change structure as a consequence of alteration in species distributions (Webb, 1992; Graham, 1992; Smith and Tirpak, 1989).

The potential rates of migration may not be attainable due to barriers imposed by nature and by development. Man-made barriers may block migration and exacerbate species losses (Smith and Tirpak, 1989; Peters, 1992). Climate change adds further pressure on species in reserves, which have stress due to isolation and size. Furthermore, reserves may not be at an appropriate site for future climatic conditions (Peters, 1988, 1992), which indicates that these conditions should be taken into account when designing and planning reserves (Davis and Zabinsky, 1992; Harris and Cropper, 1992). Extinction will occur when the contraction rate in one direction exceeds the expansion rate in the favorable areas; for example, species extinction during the rapid warming of the early Holocene (Adams and Woodward, 1992; Graham, 1992).

Population expansion does not necessarily occur after migration, since once a site is colonized, population growth depends on multiple factors. Doubling times of 158 years have been reported for both Britain and Queensland during postglacial periods in the early Holocene. It has been suggested that temperate tree floras will lag behind climate shift (Adams and Woodward, 1992).

Effects on High Diversity Tropical Forests

Since tropical rain forests contain more than half of the total world biota (Wilson, 1988), this section will examine with more detail some of the ideas presented in previous sections, as they may apply to tropical rain forests.

Inferences on potential direct effects of enhanced atmospheric concentrations of greenhouse gases on the tropical forests would require more experimental data (Medina, 1991) on the effects of CO_2 enrichment and a better understanding of the interactions of trace greenhouse gases and tropical forests (Vitousek and Matson, 1992); therefore, no attempt to suggest impacts of these direct effects will be made here.

Explanations for the generation and maintenance of tropical diversity have been numerous. Some of these have emphasized the role of climate, while others have relied on habitat heterogeneity, disturbance and biotic interactions. Contraction and expansion of geographical ranges during past climatic cycles has led to the postulation of refugia in Amazonia where plant species concentrated during dry periods correlated with glacial periods. The location of Pleistocene refugia have been challenged due to the bias introduced by sampling density correlated with postulated refugia; in this manner, rare species have not yet been sampled enough to demonstrate range discontinuities (Nelson et al. 1990). Colinvaux (1989) observed that not enough evidence exists for arid conditions in the Amazon during glacial periods, and therefore questions the refugial hypothesis that forest areas were confined to higher elevations with more rainfall.

The postulated location of refugia on table top mountains (tepuis) of Venezuela is based on the current abundance of endemism of the flora (Steyermark, 1979). However, measurements of the age of peat bogs in these mountains (Schubert, 1986), indicate arid conditions during the late Pleistocene and early Holocene. Substantial botanical collection has been made at these areas, but data for plant diversity at the community level is scarce (Acevedo, 1989).

Even though latitudinal gradients in diversity have received much attention, diversity variability within the tropics has been less studied (Gentry, 1982, 1988). Strong positive correlation between annual precipitation and neotropical plant species richness was suggested by Gentry (1982), using dry, moist and wet sites and richness data from 0.1ha plots. Plant diversity is strongly correlated with annual precipitation, but large tree (>10cm) richness does not show a significant correlation with annual precipitation, possibly due to inadequate sample size. Structural, floristic, dispersal and pollination strategies also show a dependence on precipitation. Wind-dispersal decreases with precipitation, whereas mammal and bird dispersal increases with precipitation (Gentry, 1982).

Biotic interactions seem to play a role in determining the number of species per genera, for epiphytes and shrubs. In this last case, a partitioning of the birds as dispersal agents indicates biotic effects in establishing diversity. This also happens among the lower number of species of *Piper* in dry forests. Low seasonality of wetter sites may allow for biotic tuning to occur. Sensitivity of different synusia to changes in precipitation differs: epiphytes increase most rapidly with precipitation, trees least rapidly, and lianas and shrubs have an intermediate sensitivity. Parameters for the log-log species-area curve are also dependent on precipitation (Gentry 1982).

Extension of these analyses to more sites in the neotropics plus incorporation of paleotropical sites has shown (Gentry, 1988) that the simple relationship of richness to total annual precipitation is more complicated. It seems to be valid in the neotropics where annual precipitation is correlated with strength of the dry season, but not in the paleotropics. An evenly distributed rainfall in one site in the neotropics makes diversity higher than would be expected from the low annual precipitation. A saturation of the richness *vs* precipitation curve is obtained when the extreme wet sites are included. Many species have habitat preferences, demonstrating the potential for habitat spatial heterogeneity to determine landscape diversity. More recently, Gentry (1990) proposed that habitat specialization and biotic interactions are major determinants of tree-species diversity in Amazonia.

Measures of energy and other environmental variables are not included in Gentry's work. Precipitation gradients will be good indicators of AET gradients if PET remains relatively constant. The relationship of productivity to precipitation is similar to AET (Lieth, 1975). The effects of temperature on tropical forests are not available and obtaining experimental data is likely to be difficult (Medina, 1991). Under climate change scenarios, the impact on neotropical plant diversity could be assessed by its relationship with total precipitation, if the rainfall seasonality preserves a correlation with annual precipitation. Hartshorn (1992) predicts that changes in total precipitation should not have a major impact on diversity, but that changes in rainfall seasonality could affect diversity patterns in humid tropical forests. Such impacts will be due to immeditate effects of seasonal changes on phenology.

Current GCM output predicts a precipitation increase in northern South America (Robock et al. 1992b), but the GCM results for changes in the rainy season pattern are uncertain. Even if the slope of the correlation line between diversity and precipitation were preserved, the increase in temperature or PET (Rind et al. 1990) could reduce water availability, potentially lowering the number of species per mm of precipitation. Species more sensitive to change in precipitation, such as epiphytes and shrubs, could potentially account for diversity changes.

The large scale distribution of vegetation in southern Venezuela was analyzed as a function of several climatic variables (Fariñas et al. 1990). Mean annual temperature and precipitation explain a large fraction of the variance; the axis strongly correlated with precipitation separates forests from savannas. Introduction of additional factors representing seasonal variability of precipitation allows for separation of different forest types (Fariñas et al. 1990). Using the scenarios of climate change proposed for Venezuela (Robock et al. 1992b), an examination of the potential changes in distribution suggests a tendency for forests to increase in savanna areas (Fariñas et al. 1990).

Distribution of NPP in Venezuela, calculated using the Lieth (1975) protocol, and its potential changes for the scenarios of climate change (Robock et al. 1992b) indicate a potential increase in NPP (work in progress in the PAN EARTH Project by C. Maytin and M. Rodriguez). The balance of PET/precipitation is not explicitly considered in these analysis.

Geographical range shifts are possible due to shifts in rainfall patterns, but little is known about potential migration rates and population expansion rates for tropical trees (Adams and Woodward, 1992).

CONCLUSIONS

A comprehensive assessment of the potential impacts of climate change on biodiversity would require unambiguous relationships of climate and diversity at different scales as well as plausible scenarios of climate change due to anthropogenic enhancement of the greenhouse effect. The state-of-the-science for both requirements is limited in at least several respects:

- uncertainties about sensitivity of climate to build-up of atmospheric concentrations of greenhouse gases,
- inadequate GCM soil hydrology for the purposes of estimating AET changes,
- the coarse resolution of GCM's makes difficult the specification of climate change scenarios for assessment at the landscape and local scale,
- empirical relationships between climate and diversity are available for large scale diversity patterns, but other factors related to topography, soils and biotic interactions play an important role at the landscape and local scale,
- not enough information is available, especially from the intertropical areas, to reliable infer adaptative and migration responses to climate change,
- geographical ranges patterns from paleorecords are still being debated and information is scarce for high diversity tropical areas.

Under these limitations, few broad inferences can be made. Large scale predictions indicate a potential reduction of plant species richness, due to a potential decrease of water availability in areas where precipitation will not compensate for increasing PET. A concomitant decline of faunal richness is possible, caused by the reduction of habitat due to the loss of plant species richness. However, empirical relations showing that fuanal richness increases with PET will have to be considered in the analysis. Landscape and local scale predictions are more uncertain. Those landscapes in areas where water availability is a limiting factor would seem to be more sensitive.

For tropical forests, there is even more uncertainty about diversity impacts from climate change, due to higher uncertainty about PET and precipitation changes. Effects on plant richness could be more rapid for epiphytes and shrubs. The wettest tropical forests where the species richness curve shows asymptotic behavior with precipitation would seem less sensitive.

REFERENCES

Abramsky, Z. and M.L. Rosenzweig. 1984. Tilman's predicted productivity-diversity relationship shown by desert rodents. *Nature* 309:150-151.

Acevedo, M.F. 1989. Diversidad de plantas en los cerros Marahuaka, Huachamacare y Duida. *Acta Terramaris* 1:21-32.

Adams, J.M. and F.I. Woodward. 1992. The past as a key to the future: the use of palaeoenvironmental understanding to predict the effects of Man on the Biosphere, pp. 257-314. In: F.I. Woodward (Ed.) *The Ecological Consequences of Global Climate Change.* Advances in Ecological Research Vol. 22 Academic Press, pp. 337.

Alexander V. 1992. Arctic marine ecosystems, pp. 221-232. In: R.L. Peters and T.E. Lovejoy (Eds.) *Global Warming and Biological Diversity.* Yale University Press, pp. 386.

Bonan, G.B., H.H. Sugart and D.L. Urban. 1990. The sensitivity of some high-latitude boreal forests to climatic parameters. *Clim. Change* 16:9-29.

Botkin, D.B. and R.A. Nisbet. 1992. Projecting the effects of climate change on biological diversity in forests, pp. 277-293. In: R.L. Peters and T.E. Lovejoy (Eds.) *Global Warming and Biological Diversity.* Yale University Press, pp. 386.

Brown, J.H. 1981. Two decades of homage to Santa Rosalia: toward a general theory of diversity. *Am. Zool.* 21:877-888.

Brown, J.H. and P.F. Nicoletto. 1991. Spatial scaling of species composition: body masses of north american land mammals. *Am. Nat.* 138(6):1478-1512.

Cawker, K. 1980. Evidence of climatic control from population age structure of *Artemisia tridentata* Nutt. in southern British Columbia. *J. Biogeogr.* 7:237-248.

Colinvaux, P. 1989. Ice-age Amazon revisited. *Nature* 340:188-189.

Colwell, R.K. 1992. Making sense of ecological complexity: a personal and conceptual retrospective. *Biotropica* 24(2b):226-232.

Currie, D.J. 1991. Energy and large-scale patterns of animal- and plant- species richness. *Am. Nat.* 137:27-49.

Currie, D.J. and V. Paquin. 1987. Large-scale biogeographic patterns of species richness of trees. *Nature.* 329:326-327.

Davis, M.B. and C. Zabinsky. 1992. Changes in geographical range resulting from greenhouse warming: effects on biodiversity in forests, pp 297-308. In: R.L. Peters and T.E. Lovejoy (Eds.) *Global Warming and Biological Diversity.* Yale University Press, pp. 386.

Dawson, W.R. 1992. Physiological responses of animals to higher temperatures, pp. 158-170. In: R.L. Peters and T.E. Lovejoy (Eds.) *Global Warming and Biological Diversity.* Yale University Press, pp. 396.

Fariñas M., R. Jaimez, R. Andressen and M. Acevedo. 1990. Análisis de la relaciones entre variables climáticas y tipos de vegetación en treinta localidades de la Guayana Venezolana. Manuscript.

Ford, M.J. 1982. *The Changing Climate: Responses of the Natural Fauna and Flora.* London George Allen and Unwin, pp. 190.

France R. 1992. The North American latitudinal gradient in species richness and geographical range of freshwater crayfish and amphipods. *Am. Nat.* 139:342-354.

Franklin, J.F., F.J. Swanson, M.E. Harmon, D.A. Perry, T.A. Spies, V.H. Dale, A. McKee, W.K. Ferrell, J.E. Means, S.V. Gregory, J.D. Lattin, T.D. Schowalter and D. Larsen. 1992. Effects of Global Climatic Change on Forests in Northwestern North America, pp. 244-257. In: R.L. Peters and T.E. Lovejoy (Eds.) *Global Warming and Biological Diversity.* Yale University Press, pp. 386.

Gentry, A.H. 1982. Patterns of neotropical plant-species diversity. *Evol. Biol.* 15:1-85. Plenum Press.

Gentry, A.H. 1988. Changes in plant community diversity and floristic composition on environmental and geographical gradients. *Ann. Mo. Bot. Gard.* 75(1):1-34.

Gentry, A.H. 1990. Floristic similarities and differences between southern Central America and upper and central Amazonia, pp. 141-157. In: A.H. Gentry (Ed.) *Four Neotropical Rainforests.* Yale University Press, pp. 627.

Goodess, C.M. and J.P. Palutikof. 1992. The development of regional climate scenarios and the ecological impact of greenhouse gas warming, pp. 33-62. In: F.I. Woodward (Ed.) *The Ecological Consequences of Global Climate Change.* Advances in Ecological Research Vol. 22 Academic Press. pp. 337.

Graham, R. 1992. Late Pleistocene faunal changes as a guide to understanding effects of greenhouse warming on the mammalian fauna of North America, pp. 76-87. In: R.L. Peters and T.E. Lovejoy (Eds.) *Global Warming and Biological Diversity.* Yale University Press, pp. 386.

Harris, L.D. and W.P. Cropper, Jr. 1992. Between the devil and the deep blue sea: implications of climate change for Florida's fauna, pp. 309-324. In: R.L. Peters and T.E. Lovejoy (Eds.) *Global Warming and Biological Diversity.* Yale University Press, pp. 386.

Harte, J., M. Torn and D. Jensen. 1992. The nature and consequences of indirect linkages between climate change and biological diversity, pp. 325-343. In: R.L. Peters and T.E. Lovejoy (Eds.) *Global Warming and Biological Diversity.* Yale University Press, pp. 386.

Hartshorn, G.S. 1992. Possible effects of global warming on the biological diversity in tropical forests, pp. 137-146. In: R.L. Peters and T.E. Lovejoy (Eds.) *Global Warming and Biological Diversity.* Yale University Press, pp. 386.

Houghton, J.T., G.J. Jenkins and J.J. Ephraums. 1990. *Climate Change. The IPCC Scientific Assessment.* Report from Working Group I .WMO-UNEP. Cambridge Univ. Press. pp. 365.

Houghton, J.T., B.A. Callander and S.K. Varney. 1992. *Climate Change 1992. The Supplementary Report to the IPCC Scientific Assessment.* Cambridge University Press. pp. 200.

Hutchinson, G.E. 1959. Homage to Santa Rosalia, or why are there so many kinds of animals? *Am. Nat.* 93:145-159.

Lieth, H. 1975. Some Prospects beyond production measurement, pp. 286-304. In: H. Lieth and R. Whittaker (Eds.) *Primary Productivity of the Biosphere.* Springer Verlag, pp. 339.

Lieth, H. 1975. Modeling the primary productivity of the world, pp. 238-263.

Magurran, A.E. 1988. *Ecological Diversity and its Measurement.* Princeton University Press, pp. 179.

Medina, E. 1991. Deforestation in the tropics: evaluation of experiences in the Amazon basin focusing on atmosphere-forest interactions, pp. 23-43. In: H. Mooney et al. (Eds.), *Ecosystems Experiments.* SCOPE. Wiley.

Mitchell, J.F.B., S. Manabe, V. Meleshko and T. Tokioka. 1990. Equilibrium climate change and its implications for the future, pp. 131-172. In: J.T. Houghton, G.J. Jenkins and J.J. Ephraums (Eds.) *Climate Change, The IPCC Scientific Assessment.* WHO/UNEP. Cambridge University Press. pp. 365.

Moran, J.M. 1992. The climatic future and lessons of the climatic past, pp. 379-391. In: D.A. Dunnette and R.J. O'Brien (Eds.) *The Science of Global Change: The Impact of Human activities on the Environment.* Ame. Chem. Soc. Symposium Series 483. pp. 498.

Refer to page 453 for remainder of references.

Biological Diversity: Problems and Challenges. Edited by S.K. Majumdar, F.J. Brenner, J.E. Lovich, J.F. Schalles, and E.W. Miller.

Chapter Thirteen

BIODIVERSITY AT RISK: Deforestation

MARK S. BOYCE*

Centre for Ecological Sciences
and
Centre for Theoretical Studies
Indian Institute of Science
Bangalore 560 012
India

INTRODUCTION

Recent scientific and popular literature is replete with alarming data on the drastic changes in the world's biota due in large part to increasing tropical deforestation. It would be redundant to detail current deforestation rates, and few data exist beyond those reviewed by Sayer and Whitmore (1991) and Myers (1992). Estimates vary depending upon the source of data and definitions for what constitutes deforestation (Sayer and Whitmore 1991). Recent applications of remote sensing have greatly enhanced the accuracy of estimates, but data are consistently daunting (Nair 1991, Dirzo and Garcia 1992). It is sufficient to say that each year approximately 1.8% of the world's remaining tropical forests are being cut or burned, approximately twice the area deforested annually ten years ago.

Because moist tropical forests probably host more than one half of the earth's species (Reid and Miller 1989), the biodiversity consequences of deforestation are astounding, with an estimated two million species projected for extinction during the next 30-50 years (National Research Council 1992). From a strictly utilitarian

*present address: Department of Zoology & Physiology, University of Wyoming, Box 3166, Laramie, WY 82071-3166.

perspective, we stand to lose opportunities for developing new medicines, industrial materials, and genetically disease-resistant strains of domestic organisms (Myers 1992, Oldfield 1989). If these consequences were not sufficient, an estimated 30% of our accumulating atmospheric CO_2 can be attributed to deforestation activities (Myers 1992), not to mention the values in other ecosystem services such as water storage and erosion prevention. Other arguments have been raised that biological communities afford benefits which transcend values of individual species (Gee 1992), but these may be too esoteric to convince developing countries (Boyce 1991). Nevertheless, given the irreversible consequences of extinctions, one may easily argue that deforestation is *the* most serious environmental issue on earth today.

At stake is the sanctity of our planet. The obstacle which we face is the very Darwinian basis for our existence, and its imprint on our social and economic systems. The driving forces behind deforestation are the fundamental afflictions of the developing world including rapid population growth, oppressive poverty, desire for development, agricultural inefficiency, politics, and economic pressures which can often be traced to the developed world.

The world's conservation community is making distressingly little progress at curbing these devastating rates of deforestation, and I fear our approaches are often misdirected, counterproductive, or ineffective. We dwell on issues of the appropriate design for parks and nature preserves (Soulé and Simberloff 1986), yet in developing countries parks and preserves are repeatedly shown to be ineffective and to alienate local people (Brockelman 1990, Alcorn 1991). We focus energies and squander conservation resources for the preservation of ancient forests in the Pacific Northwest (Boyce and Irwin 1990) yet this increases demand for tropical timber where consequences to biodiversity are much more serious (Sprugel 1990). And we invest in detailed population viability analyses for single species (Burgman et al. 1988, Boyce 1992) when it is clear that such efforts are hopeless given the two million extinctions projected over the next 30-50 years (National Research Council 1992).

I will discuss how new economic incentives and policies are probably the most urgent needs in our mission to deflect the current trajectory in tropical deforestation. Regrettably, ecologists seldom have the tools or knowledge to contribute in such arenas. But we can contribute to the design of much-needed conservation programs which are both economically and environmentally sound. For example, research on wise-use programs and restoration ecology can foster improved resource management in developing countries which can help immensely in our efforts to protect global biodiversity.

TARGETING THE DEFORESTATION CRISIS

The deforestation crisis is a global issue, although deforestation in most developed countries does not seriously threaten biodiversity. In some areas, such as the northeastern United States, the period of exploitative deforestation has already past and a greater area is forested now than a century ago. Furthermore, temperate forests have many fewer species than tropical forests, and therefore the consequences to biodiversity attributable to deforestation are much less in these areas. Although there is heated controversy regarding logging of old-growth forests in the Pacific Northwest of the United States (Boyce and Irwin 1990), the ramifications to biodiversity pale in comparison to the effects of deforestation in tropical developing countries of the world.

Resources for conservation are limited, therefore it makes sense to direct a high fraction of these resources to regions of the world where extinction rates are projected to be the highest. Regions of highest diversity and endemism clearly should be of greatest interest and concern, and many areas of special interest are revealed by the global maps depicted by Reid and Miller (1989). For example, Colombia, Peru, and Ecuador contain one sixth of the world's diversity of flowering plants (National Research Council 1992), and these are clearly areas where conservation efforts may have dramatic implications for preserving genetic diversity. Previous and projected patterns of development also help to establish priority areas for conservation. For example, areas valuable for agriculture are often the most heavily developed and therefore possess the smallest remnants of pre-development biota.

Myers (1992) conducted a review of areas of high diversity where concern over deforestation is most urgent, identifying several "hot spots." In alphabetical order these are: (1) northeastern Borneo, (2) the Atlantic-coast forest of Brazil, (3) the Choco strip of Pacific-coast forest of Columbia, (4) western Ecuador, (5) peninsular Malaysia, (9) New Caledonia, (10) Philippines, (11) montane forests of Tanzania, (12) Western Amazonia, and (13) the Western Ghats of India. Although these targeted regions constitute only 0.2% of the earth's land surface, collectively they contain over 37,000 endemic plant species and over 750,000 endemic animal species.

There may be danger in focusing attention only on moist tropical forests because there are many other diverse communities which may be even more threatened by development. For example, native grassland and savannah types in many parts of the world have been totally usurped by agriculture. In peninsular India, moist tropical forests are relatively secure in parks and sanctuaries, whereas undisturbed shrub/grassland habitats are exceedingly rare and have virtually disappeared. Some specific examples include the dry Albizia amara zone of India, where Tungabhadra, Point Calimere, and Rannebennur need immediate protection or restoration, and the Acacia-Capparis scrub only represented in Velavadar National Park (Gadgil and Meher-Homji 1990), and the dry tropical forests of Costa Rica (Janzen 1988).

To seek solutions to the extinction crisis, we must identify root causes. It is commonly observed that population growth and concomitant pressures on the land are ultimately to blame. Indeed, it is clear that there exists an inverse correlation between population density and proportion of an area in forests (Palo 1987). But it seems misguided to dismiss conservation priorities in deference to human population pressures (Radulovich 1990); rather, the existence of such human demands on limited resources means that it is imperative to link conservation to development! Some of the exemplary conservation programs in India and Japan show that biodiversity can be sustained even in areas of extraordinary human density. It would be outrageous to imply that human population growth is not one of the most serious environmental issues facing our planet. But perhaps even more urgent and more easily rectified is the manner in which current development policies and economic practices foster deforestation and consequent threats to biodiversity (Lélé 1991).

Governments of many countries possessing rich forest resources have economic incentives which accelerate the rate of deforestation (Repetto and Gillis 1988). Perhaps the most serious are those which encourage shifting cultivation practices (Myers 1992). Opportunity exists to use incentives to reverse the effect (McNeely 1988). One approach may be more economically productive as forests than for other land uses (Peters et al. 1989, Tobias and Mendelshon 1990, Balick and Mendelsohn 1992). Such economic arguments carry considerable weight in developing countries where conservation may be viewed as a luxury. Conservation programs have the greatest chance for success when they are both economically and environmentally sound (Silver and DeFries 1990).

WISE USE VERSUS PRESERVATION

How does one design conservation programs that accomplish biodiversity preservation and are still consistent with development objectives? The "Western" approach to nature conservation is to establish protected areas in the form of parks and preserves. But in tropical regions with high human populations, such parks and preserves may be ineffective (Brockelman 1990). It is difficult, if not unethical to cavalierly impose protected area boundaries in localities where people's traditional livelihoods depend upon resources extracted from the forest. Such approaches invariably lead to conflicts which may thwart conservation efforts (Dang 1991).

Moreover, such conflicts may not be necessary if certain forms of resource use can be accommodated without seriously jeopardizing the conservation effectiveness of an area. Forests offer resources which can be harvested sustainably while maintaining or even enhancing forest diversity. Conservation programs permitting such harvests thus meet the criterion of being both environmentally and economically sound. Managing forests to provide sustainable supplies of fuelwood is better than encouraging use of non-renewable energy resources of coal and oil; likewise, renewable wood resources make more ecological sense than use of metal and plastic. Managing forests for fuelwood production in India would appear to be wiser than encouraging cattle dung production, which might indirectly encourage increases in cattle numbers to meet demands for dung for fuel and fertilizer. (This argument does not imply, however, that more efficient use of dung should not be encouraged, e.g., by generating gobar gas [Ghorpade 1991]). Indeed, I believe that notions such as "wise use" and "sustainable development," although perhaps shibboleths (Lele 1991), are essential for the future effectiveness of global conservation.

Many types of resource exploitation are sustainable, and can take place whilst maintaining relatively natural landscapes. Selective timber harvest, hunting for meat and hides, fishing, collection of non-timber forest products, fuelwood collection, etc. all can take place without threatening biodiversity. Advocating such "wise use" affords benefits at several levels, and can be exactly the sort of management approach which will meet with the satisfaction and support of the local human population (Brockelman 1990). All such resource use must be monitored and regulated to prevent overuse. Poaching is difficult to factor into a resource management program, whereas hunting can be regulated very effectively. Likewise, stealing timber and other products from the forest would be actively discouraged by locals if they knew that they would have a future stake in the resource. This would not necessarily prevent individuals from taking advantage of opportunities for selfish gain, but it seems likely that there would be better monitoring and regulation if local populations had the incentive (Dang 1991).

Development planners consistently undervalue the economic significance of local use of wild products which may never make it to market, e.g., vegetables, meat (Malhotra et al. 1983), hides, fibers, bamboos, canes, grasses, medicines, spices, seeds for oil and propagation, gums and resins, dyes, honey and wax, and wood. The value of these items can be higher than that of timber harvest or conversion to pasture or agriculture. Over 50 years, harvesting fruit and latex in one forest in Peru would produce twice as much money as either cattle ranching or conversion to timber production (Peters et al. 1989). In India, minor forest products account for 75% of net export earnings from forest products (Miller et al. 1992).

To develop sagacious resource management programs, we need to understand the consequences of different types of resource extraction and use in the forests. Indeed, developing exploitation strategies for resources while retaining biodiversity is a complex new arena in resource management. But there is great potential. For example, various logging practices can be used to simulate natural disturbance patterns in ways which may have minimal consequences to forest structure (Hartshorn 1989). Likewise, there is a need to evaluate the efficacy of various sustainable use options, e.g. consequences of collecting non-wood forest products, hunting, etc., in the context of biodiversity preservation. Although limited resource extraction may be compatible with forest protection and preservation of biodiversity, such harvests are not without costs. For example in French Guiana selective logging at the level of 3 trees ha^{-1} has been shown to cause significant reductions in avian diversity (Thiollay 1992). Certain spatial patterns of logging can degrade the forest by fragmenting various forest types (Nair 1991). Yet, logging can be used to increase the spatial heterogeneity of landscapes by creating a variety of seral stages and thereby increasing overall species diversity. Historically, our approach to resource management often targeted efficiency and maximum sustained yield. We may expect quite different management prescriptions now that sustaining biodiversity is a further objective.

All landscapes on earth are subject to natural disturbance regimens, and given sufficient time and resources virtually all communities can be restored so long as constituent species have not gone extinct. Restoration ecology is an arena in which ecologists may play an exceedingly important role in the future, because much research is needed to develop methods for reclamation and afforestation in tropical regions. Restoration ecology presents great opportunities (Jordan et al. 1988).

For example, in countries with extensive areas of degraded lands, such as India and Pakistan, afforestation technology desperately needs attention. Returning these lands to productive forests can function to preserve biodiversity, maintain ecosystem function, as well as divert pressures away from national parks and other protected areas. Because there exists such a large area of degraded land, the effects of afforestation on atmospheric CO_2 may be significant. In India, for example, economic rationale can easily justify returning in excess of 20 million ha to forest cover (Vohra 1991). If this area could be placed into productive forest (see Myers 1992) this could remove some 100-200 million metric tons of CO_2 from the earth's atmosphere every year! In all fairness, developed countries should consider reducing CO_2 emissions by 100-200 million metric tons to balance 20 million ha of deforested landscape.

GLOBAL PERSPECTIVES

Development agencies can play a major role in shaping economic programs for developing nations which can rein the trajectory of tropical deforestation (National Research Council 1992). Although it is natural to target developing countries because they host the greatest wealth of biodiversity, the economic forces that affect biodiversity are often transnational. Indeed, the world's global economy is largely shaped by developed countries, and high rates of resource consumption, especially in developed nations, play a major role in deforestation. Unfortunately, the relationship between these transnational forces and natural resource exploitation are poorly understood (National Research Council 1992). It is fundamental that we must gain global perspective in the formulation of conservation strategies.

Of the biomes on earth, those which are best protected by parks and preserves paradoxically tend to be among those with the least diversity. For example, the Arctic has extensive parks, especially in Alaska. Yet Alaskan fauna and flora is remarkable for its low diversity and broad geographic distributions, and there are few if any species protected against extinction by this extensive park system. Rather than biotic preservation, the perceived function of most of these parks and preserves is largely to prevent energy and mineral development and to protect the pristine wilderness character of these remote areas. Indeed, protection of oil and mineral reserves for national security was one of the compelling arguments offered in the late 1970s by President Carter's administration for the need to create such long-term preserves of energy resources in Alaska.

From a global perspective the consequences of these Alaskan parks is to divert extraction of energy and mineral resources to other parts of the globe. So while the depauperate biota of Alaska is protected, developed countries turn to developing countries to meet the demand for these energy and mineral resources. For example, the second largest oil-producing nation for US imports is Nigeria, a country hosting diverse moist tropical forests. We would expect that the biodiversity consequences of oil development would be far more serious in Nigeria than in Alaska.

But this analysis is too simplistic. What if the United States stopped buying oil from Nigeria? It would not necessarily have any positive ramifications to conservation. Indeed, without the oil revenues coming into Nigeria, there may be greater demand on biological resources from intensification of forest and agricultural practices. Understanding such complex economic issues is essential to be able to predict the outcome of conservation policy at international levels.

A simpler case may be illustrated by the large set-aside for the northern spotted owl in the Pacific Northwest of the United States. By curtailing logging in old-growth forests, there has been a substantial reduction in timber marketed from northern California, Oregon and Washington (Sprugel 1990). Yet, demand for timber has not subsided and current tropical timber production is largely stimulated by demands from consumers in developed countries (Radulovich 1990). The tragic prospect is that such sustained demand for timber may be met by multi-national or foreign timber firms exploiting tropical forests where the consequences to biodiversity will be astronomically higher than in the Pacific Northwest (Sprugel 1990). Indeed, so far as we know, there are only a couple of taxa which are potentially threatened by old-growth logging in the Pacific Northwest. When conservation programs have substantial economic consequences, our failure to think globally may have serious consequences to biodiversity on a larger scale. We do not understand the economics of such transnational interrelationships well enough to predict the outcome, thus such topics clearly need to be focus for future research.

I presented these arguments regarding transnational consequences of Alaskan parks and old-growth for spotted owls to Dr. Thomas Lovejoy (Smithsonian Institution, pers. comm.), who replied that it was important that the United States set a good example for developing countries. If the United States cannot forego economic development for the sake of conservation, it will be impossible to expect greater commitment from developing countries where the need for development seems more urgent. However, I believe that the manner in which we are going about it is sending entirely the wrong message, i.e. we do not want to demonstrate that conservation involves a dichotomy between preservation and development. If developing countries are presented with such a choice, I suspect that conservation will fare poorly. It would be far better to demonstrate how we can accomplish development without jeopardizing biodiversity through careful planning and enforcement of regulations.

Another facet to the conservation consequences of global trade is raised in an unpublished draft of the *Global Biodiversity Strategy* (Miller et al. 1992): international trade in general may ultimately be detrimental because it leads to specialization of agriculture and consequently intensively managed monocultures. And from this develop all of the associated risks and consequences associated with intensive agriculture: soil depletion, changes in soil structure, top-soil erosion, pollution from agrochemicals, loss of traditional modes of agriculture, etc. If India, for example, were to maintain its currect policy of being largely self-sufficient, the country would maintain diversity of economy, agriculture, land use, etc. However, if such inefficiency requires that virtually all lands be put into production, I believe such self-reliance may be counter to global conservation strategy.

The United States currently has a policy that global trade barriers should be relaxed because of a belief that protectionist trade barriers damage both local and global economies. If the standard of living and per capita economy are any indication, there would appear to be strong support for such a contention, e.g., Japan, South Korea, Malaysia, Indonesia have dwarfed India in economic growth during the past two decades. Enhanced economies offer many advantages for conservation. For one, people who are more affluent can afford the "luxury" of conservation, and we may see more active programs to set aside lands for conservation practice. Furthermore, trade allows specialization of production so that regions of the world which are most efficient at producing a particular product will be able to capitalize on this production. This increases the efficiency of resource production from a given land base, and thereby reduces the total land which must be allocated to agricultural production. Such decrease in the demand on land for production can facilitate sound conservation measures on remaining lands. Unfortunately, there is not assurance that the economic system would stop with meeting basic human needs.

CONCLUSION

During the past decade, tropical deforestation has increased from 75,000 km^2 (1979) to 142,200 km^2 (1989) per year (Myers 1992), an increase of 89%. Instead of curbing tropical deforestation, efforts from developed countries to intervene in tropical countries, e.g., Brazil, has met hostile reaction to protect national sovereignty. Resolutions on biodiversity at the June 1992 United Nations Conference on Environment and Development in Rio de Janeiro, Brazil, failed to secure the support of the United States, signifying to many that the developed world was unwilling to sacrifice its economic stature for the sake of conservation. If this decade has been any indication, the prospects for the diversity of life on this planet appear bleak.

Despite these facts, there is reason to believe that global deforestation rates will decline (Myers 1992). A number of non-government organizations, including the Nature Conservancy, the World-Wide Fund for Nature, and the Rainforest Action Group, are aggressively pursuing tropical forest preservation in a number of places in the world, and heightening public awareness. The fact that some 150 nations were represented at the Rio Conference made a clear statement that governments are beginning to acknowledge the need to foster environmentally responsible development (Zuckerman 1992).

The message that there may be economic rewards to protecting genetic diversity has clearly reached developing countries as evidenced by the stance taken by G-77 nations at the Rio Conference. Not surprisingly, developing countries expect some economic return from their genetic resources. Developed countries are clearly reluctant to pay, yet expect economic protection for patents on how to make use of these genetic resources. It cannot go both ways. For pharmaceutical firms to explore the utility of genetic resources, their R&D investments must be protected by patents or other security for intellectual property rights. Similarly, royalties paid to developing countries for genetic resources may be an excellent way to offer incentives for forest protection.

Caveats are necessary here. Patents do not guarantee incentives for biodiversity preservation. Indeed, one can envisage that upon discovering a valuable biochemical product a pharmaceutical company may wish to destroy remaining natural stocks of a rare plant or animal to protect their pecuniary interests. Offering incentives for biodiversity preservation to local communities and tropical countries is an important strategy in biodiversity preservation, but it is necessary to develop mechanisms to ensure that such incentives are not counter-productive. Although agreement on this issue was not finalized in Rio, there are opportunities for attractive compromises. Clearly both the developed country technology and the developing country genetic resources are of value.

To be effective on a global scale, our approaches to nature conservation must change. Preservation is not the only way to protect biodiversity, and in many developing countries it is not a realistic approach. We have much to learn about interfacing resource use with conservation for biodiversity (Fitter 1986). It is worth reiterating the conclusion of the National Research Council that conservation programs have the greatest chance for success when they are both economically and environmentally sound (Silver and DeFries 1990).

ACKNOWLEDGEMENTS

I am grateful for support from the American Institute of Indian Studies, the U.S. Educational Foundation in India (Fulbright program), the Indian Institute of Science, the National Council for Air and Stream Improvement, the World Conservation Fund, and the College of Arts and Sciences and the Office of Research at the University of Wyoming. I thank Jerrold Dodd, Charles Fernandes, Madhav Gadgil, Larry Irwin, Tom Lovejoy, Evelyn Merrill and S.A. Mohnot for discussion.

REFERENCES

Alcorn, J.B. 1991. Ethics, econonics, and conservation, pp. 317-349. In: M.L. Oldfield and J.B. Alcorn (Ed.) *Biodiversity: Culture, Conservation, and Ecodevelopment.* Westview Press, Boulder, CO, pp. 349.

Balick, M.J. and R. Mendelsohn. 1992. Assessing the economic value of traditional medicines from tropical rain forests. *Conserv. Biol.* 6:128-130.

Boyce, M.S. 1991. Natural regulation or the control of nature?, pp. 183-208. In: R.B. Keiter and M.S. Boyce (Ed.) *The Greater Yellowstone Ecosystem: Redefining America's Wilderness Heritage.* Yale University Press, New Haven, CT, pp. 425.

Boyce, M.S. 1992. Population viability analysis. *Ann. Rev. Ecol. Syst.* 23:481-506.

Boyce, M.S. and L.L. Irwin. 1990. Viable populations of Spotted Owls for management of old growth forests in the Pacific Northwest, pp. 133-135. In: R.S. Mitchell, C.J. Sheviak and D.J. Leopold (Ed.), *Ecosystem Management: Rare Species and Significant Habitats.* Proc. 15th Ann. Natural Areas Conf., New York State Museum Bull. 471.

Brockelman, W.Y. 1990. Wildlife conservation in Thailand: A strategic assessment. pp. 300-304. In: J.C. Daniel and J.S. Serrao (Ed.) *Conservation in Developing*

Countries: Problems and Prospects. Oxford Univ. Press, Bombay, India, pp. 656.
Burgman, M.A., H.R. Akçakaya and S.S. Loew. 1988. The use of extinction models for species conservation. *Biol. Conserv.* 43:9-25.
Dang, H. 1991. *Human Conflict in Conservation. Protected Areas: The Indian Experience.* Har-Anand Publ. and Vikas Publ., New Dehli, India, pp. 255.
Dirzo, R. and M.C. Garcia. 1992. Rates of deforestation in Los Tuxtlas, a neotropical area in southeast Mexico. *Conserv. Biol.* 6:84-90.
Fitter, R. 1986. *Wildlife for Man: How and Why We Should Conserve Our Species.* Collins, London, U.K., pp. 223.
Gadgil, M. and V.M. Meher-Homji. 1990. Ecological diversity. pp. 175-198. In: J.C. Daniel and J.S. Serrao (Ed.) *Conservation in Developing Countries: Problems and Prospects.* Bombay Natural History Society and Oxford University Press, Bombay, India, pp. 656.
Gee, H. 1992. The objective case for conservation. *Nature* 357:639.
Ghorpade, M.Y. 1991. *Development Ethos and Experience.* Southern Economist Publ., Bangalore, India, pp. 281.
Hartshorn, G.S. 1989. Application of gap theory to tropical forest management: Natural regeneration on strip clear-cuts in the Peruvian Amazon. *Ecology* 70:567-569.
Janzen, D.H. 1988. Tropical dry forests: the most endangered major tropical ecosystem. pp. 130-137. In: E.O. Wilson and F.M. Peter (Ed.) *Biodiversity.* National Academy Press, Washington, D.C., pp. 521.
Jordan, W.R. III, R.L. Peters II and E.B. Allen. 1988. Ecological restoration as a strategy for conserving biological diversity. *Environ. Manage.* 12:55-72.
Lélé, S.M. 1991. Sustainable development: A critical review. *World Develop.* 19:607-621.
Malhotra, K.C., S.B. Khomme and M. Gadgil. 1983. Hunting strategies among three non-pastoral nomadic groups of Maharashtra. Man in India 63(1):21-39.
McNeely, J.A. 1988. *Economics and Biological Diversity: Developing and Using Economic Incentives to Conserve Biological Resources.* International Union for the Conservation of Nature, Gland, Switzerland, pp. 236.
Miller, K., et al. 1992. *Global Biodiversity Strategy.* World Resources Institute, The World Conservation Union, United Nations Environment Program, Washington, D.C., pp. 244.
Myers, N. 1992. *The Primary Source: Tropical Forests and Our Future.* W.W. Norton & Co., New York, NY, pp. 416.
Nair, S.C. 1991. *The Southern Western Ghats: A Biodiversity Conservation Plan.* Indian National Trust for Art and Cultural Heritage, New Dehli, pp. 92.
National Research Council. 1992. *Conserving Biodiversity: A Research Agenda for Development Agencies.* National Academy Press, Washington, D.C., pp. 127.
Oldfield, M.L. 1989. *The Value of Conserving Genetic Resources.* Sinauer, Sunderland, MA, pp. 379.
Palo, M. 1987. Deforestation perspectives for the tropics: A provisional theory with pilot applications, pp. 57-90. In: M. Kallio, D.P. Dykstra, C.S. Binkley (Ed.) *The Global Forest Sector: An Analytical Perspective.* J. Wiley, New York, NY, pp. 703.

Peters, C.P., A.H. Gentry and R.O. Mendelsohn. 1989. Valuation of an Amazonian rainforest. *Nature* 339:655-656.

Radulovich, R. 1990. A view on tropical deforestation. *Nature* 346:214.

Reid, W.V. and K.R. Miller. 1989. *Keeping Options Alive: The Scientific Basis for Conserving Biodiversity.* World Resources Institute, Washington, D.C., pp. 128.

Repetto, R. and M. Gillis. 1988. *Public Policies and the Misuse of Forest Resources.* Cambridge Univ. Press, Cambridge, U.K., pp. 432.

Sayer, J.A. and T.C. Whitmore. 1991. Tropical moist forests: Destruction and species extinction. *Biol. Conserv.* 55:199-213.

Silver, C.S. and R.S. DeFries. 1990. *One Earth, One Future: Our Changing Global Environment.* National Academy of Sciences, Washington, D.C., pp. 196.

Soulé, M.E. and D. Simberloff. 1986. What do genetics and ecology tell us about the design of nature reserves? *Biol. Conserv.* 35:19-40.

Sprugel, D.G. 1990. Ancient forests: Valuable data, questionable conclusions. *Conserv. Biol.* 4:461-462.

Thiollay, J.M. 1992. Influence of selective logging on bird species diversity in a Guianan rain forest. *Conserv. Biol.* 6:47-63.

Tobias, D. and R. Mendelsohn. 1990. The value of recreation in a tropical rainforest reserve. *Ambio* 20:91-93.

Vohra, B.B. 1991. Land and water. *Sanctuary Asia* 11(3):65-80.

Zuckerman, L. 1992. Between Stockholm and Rio. *Nature* 358:273-276.

Biological Diversity: Problems and Challenges. Edited by S.K. Majumdar, F.J. Brenner, J.E. Lovich, J.F. Schalles, and E.W. Miller. © 1994, The Pennsylvania Academy of Science.

Chapter Fourteen

MANAGING BIODIVERSITY: Conservation vs. Preservation

ROBERT L. BURGESS

Department of Environmental and Forest Biology
SUNY College of Environmental Science and Forestry
Syracuse, New York 13210

INTRODUCTION

Since the closing decades of the Nineteenth Century and the multiple activities of President Theodore Roosevelt, John Muir, Gifford Pinchot, John Wesley Powell, and others, there has been a strong and continuous movement in the United States to set aside areas of land (primarily) for education, scientific study, outdoor recreation, and aesthetic beauty. Beginning with Yellowstone National Park in 1872, the U.S. government has created a system of areas, of various sizes and configurations, to attempt to preserve (conserve) aspects of nature. Such conservation (preservation) measures have largely been successful, and we now take great pride in the National Park Service, the system of National Forests and Grasslands, a huge series of National Wildlife Refuges, a national system of Wilderness Areas, and numerous other kinds of areas, preserved (conserved) by a variety of agencies and organizations for myriad purposes. These are all bound together by a powerful underlying thread of concern for natural areas and the enhancement of the quality of life in America.

SOME DEFINITIONS

Initially, John Muir (and Thoreau, Gilbert White, and George Perkins Marsh) spoke (and wrote) of conservation as synonymous with preservation. Webster's

Unabridged Dictionary still treats them as synonyms. Conservation is "1. A conserving, preserving, guarding, or protecting; a keeping in a safe or entire state; preservation. 2. Official care or keeping and supervision, as of a river or forest; conservancy; also, a division, as of a forest, under such conservation." Preservation reads "1. Act or process of preserving, or keeping from injury or decay; state of being preserved, as *preservation* of life, fruit, game, etc.; a picture in good preservation. SYN. Safekeeping, conservation, saving."

During the administration of Teddy Roosevelt, conservation was officially defined as "wise use," a definition that persists and perhaps has led, either directly or indirectly, to many of the problems we face today. In particular, the wise use doctrine meant, in a very real and economic sense, that no areas were above the needs of the human population, for raw materials, food and fiber, water and wildlife, and energy sources, especially coal and oil, and later, uranium. National Parks have largely been spared from "wise use," while many national forests have been repeatedly assaulted. In addition, the doctrine of "multiple use," a touchstone for many land management agencies (e.g., Forest Service, Bureau of Land Management, Tennessee Valley Authority, and even the Army Corps of Engineers), came to mean that all areas must be eligible for all uses. To scientists, educators, and much of the lay public, it was obvious that in many cases, areas were best conserved through single-use management. Areas that were great for growing trees should be used for growing trees; areas that were suited for grazing cattle should be used for grazing cattle; and forested areas that protected the integrity of critical watersheds should be maintained, inviolate, in healthy forest vegetation.

Through much of the Twentieth Century, however, two significant and unrelenting pressures began to impact on "conservation." The human population kept growing at geometric rates, and the steady decline in the quality of the environment became noticeable to a majority of the American people. In 1969, a triumvirate of events triggered the first "Earth Day" in 1970. These were a massive oil spill in the Santa Barbara Channel off the coast of California, the reported death of western Lake Erie, and the burning of the Cuyahoga River in downtown Cleveland. Almost overnight, the mood of the nation changed, and environmental quality has been a major issue in the 23 years since 1970. Many prophets predicted that environmental issues would guide the election process, at national, state, and local levels. Thus far, this has not happened, although environmental concerns reappear at each election time.

GLOBAL PERSPECTIVES

In addition, environmental problems are now global in scope, and the American public is well aware of tropical deforestation, the ozone hole, acid precipitation, and the potential for global warming. Along with this cognizance, the public was introduced to a new term, "biodiversity," a short-hand coinage for biological diversity, the kinds of different organisms, communities, ecosystems, and landscapes that occur on Planet Earth. Biodiversity is what creates the rich and varied tapestry of the world's biota, and not coincidentally, results in the beautiful patterns of colors that appear on many maps of natural resources.

Conservation and preservation are ostensibly synonymous, but recent usage has confused the two terms, and in many cases has almost made them antithetical. The recent controversy over the northern spotted owl in the Pacific Northwest is a case in point. So-called conservationists contend that cutting of old-growth forest is "wise use," and damn the ecologists and environmentalists as "preservationists." Preservation has almost become a dirty word. Yet for something to be conserved, it often must be done through preservation. If the spotted owl is to remain, old-growth forest habitat is required. Such examples are often short-sighted, as economists and politicians look only to the next election. If the old-growth habitat is destroyed, there will be nothing for the next generation of our children to cut, and the question will be moot. Throughout the polemic, sacrifice of long range values is proposed as preferable, and short-term expediency overcomes. We have become a nation totally driven by economics, thinking that the frontier of Frederick Jackson Turner still exists, and that the biblical admonition to conquer and subdue the earth is still the mandate of humankind.

Eons ago, when human populations were small, the impact on natural resources was minimal. Now, however, with continued population growth, the demand for the world's natural resources is outstripping their supply. Hence, the continued controversy over "conservation" versus "preservation." The problem is largely one of semantics, but that it makes no less important or less critical for the survival of the planet and all of its inhabitants. In many parts of the world, damaged ecosystems may have already lost the capacity to recover to some semblance of long term productivity, and the science of ecosystem rehabilitation is still in its infancy.

Preservation is often interpreted as the "locking up forever" of usable lands, forests, wildlife, and water resources, while many ecologists view such preservation as simply maintaining our options for the future. What we destroy in the Twentieth Century will not be available for our grandchildren in the Twenty-first century, and the world, at some point, must pay attention to this basic fact. The world of the future may well be one with minimal forests, degraded grasslands, and drained wetlands, and one devoid of condors, cougars, whooping cranes, and rhinos.

The point that must be stressed is that while politicians and ecologists assail each other in the most pejorative terms, objectively there is only one choice. If **(IF)** we are to "conserve" some of the resources that many deem crucial for our continued existence, then some degree of "preservation" must be accepted.

A FEW EXAMPLES

During the latter half of the Nineteenth Century, the Adirondack Mountains lay stripped and raw, ravaged by timber harvest, the advent of the railroad, and widespread fires. In the 1880's, government officials in New York created the Adirondack Park, over 6,000,000 acres of protected land. Today, about half is in state ownership, part of the "New York Forest Preserve," while the rest is privately owned or is part of local government units such as hamlets and villages. The Constitutional Convention of 1894 proposed and ratified Article XIV of the New York State Constitution. This Article includes the famous "forever wild" clause, essentially PRESERVING, in perpetuity, the state lands of the Adirondack Park. For almost a century, despite repeated challenges, this constitutional article seems to offer the most iron-clad protection possible to the region and its natural resources.

The Article precludes the cutting of a tree in the Forest Preserve, under strict penalties. The concept is flawed in ecosystem terms, however, as fishing and hunting are allowed on state lands in the Forest Preserve. To point out that from an ecosystem perspective, shooting a white-tailed deer is no different than cutting a tree, continues to carry no argumentative weight. More importantly, the "forever wild" protection precludes management activities that somehow would mitigate against the original protection. Despite needs (such as fire suppression), the state and its agencies are effectively prevented from taking action,. It has been discussed many times that some degree of "management" may well be required in the Forest Preserve. If, as an example, scientists succeed in eliminating the spruce budworm from the northeastern forests, but the Forest Preserve still harbors a population from which surrounding lands in New England and Canada continually become reinfected, what will the state do? Similarly, it is not inconceivable that the Adirondack Park may yet be a final reservoir for the virus-carrying mosquitos that spread eastern equine encephalitis. If we succeed in eradicating this scourge elsewhere in the Northeast, will the state not control mosquito populations in the vast wetlands of the Adirondacks? These are questions of immense importance, as we continue to try to maintain biodiversity, preserve natural areas, conserve our natural resources, and provide a quality environment for the United States human population. There is still time to plan ahead, but only if governments and regulating agencies are cognizant of the critical nature of these impending decisions.

The Ecological Society of America (ESA) was incorporated in 1915, and one of its first official acts was to institute a Committee for the Preservation of Plant and Animal Communities. Under the leadership of Victor Ernest Shelford at the University of Illinois, the ESA began to inventory the remaining natural areas in the western hemisphere. The thrust was to eventually set aside (i.e., preserve) vestiges of the natural landscape for scientific, educational, and aesthetic purposes. By enlisting the cooperation of ecologists in many states (and/or regions), this committee succeeded in publishing a major compendium, *"Naturalists' Guide to the Americas"* (Shelford 1926) giving information on over a thousand "natural areas" that were candidates for some kind of "preservation." In my view, this is still the best work of its kind, although it is almost 70 years old, and many of the areas have long since been lost or compromised. On the other hand, many of the listed areas were eventually preserved to some degree.

In various forms, ESA committees continued to be active in the preservation arena, although the Society never could bring itself to purchase or otherwise acquire land for scientific purposes. Despite repeated attempts, Shelford failed to get ESA to become an active player. In 1946, the Society did, however, form (from its own membership), an organization called "The Ecologists Union." (Dexter 1978,

Croker 1991). This group was initially formed to do what the full Society was not prepared to undertake — namely, the identification and acquisition of "natural areas." The Ecologists Union was short lived, giving way in 1950 to the incorporation of The Nature Conservancy, the single most effective land acquisition organization the world has ever seen. Today, The Nature Conservancy, with over 650,000 members and multi-million dollar budgets, is operating worldwide on behalf of biodiversity.

National Forests were being delineated as early as 1905, under the administration of Theodore Roosevelt and with the constant urging of Gifford Pinchot. Yellowstone National Park had been established in 1872, to be followed shortly by Yosemite. Nevertheless, it was not until 1916 that the National Park Service was inaugurated and additional lands began to be set aside as parks and monuments. The process continues today, with the new Marsh-Billings National Historical Park in Vermont established in 1992, commemorating the wisdom, philosophy, and writings of George Perkins Marsh. In the early 1920's, under the leadership of Aldo Leopold, the Forest Service began to designate "Primitive Areas," forerunners, in truth, of today's wilderness areas. Through the 1950's, the Forest Service continued to identify such areas, using a variety of names.

ENVIRONMENTAL LEGISLATION

The Wilderness Act was passed in 1964, followed in 1969 by the National Environmental Policy Act (NEPA), and the Endangered Species Act in 1972. Each of these has been fraught with controversy, primarily (again) over the semantic distinctions between "conservation" and "preservation." The Wilderness Act permitted Congress to set aside large tracts of essentially roadless areas, predominantly in the west where such areas still existed. Original wilderness areas in the eastern states were virtually non-existent. Most of the areas were carved out of public lands, particularly National Parks, National Forests, and holdings of the Bureau of Land Management, a primary force west of the Mississippi. In most cases, arguments over boundaries were the rule, but often the underlying problems rested on the "locking up" of public land "forever." Most of this activity took place during a time when the term "biodiversity" was almost never heard, with no indication that it would become a part of headline jargon, a source for magazine editorials, and intermittent commentary on the network evening news.

Partly in response to increasing concern over the general loss of environmental quality worldwide and the continued evidence for massive extinction of species, a number of other conservation programs arose. These included the concept of Biosphere Reserves, under the auspices of the United Nations Man and the Biosphere (MAB) program. Internationally, over 100 areas in more than 40 nations are now designated as UN Biosphere Reserves and enjoy a measure of protection and continuity that would otherwise not be available. In 1976, the National Science Foundation instituted a program of Experimental Ecological Reserves, a designation to identify areas that already had a degree of protection and management, but which were amenable to certain experimental manipulations of both biota and environment that could elucidate complex ecosystem processes that were being impacted by the continued environmental insults of modern technological society. In addition, the U.S. Department of Energy established "National Environmental Research Parks" at several of its national laboratories, including those at Hanford, Washington, Savannah River, South Carolina, Oak Ridge, Tennessee, and Brookhaven National Laboratory on Long Island. These too, were a combination of some protection for the native flora and fauna coupled with the ability to perform necessary experiments that were either large-scale, long-term, or both.

The Wilderness Act gave an impetus to the preservation of natural diversity by providing large areas with a degree of protection heretofore unknown. Other programs also assisted in the identification and subsequent designation of "natural areas." However, major impact on the biodiversity question has sprung primarily from the Endangered Species Act. Almost overnight, the Act gave focus, reason, and funding to preserve individual species that were rare, threatened, or endangered. Like many other pieces of legislation, the Endangered Species Act was replete with bureaucratic hurdles. Species proposed for listing (as rare, threatened, or endangered) had to undergo extensive review by a suite of federal agencies, a process that took a long time. The early ones were relatively easy; the California Condor, the Whooping Crane, the Gray Wolf, the Florida Panther, and the American Alligator. Virtually all those involved knew, *a priori,* that each of these was, at the time, probably on the verge of extinction, and only overt action by the Federal Government could enhance their potential for survival. For some, it appeared to work. With a ban on hunting, the alligator rebounded beyond all expectations. Healthy populations of the wolf exist in some of the lower 48 states. The Whooping Crane status continues to improve, albeit slowly. On the other hand, the condor is now extinct in the wild, and only time will tell if the captive breeding program will succeed in reintroducing this magnificent bird to the mountains of California.

The Endangered Species Act generated thousands of searches for species that could (or should) be listed on Federal registers. Biologists with many organizations were put to work to ferret out "endangered species" that might then be offered some degree of protection. "Endangered species" appeared everywhere. Two problems were immediately apparent. Some species, rare in one state because of general range limits, were healthy and widespread in adjacent states. In addition, the taxonomy was often called into question. Populations that for years had been treated as subspecies (or varieties, or forms) were now proposed for specific status, so that they, too, would come under the aegis of the Endangered Species Act.

As early as 1973, the Endangered Species Act was being used to prevent, or at least forestall, Federal action. Discovery of the snail darter in the Tellico River held up completion of work on the Tellico Dam, a Tennessee Valley Authority project. While of questionable value, the Tellico Dam project included large areas for industrial park development where major national and international firms had already purchased land and pledged their occupancy. Consequently, it was unthinkable that the project would (or could) be stopped because of the discovery of a tiny, seemingly insignificant, little fish. Yet the mandates of the Endangered Species Act were clear. The Federal government could take no action that would threaten or endanger the continued existence of a living species.

However, at that time Howard Baker of Tennessee was the Senate Majority Leader, and he was convinced that a way could be found to complete the Tellico Dam despite the Endangered Species Act. In the end, the solution seemed almost too simple: a law was passed in Congress exempting the Tellico Dam project from the provisions of the Act. Once that was done, the project could move ahead to completion.

The dam was completed and the reservoir began filling in the early 1980's. Thousands of acres of farmland and forest were purchased as part of the Tellico project, and hundreds of families were displaced by TVA to provide for reservoir construction and industrial park development. As of this writing, not a single occupant of the industrial park is evident, but another valley has been drowned beneath a giant reservoir, despite the Endangered Species Act.

In many cases, scientists (and others) have continued to locate populations of species that were presumed (or known) to be rare or endangered. Given the appropriate program, it is sometimes possible to purchase lands surrounding their occurrence, usually under the guise of a "natural area." This is backwards, as an endangered species does not a natural area make. If we preserve natural areas, we may, in many situations, preserve one or more rare species. However, the conservation and preservation programs that are currently in place, both in the United States and throughout the world, should not be invoked to draw imaginary lines around tiny populations (or sometimes single occurrences) of rare species for purposes of preservation.

Extinction is a natural process, and has been going on since the beginning of life on earth. Throughout the history of evolution, species have arisen, flourished, and died out. Today, we are concerned not so much with the fact that species continue to become extinct, but with the incredible acceleration in the RATE of extinction, largely through human activity. Estimates vary, but many scientists believe that the world is losing between 100 and 1,000 major species each year. The point is that it is impossible to save every species that is threatened or in danger of becoming extinct. We could not save the moa or the dodo anymore than we could have saved the woolly rhinoceros or the sabre-toothed tiger. We have already spent millions of dollars on attempts to save the California condor, actions that may yet be significant for their futility. If biodiversity is important to the human species, then natural habitat must be PRESERVED in which communities of species can flourish. Piecemeal, single species approaches to the conservation of biotic diversity are largely doomed to failure.

EPILOGUE

Millions of people are concerned with preserving biodiversity. These range from scientists to engineers, from politicians to school children, all with a vision, sometimes unrecognized, that biodiversity is important to human well-being.

Except in rare cases, maintaining biodiversity means that habitats, communities, and ecosystems must be preserved in some semblance of working condition. We need a mix of options, where preserves are a part of a larger managed landscape. Thus, it should be possible to keep needed products and services in sustainable condition, including the richness of biological species. Preserving as much as we can also insures that the human species maintains its options for the future. As civilization as we know it seemingly rushes headlong toward oblivion, it may still be possible to cherish the biodiversity of life, and through conservation/preservation, find ways to live in harmony on one earth.

LITERATURE CITED

Croker, Robert A. 1991. Pioneer Ecologist. The life and work of Victor Ernest Shelford 1877-1968. Smithsonian Institution Press, Washington, D.C., 222 pp.

Dexter, Ralph A. 1978. History of the Ecologists' Union — spinoff from the ESA and prototype of The Nature Conservancy. Bull. Ecol. Soc. Amer. 59:146-147.

Shelford, Victor E. 1926. Naturalists' guide to the Americas. William and Wilkins Co., Baltimore, MD. 761 pp.

Biological Diversity: Problems and Challenges. Edited by S.K. Majumdar, F.J. Brenner, J.E. Lovich, J.F. Schalles, and E.W. Miller. © 1994, The Pennsylvania Academy of Science.

Chapter Fifteen

PROTECTION OF BIOLOGICAL DIVERSITY: Missing Connection Between Science and Management

CHRISTINE SCHONEWALD-COX

National Biological Survey
Division of Environmental Studies
Wickson Hall, University of California
Davis, CA 95616-8576

INTRODUCTION

The rate at which diversity disappears is difficult to show with the existing knowledge, management and political avenues we have now. However, there is one marked problem that can be addressed constructively. Science intended for application to conservation does not reach its destination sufficiently. The management destination does not often get what it wants from science. Science and management are often mutually ignorant of the status of knowledge and experience in the other's profession. All three of these difficulties are addressed in Schonewald-Cox et al. (1991) and Buechner et al. (1992) in which the authors identify this missing connection between science and management, from both perspectives. While communication gaps existing between professions are not new, this one is particularly troublesome. Knowledge of a communication gap exists, but its negative effect on protection of biological diversity is accepted out of default by managers as the researcher's fault or by researchers as the manager's "handicap." This causes at least two professions to work at counter purposes to one another despite their common goals. When this happens, it ultimately slows the protection process. Such slow and handicapped communication feeds back as a managerial distrust and frustration with science.

This paper focuses on two subjects, that are generally dismissed as unworthy for publication. During the last 15 years, however, I have found them to be major impediments to successful protection of biological diversity. First, methods of communication between science and management are largely incompatible. Second, there is a prevailing ignorance of the state of manager's institutional memory. This ignorance is shared by both science and successive generations of managers. Both are principal factors in the missing connection.

There are many professions that span from research to an implementation of policy in management. For practical purposes in this discussion I have lumped various types of science and management. The latter includes, for example, park superintendents and policy makers (local, regional, and broader). Not all managers are professional by training, and not all scientists conduct research. I have not made the distinction here although it should be done elsewhere. The missing connection really consists of several missing connections operating at different levels, in tandem. In combination they produce the effects I discuss, and they are focused where their combined effects influence conservation of biological diversity most clearly.

While individuals profess knowledge of ultimate causes for the missing connection(s). I have found few explanations that are at once objective, easily compartmentalized, and practically addressed. Information transfer relies at the very least on an ability to understand the generalities associated with both worlds' jargons. Particularly, several identical terms are used within and between professions but have variable meanings ("management" or "wildlife," for example). *Few profession-specific dictionaries exist.* In addition, some information transfer may require specialized training for managers. Information transfer also relies on searching for and finding of useable information for application. Scientists need a point of departure and information on the status of existing knowledge. But, what if it is not clear what knowledge exists except for what is found in the scientific literature? When scientists attempt to translate research for managers, they simplify the scientific terminology, and report their results in special management symposia and documents. The scientist assumes the manager can and may read the document, or that she/he can and should attend the symposium. Additional assumptions are made in the interpretation of language. That managers may depend on different media or have different styles for primary communication has seldom been compensated for by scientists. When institutions and individuals attempt to bridge gaps between research and management, they simplify communication, but retain their own profession's principal means of communication, and rely on their own assumptions regarding the status of knowledge for others.

I assume communication develops within professions and modifies with time (historically) to match the manner in which the professional does the work. While communication may not be efficient, some optimization occurs matching time and technology. Within professions such as science and management, communication functions as a major element defining working style. In science, the primary communication modes for most professionals have been journals, newsletters, books, conferences, seminars and symposia (mostly from scientist-to-scientist).

THE FIRST PROBLEM: COMMUNICATION BETWEEN SCIENCE AND MANAGEMENT

Science Communicates

Clear examples of slow information transfer are in the assimilation of knowledge into practical applications of genetics, long term monitoring, and biogeography to the conservation of biological diversity. (Slow is relative to rate at which biodiversity is lost.) For example, Sewall Wright's 1969, 1977 and 1978 works summarizing genetics of small populations were applied in the early 1980's to conservation research, and management a few years after (Soulé and Wilcox 1980, Schonewald-Cox et al. 1983, Soulé 1986, Soulé 1987a). Lasley's (1978) work on cattle genetics was put to use in the 1980's for zoo research and management (Ralls et al. 1979). A few years later the implications were carried further into recommendations for management of wild populations (Schonewald-Cox et al. 1983). The discussions of isolation and island biogeography (MacArthur and Wilson 1963) were applied to park design in the 1980's only to be miss-applied to the realities of park design and management (subject reviewed in Shaffer 1990). Much of what was published had little or no input from resources management documents, nor management's organizational memory. It is from landscape ecology that the influences of corridor and adjacent landscape quality are being used to describe how parks function on continents (Forman and Godron 1986; Schonewald-Cox and Bayless 1986, Turner and Gardner 1991). Now it is in the form of Hurd (1903), McMichael and Bingham (1923), and Alexander et al. (1977) that the spread of urbanization is being studied in conservation.

Public media such as newspapers, magazines and television have been used for communication (scientist-to-layperson), but there is little for communication from scientist to manager. Managers are exposed to public media to varying extents. As part of their work they attend training workshops and occasionally one ($\pm$) symposium led by scientists. Mostly, functions are within-profession. Few managers at decision making levels can pull away from their daily routine for more than one day at a time, let alone for more than one seasonal workshop. The symposia and workshops designed to attract mixed audiences undergo information bottlenecks due to who can participate (based on time constraints). The sampling process restricts information flow, due to emphasis of the workshop, and the few repeated attendances by managers. This is evident in the attendance records of a biennial symposium on research in California National Parks. Records for the last decade show only one chief manager out of about 40 attended each of three symposia. Managers may not perceive repeated attendance to be necessary. Regrettably, the communication passes without reaching most of the intended audience. The research and the journals reach few practitioners and decision makers among managers.

There are academic scientists whose research has direct value for conservation of biological diversity, but they are not aware of this, for lack of contact with the applied community. Also scientists involved in research who wish to offer advice useful to management are not necessarily the best people to discuss their own work with managers, either orally or in published guidelines. (The work is lost, effectively.) Further, many scientists consider it outside the domain of science to advise in decision making. Many managers agree with them.

A different view of linkage is required to bridge the missing connection between science and management. Application of science to management would benefit from a new view of what is needed to induce communication and subsequently application of science. Previous recognition of the need for such linkage is why translation workshops occur and why scientists do write articles translating their research results. Still, the traditional assumption has been that linkage is an individual responsibility, a product of personal sensitivity, commitment or part of a job description, tacked on after a list of primary duties.

Although outreach from researchers can help bridge the communication gap between researchers and managers, this is a limited, unstable and potentially abusive solution.

Management Communicates

The management systems of communication depend on telephone, reports (oral and written) and memoranda (3 or less pages preferably). There are also committee meetings, briefings and occasional training courses, usually to improve managerial and administrative skills. As the level of management responsibility increases, the quantity of materials to be examined increases. Length of daily material (memoranda, routine briefings, etc.) decreases and the media selected for communications change in relative importance to one another. The shift in communication style, moving up the professional ladder, decreases the time spent reading or listening to any but a few high priority reports. The reading may be the daily stock market or other varied reports, to keep up with land purchases and economics in lands next to the reserve or park. The assimilation of science at each of these levels is filtered within subject, increasingly, while the subject area coverage broadens.

Generally most managers, particularly those in upper level positions, do not have time within their daily schedules to read journals, magazines or even newsletters. Their time budgets are limited and schedules restricted to essential conferences, seminars or symposia, (manager-to-managers). The scopes of subject matter coverage within the managers' jobs are usually broad. A local administrator may be equally responsible for law enforcement, commissioning research, putting research recommendations into practice, maintenance of grounds, sewage removal, attendance at local city council meetings and more. Most of all, the individual is responsible for the successes and failures of these combined efforts, within the manager's institution. A natural resources manager in the National Park Service, for example, is generally responsible for integrated pest management, water quality, air quality, inventory and monitoring, hazardous waste, climate change, acid rain, endangered and threatened species, migratory species, environmental impact statements, research contracts, briefings, and much more.

The available research within each applicable subject area of park interest is too much for the single individual (or small technical staff) to collect. Few to none in the staff may be sufficiently conversant with all the applicable disciplines to understand technical papers in each field. The financial cost of contracting annual searches in all these topics is initially prohibitive, and does not provide advisory expertise on call.

Managers that are tasked with developing solutions to problems will seek scientific advice, but initially are faced with the repeated dilemma of: 1) "What is needed?"; 2) "How do I phrase the question?"; 3) "Where do I go to get help?"; 4) "How do I convince superiors that the mysterious research is worth the cost?"; and 5) "Why aren't our agency scientists more readily available on call?" Other problems surface when the individual tries to interpret and evaluate proposals for quality and cost effectiveness. The difficulty surfaces again when the individual needs to interpret, evaluate and apply the results. In the worst case the results of the commissioned research may seem to the manager to be different from what was requested in the first place. Consequently, the management action may be implemented giving less than the intended attention to the research findings. As the responsibility of the manager increases, the difficulty in passing through these hoops also increases, especially if local help is unavailable to translate and assist in evaluating the research problem and research. When scientists are present to assist in formulating the questions for needed research they provide a filtered view of what is available. This is by virtue of their own professional exposures within science, such as sample of journals read.

The following Table 1 depicts a coarse illustration of the separation of communication channels which exist between professions.

TABLE 1

The following is a coarse illustration of the separation of communication channels existing between professions. Forcibly, it is a generalization and can be improved upon. Note, however, how little overlap exists between group 1 and others, and between 1 and 2 and others. Group 1 = scientists, 2 = field or resource managers, 3 = reserve manager or superintendent, 4 = regional manager, director or natural resources coordinator, and 5 = central secretary, director or associate level administrator.

Mode of Communication and Relative Importance in Every Day Work	1	2	3	4	5
Publications read [specialized (A) to (C) unspecialized topics]	AA	A/B	C	C	CC
Memorandum size [short (D) to long (F)]	FF	F/E	E/D	D	DD
Memorandum number [few (G) to many (I)]	GG	G	G/H	I	II
Long internal reports (r) number of publications (p) produced [few (J) to (L) many]	LLp	Lr	Kr	Jr	JJr
Subject coverage of job [specialized (M), non-specialized (N), to diverse (O)]	MM	M/N	N/O	OO	OO
Verbal communication [only lectures/seminars (P) to brief appointments and meetings (R)]	PP	Q/R	Q	RR	RR
Professional contacts [one on one, exclusively (S) to include multiple conferences and workshops (U)]	U	R	S	SS	SS
Access to published materials, books, etc. [easy (W), difficult but available (X), not likely to hear of them or obtain them (Y), strictly private reading (Z)]	WW/X	X	Y/ZZ	ZZ	ZZ

A solution for closing the missing connection between research and management cannot rest effectively with the managers.

Can Scientists Close the Gap?

The interests of park, reserve or forest scientists may be broader than for most academic scientists, but they are also specialized to varying degrees. Although scientists associated with institutions that manage parks and reserves may know how to obtain information from diverse fields, the limited access to libraries from remote sites and competition for their time between research and management, influences the relationship they share with managers. It also influences the creativity and quality of their research. For example, some managers have organization-scientists near by that can access new scientific developments, translate them, and show management staffs how to implement the results. A few good examples in the U.S. National Park System are Everglades, Sequoia-Kings Canyon and Glacier National Parks. Having this in-house scientific capability in the U.S. characterizes mostly large parks or reserves (500,000 + ha) that have the capability of securing large budgets and justifying a localized research program, in place. The majority of other sites rely on fractions of work schedules of regionally located scientists who serve multiple parks and other offices or universities. The U.S. National Park Service is not alone in facing the dilemma of park scientists gravitating toward either the esoteric research (the minor extreme) or engaging in resources management practice or administration at the other. Similar difficulties have been reported for other federal agencies. Tight funding accentuates the problem. Geographic and intellectal isolation of scientists and loss of contact with developments in their fields contribute to the gap. Salary and promotion in the U.S. Federal research positions are based primarily on research performance and professional standing in science, not usually in management or communication between professions. Were it the other way around, research competence and productivity would have to give way; time available for work will not increase. Consequently time devoted to information transfer, translation and local consultation to management compete directly with professional responsibilities. All the difficulties mentioned above are problems that provide direct negative feedback to the link in communication. These are good examples of how science and management can be forced into working at cross purposes to one another. I see no evidence that science is able to close the missing connection, and in the mean time the communication and application of science to conservation of Biological Diversity is slowed.

Some have attempted stop-gap measures with indefinite success. There have been appeals to modify the evaluation process for scientists within Federal organizations to increase their advisory responsibilities and credit for advising. And some have suggested scientists should view themselves differently, especially within Federal organizations. There is a continuing conflict that for most, the crossing over doesn't provide for specialization required in research. Quality of research and translation and advising for management are not entirely compatible. Both are time consuming and do not necessarily call upon the same talents and personalities. Automatic responsibity for both can confer unnecessarily and misplaced frustration and guilt.

Proposed Solution: The Value of Change

With the information explosion that has occurred it is already difficult for a scientist to keep up with the available data bases, reference services, and relevant conferences. It is difficult to read all the abstracts, let alone read the complete manuscripts pertinent to a scientist's specialty and conduct research. The less specialized (within traditional discipline) the scientist the more expanded and diverse the reading load becomes. Transmitted information is destined to be available only piece meal to managers. They depend on windows of opportunity for obtaining information, rather than on a *systematic and comprehensive* information transfer process. The manner in which we (researchers, administrators, ... and policy makers, alike) presently recognize professions and professional responsibilities does not lead to a solution and appears erosive of the good will and common ideals shared by the diverse professions interested in conservation of biological diversity.

There is value in initiating change. It seems that the communication gaps which develop between researchers and managers are not restricted to land managing institutions. This problem is not restricted to the Federal system, the U.S. nor even North America. I have observed similar difficulties in western Europe and India, for example. If the problem is widespread, if the information explosion exacerbates it, and if it affects more professions than just research and management, the difficulty may require a remedy that is not simply a "technique." The remedy may be a full-time and permanent professional occupation. If the problem were only the product of ivory tower behavior, researchers may be persuaded to convey their knowledge for their field specialties, but not necessarily anything more broad than that.

In Other Worlds

At a historic moment in commerce and trade there must have been an analogous problem to the communication gap in conservation that instituted the "broker," a bona fide profession. Brokers do not usually involve themselves in direct sales with the public (consumers), nor in product development. But the broker finds the suppliers, promises to match them to buyers and promises the buyers that suppliers can be found to fill a need.

I have only found informal recognition of a missing profession in conservation and natural resource management that would help fill the missing connection. While there have been several starts by institutions to recognize such needs (such as the new National Biological Survey), these attempts have been few and incomplete. For example, the wildlife extension (in wildlife management) is similar to the types of translators that need definition, increased recognition and a standardized course of professional development. There are other isolated cases in which the intermediary is recognized and acts as the connection providing the many types of services necessary to bridge the gap. That which is in existence today in independent cases needs to be solidified into a recognized *profession* requiring qualifications.

The Linkage Profession

The profession of translator/advisor needs a focus and definition, including statements of breadth of subject expertise. Operating in a fashion analogous to brokers, representatives of such a profession have the potential to vary in degree of specialization, to be broader in subject coverage than most scientists can afford to be, and be more in tune with what science has to offer than any manager can afford to be. Such professionals, acting within an institution (or as private contractors) functioning as intermediaries, investigators and translators, can connect specialized information to targeted needs. They also can search broad topics for application. Functioning as a series of articulated links of varied specialities its professionals should be flexible enough to adjust in degree of specialization, and adjust in degree of involvement in research/or management. They could interface well with applied scientists and together with the researchers effectively convey advice. In the medical profession, for example, there is a variant strategy. It is the nurse-practitioner or doctor-practitioner that reaches out, reads, evaluates and apply most of new medical knowledge. Because the liability is generally on the individual this makes sense. When as with business, the liability is on the company or institution the responsibility appears to shift to specialists. Having such specialists within the company or contracting with specialist brokerages, the company reduces risk of financial loss and increases probability of increasing profits. In the government (public) institution, responsibility rests with the institution, though some responsibility may be localized. It is the Director or Secretary of the institution that is ultimately accountable to Congress (U.S. example). Translators reduce risk of failure, increase chances for the institutional successes, by being all at once familiar with the subject matters, the institution's information needs, and opportunities for cross-communication. The translator can articulate what she/he knows, and is suited to share in the institution's responsibility for achieving its mandated objectives.

Although individuals have attempted to develop institutional translation, the initiative has dealt only with data bases. The profession has not been institutionalized, and systematically integrated into the functioning of conservation applications. Contractors are sometimes asked to fill such intermediary communication/translation roles. While exceptions need to be recognized, most contractors do not possess intimate knowledge of the organization's functioning, nor are they necessarily committed to the organization's objectives. While contractors *are* also essential players in accomplishment of research and management, the close familiarity of the "translator/intermediary" with the organization gotten from inhouse training superimposed on an academic background is particularly crucial for public institutions. This argues for having a cadre of translator/intermediaries within institutions that can act and coordinate information flow among scientists, managers, private contractors and affiliates. Only rarely does the purchase order, contract or memorandum of understanding specify that the contractor shall discover potential but still unidentified pertinent topics and applicable technologies, or specify the contractor discover information of potential use, for which the needs are not yet identified. "Think tanks" are sometimes improvisational responses to these exploratory and shifting needs, as are advisory panels. The point is that present approaches to crossing the communication gap do exist, but in subjects of environment and conservation they are generally disarticulated and improvisational. Despite the promising activities of the US National Biological Survey, there is little to tie all these functions together and no recognition of the unique function of tying it all together. Consequently, efforts of most institutions currently underway are frequently redundant, unintentionally competitive, and taxing on the time and abilities of both scientists and managers. Just as interpretive and public relations professions are recognized as valuable connections between institution and "public," the professional translator/intermediary also would be valuable. (I do not use the word translator advisor outright, because the work involves much more than just translation.)

To search out, translate and produce available scientific material for use by management is labor intensive, and requires a peculiar talent. There is a direct need for formal recognition such as a profession that competently searches for and conveys knowledge, a profession that is recognized for its awareness and access to information at various levels of detail. These may be advisors but also may assist scientists to advise more effectively than alone, or may help managers to better comprehend advice. This is where I think one major solution lies.

The Second Problem: Knowledge Lost

The second poorly recognized contributor to the missing connection between science and management, is more localized and easier to remedy. It is the lack of a systematic documenation of manager's knowledge and experience. When scientists report findings or observations these are documented and archived in published literature. When managers and administrators of parks (such as with our U.S. parks) make observations and learn from their work, theirs are mostly oral reports of findings delivered to immediate associates. Writing publications is not part of the manager's professional communication system. Memos, incidence reports, etc. are all too superficial to do justice to the collective (including historical) memory and technical knowledge of an organization. Having no systematic documentation of institutional memory at their disposal for more than the extroverts or leaders of the system, individuals not in the system (such as scientists outside) make default assumptions regarding the knowledge and sophistication of the institution and its managers. I have observed managers to be underestimated; there is no source (publication) that documents their knowledge systematically and consistently. Internal newsletters do not suffice, no matter how expert they are at reporting news. This non-registry of institutional experience contributes significantly to the missing connection between science and management; it isolates managers from the pool of available knowledge. Their contributions are not evaluated or considered in the research of new and workable techniques unless the scientist happens to be personally familiar with the manager and aware of his/her contributions. By ignoring the manager's (including upper level administrators') contribution or their understanding of a subject, scientists offering research or advice easily offend and cross the very individual they seek to assist.

For example, in Washington, D.C., in Parks and Regional offices of the U.S. National Park Service, in other U.S. Federal and State agencies, and in other countries, I have seen the manager's knowledge transferred mostly by word of mouth and from one individual to another, rather than from one individual to groups at a time. Occasionally conferences and workshops allow the more eloquent and extroverted to share their experiences (superintendent workshop), but the remaining majority contributes little by comparison. Experience and knowledge are predisposed to disappear with retirement. Some individuals publish memoirs after retirement, but the information is only a small sample (and a potentially distorted rendition) of their accumulated professional insights. I have heard of no central place nor consistent method in place designed to collect this information analogous to a professional debriefing.

Scientists that are interested in looking for the knowledge and experience contained within an organization are going to look for published material, first. My personal experience with scientists interested in involving themselves in park research, and those that publish on conservation topics is that they have little to no access to institutional experience. Talking to managers is difficult, due to unfamiliarity, difficulty of locating appropriate knowledgeable individuals. (People in public agencies are apt to move frequently.) Before the scientist designs the study, interprets research results, or offers advice, a means is needed for making essential connections to the accumulated knowledge of managers.

The best examples of missed connections are evident among the last decade's articles on conservation of biological diversity and reserve design. Applications of island biogeography and population biology to conservation in continental habitats have lacked semblance to real systems. While some of this has resulted from a trend to pit reality to models (as discussed for ecology by Peters 1991), the published interpretations and models are impacted by a lack of infusion of the managers' collective knowledge. Most obviously missing is knowledge of practical constraints on design and management. These miscommunications and misinterpretations lead to conflicts and distrust between managers and scientists. While this difficulty is simple to describe, the solution is not as accessible.

Proposed Solution: Recording Organizational Memory of the Manager

To cope with the continuous loss of institutional knowledge the events such as workshops, promotions, and retirements, etc. should be targeted for extensive rather than superficial interviews. The documentation of the interviews should be organized, archived and their subject/source indices regularly published. An oral history would be valuable if gathered and focused on protection and management of natural resources, to include the various topics of biological diversity, ecosystem processes, through to mismanagement of alien species. Ideally, individuals should be interviewed at least on leaving each job for their memories and insights. But, realizing that the publication of such documents could be expensive due to the volume and problems of managing confidential material, an inexpensive and methodical approach is preferable. The Indices to the archives could be published, periodically. Internal magazines and newsletters, such as the U.S. National Park Service's *Park Science* or *Courier*, and U.S. Fish and Wildlife Service's *Endangered Species* would be good outlets for such institutional indices. Many managers write articles that document their memory but these are a biased sample and not necessarily complete sample of the subject matters covered by the organization. They are non-systematically collected. Publications such as these definitely contribute to such an effort, but they do not have the same purpose, nor are they equipped with the funds and facilities to undertake the broad task. Besides, the readers of internal magazines, journals, and newsletters may not find most of the reading of archived institutional knowledge too exciting. I imagine the material would be like reading any court record or Federal Code, intended for reference mostly. If the task is left to individuals to contribute material, I suspect that all but a few people are too rushed, too shy or for other reasons unable to contribute their written experience.

In their efforts to be responsive to their own needs to share "news," specialist professions develop internal newsletters. If several of these exist within each institution they do increase information flow within professions but, I expect they are likely to decrease information flow between professions. Reading time is limited and, where publication efforts are scattered and tailored to a specific group, the cohesiveness is supported (a good thing), but isolation of that group from others is the drawback (not so good). Summarizing this issue: some structure and visibility of the newsletters already provides documentation of institutional knowledge by drawing attention to history in progress, by publishing selected experiences, views, attitudes, etc. Once archived and referenced the material is accessible to researchers in a communication medium likely to be recognized and accessed. However, because of how the institutional memory is sampled, these specialized newsletters cannot replace a reference that specializes in collecting, archiving and indexing collections of institutional memory representatively and objectively.

Historians (and translators) are both in a position to analyze what the institution knew, and how it used what it knew.

Whatever method is finally settled upon to record and preserve organizational knowledge, there is an urgent need to deal with the drain of previous investments caused by loss of information as individuals forget, leave or retire. Such information not only has the potential to communicate to science and to the future, but has the potential to communicate within the institution, building the qualifications of its professionals.

To not keep records of institutional memory is much the same as not keeping records of floor discussions in Congress (or legal profession). Such published records are essential for evaluating the dynamics of the congressional and legal systems, and are frequently used in the future for historical reference.

CONCLUSIONS

Earlier I mentioned two principal factors necessary to bridge the missing connection between science and management. The first was the incompatible modes of communication between science and management, and the second was the unavailability of accumulated organizational memory of managers involved in protection of biological diversity. I proposed that facilitating communication would both speed and increase the amount of research findings that would be brought to assist in conservation of biological diversity. The reason we should be concerned with this is made evident every time potential losses and rate of losses of diversity are mentioned. This is not only a tropical matter but is as much a crisis in the rest of the world. There are several factors contributing to the speed with which we can respond to threatened losses of diversity, but I have chosen to focus on these because I hear or read about them the least, but have confronted them the most at national, regional and local levels.

The first recommendation made is to formally recognize and define a profession with responsibility to translate and communicate scientific findings to managers. The profession should be diverse in it types of expertise, but anyone should have the capability to conduct cross-disciplinary searches for data that can be used to guide management, and be equally capable of searching out appropriate scientists to serve as advisors to managers. The reasons why current scientists and managers are inadequately accomplishing this task are evident and should not be used to blame existing professions but to recognize that brokerages are necessary in conservation applications also. The proposed job tackles the communication challenges brought on by the explosion of information, needs for review and synthesis, and the serious and complex nature of contemporary environmental problems. The institutions using such a profession stand to benefit from having a variety of specializations to 1) meet the increased specialization of disciplines (and their publications) as these expand and subdivide, 2) to conduct the detective work (not necessarily just research) that is required to collect obscure information, and to 3) synthesize, 4) translate, 5) transmit and 6) make available information useful for application to institutional use. The standardization to which I refer above would enable us to talk more than vaguely about what such work constitutes, and would enable agencies and organizations to communicate their needs for such individuals more effectively than presently. The notion that such tasks exist, subsumed under responsibilities of other professions such as research or resources management or contractor, for example, assumes too little about the task.*

The second recommendation made is that institutions involved in land management consider managerial experience and other institutional knowledge to have value for guiding research and informing successive generations of managers about the organization memory of management. By recording this accumulated experience they can assist their own goals of protection of biological diversity by 1) documenting institutional history; 2) having positive feedback in matters of research funding and incorporation of research into institutions' managerial programs; 3) increasing respect for the accumulated knowledge and cultural wisdom of managerial professions; 4) increasing the rate of infusion of newly acquired scientific knowledge into management; and 5) in other less obvious ways effectively bridging, potentially closing the missing connection. Accomplishing this satisfactorily not only

---

*The institutionalization of the US National Biological Survey (NBS) after this paper was written does not negate what is said here. But, the translations function within the NBS has opened the door for the translation activity to develop for the US Department of Interior.

depends on collection of the knowledge, but also on a systematic collection, the curation of archives, and the publication of indices accessible to scientists and other interested professions in academic and other major libraries.

ACKNOWLEDGEMENTS

The author's comments do not necessarily reflect the policies or points of view held by the National Park Service or National Biological Survey. This work was supported by the Western Region Science Program of the National Park Service. The author is grateful to Drs. Marybeth Buechner and Bruce Wilcox and Ray Sauvajot for their collaboration on projects contributing to this work. None of this could have been prepared without the supportive cooperation of National Park Service Superintendents, Resource Managers and their staffs. Special thanks for their comments on earlier versions of this manuscript go to Drs. J. David Cherry, William P. Gregg, Jr., Cat Hoffmann, Jean Mathews, and Walter Rosen, acknowledging that the contents of the paper does not necessarily reflect their personal or professional views.

LITERATURE CITED

Alexander, C., S. Ishikawa, M. Silverstein, M. Jacobson, I. Fiksdahl-King, S. Angel. 1977. *A Pattern Language.* Oxford University Press. New York. 1171 pp.

Blume, S., J. Bunders, L. Leydesdorff and R. Whitley. 1987. The social direction of the public sciences; causes and consequences of co-operation between scientists and non-scientific groups. *Sociology of the Sciences* Vol. 11. D. Reidel Publishing Co., Dordrecht.

Buechner, M., C. Schonewald-Cox, R. Sauvajot and B. Wilcox. (In Press) Cross-boundary Issues for National Parks: What Works "On the Ground." *Environmental Management.*

Forman, R.T.T. and M. Godron. 1986. *Landscape Ecology.* John Wiley and Sons, New York.

Lasley, J.F. 1978. *Genetics of Livestock Improvement,* 3rd edition. Prentice-Hall, Englewood Cliffs, NJ.

MacArthur, R.H. and E.O. Wilson. 1963. An equilibrium theory of insular zoogeography. *Evolution* 17:373-387.

McMichael, S.L. and R.F. Bingham. 1923. *City Growth and Values.* The Stanley McMichael Publishing Organization. Cleveland, Ohio. 368 pp.

Peters, R.H. 1991. *A Critique for Ecology.* Cambridge University Press. Cambridge. 366 pp.

Ralls, K., K. Bruggar and J. Ballou. 1979. Inbreeding and juvenile mortality in small populations of ungulates. *Science* 206:1101-1103.

Schonewald-Cox, C., M. Buechner, R. Sauvajot and B. Wilcox. 1991. Resolving Conflicts Between National Parks and Surrounding Lands: A Review and Discussion. Environmental Management 16(2):273-282.

Schonewald-Cox, C. (submitted, invited paper) Social Issues of Landscape Fragmentation. In [will be available shortly].

Shafer, C.L. 1990. *Nature Reserves: Island Theory and Conservation Practice.* Smithsonian Institution Press, Washington, D.C., 189 pp.

Soulé, M.E. and B.A. Wilcox. 1980. Conservation biology: its scope and its challenge. Pages 1-8 in M.E. Soulé and B.A. Wilcox (eds.), *Conservation Biology: An Evolutionary-Ecological Perspective.* Sinauer Associates, Inc., Sunderland, MA.

Soulé, M.E. 1986. (ed.) *Conservation Biology: The Science of Scarcity and Diversity.* Sinauer Associates, Inc., Sunderland, MA, 233 pp.

Soulé, M.E. 1987. (ed.) *Viable Populations for Conservation.* Cambridge University Press.

Turner, M.G. and R.H. Gardner (eds.). 1991. *Quantitative Methods in Landscape Ecology.* Springer-Verlag, NY. Ecological Studies Series Vol. 82, 536 pp.

Wright, S. 1969. *Evolution and the Genetics of Populations,* Vol. 2. *The Theory of Gene Frequencies.* University of Chicago Press, Chicago, IL.

Wright, S. 1977. Inbreeding in animals: differentiation and depression. Pages 44-96 in *Evolution and the Genetics of Populations,* Vol. 3. University of Chicago Press, Chicago, IL.

Wright, S. 1978. *Evolution and the Genetics of Populations.* Vol. 4. *Variability within and among Natural Populations.* University of Chicago Press, Chicago, IL.

Biological Diversity: Problems and Challenges. Edited by S.K. Majumdar, F.J. Brenner, J.E. Lovich, J.F. Schalles, and E.W. Miller.

Chapter Sixteen

THE ROLE OF SCIENCE IN PRESERVING BIODIVERSITY IN NATIONAL PARKS

JOHN LEMONS*
Department of Life Sciences
University of New England
Biddeford, ME 04005

INTRODUCTION

The loss of the world's biodiversity is thought to be a major problem by many scientists, public policy makers, and the general public. However, no precise estimate can be made of the numbers of species being lost or the rate of loss for three reasons.[1] First, scientists do not know the numbers of species originally present in major habitats, or the number of existing species, even to the nearest order of magnitude. Estimates of the number of existing species range from five to 30 million. Of the approximately 1.4 million named species, about 750,000 are insects, 41,000 are vertebrates, 250,000 are vascular plants and bryophytes, and the remainder are invertebrates, fungi, algae and microorganisms. Second, extinction rates are usually determined indirectly from principles of biogeography.[2] Diversity reduction is a function of the size of island fragments and their distance from each other, and these factors vary enormously between countries. Third, the ranges of most species are not known and it is therefore not always possible to estimate which ones will be eliminated due to habitat alteration. Nevertheless, estimates suggest that the rate of extinction is approximately 1,000 - 10,000 times that prior to human intervention in ecosystems, and the number of extinct species in the tropics alone may exceed 17,500 per year.

---

*Dr. John Lemons is Professor of Biology and Environmental Science, and Editor-in-Chief of *The Environmental Professional.*

Loss of biodiversity is due to ecological effects of development and deforestation, conversion of natural lands to agricultural uses, chemical pollution, introduction of exotic species and artificially produced strains of existing organisms, the frequencies and types of management practices in existing parks and refuges, and their sites and sizes. Lastly, an important cause of habitat alteration and subsequent loss of biodiversity is the fact that humans presently utilize approximately 40 percent of the potential net terrestrial productivity of the planet.[3]

National parks offer one means of preserving biodiversity, since their legislative mandates and other environmental laws confer various degrees of protection to species and ecosystems. The United States' National Park Service (NPS) manages over 63 natural area units, which range between 30,000 to over 3,000,000 acres in size.[4] Included are well-known parks such as Yosemite, Yellowstone, Glacier, Grand Canyon, Great Smoky Mountains, Everglades, and Denali. Twelve of the units are Biospher Reserve Parks, which are dedicated to long-range ecosystem monitoring under the UNESCO Man and Biosphere program. Importantly, these units are considered to be ecological standards for the network of International Biosphere Reserves. Despite the fact that national parks offer the potential to preserve biodiversity, various threats are irreparably damaging biological resources.

Enhancing the protection of biodiversity in national parks requires assessment of the: (1) legislation governing use and protection of parks' resources, (2) threats to parks' species and ecosystems, and (3) use and adequacy of science in management decisions. In this article, I assess these factors for United States' national parks. However, this assessment should be relevant to parks' problems in other nations because of similarities of management goals for parks and because problems of protection of biodiversity are common.[5]

LEGISLATIVE PROTECTION OF PARKS' RESOURCES

Protection of biodiversity within parks is governed primarily under the general authority of the National Park Organic Act (16 United States Code [USC] §§1, 1(a)-1, 3, 20), as well as under authority of enabling legislation specific for each park. These legislative mandates are used as a basis for determination of management policies, as well as a basis to mitigate internal threats to parks' resources. Internal threats originate from within park boundaries, and include such impacts as those resulting from numbers of visitors and types of use of particular areas, numbers of vehicles permitted in parks, and the types, numbers, and location of visitor facilities and roads. Additional problems stem from historical NPS decisions to: eliminate predators from some parks, stocking of fish in rivers and lakes, suppress natural fires in parks' ecosystems, ignore the role of humans in influencing parks' biota, and assume that "natural" regulation of species could occur in parks despite the fact that parks' boundaries were established for political reasons and do not reflect biotic or ecosystemic boundaries.

Many of the internal threats to parks' biodiversity originate from controversies surrounding the traditional interpretation of the legislative mandates for parks.[6-9]

Title 16 USC §1 states that the fundamental purpose of national parks is:

> ... to conserve the scenery and the natural and historic objects and the wildlife therein and to provide for the enjoyment of the same in such manner and by such means as will leave them unimpaired for the enjoyment of future generations.

Title 16 USC §20 states congressional findings and purpose regarding park concessions, accommodations, facilities, and services. It provides:

> In furtherance ... of this title, ... which directs the Secretary of the Interior to administer national park system areas in accordance with the fundamental purpose of conserving their scenery, wildlife, natural and historic objects, and providing for their enjoyment in a manner that will leave them unimpaired for the enjoyment of future generations, the Congress hereby finds that the preservation of park values requires that such public accommodations, facilities, and services as have to be provided within these areas should be provided only under carefully controlled safeguards against unregulated and indiscriminate use, so that the heavy visitation will not unduly impair these values and so that development of such facilities can best be limited to locations where the least damage to park values will be caused. It is the policy of the Congress that such development shall be limited to those that are necessary and appropriate for public use and enjoyment of the national park area in which they are located and that are consistent to the highest degree practical with the preservation and conservation of the areas.

Title 16 USC §3 creates penal sanctions for violation of park rules and regulations promulgated by the Secretary of the Interior governing the use and management of parks. This section also allows the Secretary of the Interior to destroy exotic species of animals and plants that may be detrimental to native flora and fauna.

Title 16 USC 1(a)-1 states that areas of the National Park System derive their increased national dignity and recognition from their superb environmental quality:

> Congress further reaffirms, declares, and directs that the promotion and regulation of the various areas of the National Park System ... shall be consistent with and founded in the purpose established by Section 1 of this title, to the common benefit of all the people of the United States .... The authorization of activities shall be construed and the protection, management, and administration of these areas shall be conducted in light of the high public value and integrity of the National Park System, and shall not be exercised in derogation of the values and purposes for which these various areas have been established, except as may have been or shall be directly and specifically provided by Congress.

The traditional interpretation of the NPS's legislative mandates is that parks are to be managed for two conflicting purposes: maintenance of park resources in an unimpaired condition, and provision for their use and enjoyment by the public. These conflicting goals persist in official administrative policies and actions.[10,11]

However, the NPS is not entirely explicit or consistent on how it interprets the conflicting purposes; in fact, it is ambiguous on this matter. At times, the NPS has stated that biodiversity in parks should be managed according to a strict policy of preservation.[4] More frequently, the NPS has stated its proper function is to balance preservation and use equally.[6,12] Occassionally, the NPS has proposed a goal of maximizing use and enjoyment of parks.[13] The NPS has also argued that it should be free to manage parks according to what the agency perceives the public wants, and is therefore free to oscillate between preservation and use if public desires change over time.[8]

Management goals to preserve biodiversity in national parks is complicated also by disagreement about the meaning of the word "natural." The common interpretation of the phrase "unimpaired" in Title 16 USC §§1, 1(a)-1 and 20 is that biodiversity of parks should be maintained in its natural condition.[7] The NPS legislation does not define "unimpaired" or "natural," and it does not contain standards by which to guide or assess management decisions affecting biodiversity. Further, the NPS has not defined these terms in its administrative policies. However, the agency has used different concepts of "natural" in management decisions. Some definitions have included the influence of modern humans in park areas, while others exclude such influence.[8] Additionally, the NPS does not have firm policies on whether ecosystem structure or function (processes), respectively, are desired goals to promote "naturalness" in parks. For example, policies regarding prescribed burning in parks have been controversial because of disagreement about whether to attempt to recreate a pre-Columbian ecosystem structure, or alternatively, whether to allow ecosystem processes to occur with minimal influence by humans.[14,15] Unfortunately, there are no judicial decisions that have resolved the ambiguity surrounding the meaning of NPS legislative mandates. However, Lemons and Stout[7] and Lemons[8,9] argued that a reinterpretation of NPS legislative mandates supports management goals that maximize preservation of biodiversity and other national park resources.

In general, the NPS lacks explicit legal authority under either the National Park Organic Act or park-specific enabling legislation to manage external threats, which originate from outside park boundaries. Examples of external threats include air pollution from energy developments and mine smelters on lands adjacent to park boundaries, water pollution from mining operations adjacent to park lands, impacts from dam constructions, acid precipitation, global warming, geothermal development, pesticide use, encroachment by exotic species, and other forms of development and activities on lands adjacent to parks. Authority to manage external threats falls generally under legislation such as the National Environmental Policy Act, the Clean Water Act, the Clean Air Act, the Endangered Species Act, the Wilderness Act, the Wild and Scenic River Act, and various other state and federal laws pertaining to water rights, energy and mining development, and environmental protection.

Considerable disagreement exists about the efficacy of these laws to mitigate external threats to parks' biodiversity. Prevailing opinion is that although case law pertaining to the aforementioned environmental legislation can provide some protection against external threats, new and more explicit statutory protections are needed. Space limitations preclude more thorough discussion of external threats; extensive information on can be obtained from Simon[10] and Mantell.[11]

THE ROLE OF SCIENCE

National Park Service decisions to preserve biodiversity will be consistent with legislative mandates only if they are based upon adequate scientific knowledge. Regardless of which interpretations of legislation pertaining to national parks are thought to be correct, NPS decisions concerning biodiversity will be effective only if they are based on scientific information. As a matter of administrative law, NPS decisions and actions are presumed to be valid and require affirmation if the agency has a rational basis for its decisions and actions.[16] Consequently, people who argue that the NPS should promote greater protection of biodiversity are required to base their arguments upon compelling factual evidence. Thus, the role of science in preserving parks' biodiversity is extremely important, and ways should be sought to increase its use as a basis for decisions. Unfortunately, it may not be possible to meet a stict burden of proof to base decisions upon compelling factual evidence due to a shortage of both practical and theoretical knowledge.

THREATS TO NATIONAL PARKS

Threats to parks' biodiversity have been documented in various ways. The *State of the Parks* report was the first and most comprehensive attempt by the NPS to systematically identify threats to biodiversity in parks.[4] The report focused on 32 parks, and identified 73 different kinds of threats in the following categories: aesthetic, air pollution, physical removal of resources, encroachment by exotic species, visitor physical impacts, water quality pollution and water quantity changes, and park operations and planning of facilities. A total of 1954 internal threats, and 2391 external threats in these categories were reported. Categories of resoures threatened were: biological, physical, aesthetic, cultural, and park operations. Overall conclusions of the report were:

1. External threats, although serious, receive little attention because they are complex and difficult to deal with.
2. Internal threats constitute close to 50 percent of all threats. The most frequently reported internal threats are associated with heavy visitor use, visitor facilities, vehicle noise, water pollution and quantity changes, soil erosion, and exotic plants and animals. There needs to be more attention paid to the carrying capacity of park areas and ways to accommodate use without sacrificing major resources.
3. Research into the problems affecting parks is inadequate. Some 75 percent of the threats reported by park staffs lacked adequate research and scientific documentation. Greater resources need to be devoted to research and additional monitoring to meet management needs.
4. The 12 Biosphere Reserve Parks reported an average number of threats three times the NPS-norm. This is a major problem insofar as the Biosphere Reserve Parks are considered to be model ecological control areas.
5. The identified threats to park resources and needed monitoring/research actions are probably grossly understood because the NPS does not have sufficient qualified personnel to assess the threats properly.

6. The overall summary of the report stated "The results of this study indicate that no parks of the system are immune to external and internal threats, and that these threats are causing significant and demonstrable damage . . . In many cases this degradation or loss of resources is irreversible. It represents a sacrifice by a public that, for the most part, is unaware that such a price is being paid."

Although the *State of the Parks* report represents the NPS's first systematic attempt to study threats to parks and call attention to park problems, it has several serious limitations. The report was over-quantified and underqualified. The overall number of threats for each park unit were provided, but no information was given as to the type of threats or degree of severity. Threats were also counted on a system-wide basis and averaged according to category. However, the report did not provide information on the qualitative significance of the threats. Consequently, it is not possible to interpret the results in the context of whether they are consistent with the NPS's legislative mandates. The way the report counted threats, all threats were equal and insignificant problems were obscured by the overall numbers. An illustration of this problem is provided by examining how the report treated water resources. Four hundred and sixty-six threats were reported to water resources; this category ranked only sixth of the seven threat categories in the number of reported threats (11 percent of total). Yet, a closer review of NPS documents indicates that most, if not all, parks where water is a significant resource suffer problems of water quality or quantity; these are having major impacts on park vegetation and wildlife.[17]

The numbers contained in the report do not provide information on the extent of the threats. Is a threat confined to a small area within a park and therefore more susceptible to mitigation? Alternatively, is the threat pervasive throughout a park and more difficult to resolve? Nor does the report indicate whether threats affect a critical biodiversity resource or one of incidental importance.

There was little control over the quality of the information used in the report. Most threats were assessed by asking park scientists, resource managers, or superintendents to fill out a standard questionnaire. A review of the responses indicated considerable disparity in their quality. In some cases, responses were supported by research; however, for many responses this was not the case. In fact, no literature citations were provided in the report. In part, the deficiencies of data used for the report are attributable to lack of NPS support for scientific study and resource management.

Published articles represent a second source of information to describe and evaluate threats to parks. A search of peer-reviewed scientific and environmental management journals, using the *Biosis* and *Enviroline* databases, was conducted for published articles pertaining to the 32 national parks identified in the *State of the Parks* report, for the period 1970-1986.[18] Only 77 articles were identified; 50 articles focused on internal threats, while 27 emphasized external threats. The number of publications per park ranged from zero to 21. Thirty-four publications focused on threats due to physical removal of resources, fifteen emphasized threats due to water quality changes, and ten focused on the threats of air pollution. Only five studies pertaining to the impacts of concessioner developments and visitor use levels in causing impairment to park resources were identified. Fifty-two of the publications focused on biological resources, nineteen focused on physical resources, and only six publications pertained to the other resources. Although this survey was conducted in 1986 and more research on national parks has been published since that time, it is reasonable to expect that there is still too little available information on threats to parks' biodiversity.

THE STATUS OF SCIENCE

Since the creation of most national parks, significant changes have occurred in their biota, both for natural and human-induced reasons. Prior to 1963, there was relatively little use of science as a basis for management decisions concerning biodiversity in parks. In 1963, reports by Robbins[19] and Leopold[20] provided specific recommendations to improve the use of science in management decisions. The former report emphasized the need for research in national parks and made numerous recommendations for its use; the latter report focused on problems of wildlife management and made recommendations for goals and policies. However, many of the recommendations contained in these reports calling for improved scientific research and resource management were not acted upon by the NPS due to resistance to change by the agency, and uncooperative relationships between NPS scientists and scientists from outside the agency. Past problems stemming from too little use of science can be ameliorated if the NPS recognizes the importance of science as a basis for sound management decisions.

Conservation biology is a newly emerging synthetic field which addresses the dynamics and problems of perturbed species, communities and ecosystems, and thus can serve potentially as a basis for management decisions. Conservation biologists provide principles and tools to those interested in preserving biodiversity, and draw upon research from fields such as population biology, physiology, genetics, island and historical biogeography, hazard and risk assessment, veterinary medicine, environmental monitoring, environmental ethics, social sciences, and natural resource fields.

Drawing upon both theory and limited empirical data, Soulé has proposed several functional postulates for conservation biology: (1) many species that constitute natural communities are products of coevolutionary processes; (2) species are interdependent; (3) many species are highly specialized; (4) extinctions of keystone species can have important long-range consequences; (5) introductions of exotic species may reduce diversity; (6) many ecological processes have thresholds below and above which they become discontinuous, chaotic or suspended; (7) the temporal continuity of habitats and successional stages depends on size; (8) if population densities of ecologically dominant species exceed sustainable levels, they can alter prey populations and other species with whom they share resources; (9) genetic and demographic processes have thresholds below which nonadaptive, random forces prevail over adaptive and deterministic forces within populations, and (10) most nature reserves are inherently disequilibrial for large or rare organisms.[21] These functional postulates are used in an attempt by conservation biologists to suggest rules for actions to maintain structure and function of natural ecological systems, and are relevant therefore to national park management. However, the adequacy of conservation biology as a basis for management decisions is constrained by a number of factors, both theoretical and applied.

Application of conservation biology requires knowledge of biological and ecosystem attributes such as: (1) ecosystem processes; (2) productivity; (3) decomposition and nutrient cycling; (4) interactions between the biosphere, atmosphere, hydrosphere, and lithosphere; (5) communities; (6) community structure; (7) dynamic networks of interacting individuals and species; (8) symbiotic and mutualistic species; (9) populations; (10) individual organisms; and (11) overall homeostasis. However, often there is no consensus as to which attribute should be studied for a particular problem. Further, studies of numerous parameters is often required, since changes in one parameter are not coincident necessarily with changes in others or with overall ecosystem changes. Practical constraints almost always limit the number of parameters that can be studied for any theoretical or practical ecological problem.

Disagreement regarding the importance of the various biological and ecological parameters as a focus of studies has led to two opposing trends in conservation biology: (1) emphasis on autecology and the population viability of individual species; and (2) emphasis on communities, ecosystems, and regions as appropriate management units.[22] The emphasis on autecology and population viability of individual species is predicated upon the notion that biodiversity management strategies should stress ecosystem processes as opposed to patterns, as well as populations as opposed to communities. The emphasis on communities and whole ecosystems is predicated upon recent critiques of the species approach to preservation of biodiversity. In any event, it is not clear yet which approach will be most effective. One practical consequence of this disagreement is whether NPS resource managers should or will focus management efforts on all or most of the abiotic and biotic components of parks' ecosystems, or just one species or selected species.

Knowledge of species' and ecosystems' attributes is limited also by the fact that substantial natural variations of ecosystem structure and functions include successional change, regular and periodic change due to temporal cycles, and random or stochastic changes. All variations may be essential to the continuation of species which are adapted to the predictable variation patterns, and to ecosystem processes. Effective management of biodiversity must be predicted upon understanding of the natural dynamics of ecosystems, and the ability to separate change into that which would occur in the absence of human influences. In general, knowledge of the kinds and magnitudes of variations increases the more the lower levels of biological or ecosystem organization are studied; large ecosystem components may sometimes appear to be static because changes can be detected only after long study times. Unfortunately, few long-term ecological studies have been conducted.[23] Consequently, it generally is not possible to detect, explain or predict fully the causes or consequences of environmental changes.

In addition, considerable confusion has existed concerning the question of whether ecosystems tend toward a relatively static equilibrium[24] or are comparatively dynamic.[25,26] Many past NPS decisions were predicted upon the belief that ecosystems tended toward relatively static or equilibrium states, and that preservation of such states should be both desirable and feasible.[27] Some modern ecologists also propose that NPS management decisions have as a goal the maintenance of relatively static ecosystem structure. For example, Bonnicksen and Stone argue that parks be managed to restore and preserve pre-Columbian ecosystems, or ecosystems that would have existed today or at some point in the future if European settlers had not interfered with their process; such ecosystems being tantamount to being "natural" in their view.[15] However, many others recognize the importance of trophic-level and other biotic interactions, periodic natural events such as fire, and random fluctuations of environmental factors in creating comparatively dynamic ecosystems.[15,27]

The dynamic nature of ecosystems has important consequences for NPS management decisions. If NPS policies are to maintain static ecosystems, then significant and perhaps unrealistic levels of intervention in biological and ecological systems may be required of resource managers. Consequently, a more appropriate management goal might be to restore and sustain ecological processes that have shaped parks' landscapes. However, this policy should be tempered with recognition that restoring and sustaining ecological processes will not necessarily reestablish pre-Columbian or presettlement ecosystem structures. The reason is that ecosystem structure and functions are inseparable, and the presettlement structure of an ecosystem must be restored (or at least known) before management interventions will be successful. Usually, this is not possible.[15]

Ecosystems' processes operate on widely different structural, functional, temporal, and spatial scales. There is no *a priori* single correct choice of scales for a particular ecosystem. On the structural level, a focus of study may be on individual organisms, populations, or species; alternatively, a functional study may emphasize factors such as production, nature of change in population densities with predation, or energy flow. A suitable temporal scale may be defined, in part, by factors of climate, physiology, and nutrient cycling, and may include the short or long-term. Spatial scales may be locally or globally defined. Spatial scales may interlock with temporal ones, but there is not a necessary correspondence.

Management of biodiversity takes place on a variety of structural, functional, temporal, and spatial scales. Ecological studies necessarily must define bounds on the scales of interest; this operationally defines the units of study. For example, in the absence of limiting the boundaries of scales, it is not possible to unambiguously define an ecosystem. All ecosystems exchange information and material with others up to the global level. However, because ecosystems are not closed, even with boundaries established on scales ecosystems cannot be defined unequivocally. A major problem is that in many cases the scales of interest are not obvious; when scales are known, they often are known better for lower hierarchial levels of organization. Effective management of biodiversity therefore demands that structural, functional, temporal, and spatial scales be chosen carefully; yet, it is not often possible to make choices on sufficiently precise data.

Some research on spatial and temporal scales of ecosystems has focused on the biogeographic consequences of fragmentation.[28] Ecosystem fragmentation causes large changes in the physical environment as well as biogeographic changes. A result is that landscapes consist of remnant areas of native flora and fauna surrounded by land modified by human influences. Consequently, fluxes of physical and biotic inputs across ecosystem boundaries are altered, thereby affecting native species in remnant areas. Biological consequences of the isolation of protected areas due to the modification of adjacent lands is significant, and varies as a result of the time since isolation and distance from other remnant areas. Research also has indicated that the consequences of fragmentation are influenced by size, shape, and distance of remnant areas from each other. Controversy exists regarding whether one large reserve will protect more species than several smaller areas with a total equivalent area to the larger reserve. Generally, larger reserves are more buffered from adverse consequences of fragmentation. However, most research on size and shape of reserves has provided little of practical value to resource managers for the reason that managers of protected lands are dealing with ecological conditions that are a *fait accompli*. In other words, with few exceptions protection of biodiversity must occur on lands already set aside for such protection but whose boundaries were based upon political as opposed to biotic consideration. Consequently, a critical need exists to develop integrated approaches to land management that places preservation of biodiversity in national parks in the context of the overall landscape. Such an approach may be beginning with new research and management of the greater Yellowstone ecosystem, which is directed at understanding and controlling external influences on species of Yellowstone National Park.[29]

Although a number of species and ecosystem attributes are known to be important as a basis for management decisions, excessive emphasis on either structural or functional characteristics may influence managed ecosystems in a questionable manner.[30] For example, an accurate description of species composition may indicate changes due to introduction of exotic species. However, such changes may not necessarily be of consequence functionally if new species replacing the old perform similar functional roles in the ecosystem. Further, the removal of certain exotics from some park areas can result in significant perturbations in some ecosystem functions during the period of transition to native cover, at least for plants. Likewise, changes in functional parameters by themselves do not indicate whether a system has been altered by replacement or disappearance of species, changes in trophic relationships, or whether it remains structurally unchanged. Temporal responses of species to human-induced impacts may range from the instantaneous to the long-term. Consequently, short-term measurements may not predict long-term responses, and the natural variation of undisturbed systems may mask the response of a particular system to human influences. Since most ecological studies are relatively short-term, it is problematic to separate natural variation from the effects of human influences, as in the case of assessing the ecological affects of human-induced global warming.[23]

Ecosystem stability is important for the preservation of biodiversity, and is a focus of study for many conservation biologists. However, the concept of stability has stimulated considerable debate among scientists.[31] Ecologists have been successful in determining the relationship of some, but not all ecosystem attributes to stability. For example, it is not known whether stability is due to species diversity, or the cause of it. Further, concepts of stability can variously emphasize the resistance to disturbance of an ecosystem, the time an ecosystem requires to recover from damage, the zone from which an ecosystem will return to a stable state, the degree to which the pattern of secondary succession is not an exact reversal of the retrogression following environmental impact, and the degree to which a stable ecosystem established after disturbance differs from the original steady state. Consequently, management decisions based upon concepts of stability must recognize the debate and uncertainty surrounding the different concepts, as well as the practical implications of managing for one concept as opposed to another.

Obtaining adequate ecological knowledge to manage threats to biodiversity is difficult also because of the problem of cumulative effects, which is the continuing degradation of ecological systems due to repeated perturbations. Cumulative effects can occur due to time-crowded perturbations that are so close in time that the effects of one are not eliminated before the next one occurs. Space-crowded perturbations occur when perturbations are so close in space that their effects overlap. Synergisms are different types of perturbations in the same area that interact qualitatively and quantitatively to produce different responses in ecological communities. Indirect effects can be produced after or away from an initial perturbation or by a complex pathway, and are often not able to be predicted in advance. "Nibbling" involves the total effects of many smaller incremental actions in space and time. While the scientific understanding of cumulative effects is difficult, problems may be compounded because boundaries of affected environments often cross local, regional, or national jurisdictional boundaries. For example, perturbations might originate in one jurisdictional boundary, but the ecological effects might occur in another.

CONCLUSIONS

National parks can play an important role in preservation of the world's biodiversity. Traditional interpretations of national parks' legislative mandates has oscillated between preservation and use of resources. Regardless of which interpretation is thought to be correct, scientific knowledge is rarely a decisive factor in management decisions because of both a lack of scientific standards contained in the legislation, and because of inadquate scientific information to guide decision-makers. Accordingly, most decisions include three variables: park values (including biodiversity), social benefits, and financial.

All human use impairs natural resources and detracts from visitor experiences to some extent. Since park lands are public, decisions about levels of acceptable impairment are political. Because legislation does not define scientific standards nor criteria of appropriate use, the NPS traditionally has fashioned park policy by compromise to reflect different citizen constituencies with different preferences.[8,32] Scientific knowledge is required to inform us as to whether NPS policies are consistent with parks' legislative mandates, and to serve as a basis for decisions to conserve biodiversity. Unfortunately, it is questionable whether such knowledge is adequate for these purposes.

Historically, the NPS has not made rigorous attempts to base management decisions on the best scientific information available. The *State of the Parks* report identifies a broad spectrum of park threats; yet, considerable uncertainty exists about them given the nature in which information was compiled and in which data were reported. By the NPS's own admission, 75 percent of the 4345 threats identified were not documented adequately.[4] The primary value of the report is to call public and congressional attention to the serious deficiency of adequate scientific data on which to predicate management decisions. A search of the refereed scientific literature identified relatively few articles focusing on threats to parks, despite the high number of threats reported in the *State of the Parks* report. While factors such as budgetary restraints limit the quantity and quality of park research, the fact that conservation biology is a newly emerging field and is mostly theoretical as opposed to an applied science imposes additional problems. With respect to both the design of research and reserves to conserve biodiversity, there is too much uncertainty regarding species-specific and site-specific information, appropriate biological or ecological parameters, spatial and temporal scales, problems of fragmentation, and how to handle cumulative effects. Consequently, conservation biology is not adequate to serve as a practical guide for decisionmaking at this time.

The lack of adequate data about threats to parks' biodiversity and ways to mitigate them have important consequences. Policy decisions will be determined primarily by political concerns, with no clear indication of whether decisions are consistent with legislative mandates to preserve biodiversity. The Secretary of the Interior is ultimately responsible for managing parks, and bases decisions on the best information available—such as the information used in the *State of the Parks* report and the refereed literature, despite their inadequacies. Because these sources of information do not provide adequate documentation of threats, and because the status of conservation biology is primarily theoretical as opposed to applied, those seeking to overturn NPS decisions and have the agency adopt a more strict policy of preservation have little factual bases on which to argue. In this situation, NPS decisions are assumed to be valid by the court.[16] Given the NPS history of developing policies based upon compromise of public preferences, it is not likely that a more strict policy of preservation will be implemented without compelling factual knowledge about the nature and significance of threats to parks and ways to mitigate them.

Absence of factual knowledge also will constrain mitigation efforts directed at external threats. Despite the fact that some legislation such as the Endangered Species Act, the Clean Air Act, the Clean Water Act, and the National Environmental Policy Act provide some means of protecting parks' biodiversity from activities originating on adjacent lands, Congress nevertheless has given the NPS little explicit authority to regulate such activities. To the extent that the NPS lacks authority for dealing with problems of incompatible uses on adjacent nonpark lands, the agency must rely on cooperation with private landowners and representatives from local, state, and other federal agencies to resolve conflicts. Such cooperation will be based on the NPS's ability, and the ability of others concerned with park resources, to convince those responsible for activities on nonpark lands that their actions are needed to protect park resources, and to convince them to be willing to undertake those actions. Such actions will not be forthcoming unless arguments are predicated upon compelling factual knowledge.

Recommendations to improve the use of science in preservation of parks' biodiversity include: (1) reinterpret national park and other legislation in a manner which protects biodiversity against internal and external threats to the maximum extent possible; (2) maintain park areas in a condition as free from human-induced impacts as possible; (3) support the contined growth of both the theoretical and applied aspects of conservation biology, and base management decisions on the best available knowledge from the science, and (4) develop cooperative agreements with private landowners and state/federal agencies that control lands adjacent to parks.

Despite these recommendations, it should be understood that too many uncertainties exist, and are likely to exist for a considerable time, for conservation biology to yield compelling factual knowledge to serve as a basis for management decisions, especially when intensive management is required. We simply do not know enough. National Park Service managers, as well as others interested in preservation of parks' biodiversity, should explicitly acknowledge the uncertainties of scientific information about threats to parks' biodiversity and those inherent in the science of conservation biology. Given the uncertainties, it would be wise to manage parks in a conservative manner. In other words, limit to the maximum extent thought reasonable the numbers of visitors, our types of uses of parks' resources, and activities on lands adjacent to parks.

REFERENCES

1. Wilson, E.O. 1988. The current state of biological diversity, pp. 3-20. In: E.O. Wilson (Ed.) *Biodiversity.* National Academy Press, Washington, D.C., pp. 521.
2. MacArthur, R.H. and E.O. Wilson. 1967. *The Theory of Island Biogeography.* Princeton University Press, Princeton, NJ, pp. 203.
3. Ehrlich, P.R. and E.O. Wilson. 1991. Biodiversity studies: Science and policy. *Science* 253:758-762.
4. National Park Service. 1980. State of the Parks - 1980, A Report to the Congress. Office of Science and Technology, U.S. Department of the Interior, Washington, D.C., pp. 57.
5. McNeely, J.A. and K.R. Miller (Eds.). 1984. *National Parks, Conservation, and Development.* Smithsonian Institution Press, Washington, D.C., pp. 825.
6. Mantell, M. 1979. Preservation and use: Concessions in national parks. *Ecology Law Quarterly* 8:1-54.
7. Lemons, J. and D. Stout. 1984. A reinterpretation of national park legislation. *Environmental Law* 15:41-65.
8. Lemons, J. 1987. United States' national park management: Values, policy, and possible hints for others. *Environmental Conservation* 14:329-340,328.

9. Lemons, J. 1987. Title 16 United States Code §55 and its implications for management of concession facilities at Yosemite National Park, California. *Environment Management* 11:461-472.
10. Simon, D.J. (Ed.). 1988. *Our Common Lands: Defending the National Parks.* National Parks and Conservation Association, Washington, D.C., pp. 567.
11. Mantell, M.A. (Ed.). 1990. *Managing National Park System Resources.* The Conservation Foundation, Washington, D.C., pp. 270.
12. National Park Service. 1975. Management Policies. United States Department of Interior, Washington, D.C.
13. Shabecoff, M. 1981. Administration seeks greater role for entrepreneurs at federal parks. New York Times, March 29, §1, at 1, col. 1.
14. Parsons, D.J., D.M. Graber, J.K. Agree and J.W.V. Wagtendonk. 1986. Natural fire management in national parks. *Environmental Management* 10:21-24.
15. Bonnicksen, T.M. and E.C. Stone. 1985. Restoring naturalness to national parks. *Environmental Management* 9:479-486.
16. Administrative Procedure Act §10, 5 U.S.C. §706 (Supp. III, 1965-67).
17. Shands, W.E. 1982. Testimony before the Subcommittee on Public Lands and National Parks, February 4, 1982, U.S. House of Representatives, Washington, D.C.
18. Lemons, J. 1986. Research in the national parks. *The Environmental Professional* 8:127-137.
19. Robbins, W.J. 1963. A Report by the Advisory Committee to the National Parks Service on Research. National Research Council, National Academy of Sciences, Washington, D.C.
20. Leopold, A.S. 1963. Wildlife management in the national parks. *American Forests* 68:32-35, 61-63.
21. Soulé, M.E. 1985. What is conservation biology? *BioScience* 35:727-734.
22. Noss, R.F. 1990. Can we maintain biological and ecological integrity? *Conservation Biology* 4:241-243.
23. Franklin, J.F., C.S. Bledsoe and J.T. Callahan. 1990. Contributions of the long-term ecological research program. *BioScience* 40:509-523.
24. Chase, A. 1990. What Washington doesn't know about the national park system, pp. 139-149. In: J.A. Baden and D. Leal (Eds.) *The Yellowstone primer: Land and Resource Management in the Greater Yellowstone Ecosystem.* Pacific Research Institute for Public Policy, San Francisco, CA.
25. Houston, D.B. 1982. *The Northern Yellowstone Elk: Ecology and Management.* Macmillan, New York, NY.
26. Romme, W.H. and D.G. Despain. 1989. Historical perspectives on the Yellowstone fires, 1988. *BioScience* 39:695-699.
27. Boyce, M.S. 1991. Natural regulation or control? pp. 183-208. In: R.B. Keiter and M.S. Boyce (Eds.) *The Greater Yellowstone Ecosystem.* Yale University Press Press, New Haven, CT, pp. 428.
28. Saunders, D.A., R.J. Hobbs and C.R. Margules. 1991. Biological consequences of ecosystem fragmentation—A review. *Conservation Biology* 5:18-32.

29. Keiter, R.B. and M.S. Boyce (Eds.) 1991. *The Greater Yellowstone Ecosystem.* Yale University Press, New Haven, CT, pp. 428.
30. Westman, W.E. 1990. Park management of exotic plant species: Problems and issues. *Conservation Biology* 4:251-260.
31. Westman, W.E. 1990. Managing for biodiversity: Unresolved science and policy questions. *BioScience* 40:26-33.
32. Sax, J. 1980. *Mountains Without Handrails: Reflections on the National Parks.* The University of Michigan Press, Ann Arbor, MI, pp. 160.

Biological Diversity: Problems and Challenges. Edited by S.K. Majumdar, F.J. Brenner, J.E. Lovich, J.F. Schalles, and E.W. Miller. © 1994, The Pennsylvania Academy of Science.

Chapter Seventeen

ETHICAL AND PHILOSOPHICAL ASPECTS OF BIODIVERSITY CONSERVATION

WILLIAM H. MURDY

Oxford College
Emory University
Oxford, Georgia 30267

INTRODUCTION

The idea that the world's diversity of species was created expressly for the benefit of humans who were separate from nature has been prevalent throughout most of Western history. In this view, nature was simply a theatrical background against which the meaningful human drama was played out. Thomas Malthus and Charles Darwin provided evidence which undermined this narrow form of anthropocentrism. Malthus, in his *Essay on Population,* published in 1798, argued that species have a natural tendency to overreproduce. This, coupled with the earth's limited power to provide subsistence, results in "a strong and constantly operating check on population." He concluded that humans are not apart from nature but subject to "the weight of this law which permeates all animate nature."[1] Charles Darwin, in the *Origin of Species*[2] and the *Descent of Man*[3] maintained that all species were the products of an evolutionary process guided principally by the laws of natural selection and that the human species was no exception. He argued convincingly that species exist as ends in themselves and not for the exclusive benefit of any other species. The science of ecology, which emerged only recently as a respectable and legitimate scientific discipline, has provided abundant evidence that no species can exist apart from a complex matrix of relationships involving other species and the physical environment.

For most of its history, the human species consisted of small populations whose impact on nature, consisting mainly of hunting, fishing, gathering and burning, was slight. A major change in the impact of human activities on the environment occurred with the domestication of plants and animals about 10,000 years ago. Populations enlarged, cities arose and ancient civilizations came into existence. The tapping of fossil fuels beginning in the last century, created the industrial revolution which again led to an increase in human populations and in their power to alter earth's ecosystems. Vast areas of the earth are presently occupied by human-managed ecosystems resulting in a loss of the earth's original biodiversity. Moreover, biodiversity is declining as natural ecosystems become converted to managed ones in order to feed, clothe and shelter expanding human populations. While extinction is the fate of most species, the reduction of biodiversity from human-generated extinction seems destined to approach the mass extinctions at the end of the Paleozoic and Mesozoic Eras.[4]

ENVIRONMENTAL ETHICS

It is our fate to live at a time when the human species has reached a critical stage in its evolution. Within a relatively short period of time, humans have acquired enormous power to consume natural resources and to reduce the earth's biodiversity. Perhaps cultural species have arisen on other planets in the universe and become extinct because they were unable to manage their potentially destructive powers. Chaisson writes: "All technological beings on any planet, must evolve a planetary ethic, lest they be unprepared to endure the by-products of technoculture." Chaisson envisages a planetary ethics that "will guide our attitude and behavior toward what is best for all humankind."[5]

Nash describes how our environmental consciousness has changed from the conservation movement of the Teddy Roosevelt era, which he characterizes as being "utilitarian, anthropocentric and resource oriented," to a new environmental ethics that developed after World War II and which attributes value to non-human species as ends in themselves rather than simply means to human ends. Passage of The Endangered Species Act in 1973 was for Nash a momentous event which marked the "beginning of the institutional extension of rights to non-humans."[6] Today, environmental ethics includes a broad spectrum of ideas ranging from those of anthropocentrists, who "must save the environment to save humanity" to radical biocentrists or "deep ecologists," who view the human species "as an obstacle to saving the environment."[7]

Beginning with Aldo Leopold's "Land Ethic," there has developed a growing interest to construct an environmental ethic that goes beyond the traditional ethics of human personal relationships. The movement has raised our ecological consciousness by challenging us to consider cosmic relationships that transcend human-centered values. At the same time, the movement has raised questions that are not easily resolved. In what sense do species and other items in nature have inherent value independent of sentient beings having the capacity to value things? If taking the biocentric position, we accord to nature an "ethical status at least equal to that of humans,"[8] how should if affect our behavior toward nonethicizing forms of life which cannot reciprocate? Leopold writes: "A thing is right when it tends to preserve the integrity, stability, and beauty of the biotic community. It is wrong when it tends otherwise."[9] If we value biospheric stability as the highest good, how can we escape the conclusion that the planet would be better off without the one species which threatens it?

The ecological ethics discussed in this paper is human-centered and similar to Potter's bioethics, which he describes as "a system of morality based on biological knowledge and human values, with the human species accepting responsibility for its own survival and for the preservation of the natural environment."[10] This position is consistent with a philosophy that affirms the essential interrelatedness of things and that values all items in nature since no species or event is without some effect on wholes of which we are parts. "The key to contemporary ecological consciousness," writes Sessions, "is to see the diminishment of man and the diminishment of the planet and its non-human inhabitants as essentially one and the same problem."[11]

THE IMPORTANCE OF BIODIVERSITY

Biodiversity conservation is obviously important because we depend upon genetic diversity both within and between species to sustain and improve agricultural productivity and as a source of useful medicines. Not so obvious, but of equal importance, is the role of biodiversity in the function of ecosystems which comprise our life support system.

Biodiversity and Agriculture

An essential factor in the preservation of the world's food resources is the genetic variation within and among species that can be used by the plant breeder or genetic engineer in a never-ending search for beneficial genes that can be incorporated into food crops. The genetic variation currently residing in the world's biodiversity is vitally important for a sustainable agriculture of the future because of the continuing need to counter the evolutionary adaptations of diseases and insect pests and to find ways to produce food more efficiently on a sustainable basis.

The perennial battle between farmer and crop pest follows a now predictable pattern. Ever greater amounts of pesticide are needed to achieve an adequate level of protection. Within a few years, the pest species evolves to become resistant to the pesticide and a new one must be applied. Meanwhile, beneficial insects are killed off, soils, streams and lakes become polluted with toxic chemicals, our crops are decimated and the pest lives on.

The finding of a rare relative of corn provides an example of the importance of biodiversity in maintaining potentially useful genes. Vietmeyer[12] relates the discovery of a new species on the verge of extinction, which could revolutionize the cultivation of corn throughout the world. While searching for a perennial teosinte *(Zea perennis),* which had not been seen since the 1920's, a botany student at the University of Guadalajara discovered a diploid species of teosinte *(Zea diploperennis)* in a remote, 3-acre site at 8,000 feet in the Mexican State of Jalisco. Whereas *Zea perennis* is a tetraploid perennial not crossable with corn, *Zea diploperennis* is a diploid perennial fully compatable with corn. This rare species may provide genes for converting corn from an annual which has to be planted and harvested annually, to a perennial which could yield grain from the same plants year after year.

Biodiversity and Medicine

The Pacific Yew is an example of a species recently found to be the source of a "miracle drug", taxol, used in cancer therapy. Taxol is a complex chemical effective against ovarian and breast cancer. This remarkable discovery overnight converted a species from a "trash" tree considered fit only for burning to make way for more Douglas fir, to one in very high demand.

A renewed interest in phytochemical prospecting is being fueled by the fear that thousands of tropical species with a potential benefit to medicine may soon be lost forever.[13] It is a welcome initiative that may encourage countries to mitigate the destruction of their tropical forests where most of the earth's biodiversity is located and to value forests for reasons other than either conversion to agriculture or timber harvesting. Merck "is paying one million to Costa Rica's National Institute of Biodiversity for the right to analyze hundreds of indigenous plant and animal extracts for possible drugs or other commercial products."[14] If marketable drugs are discovered, Costa Rica will share in the royalties.

Biodiversity and Ecosystem Integrity

There is nothing more fundamental to our survival and well-being than the earth's ecosystems which affect the quality of air, climate, the recycling of nutrients, the accommodation of wastes and the generation and maintenance of soils.[15]

The biodiversity of ecosystems is the result of an extended process of evolution whereby selection favors the maximum number of organisms that can exist in a particular set of environmental circumstances. Darwin was one of the first to note that the most severe competition is among individuals of the same species and that diversity allows a greater number of individuals to occupy the same site. "In the general economy of any land," he writes, "the more widely and perfectly the animals and plants are diversified for different habits of life, so will a greater number of individuals be capable of there supporting themselves."[16] Thus, greater diversity results in a greater biomass per unit area. "Mixed forests have a higher rate of biomass production than a homogeneous stand of trees." (Ewel, 1986 cited in Pimental, et. al., 1992). "This difference occurs because each tree species has a specific set of nutrient requirements, and the mixed planting of trees makes effective use of the complex mix of nutrients present in the soil."[17]

Darwin was struck by how unrelated species "are bound together by a web of complex relationships." He notes that in England only humble-bees pollinate the common red clover, "as other bees cannot reach the nectar." He concludes that if the species of humble-bees become extinct, red clover would also become extinct.[18] In a contemporary sequel to Darwin's illustration of co-dependency, honey bees in the United States currently are threatened by parasitic mites. Since honey bees pollinate most fruit and vegetable crops, their regional extinction could have a devastating effect on both agriculture and wild plants.

We simply do not know how many species can be lost from an ecosystem before it becomes disfunctional. Once key species become rare or extinct, the ability of an ecosystem to function may be impaired with an attendant decrease in biomass, increase in soil erosion and a diminished capacity to retain moisture, recycle nutrients and exchange gases. In marine ecosystems, the loss of a key bacterium or alga could diminish ecosystem function with a resulting loss of food resources as well as the capacity of the oceans to absorb carbon dioxide. A recent study of Antarctic marine ecosystems revealed that as the ozone layer thinned, ultra-violet radiation to the planktonic layer increased, which in turn resulted in a 6-12 percent reduction in primary productivity.[19]

In Daniel Quinn's recent novel, Ishmael, the main character, a gorilla, asks: What's wrong with "a global community that consists of grass, gazelles, and lions? Or a global community that consists of nothing but rice and humans?" He answers:

> "Diversity is a survival factor *for the community itself.* A community of a hundred million species can survive almost anything short of total global catastrophe. Within that hundred million will be hundreds of thousands that could survive a global temperature drop of twenty degrees... [or a] rise of twenty degrees. But a community of a hundred species or a thousand species has almost no survival value at all."[20]

THE INCLUSIVE COMMUNITY

Aldo Leopold writes:

> All ethics so far evolved rest upon a single premise: that the individual is a member of a community of interdependent parts. His instincts prompt him to compete for his place in that community, but his ethics prompt him also to co-operate (perhaps in order that there may be a place to compete for).[21]

Richards[22] considers community welfare to be the highest moral good. While this remains constant, concepts of what particular acts promote community well-being will change as the rational and moral capacities of the human species change. For example the biblical injunction "Be fruitful and multiply and subdue the earth" was appropriate for most of human history when communities consisted of small groups of related individuals struggling to survive, but in the modern world it is an anachronism. A central question given the global effects of our behavior is how to develop a more inclusive community.

Throughout most of human history the successful genes belonged to those hunter-gatherer groups who combined amity and cooperation among members with enmity and hostility toward other groups. Darwin noted that humans and other animals endowed with the social instincts "take pleasure in one another's company, warn one another of danger, defend and aid one another in many ways. These instincts do not extend to all the individuals of the species, but only to those of the same community."[23]

The scientific concepts of kin selection and inclusive fitness[24] explain why individuals who share many genes in common care for one another. Science may also explain why unrelated individuals within a group would behave decently toward one another based upon the concept of reciprocal altruism.[25] However, one cannot deduce from the data of science how far and to whom our moral obligation should extend. Should it extend to persons who differ by race, creed or ideology? To everyone on the planet? To generations yet unborn? Here philosophy and religion come into play.

An important aspect of all religions is a relational one: the ability through myth, metaphor and symbol to provide a framework of connectedness to greater wholes. The common ground "between all the major systems of different religions and groups of mankind," writes Waddington include "such values as truth, respect for self, respect for others, and respect for something larger and more embracing than one's immediate experience. (God? mankind? the whole creative universe?)"[26] Joseph Campbell anticipated the emergence of a new global myth for our time. Its symbol would be the picture of the earth taken from the moon and the community would include all inhabitants of the planet.[27]

A critical need is to extend our concept of community beyond family and friends to others who share neither our genes nor our values. Whatever extension of our moral obligation takes place must begin with the family and local community, because these are essential elements of our human nature. The capacity to extend "the sense of 'we' to include 'them' ", which Cobb calls the capacity for transcendence, "has an important role to play, but its task is not to suppress or displace family or community. It is to call attention to the need for extending these and to ways of overcoming obstacles to that extension."[28]

Some philosophers such as Singer would have us extend moral obligation not only to all humans, but to all species "whose welfare can be affected by our actions." This would include non-human animals "with the capacity to feel pleasure or pain."[29] Leopold would have us extend ethics to the land, which for him was not merely soil but "a fountain of energy flowing through a circuit of soils, plants and animals."[30] He saw this enlargement of the boundaries of community as a logical progression in our social evolution and an ecological necessity for human survival.

Intergenerational Justice

Until recently, we did not realize that the activities of one generation could lower drastically the quality of life for the next. Traditionally, most people believed that their activities were building a better world for their descendants. Nineteenth-century scientists saw the growth and application of scientific knowledge "leading infallibly upward to an empyrean noon hour for mankind," writes Monod, "whereas what we see opening before us today is an abyss of darkness."[31]

Evolution has programmed us to promote the propagation of our genes by caring for our offspring, the carriers of our genes, and to promote the common welfare of our social group. Once we understand that unbridled self-indulgence on the part of one generation will reduce the quality of life for the next, we must deal with the moral question of intergenerational justice. Most ethical systems would consider it morally wrong to undermine the well-being of others simply for personal gain. Heilbroner holds that the willingness of one generation to sacrifice in order not to jeopardize others, requires that we "form a collective bond of identity with future generations."[32] Rolston reminds us that "self-interest and benevolence are not necessarily incompatible especially where one derives personal fulfillment from the welfare of others."[33] Hartshorne explains how concern for the future of others can be grounded in our own self-interest. We don't only value others for their usefulness to ourselves, but we also value ourselves for our usefulness to others. Regard for the future of others may be as important in providing our lives with meaning, right now, as regard to our own future.[34]

DISCUSSION AND CONCLUSIONS

In light of our knowledge that humans, like all other species, are dependent upon the proper function of earth's ecosystems, which in turn are dependent upon a minimum level of species diversity, is it rational to degrade or destroy ecosystems and extinguish millions of species in order to expand human numbers and increase our power to consume resources and generate wastes? It has become clear that a continuation of our treatment of nature in pursuit of the expansionist economic vision cannot be sustained. At the recent United Nations sponsored Earth Summit in Rio de Janeiro, Maurice Strong, Secretary-General of the Conference, said: "We have been the most successful species ever; but ladies and gentlemen, we are now a species out of control." A commitment to biodiversity conservation demands that we adopt a concept of progress more appropriate to the age; one which would be measured by "our ability to move toward less consumption, less depletion of resources, and lower rates of population growth."[35]

Metaphorically, the human species is to the biosphere as a virulent parasite is to its host. Since extinction of the host causes extinction of its parasite, it is adaptively advantageous for parasites to evolve into less virulent forms in order to achieve a more balanced association with their host—one where the parasite benefits by its relationship, but the host is not destroyed. If we human parasites are to have a future we must develop a more benign association with our host, planet earth.

The kind of human-centered environmental ethics described in this paper is justified because it promotes the survival of the human species. However, perpetuation of the human species is a necessary, but not sufficient goal for human beings, who need to preserve and enhance cultural values such as truth, justice, compassion and beauty, because they define us as humans. We need to stabilize world population not only because of the pressure of population growth on biodiversity, resources and natural areas but because it threatens our human values regarding the dignity of life. Traditional, moral values have conditioned us to regard famine, war and epidemic disease as human calamities to be ameliorated by compassion and material aid. In a world where poverty-stricken, politically-explosive billions seethe with resentment against the privileged minority, the latter, it is feared, will come to view such catastrophic events not as tragedies, but as "solutions" to the problem of an overcrowded earth.

Humans are social animals, who can function properly only within the context of a family and social group. In a world that increasingly resembles a global village, it is important that we view ourselves as members of inclusive communities that do not displace, but transcend the family and local group.

The test ban treaty in the 1960's, an agreement among nuclear powers to stop above ground testing of atomic weapons, demonstrates that diverse, even adversarial, human groups can view themselves as part of a single, global community when it comes to a common health threat (radionuclides entering the human food chain) to ourselves and our children. The Vienna Convention for the Protection of the Ozone Layer in 1985, the Montreal Protocol on Substances that Deplete the Ozone Layer in 1987 and the Earth Summit at Rio de Janeiro, where, among other agreements, most diplomats signed a biodiversity treaty, are hopeful signs that international cooperation is possible when dealing with ecological issues that transcend political boundaries.

As temporal manifestations of an evolutionary process, we are the links between its past and future. Billions of years of atomic, chemical and biological evolution were required to produce a biosphere hospitable toward *Homo sapiens.* Our behavior will determine in large measure the fitness of the biosphere to support our descendants. The earth has finite resources (high entropy fuels, highly ordered minerals, biodiversity and functional ecosystems) which should be valued as a precious heritage to be shared with future generations. Utilization of resources by each generation is of course necessary, but squandering them in excessive or "conspicuous" consumption and destroying natural ecosystems together with their diversity of species is immoral if our concept of community extends to future generations as it knowingly deprives others of their essential need for a livable world. The preservation of sufficient biodiversity to sustain human life and cultural values, in the final analysis, may depend on extending our sense of community to include others with whom we share the earth and a willingness to assume a moral obligation to the next generation.

REFERENCES

1. Malthus, T.R. [1798] 1959. *Population: The First Essay.* The University of Michigan Press, Ann Arbor, MI, pp. 139.
2. Darwin, C. 1859. *The Origin of Species.* The Modern Library, New York, NY, pp. 1-386.

3. Darwin, C. 1871. *The Descent of Man.* The Modern Library, New York, NY, pp. 389-924.
4. Wilson, E.O. 1985. The biodiversity crisis. *Bioscience* 35: 700-706.
5. Chaisson, E.J. 1988. Our cosmic heritage. *Zygon* 23: 469-479.
6. Nash, R.F. 1984. Rounding out the American revolution: ethical extension and the new environmentalism, pp. 170-1818. In: M. Tobias (Ed.) *Deep Ecology.* Avant Books, San Diego, CA, pp. 296.
7. Watson, R.A. 1992. Comment: Misanthropy, humanity and the eco-warriors. *Environ. Ethics* 14:95.
8. Nash, R.F. 1989. *The Rights of Nature.* The Univ. Wisconsin Press, Madison, WS, pp. 290.
9. Leopold, A. [1949, 1953] 1970. *A Sand County Almanac with Essays on Conservation from Round River.* Ballantine Books, New York, NY, pp. 295.
10. Potter, V.R. 1988. *Global Bioethics.* Michigan State University Press, East Lansing, MI, pp. 203.
11. Sessions, G. 1984. Ecological consciousness and paradigm changes, pp. 28-44. In: M. Tobias (Ed.) *Deep Ecology.* Avant Books, San Diego, CA, pp. 296.
12. Vietmeyer, N.D. 1979. A wild relative may give corn perennial genes. *Smithsonian,* December: 68-76.
13. Joyce, C. 1992. Western medicine men return to the field. *Bioscience* 42: 399-403.
14. Stone, R. 1992. The biodiversity treaty: Pandora's Box or a fair deal? *Science* 256:1624.
15. Tobin, R.J. 1990. *The Expendable Future: U.S. Politics and the Protection of Biological Diversity.* Duke University Press, Durham, NC, pp. 325.
16. Darwin, C. 1859, op. cit.
17. Pimental, D., et. al. 1992. Conserving biological diversity in agricultural/ forestry systems. *Bioscience* 42: 354-362.
18. Darwin, C. 1859, op. cit.
19. Smith. R.C., et. al. 1992. Ozone depletion: ultraviolet radiation and phytoplankton biology in Antarctic waters. *Science* 255: 952-958.
20. Quinn, D. 1992. *Ishmael.* A Bantam/Turner Book, New York, NY, pp. 266.
21. Leopold, A. 1970, op. cit.
22. Richards, R.J. 1987. *Darwin and the Emergence of Evolutionary Theories of Mind and Behavior.* University of Chicago Press, Chicago, IL, pp. 700.
23. Darwin, C. 1871, op. cit.
24. Hamilton, W.D. 1964. The genetical evolution of social behavior. Jour. Theor. Biol. 7:1-52.
25. Trivers, R.L. 1971. The evolution of reciprocal altruism. *Quart. Rev. Biol.* 46:35-57.
26. Waddington, C.H. 1971. *Biology, Purpose and Ethics.* Clark University Press with Barre Publishers, pp. 54.
27. Campbell, J. 1988. *The Power of Myth with Bill Moyers.* Doubleday, New York, NY. pp. 229.
28. Cobb, J.B. 1988. Befriending an amoral nature. *Zygon* 23:431-436.

29. Singer, P. 1981. *The Expanding Circle: Ethics and Sociobiology.* Farrar, Straus and Giroux, New York, NY pp. 190.
30. Leopold, A. 1970, op. cit.
31. Monod, J. 1971. *Chance and Necessity.* Knopf, New York, NY, pp. 198.
32. Heilbroner, R.L. 1974. *An Inquiry into the Human prospect.* Norton, New York, NY, pp. 150.
33. Rolston, H., III. 1975, Is there an ecological ethic? *Ethics* 85:93-109.
34. Hartshorne, C. 1974. Beyond enlightened self-interest: A metaphysics of ethics. *Ethics* 84:201-216.
35. Brown, G.E., Jr. 1992. Reorienting scientific and technological inquiry to tackle the global crisis facing humanity. *Chronicle of Higher Education,* April 22: B1-B2.

Biological Diversity: Problems and Challenges. Edited by S.K. Majumdar, F.J. Brenner, J.E. Lovich, J.F. Schalles, and E.W. Miller. © 1994, The Pennsylvania Academy of Science.

Chapter Eighteen

THE ROLE OF WILDLIFE MANAGEMENT IN THE MAINTENANCE OF BIODIVERSITY

WILLIAM E. MAGNUSSON
Depto de Ecologia, INPA, CP478
69011 Manaus AM, Brazil

INTRODUCTION

This is not a chapter about how to manage wildlife. There are many good books that cover techniques and examples in detail (e.g. Schemnitz and Toschik 1980, Teague and Decker 1979, Eltringham 1984, Robinson and Redford 1991: see also Nichols and Applegate 1987). There are also many relevant books on biodiversity and conservation (e.g. Soulé and Wilcox 1980, Prance 1982, Western and Pearl 1989). Here I will present a highly personal view about the direction of wildlife management in the future and the political/administrative ground work that needs to be done before management can be implemented. Biodiversity and conservation are now so fashionable that it is unlikely that I will say anything truely original. Rather, I will use examples from the management literature as well as from primarily ecological sources to try to bring about a marriage between the two groups of disparate professionals. One group that is quickly developing theory about what should be done and another group who will probably have to take responsibility for getting it done.

Wildlife management is the manipulation of populations or communities of animals to obtain a desired effect. The manipulation may be direct or may be brought about by modification of habitat or by the control of competitors and predators such as humans. The aim of wildlife management is always conservation, whether it be conservation of a species, conservation of an ecosystem or even conservation of a crop being damaged by a wildlife species. However, different forms of management may have vastly different effects on the maintenance of biological diversity. It is not uncommon for wildlife managers to lament the reversion of disturbed habitats to a more natural state (e.g. Brennan 1991).

To understand the effects of management we have to realize that there is no plan that has only positive effects and no plan that has the same effect in different geographic areas or at different points in time. We are talking about trade offs. At each decision we ask: "How much are we willing to lose and will the amount we gain compensate for this?". Decisions will be made on moral, ethical or social grounds (Decker et al. 1991); the science merely puts probabilities on potential outcomes.

It should be obvious that I am referring to wildlife management in its modern sense. In some parts of the world, especially in some states of the U.S.A., wildlife management is considered synonymous with game management. Those attitudes are changing (Herberlein 1991; Gentile 1987) but, as recently as 1990, one of the leading state game biologists in the U.S. said to me "Why do you need to create National Parks? We have reserves all over my state and we hunt on every one of them. If you do not hunt, the populations will be decimated by disease or natural disasters. Hunting is necessary to keep them healthy" (See MacDonald [1987] and Kennedy [1985] for other examples of this philosophy).

Some animal rights groups believe that any intervention imposes unnecessary unnatural suffering on animals. A friend told me of a conversation with the head of such an organization who told him "I do not care if the last individual of a species is killed as long as it dies humanely" (See Decker and Brown [1987] for some wildlife managers views). Obviously, many persons deeply committed to management or non-management of biological diversity do not hold the maintenance as a high priority. Managing one species and managing for diversity usually require very different strategies (e.g. Graul and Miller, 1984). Shepherd and Caughley (1987) give more detailed discussion of the problem of defining strategies for decisions made on different moral criteria.

Supposing that we convice the relevant politicians that the conservation of biological diversity deserves high priority, how do we define "diversity"? The number of species is not sufficient. Most people would rank an assemblage of 10 species of grasshoppers as less diverse than one with only 4 species but that included a grasshopper, a snail, a fish and a mammal. Some people are even thinking of potential diversity, trying to preserve the evolutionary potential to create new life forms in the future (Soulé and Wilcox 1980). I do not want to get bogged down in a discussion of these issues here but they are issues that must be faced by would-be managers. Management is a tool, and tools are only useful when we know what we want to fix.

Wildlife management grew out of game management. If it were not for the private hunting preserves of royalty, places like Europe and India would have far less biodiversity than they do today. The recognition that habitat is necessary to preserve game species led to the first unintentional link between wildlife management and the maintenance of biological diversity. However, game management is not always the alternative that leads to the maintenance of highest biological diversity and I will discuss the relationships between game management and biodiversity in more detail later.

In the late 1950s and early 1960s, the science of ecology became popular. Researchers tried to determine the causes of the distribution and abundance of organisms. Some worked at the level of species, others with communities, and some at the level of energy flow. All of these studies have influenced wildlife managers but it was probably the book by MacArthur and Wilson (1967) on island biogeographic theory that brought home most forcefully to wildlife managers the link between ecology and management. Fragmentation of habitats leads to a loss of biodiversity and it is often the large anmials that humans consider as game that go first.

Concommitantly, increasingly urbanized populations started to put more value on non-consumptive uses of wildlife such as photography, birdwatching and ecotourism. These urbanites frequently agitated and voted to save species with which their only contact was through the television. The camera had made valuable the life of a spider that would go unnoticed beneath a hunters boot. Wildlife became available to people who would never feel the chill of the dawn wind in a treeless marsh or the blaze of an open fire in a moonlit forest.

The definition of wildlife was rewritten in legislation in countries throughout the world and management took on a wider meaning. However, the increased awareness of wildlife needs reflected an increased awareness of human needs. Despite the enormous social injustices that exist in the world today, never before in the history of civilized man have there been so few wars and such a small fraction of the world population living under direct military tyranny. No longer is it acceptable to preserve wildlife at the whim of the ruling classes. Wildlife management will only be acceptable if it can be shown to increase the quality of human life, whether it be for economic, aesthetic or religious reasons. The greater part of wildlife management in the greater part of the world is mediated throught the control of a single species: *Homo sapiens.*

Human intervention in the loss of biodiversity started with the game parks; today many of these have become national parks. However, even the largest parks are probably too small to preserve many of their constituent species (Western and Pearl, 1989). The human populations are increasing and species that cannot use human-impacted environments, at least for dispersal, are probably doomed to extinction. Also, many of the areas with the highest remaining biological diversity are inhabited by indigenous peoples or are being colonized by landless peasants. In some cases, those interested in cultural conservation and those interested in conservation of biodiversity have similar short-term goals but they will not be able to reconcile their long-term objectives unless the trade offs are understood by both parties and strong planning (management) is initiated now.

I will try to keep the following discussion of options for terrestrial wildlife as simple and general as possible but the concentration of biodiversity in third world countries and my experience will undoubtedly lead to a third world bias.

There is a natural progression in the management options with the scale of the problem. Unfortunately, few third world governments realize this and there has been considerable money wasted by governments importing inappropriate technology. There is also considerable waste of resources by dedicated conservationists in first-world countries trying to export high-technology management practices to countries which do not have the infrastructure to use them. In this grossly-simplified overview I will try to link options to situations. There are many exceptions to the "rules" I will present but someone incapable of seeing the generalities is unlikely to be able to understand the exceptions.

What options for management are appropriate in any situation? Probably any active management is better than none at all, simply because it involves a scientific presence. Habitat loss is the major cause of reduction in biodiversity everywhere in the world. Without basic monitoring and long term studies we may not even realize what we are losing (Wilson 1989). This is the reason that researchers doing "pure" science, especially taxonomists, are often at the forefront of the fight to save biodiversity. Unfortunately, most third-world countries do not have the resources to maintain "pure" scientists in the field. Universities are understaffed and professors underpaid. In this situation, wildlife managers may offer the only option to maintain professionals in the field. In most countries, it is possible to find persons responsible for issuing permits, collecting taxes and advising the government. However, it is difficult to find anyone responsible when the resource is destroyed. Importing management technology is of no use unless there are mechanisms for implementation. Wildlife managers are usually not trained in community organization, political administration or the other social sciences that are necessary to create viable long-term programs in the third-world countries where there is no tradition of resource management. Brazil is a good example. It is large and has more biodiversity than any other country in the world. Individual states within Brazil are similar in size to, or larger than, most European countries. Nevertheless, the number of persons responsible for wildlife would fit in a closet. No state has a wildlife department and all decisions are made at the national level. The persons who write permits and make decisions about wildlife have often never seen the species and areas about which they legislate. What use is it to import technology if there is no to one use it?

Mismanagement generates controversy but at least when we have mismanagement we have something to target and improve. In wildlife management there are no good options but some are better than others. Given that a political/administrative system is in place, what are the common options for common situations?

SITUATION 1: Remnant habitat in areas that are totally alienated in terms of survival of most wildlife.

Few species will persist for long periods (hundreds of years) if confined in small islands (Samson et al. 1985). Intensive, active management is the only option for perserving most species in this situation which is typical of most of the developed regions of the earth. Simply designating the remnant patches as reserves is not enough; manipulations must be done to slow the rate of extinctions. Essentially, every species needs a survival/recovery plan and it is obvious that even the richest nations do not have the resources for this. Usually, a large vertebrate species with an extensive home range that includes habitat suitable for other wildlife and which is attractive to the public is targetted (Conway 1989). They are the so-called flagship species which attract conservation money and political good will. Intensive study of the life history of the species leads to demographic models which predict the probability of extinction due to random environmental (e.g. Dennis et al. 1991) or genetic fluctuations (e.g. Soulé 1980; Vrijenhoek 1989; Woodruff 1989; Hedrick and Miller 1992). Such models have been criticized because they can give the impression that there is a minimum acceptable size of reserve. Of course they cannot do this and minimally viable populations may not be sufficient to maintain ecosystem processes (Conner 1988). However, they can show which species are obviously at immediate risk.

Increasing the area of habitat for the species usually is not a viable option but simple habitat manipulations may increase the size of populations and help guard against random extinctions (e.g. Walters 1991). This is the most elegant and effective means of conserving species. However, in many cases we do not know what are the factors causing the decline or the cause may be something that we cannot control in the short term. In this case, captive propagation may be the only alternative to save the species. Successful examples of the use of this life-boat approach such as the Arabian Oryx, the Indian Gharial, the Golden Lion Tamarin and the Hawaiian Goose (Conway 1989) have attracted much attention. However, the approach is extremely expensive, success rates may be low (Dodd and Seigel 1991), and the vast majority of less spectacular species will never be considered for such programs.

Economic exploitation of species in remnant habitat is not usually a viable option. In most cases, the risks far outweigh the benefits. Even non-consumptive uses such as tourism must be strictly controlled (e.g. Boyle and Samson 1985; Pomerantz et al. 1988; Anderson 1988). The "hands-off" option is also not viable. Without direct manipulative management by qualified wildlife managers, most species in situation 1 will be lost.

SITUATION 2: Natural habitat of various sizes in a sea of partially degraded habitat.

Here I will define degraded habitat as that which contains humans or their domesticated hangers on. This will raise the eyebrows of some biologists and most anthropologists. However, human cultural evolution seems always to outpace the genetic adaptations of multicellular organisms. Cultural evolution goes through many phases but agriculture always seems to be its end point. More energy per unit area or unit time can be extracted from the land by agriculture than from any other form of renewable resource use. The industrial revolution was based on the agricultural revolution. If 5-10 percent of the population were not able to provide food for the rest, no one would have time to invent machines. Agriculture allows humans to degrade habitat that is not cultivated because it gives them time for hunting and activities such as gold mining. Williams (1987) described agriculture as "our greatest mistake".

Wildlife managers interested in the maintenance of biodiversity often share short-term goals with anthropologists and social scientists interested in cultural conservation. However, biologists should be honest when they say that they want to conserve the potential for organic evolution that will take tens of thousands or millions of years. Cultural evolution in tens or hundreds of years may lead to a divergence between the biologists and the people they now support. Anthropologists often do not recognize the difference between alpha and beta diversity. Some cultural landscapes exhibit high diversity of cultivated plants, domesticated animals and anthropophyllic wild species. However, there is far less difference among cultural landscapes in different areas than there is between natural communities separated by the same distances. The differences among cultural landscapes will tend to decrease as barriers to communication and transportation break down. Therefore, although cultural landscapes have an important role in increasing local diversity, continentwide and worldwide diversity will depend on representative natural areas also being conserved. Wildlife managers will have to have the courage to take a hand in the direction of cultural evolution. Huntley et al. (1992) present a succint overview of the interrelationships and trade offs.

Small numbers of people and small changes in culture can lead to dramatic effects on wildlife. When the nomadic aborigines that inhabited the central Australian deserts were attracted to stations and towns by the more regular supplies of food and water (and more recently, medicine and education), the rate of extinctions of mammals in the desert skyrocketed. It appears that the patchy fire regime of the desert aborigines had maintained a mosaic of habitats for species that could not survive the extensive and relatively homogeneous burns of the natural fire regime (Burbridge et al. 1988). However, the aborigines may have had a hand in the demise of the megafauna which maintained a vegetation mosaic before they and their fire came to dominate the land (Caughley, 1987). Changing the density, distribution or cultural habits of people will lead to large and unpredictable changes in biodiversity. Wildlife managers must be prepared to study such changes.

Subsistence hunting induces drastic changes in the fauna of the larger vertebrates (Redford and Stearman 1989; Redford, in press) and nobody knows what long-term effects the removal of these large seed dispersers, herbivores and carnivores will have on the structure and composition of the vegetation. Subsistence hunting is almost impossible to control except where religious taboos reinforce the wildlife legislation such as in India (e.g. Krausman and Johnsingh 1990).

Contented humans, with zero population growth, living in complete harmony with the natural world, do not exist. Humans are reproductively profligate, extremely dangerous animals and every one of them has as much right to see the world the way he wants it as a wildlife manager. How can the wildlife manager use these creatures to make his life easier? He must convince them that he is able to increase the quality of their life and at the same time reduce the total amount of primary productivity that is channelled into producing more humans. Life-boat strategies are extremely expensive (Western 1989b). Habitat that is only partially degraded can support much biological diversity (Western, 1989a) and many habitat specialists can use it for disperals, and hence recolonization and genetic interchange, even though they cannot survive indefinitely there (e.g. Stacy and Taper, 1992). This allows managers to concentrate on a smaller number of sensitive species in natural reserves. However, in some cases, large areas of regenerating native vegetation may have negative effects (e.g. Janzen 1983). Decisions about the appropriate proportion of natural and disturbed habitats and their distributions require complex but testable criteria (e.g. Murphy and Noon 1992; Bodmer and Ayres 1991). There are often strong economic reasons to maintain parts of human impacted landscapes under natural cover, the most common being soil conservation and flood mitigation. Wildlife managers are usually associated with another activity however; the economic exploitation of wildlife in semi-natural environments. As outlined by Redford (in press), Mares and Ojeda (1984) and Ryman et al. (1981), economic exploitation is likely to have some negative impacts on biodiversity. It is justified only when it can be shown to lead to an increase in the amount of wildlife habitat available or funding for wildlife management. In many places, wildlife management almost pays for itself (Heberlein, 1991) and many innovative schemes have been suggested to obain revenues from nonconsumptive users in developed countries (e.g. Applegate 1984; Anderson et al. 1985; Manfredo 1980; Matlich and Adams 1987; Purdy and Decker 1989). It is not sufficient that wildlife be considered a national resource. Exploitation will only lead to conservation if the economic benefits acrue to the land holder and he can see a direct link between habitat conservation and wildlife production. However, privatizing wildlife could turn it into a force as destructive as agriculture (Geist 1985, 1990).

In some countries, such as the U.S.A., large tracts of land are maintained under seminatural conditions by the federal and state governments, hunting clubs, and private landholders because the community puts great (but declining) value on sport hunting. Much wildlife is also maintained on private property for sport hunting in South Africa (Luxmoore 1985; Benson 1991). However, in most parts of the world, subsistence hunters take most of the wildlife crop and they often have no legal claim to the land on which they live (e.g. Bucher 1989). Wildlife must be managed for "the common good". It is going to take a lot of political juggling to convince the political leaders of third-world countries that the wellbeing of landless peasants is related to their wellbeing; and just as much work is needed to convince the peasants that the maintenance of biological diversity is in their best interest (See Dicker and Purdy [1988] for a discussion of these aspects in relation to a developed country). If wildlife managers cannot do this they are going to lose the fight (Kellert 1985). By the time all the peasants have televisions and can be convinced to preserve wildlife for purely aesthetic reasons there will be nothing left to protect. (The Wildlife Institute of India [Krausman and Johnsingh, 1990] shows how remarkable successes can be achieved against overwhelming population growth and poverty).

However, maintaining people to maintain wildlife is a double-edged sword (e.g. Redford and Stearman 1989). While one species may be resistant to exploitation, other species may be driven to extinction (e.g. Beddington and May 1982) even when they are closely related phylogenetically and occupy similar ecological niches (e.g. Magnusson and Rebêlo 1983). Every success will lead to a thousand problems but often, management gives us the only opportunity of learning "by doing" (e.g. Macnab 1983; Anderson et al. 1987; Walters and Holling 1990). Long-term studies are necessary if we are to maintain what we have (e.g. Davis and Ehorn 1988). It is not enough to just start, we must create the tradition of wildlife management in every political and zoogeographic region of the earth. If not, conservation will exist only on paper (e.g. Barriga 1982) and, there will be no one to face the crises that we did not predict and in all probability helped to create with our good intentions. (See Janzen [1983] and Howarth [1991] for examples of how "safe" options such as buffer zones and the use of biological control agents instead of insecticides may backfire).

SITUATION 3. Enormous areas untouched by humans.

This is the situation in which the "hands-off" option is appropriate. Unfortunately, we missed the boat 200-2000 years ago. McNeely (1989) even goes so far as to suggest that by the year 2100 National Parks will be a thing of the past.

Table 1 presents the three broad options and three broad situations that I have discussed. In my opinion, governments rarely consider the situation before reaching for an option. In this computer age, when we can simulate processes that have never existed and biologists that go out in the field and get their hands dirty are considered clumsy oafs, few third world biologists are likely to recommend years of hard slogging in the field. Many or most are in situation 2 but they reach for situation 1 options. They can get all the results they need by commuting between the genetics laboratory and the local zoo, but who is going to go out in the field and teach class 2 options for class 2 situations? Nobody, unless first world countries are going to be willing to invest in the training of third-world wildlife managers to conserve all-world biodiversity. The world is getting smaller. There are international treaties on the trade in wildlife, on the conservation of migratory species, and on economic development. Wildlife Managers are realizing that organizational and political factors affect the efficiency of management as much as the biological effects (e.g. Langenau and Ostrom 1984). Wildlife management is inherently socialist and it is quickly becoming international socialism. The old concept of a wildlife manager as a redneck in a pickup with a rifle strapped across the rear window, whose vision stops at the state border, is dying. The new concept of a wildlife manager as a man in a white coat who understands all the latest computer programs cannot last. The future wildlife manager will be a hybrid or a team and he will probably end up not respecting national frontiers, much less state borders.

TABLE 1

Summary of Situation and Management Options to Protect Biological Diversity.

	OPTIONS		
	Captive Propagation Genetic Manipulations	Consumptive Exploitation	Hands Off
SITUATION			
Islands of natural habitat in a sea of totally alienated habitat.	+	—	—
Islands of natural habitat in a sea of partially degraded habitat.	+—	+	—
Greater part of the enormous area under natural habitat without humans.	—	—	+

He will have to accept that management is dangerous and an awesome responsibility but that without management we will lose all. He must accept that the most pressing problem in the conservation of biological diversity is the construction of self-sustaining management units in developing countries (e.g. Krausman and Johnsingh 1990; Lewis et al. 1990) and not the export of the technology that works in his country. Those management units need political respect, long term funding, social relevance and the development of a wildlife tradition. The setting up of those structures will only marginally be associated with the estimation of maximum sustainable yields or minimum critical sizes of populations even though it is on such aspects that the utility of the wildlife profession is most often evaluated (Gill 1985).

LITERATURE CITED

Anderson, D.W. 1988. Dose-response relationship between human disturbance and brown pelican breeding success. Wildl. Soc. Bull. 16:339-345.

Anderson, D.R., K.P. Burnham, J.D. Nichols and M.J. Conroy. 1987. The need for experiments to understand population dynamics of black ducks. Wildl. Soc. Bull. 15:282-284.

Anderson, M.W., S.D. Reiling & G.K. Criner. 1985. Consumer demand theory and wildlife agency revenue structure. Wildl. Soc. Bull. 13:375-384.

Applegate, J.E. 1984. Nongame tax check-off programs: a survey of New Jersey residents following the first year of contributions. Wildl. Soc. Bull. 12:122-128.

Barriga, E. 1982. Colombia's "Paper Parks". Traffic (U.S.A.) 4:3.

Beddington, J.R. and R.M. May. 1982. The harvesting of interacting species in a natural ecosystem. Scientific American 247:42-49.

Benson, D.E. 1991. Values and management of wildlife and recreation on private land in South Africa. Wildl. Soc. Bull. 19:497-510.

Bodmer, R.E. and J.M. Ayres. 1991. Sustainable development and species diversity in Amazonian forests. Species 16:22-24.

Boyle, S.A. and F.B. Samson. 1985. Effects of nonconsumptive recreation on wildlife: a review. Wildl. Soc. Bull. 13:110-116.

Bucher, E.H. 1989. Conservacion y desarrollo de alternativas. Vida Silvestre Neotropical 2:3-6.

Brennan, L.A. 1991. How can we reverse the northern bobwhite population decline? Wildl. Soc. Bull. 19:544-555.
Bucher, E.H. 1989. Conservacion y desarrollo de alternativas. Vida Silvestre Neotropical 2:3-6.
Burbidge, A.A., K.A. Johnson, P.J. Fuller and R.I. Southgate. 1988. Aboriginal knowledge of the mammals of the central deserts of Australia. Aust. Wildl. Res. 15:9-39.

Caughley, G. 1987. Introduction to sheep rangelands. Pp. 1-13 *In* G. Caughley, N. Sheperd and J. Short (eds.). Kangaroos: their ecology and management in the sheep rangelands of Australia. Cambridge Univ. Press, Cambridge.

Conner, R.N. 1988. Wildlife populations: minimally viable or ecologically functional? Wildl. Soc. Bull. 16:80-84.

Conway, W.G. 1989. The prospects for sustaining species and their evolution. Pp. 199-209 *In* D. Western and M.C. Pearl (eds). Conservation for the twenty-first Century. Oxford Univ. Press, N.Y.

Davis, G.E. and W.H. Ehorn. 1988. Science and management in U.S. National Parks. Bull. Ecol. Soc. Amer. 69:111-114.

Decker, D.J. and T.J. Brown. 1987. How animal rightist view the "Wildlife management-hunting system" Wildl. Soc. Bull. 15:599-602.

Decker, D.J., R.E. Shanks, L.A. Nielsen and G.R. Parsons. 1991. Ethical and scientific judgements in management: beware of blurred distinctions. Wildl. Soc. Bull. 19:523-527.

Dennis, B., P.L. Munholland and J.M. Scott. 1991. Estimation parameters for endangered species. Ecol. Monogr. 61:115-143.

Dicker, D.J. and K.G. Purdy 1988. Toward a concept of wildlife acceptance capacity in wildlife management. Wildl. Soc. Bull. 16:53-57.

Dodd, C.K. and R.A. Seigel 1991. Relocation, repatriation and translocation of amphibians and reptiles: are they conservation strategies that work? Herpetologica 47:336-350.

Eltringham, S.K. 1984. Wildlife Resources and Economic Development. John Wiley and Sons, Chichester.

Geist, V. 1985. Game ranching: Threat to wildlife conservation in North America. Wildl. Soc. Bull. 13:594-598.

Geist, V. 1990. Review of wildlife production systems: economic utilization of wild ungulates. Quart. Rev. Biol. 62:246-247.

Gentile, J.R. 1987. The evolution of antitrapping sentiment in the United States: a review and commentary. Wildl. Soc. Bull. 15:490-503.

Gill, R.B. 1985. Wildlife Research an endangered species. Wildl. Soc. Bull. 13:580-587.

Graul, W.D. and G.C. Miller 1984. Strengthening ecosystem management approaches. Wildl. Soc. Bull. 12:282-289.

Heberlein, T.A. 1991. Changing attitudes and funding for wildlife-preserving the sport hunter. Wildl. Soc. Bull. 19:528-534.

Hedrick, P.W. and P.S. Miller. 1992. Conservation genetics: techniques and fundamentals. Ecol. Monogr. 2:30-46.

Howarth, F.G. 1991. Environmental impacts of classical biological control. Annu. Rev. Ecol. Syst. 36:485-509.

Huntley, B., E. Ezcurra, E.R. Fuentes, K. Fujii, P.J. Grubb, W. Haber, J.R.E. Harger, M. Holland, S.A. Levin, J. Lubchenco, H.A. Mooney, I. Noble, V. Neronov, H.R.H. Pulliam, P.S. Ramakrishnan, P.G. Risser, O. Sala, J. Sarukhan and W.G. Sunbroek. 1992. A sustainable biosphere: the global imperative. Bull. Brit. Ecol. Soc. 23:5-17.

Janzen, D.H. 1983. No park is an island: increase in interferences from outside as park size decreases. Oikos 41(3):401-410.

Kellert, S.R. 1985. Social and perceptual factors in endangered species management. J. Wildl. Manage. 49(2):528-536.

Kennedy, J.J. 1985. Viewing wildlife managers as a unique professional culture. Wildl. Soc. Bull. 13(4):571-579.

Krausman, P.R. and A.J.T. Johnsingh. 1990. Conservation and wildlife education in India. Wildl. Soc. Bull. 18(3):342-347.

Langenau, E.E. and C.W. Ostrom. 1984. Organizational and political factors affecting state wildlife management. Wildl. Soc. Bull. 12(2):107-116.

Lewis, D.M., A. Mwenya and G.B. Kaweche. 1990. African solutions to wildlife problems in Africa: insights from a community based project in Zambia. Unasylva "161"41:11-20.

Luxmoore, R. 1985. Game ranching in South Africa as a force in conservation. Oryx 19:225-231.

MacArthur, R.H. and E.O. Wilson. 1967. The Theory of Island Biogeography. Harvard University Press, Princeton.

MacDonald, D. 1987. Hunting-an exercise in pluralistic democracy. Wildl. Soc. Bull. 15(3):463-465.

Macnab, J. 1983. Wildlife management as scientific experimentation. Wildl. Soc. Bull. 11:397-401.

Manfredo, M.J. 1988. Second-year analysis of donors to Oregon's nongame tax checkoff. Wildl. Soc. Bull. 16(2):221-224.

Magnusson, W.E. and G.H. Rebelo. 1983. Brazilian crocodiles: the problems of conservation in a multispecies system. Zimbabwe Science News 17(3):56-57.

Mares, M.A. and R.A. Ojeda. 1984. Faunal commercialization and conservation in South America. BioScience 34:580-584.

Matulich, S.C. and R.M. Adams 1987. Towards more effective wildlife policies: an economic perspective of wildlife management research. Wildl. Soc. Bull. 15:285-291.

McNeely, J.A. 1989. Protected areas and human ecology: how national parks can contribute to sustaining societies of the twenty-first century. Pp 150-157 *In* D. Western and M.C. Pearl (eds). Conservation for the twenty-first century. Oxford Univ. Press, New York.

Murphy, D.D. and B.R. Noon. 1992. Integrating scientific methods with habitat conservation planning: reserve design for northern spotted owls. Ecol. Applications. 2:3-17.

Nichols, T.C. and J.E. Applegate. 1987. The essential volumes for a wildlife professional's library. Wildl. Soc. Bull. 15:584-591.

Pomerantz, G.A., D.J. Decker, G.R. Goff and K.G. Purdy. 1988. Assessing impact of recreation on wildlife: a classification scheme. Wildl. Soc. Bull. 16:58-62.

Prance, G.T. (ed) 1982. Biological Diversity in the Tropics. Columbia Univ. Press, New York.

Purdy, K.G. and D.J. Decker. 1989. Applying wildlife values information in management: the wildlife attitudes and values scale. Wildl. Soc. Bull. 17:494-500.

Redford, K.H. in press. The empty forest. BioScience.

Redford, K.H. and A.M. Stearman. 1989. Local peoples and the Beni Biosphere Reserve, Bolivia. Vida Silvestre Neotropical 2:49-56.

Robinson, J.G. and K.H. Redford (eds). 1991. Neotropical Wildlife Use and Conservation. Univ. Chicago Press, Chicago.

Ryman, N., R. Baccus, C. Reuterwall and M.H. Smith. 1981. Effective population size, generation interval and potential loss of genetic variability in game species under

different hunting regimes. Oikos 36:257-266.
Samson, F.B., F. Perez-Trejo, H. Salwasser, L.F. Ruggiero and M.L. Schaffer. 1985. On determining and managing minimum population size. Wildl. Soc. Bull. 13:425-433.
Schemnitz, S.D. and L. Toschik 1980 (eds). Wildlife Management Techniques Manual. The Wildlife Society, Washington: 686 pp.
Shepherd, N. and G. Caughley. 1987. Options for management of kangaroos. Pp 188-219 *In* G. Caughley, N. Shepherd and J. Short (eds). Kangaroos: their ecology and management in the sheep rangelands of Australia. Univ. Cambridge Press, Cambridge.
Sorg, C.F. and J. Loomis 1985. An introduction to wildlife valuation techniques. Wildl. Soc. Bull. 13:38-46.
Soulé, M.E. 1980. Thresholds for survival: maintaining fitness and evolutionary potential. Pp 151-170 *In* M.E. Soulé and B.A. Wilcox (eds). Conservation Biology: An Evolutionary-Ecological perspectives. Sinauer Assoc., Sunderland.
Soulé, M.E. and B.A. Wilcox (eds). 1980. Conservation Biology: An evolutionary-ecological perspective. Sinauer Assoc. Inc., Sunderland.
Stacey, P.B. and M. Taper. 1992. Environmental variations. Ecol. Applications 2:18-29.
Teague, R.D. and E. Decker (eds) 1979. Wildlife Conservation: Principles and Practices. The Wildlife Society, Washington. D.C.
Vrijenhoek, R.C. 1989. Population genetics and conservation. pp 89-98 *In* D. Western and M.C. Pearl (eds). Conservation for the Twenty-first century. Oxford Univ. Press, N.Y.
Wallace, M.S., H.L. Stribling and H.A. Clonts. 1991. Effect of hunter expenditure distribution on community economics. Wildl. Soc. Bull. 19:7-14.
Walters, C.J. and C.S. Holling. 1990. Large-scale management and learning by doing. Ecology 71:2060-2068.
Walters, J.R. 1991. Application of ecological principles to the management of endangered species: the case of the red-cockaded woodpecker. Annu. Rev. Ecol. Syst. 22:505-523.
Western, D. 1989a. Conservation without parks: wildlife in the rural landscape. Pp 158-165 *In* D. Western and M.C. Pearl (eds). Conservation for the Twenty-first Century. Oxford Univ. Press, New York.
Western, D. 1989b. Why manage nature? Pp. 133-137 *In* D. Western and M.C. Pearl (eds). Conservation for the Twenty-first Century. Oxford Univ. Press, New York.
Western, D. and M.C. Pearl (eds). 1989. Conservation for the Twenty First Century. Oxford University Press, New York. 365 pp.
Williams, R. 1987. Our greatest mistake? Aust. Nat. Hist. 22:312.
Wilson, E.O. 1989. Conservation: the next hundred years. Pp 3-7 *In* D. Western and M.C. Pearl (eds). Conservation for the twenty-first century. Oxford Univ. Press, New York.
Woodruff, D.S. 1989. The problems of conserving genes and species. Pp 76-88 *In* D. Western and M.C. Pearl (eds). Conservation for the Twenty-first Century. Oxford Univ. Press, N.Y.

Biological Diversity: Problems and Challenges. Edited by S.K. Majumdar, F.J. Brenner, J.E. Lovich, J.F. Schalles, and E.W. Miller.

Chapter Nineteen

POSSIBLE ROLE OF REPETITIVE DNA ELEMENTS IN PLANT DIVERSITY

R.K. CHAUDHURI AND ILA CHAUDHURI

Molecular Biology Laboratory
Botany Department, Calcutta University
Calcutta 700019, India

INTRODUCTION

Phenotypic plasticity can occur in any feature of plant growth and development. The phenomena such as plasticity, selection pressure and diversity can be best understood if we analyze the molecular mechanism involved in these processes. It would be ideal to cite Lima-de-Faria (1983) "all organic forms are ways of evading selection". In plant cells, there are three genomes: one at the nucleus, another at the mitochondrion and a third at the chloroplastid. In addition, sometimes, alien genomes may invade plant cells and live therein as parasites. During growth and development of plants, interactions of these genomes are key factors for diversity. However, of the three, the nuclear genome plays the key role. For decades the plant breeders had taken the advantage of germplasm diversity in creating new plants. The observations find agreement with the results obtained from animal breeding programmes (Falconer 1981). Pattern of plasticity may differ even in closely related plant species. In natural selection, some individuals are favoured as they are best suited to survive (Dobzansky 1969). It is assumed that sensor mechanism in plant chromosome is very sensitive, and plant chromosomes are more elastic than animal chromosomes. Polyploidy is a common occurrence in plants. Under selection pressure they may acquire flexible traits through inherent mechanism lying at their chromosomes. Frequent interactions between different genotypes in a population have been detected by different authors (McClintock 1984, Primac and King 1989). In many cases different genotypes can converge to identical phenotypes or the same genotype can differentiate into different phenotypes in diverse environments (Jeffries 1984, Schlichting 1986).

SOMATIC MUTATION

Mutations in somatic cells of plants may result in genetic variability in a population (Klekowski 1988). However, most of such mutants are unstable and not heritable as most of them may not reach the germlines. Still it is an important mechanism for phenotype variability. The apical meristem of a shoot-tip contains a cluster of cells that have divided mitotically a large number of times to produce the cells of the stem, leaves, buds, flowers and fruits. During these repeated divisions, a cell in the apical meristem may mutate. If this process occurs throughout the plant body, then individual shoots will begin to divide genetically. As a result of these somatic mutations, the mutant cells are in a favourable position within the apical meristem (Gill 1986). Somatic mutations are identified in clonally propagated plants and are well known horticulturally (Grant 1975). This genetic variability within individuals should increase over time and be particularly evident in long-lived plants, such as trees (Klekowski 1988). Sharma (1985) reviewed the variation of chromosome complements in somaclonal populations. Even, in the same individual such variation in chromosome numbers could be observable. It might be pertinent to mention here that most of these mutations may not be heritable or they might be unstable too. However, somaclonal variation and somatic mutation are now well identified mechanism for generating diversity in vegetatively propagating plants. The chromosomal variations associated with the process of tissue culture and regenerations are well known. DNA samples from leaf and callus tissue from the same lines showed that almost all of the repetitive sequences showed variations. The only sequence with a constant copy number was the clone representing the major part of a light satellite DNA (Chaudhuri, unpublished data).

HETEROCHROMATIN

Heterochromatin segments are responsible for chromosome replication, recombination and conservation (reviewed by Bhattacharyya 1991, Mallik 1992). Heterochromatin harbours repetitive DNA sequences which are late replicating and responsible for allocyclic behaviour of chromosome segments. Repetitive DNAs are now being considered as important chromosome constituents with unknown gene function. Their transcripts were identified in total ceullular RNAs by DNA-RNA hybridization as well as by in situ hybridization (Dutta and Chaudhuri 1975, Kedes 1978, Long and Dawid 1980, Varley et al. 1980, Stephenson et al. 1981, Jelenik and Schmid 1982). An excess of DNA in higher organisms had puzzled cytogeneticists when it was first discovered by cytophotometry. The tendency at that time was to consider them as 'junk' or senseless DNA elements. Repeated DNAs are common in eukaryote genomes, except in fungi (Timberlake

1978). In plants, their amounts may go up to 80 percent (Rees and Bennett 1969). In some plants their amounts may be very low (Leutwiler et al. 1984, Bhattacharyya et al. 1986; Pruitt and Meyerowitch 1986); on the other hand, in plant species like *Vicia faba* and *V. sativa* there might be very few non-repeated DNA. Like other eukaryotes, in plants too, repetitive DNA sequences may occur as tandem repeats — in short or long clusters or they might be interspersed with non-repeated DNA sequences within a genome (Gupta 1980, Bhattacharyya 1991, Mallik 1992). Often they might be very relic, preserved for million years of evolution (Flavell 1982). In addition, in different tissues of the same plant amount of DNA was found to vary (Sharma 1985, Sharma et al. 1986). These findings as well as presence of their transcripts in plant and animal cells suggest that these DNA sequences might serve useful functions. However, in some organisms most repetitive DNAs can be experimentally removed with no apparent fitness effect on the organism (Li and Graur 1991). However, data are scanty to arrive at a generalized conclusion on the role of repetitive DNA in plants.

Among the first to realize the role of eukaryotic chromosome to evade selection pressure were Britten and Khone (1968), who had argued that during organic evolution eukaryotic chromosomes have accumulated hundred and thousand of copies of identical DNA pieces — the repetitive DNA elements. In the same year, Kimura (1968) provided a population genetic rational by proposing his 'neutral mutation' hypothesis, where he had argued that most differences in the rate of evolution of molecules or parts of genome are accounted for by the conservation of significant sequences. By repetition of a particular DNA sequence harmful effect of a mutation could be avoided, and these sequences can evade selection pressure. What Kimura had visualized was explained by Britten and Khone. Variation in repetitive DNAs would perhaps explain the fluid chromosome model.

The evolution of repetitive DNA sequences coincides with the organization and evolution of eukaryotic chromosomes vis-a-vis evolution of plant and animal species. Lima-de-Faria (1980) proposed a 'chromosome gradient hypothesis' for chromosome evolution, where he had argued that from centromere to telomere there is a size gradient of chromomeres — largest at the centromere and smallest at telomeres. Flavell (1982) explained that the origin of repetitive DNA sequences is from unique DNA sequences by multiplication of existing copies and their conversion into unique sequences by series of mutations. Their presence in constitutive heterochromatin is confirmed (Arrighi et al. 1970, Jones 1970, Pardue and Gall 1970). Depending on cell function and taxonomic level of species, their amount may vary (reviewed by Charlesworth et al. 1986, Chaudhuri 1975, Flamm 1972, Mallik 1992, Sharma 1985, Walker 1971). Satellite DNAs were reviewed by Commings (1972), Pardue and Gall (1972), Rae (1972), Peacock et al. (1977). Being a part of constitutive heterochromatins these DNA pieces could be moulded effectively without any serious damage to chromosome functions.

ECOLOGY AND GENOME CHANGES

Although genomic fluidity has been proposed at various times, since the discovery of Mendel's laws, most of the proposals have not been widely appreciated. The important discovery of Barbara McClintock in 1940s, on changes on gene loci, could not be appreciated until the transposable elements were discovered in maize in 1970s. Such changes are found to be operative by environmental pressure. Similarly, in cultured animal cells, gene amplification occurs under selection pressure (Walbot and Cullis 1985). These findings suggest that a rapid genomic change can occur in response to particular environmental stresses. Transposition, unequal crossing over, gene rearrangement and gene amplification are being studied extensively, and it is likely that we will soon find a molecular explanation of the mechanisms underlying genomic changes.

M.D. Bennett (1972), and later on Levin and Funderburg (1979), observed that there is a conspicuous variation in the genome sizes of annual and perennials. Most attempts to explore the significance of variation in genome size have been involved comparisons with respects to life history, taxonomy and evolutionary affiliations and geographic distribution (reviewed by Price 1976, Hinegardner 1978, Cavallier-Smith 1989, Gillespie 1991, Li and Graur 1991, Thompson 1991, Scheiner and Lyman 1991). Data show that ecology plays a major role in genome size fixation during evolution, and, contrary to earlier observations, no consistent difference is apparent within 162 common annuals and perennials of the British Isles. Irrespective of their taxonomic levels, many perennials, particularly woody plants, are found to have small genomes. Annual grasss often have large genomes (2C = 10 picogram). Average genome size in dicotyledons is 2C = 6.86 picogram (pg) and that in monocotyledons is 2C = 17.28 pg (Grime and Mowforth 1982). Therefore, in monocytledons there is three times multiplication of nuclear genomes.

In angiosperms, DNA content per nucleus varies 500-fold (Rees and Bennett 1972, Jones 1972, Sparrow et al. 1972). Polyploidy is responsible for some variation, but at least 100-fold increase in DNA contant might occur in diploid species of a taxon. For herbaceous angiosperms, the variation is 3000-fold - 0.07 pg for *Arabidiopsis thaliana* to 295 pg for *Fritillaria davisi.* The polyploidy alone cannot explain this variability. Even in a single plant, different tissues were found to contain different amount of DNAs (Sahu et al. 1980). Nagl (1978), Sharma (1985), Sen (1974) and Chaudhuri (1975) had reported endomitotic replication of chromosome during cell differentiation.

The correlation between large genome size and capacity for growth at low temperature is consistent with genomes with cereals, grasses and legumes (Levin and Funderburg 1979). The massive genomes reported in *Fritillaria* (2C = 95.5 to 295 pg) and *Tulipa* (2C = 39.9 to 90.6 pg) are found to be associated with capacity for early and rapid expansion of short-lived shoots. It is suggested by Levin and Funderburg (1979), after comparing genomes sizes and time of shoot expansion of 24 angiosperms, that the selective season of growth arises from a differential effect of low temperature on cell division and cell expansion. As temperature rises, the growth is predominantly by cell division. As the minimum duration of mitotic and meiotic cycles is inescapably long in large cells (Bradley 1954, Commoner 1964, Darlington 1978, Bennett 1981, 1992), the advantage of cell enlargement would not be expected to pertain at warm temperature. The small genomes are observed in warm tropical climates, and also in plants in which growth is restricted in summer. Britten and Davidson (1971) also reported similar observations for animals. The process is not yet clear. Biochemical evidences and electron microscopic studies showed highly looped uninemic chromatids in wide array of animals taxa (Dupraw 1972).

It is polynemy which is responsible for DNA increases in animals, and perhaps in plants too. Perhaps, unequal crossing over mechanism or duplication of a single chromosome (such as B chromosome in *Drosophila*) might be alternate mechanism for DNA amplification within an eukaryotic nucleus. It might be possible that there could be tandem duplication of chromosome segments or discrete multiplication of favoured DNA sequences within a chromosome. Sparrow and Nauman (1974) had suggested DNA doubling during evolution of 76 grass species. Price et al. (1974) observed similar mechanisms operating in conifers, and hypothesizes that at least for conifers the selective advantage is towards increase in DNA content, by small duplications along the whole chromosome surface.

Part of the variation in DNA content apparently results from highly reiterated sequences of genome. Bacteke et al. (1967) analyzed 23 angiosperm genomes and correlated them with their nuclear volume, cell volume and repeated DNA percentage. They concluded that higher DNA content is due to accumulation of higher amount of reiterated sequences within nuclei. Britten and Davidson (1969) proposed that a certain amount of growth of the vertebrate genome was due to the addition of redundant sequences. Mirksche and Hotta (1973) observed similar phenomenon in confers and Sharma et al. (1986) for *Lathyrus*. It is also believed that 15 to 38 percent of the *Vicia* species have accumulation of such redundant sequences (Flavell 1980) to an approximate 7-fold span in halpoid nuclear DNA (1C = 2.0 to 13.9 pg) (Chooi 1971, Bennett and Smith 1976, Raina and Bisht 1988). According to Rowland (1980), 85 percent of *V. faba* genome is composed of repetitive sequences, and most of the highly repetitive sequences (i.e. those copy number exceeds several thousand copies) are distributed through the whole genome. In *V. faba*, one repeated DNA class (Fok I family) could be visualized in agarose gel after digestion with Fok I restriction endonuclease. In such digestion, the DNA copies fall in one size category and could be visualized as a band in agarose gel after electrophoresis and ethedium bromide staining. This repeat family was found to be confined to the median arm regions of the large metacentric chromosomes as inverted repeated (Bassi et al. 1982, Yakura et al. 1987).

Amongst gymnosperms, cycads and conifers have the largest genomes, with an average nuclear DNA content of 80 pg. For different groups, the range is from 5.3 to 78.5 pg. But, within each group range of variation is narrow, only 2 to 4-fold. Due to limited number of observations for a particular genus or species, it could not be worked out with certainty the significance of this nuclear DNA variation. However, working on *Picea* Miksche (1970) had noticed that nuclear DNA content of the northern populations of each species (*P. glauca* and *P. sitchensis*) is greater (1.5-fold) than that of southern populations (in southern hemisphere). Grime and Mowforth (1982) also reported that plants in South Britain (warmer condition) have markedly large genomes (mean 2C = 13.31 pg) in comparison to those of northern

Britain (colder region) (mean 2C = 2.55 pg). We, therefore, find temperature plays an important role in nuclear genome amplification.

NUCLEAR GENOME

Eukaryotic genome organization was reviewed by Flavell in several articles (1980, 1982). An ecological approach to the problem was addressed by Rose and Doolittle (1983). Because coding sequences are conserved, concerted evolution involving recombination or gene conversion within the family has been proposed to preserve homogeneity. Recognition of "selfish DNA" focuses attention on the evolution of the genome due to DNA amplification, and role of repetitive DNAs in such evolution is gaining momentum (Rose and Doolittle 1983). Repetitive DNA sequences, shared among different organisms, may be flanked by species-specific less reiterated sequences (Bedbrook and Flavell 1982). Heterogeneity in length and character is noticeable more in those flanking sequences, but high repeate sequences too may be variable in character. By comparing those in closely related species the mode and rate of genome divergence could be ascertained. Bedbrook et al. (1980) demonstrated that highly repetitive DNA sequences of rye plant are also found in other cereals, though their amount and organization can vary, even within different rye cultivars.

The Mendelian principles, that genes have a fixed linear order on the chromosomes and that genetic loci are inherited unchanged from parents to offspring, are violated when we find DNA sequence copy number and position have changed rapidly, even within the lifetime of an organism. In 1975, McGhee and Engel presented the nucleosome ladder model, generated with plant material, and estimated a repeat size of 170 + /-30 base pairs of DNA for pea chromoatin. Since then a number of workers have shown that plant chromatin's have a repate size of 170 base pairs (Spiker 1985). In most of these cases, the repeat values have not been shown to differ (Leber and Hemleben 1979, Leweke and Hemleben 1982, Murray and Kennard 1984). However, Murray and Kennard (1984) have documented a striking contrast in the repeat size in bean leaves (191 base pairs). Weintraub (1978) observed a similar change in repeat length during erythrocyte development in chick and concluded that such changes are accompanied by dramatic changes in the content of Hl-like histones (H5) in the chromatin. But whether in beans too similar phenomenon occurs is yet to be discovered. If so, role of one repeat element in plant development may be documented with.

Murray and Kennard (1985) had investigated the chromatin structure of the phaseolin gene family in French bean. Expression of these is under strict developmental control, by the elements up to 1400 base pairs upstream, from 5' end of the gene, or 1000 base pairs downstream, from the 3' end of the gene. These workers demonstrated these genes have DNase hypersensitive sites within the controlling elements of the genes. Generally, by the presence of DNase hypersensitivity it could be inferred that near these sites there are sequences which could bind to functional proteins, or, in other words, presence of DNase hypersensitivity indicates the presence of controlling elements. For binding to protein, the DNA molecule assumed a tertiary structure, and small repetitive base pairs help to fold the DNA in its characteristic fold. So, small repeat elements may be prerequisite for gene functions, and their evolution may be dictated by functional proteins.

Plant ribosomal RNA genes vary with regard to their copy number, sequence organization and chromosome localization. The copy number of these genes vary considerably within and between plants. Rogers and Bendich (1987) observed a 97-fold difference in rRNA gene amount among individuals of *V. faba* population. In general, the species with smaller genomes contain fewer copies of ribosomal RNA genes than do those with larger genomes, but this relation may not be proportional (Maher and Fox 1973). 18S and 25S rRNA gene sequences were found at the nucleolus organizing region (NOR) in the appropriate chromosomes of *V. faba* and *V. barborensis* by in situ hybridization, while the 5S rRNA genes have been found to be located at two sites on the satellite segment of large metacentric chromosome in *V. barborensis* (Knalmann and Burger 1977). A detailed discussion on these gene-amplification will be taken up later.

Chromosome evolution is assumed to be dictated by natural selection (White 1973, Ohno 1974). Since the discovery of eukaryotic chromosome, we have not progressed very far to the molecular architecture of the plant chromosome and its function. Genes, their rearrangements and mutations occur as random. Almost nothing is predictable. Lima-de-Faria (1983) however argued that this explanation was used to cover up our ignorance on molecular evolution of eukaryotic chromosomes. Watson (1976) thought the eukaryotic chromsome as a magnified bacterial chromosome. The inherent dynamism and flexibility permit the nuclear chromosome to stand for genetic manipulation and chromosome evolution. However, due to lack of co-ordination between cytologists, geneticists and molecular biologists, interrelationships between origin of DNA replication, termination of replication and position of specific genes in eukaryotic chromosomes are still obscure. For most plants the knowledge remains dispersed and the chaotic situation persists (Sharma 1985, Watson 1976).

The distribution of repetitive DNA elements on chromosome will be clear if such element can be correlated with important features, like centromere, telomere, secondary constriction of the chromosome. Of which we know the role of some secondary constrictions which are found to contain NOR regions and supposed to contain highly amplified ribosomal RNA gene sequences. In chromosome evolution, these features change their relative position. In some chromosome segments like telomere and centromere, bands represent position of highly repetitive DNA sequences (Pardue and Gall 1970, Jones 1970, Bedbrook et al. 1980, Deumling and Greilhuver 1982, Walbot 1985). In addition, parts show chromomere organization where DNAs remain as highly packed and condensed, and chromomeres often assume heterochromatic appearance after staining. So, in those chromosome segments too, there may be some common sequences which help the chromatid to be in condensed or decondensed state. Perhaps, study of chromomeres may reveal chromosome evoluation and accumulation of repetitive DNA sequences in chromosomes. In *Ornithogalon, Secale cereale* and *Agapanthus,* an analysis of chromomere number and shape shows that the same chromosome decreases or increases in size in different tissues. In those plants there is complete reshaping of chromomere gradients during differentiation (Lima-de-Faria 1983). In *Drosophila,* chromomere in salivary gland differs from that in spermatocytes (Hess 1971).

Chromomere gradient indicates the presence of the chromosome field where large chromomeres are concentrated near the centromere and small at the telomeres. (In some plants like maize, the distal chromomeres may be very large and assume the appearance of knob on chromosome.) In mitotic prophase the gradient is not clear, but in meiosis (prophase) it is clear, perhaps due to prolonged meiotic prophase duration, which may often continue up to 2 days in some plants. In *Agapanthus,* this gradient is observable in second meiotic prophase, and even in mitotic prophase of *Ixerix* (Lima-de-Faria 1954). A total of 264 knobs had been identified in 15 species belonging to 4 angiosperm families. A pattern has emerged. In wild species, knobs are found at telomeres. In cultivated species, they become subtelocentric (Lima-de-Faria 1976). The centromere seems to exert an inhibitory effect on knob formation. If the knob contains satellite DNA sequences in wild species, then during domestication they might shift their position on the chromosome (Lima-de-Faria 1983). The presence of telomeres near the centromere leads of the disappearance of some DNA sequences, as for example, the knobs of *Salvia, Capsicum* and cold-induced regions of *Trillium* and *Fritillaria* (Lima-de-Faria 1976). Therefore, a particular DNA sequence, i.e. repetitive sequences at the centromere, may have an inhibitory effect on some repetitious sequences within eukaryotic chromomeres. In maize, rye, *Ornithogalom* and few other plant species, there are large distal chromomeres known as knobs (Longley 1974, Lima-de-Faria 1952, 1959). Information is now available on the nature of DNA present in the knobs. Complimentary RNA synthesized from satellite DNA sequences were found to hybridize at knob regions (Peacock et al. 1977).

GENE AMPLIFICATION

Preferential replication of a single genetic locus results in gene amplification. Amplification has been assumed to occur to provide large number of gene copies to support a high requirement for a gene product. The first gene amplification event was studied in the ribosomal RNA genes, during oogenesis of *Xenopus laevis* in 1960s. Ribosomal DNA amplification is now known to occur in protists, invertebrates, vertebrates and plants (Long and Dawid 1980). The amplified rRNAs may exist as extra chromosomal elements, which may be lost during subsequent cell divisions (Walbot and Cullis 1985). Ribosomal RNA gene sequences were found to be amplified in knob like structures of chromosome 6 and chromosome 11 of *Acheta* (Cave 1972, Wimen et al. 1973). Therefore, their large chromomeres are not only physical entities of a chromosome, but they also represent amplified DNA sequences. Here DNA amplification could be visualized as distinct knobs. Amplification in knobs of *Acheta* may go up to 1000 times, consisting of 10^9 base pairs, which is highly significant. In *Acheta* some of these extras DNAs are spliced as cytoplasm DNA circles (Lima-de-Faria 1975), and they are found to transcribe RNAs.

Changes in ribosomal RNA gene copy number have been described by many authors form any species (Long and Dawid 1980, Walbot and Cullis 1985). These genes are arranged in long tandem arrays and are present as 1 percent or more of the genome. In plants, Timmis and Ingle (1973) had demonstrated that this parameter might vary widely between lines of the same species. Philips and coworkers (1978) had demonstrated this in maize lines, and Goldsbrough and Cullis (1981) in flax genome. In flax, these genes may vary nearly three-fold among genotrops; but there might not be length heterogeneity or restriction site polymorphism in those genotrops. The generation of rapid DNA changes and been described extensively in flax (Cullis 1983). These include 25S, 18S and 5S genes. They all are found to be susceptible to environmental stresses. Heritable changes were noticed in flax genotrops after they were grown in different environments for a single generation. Other repetitive DNA components, such as a light satellite DNA, comprising of 15 percent of the flax genome, were found to be affected after environmental selection (Cullis 1979). This environmentally induced change could be traced to the progenies (Cullis and Charlton 1981). Other genes also showed such amplification phenomenon in stress condition. Schimke et al. (1978) observed that human cell lines with stable drug (methotrexate) resistance show amplified genes in the chromosomes, where cell lines quickly loose these amplified sequences and also the drug resistance if they are grown under non-stress conditions, i.e. in medium without methotrexate. Such examples of gene amplification are available in other mammalian cell cultures too (Hamlin et al. 1984).

An instance of amplification of a protein coding gene, as programmed part of normal development, is found in chorion gene of *Drosophila* (Spradling and Mahowald 1980). During follicular differentiation, reinitiation of chorion gene replication repeatedly occurs, bidirectionally, from a site within each locus of the gene family. The newly replicated strands remain part of the chromosome, creating polytenization of the chromosome (Oschieum and Miller 1983). In plants, extra-chromosomal DNAs were noticed as outcomes of gene amplification (Walbot and Cullis 1985). Extra-chromosomal DNA has been characterized in pea plants grown without an apparent selection pressure (Van't Hof et al. 1983). Perhaps, the extra-chromsomal sequences are an aspect of the normal differentiation of root cells (Krimer and Van't Hof 1983). Similar to chorion genes, plant genes too may amplify their sequences at replication points. If the amplified units integrate into the chromosomes, a new genome could be created with multigene families or pseudogene families. Sometimes, such amplification could be achieved through gene product (RNA) by an enzyme (*reverse transcriptase*). *Reverse transcriptase* plays a significant role in accumulating active or relic sequences or parsite DNAs which code for this enzyme. Unstable amplification may allow a new temporary phenotype which can withstand some selected environmental pressure, as has been demonstrated in methotrexate resistance in human cells.

SEXUAL CROSSES

Some highly imbred cultivars show a high degree of 'rogue' progenies. These rogues are hyperactive in phenotypic expression. They may occur with a frequency of 1 percent (Bateson and Pellow 1915, Pearson 1965). The cytogenetics and molecular basis of rogue is unknown. However, instability of rogue mutants could be the result of extra chromosomal DNA replication or to the presence of transposable elements (Walbot and Cullis 1985). It could be a hybrid dysgenesis phenomenon which has been observed in *Drosophila.*

McClintock (1978) has discussed the concept that wide hybridization might trigger reorganization of parental genomes. The outcome could be absolute genome change in size and copy number of repetitive DNA sequences and release of cryptic transposable elements. In *Nicotiana* spp. there is evidence of some hybrids being unstable in their DNA content. Gerstel and Burns (1976) observed chromosomes up to 15 times the normal length in some corolla, root tips and pollen mother cells from *N. tabacum* x *N. plumbaginifolia* hybrids. These megachromosomes might be the result of excessive DNA amplification, as was observed for *dihydrofolate reductase* gene in some methotrexate-resistant cell lines. The inheritance of nuclear DNA content in interspecific and intraspecific crosses is often non-Mendelan in nature. In F2 of interspecific cross, five fertile progenies were recognized which had independent 2C values (Price et al. 1983).

An extensive study in maize demonstrated that most repetitive DNA sequences could show non-Mendelean inheritance in intraspecific crosses (Rivin and Cullis 1983, Walbot and Cullis 1985). Copy numbers of highly repetitive DNA sequences (which has more than one hundred thousand copies) and low repetitive sequences (having less than thousand copies) were found to increase in such crosses. With one exception, each repetitive family varied two to four fold in copy number among the imbred lines examined. In these imbred lines of maize, an autonomously replicating sequence (*ars*) had been selected for its ability to act as an origin of replication in plasmids of yeast. It is unknown whether in maize the *ars* sequence has an origin of replication. Other repeat sequences studied here are: rRNA gene sequences, satellite DNAs, disperse repeats etc.

CHROMOSOMAL RECOMBINATION

In plants, interchromosomal recombination or intrachromosomal recombination during mitosis could be observed in cytological preparations (Lima-de-Faria 1983). However, the exact mechanism of these karyotype variations could not be traced. It is often assumed that homologous sequences within the chromosome are responsible for these arrangements. Interchromosomal recombination between naturally occuring repeated genes (such as *Ty* element of yeast, ribosomal RNA genes, duplicated copies of mating type information etc.) has been examined (Rothstein et al. 1987, Lieman and Picologlu 1988). Such phenomenon can be compared with unequal crossing over in meiosis where a particular chromsome segment is amplified or lost by such mechanism. One class of repeated genes which have a high level of intrachromosomal recombination in mitosis is rRNA genes (Szostak and Wu 1980). They showed that insertions of the LEU2 gene at the rRNA gene cluster were mitotically unstable. Their loss rate, by unequal sister-chromatid exchange (recombination), was estimated as 10^{-2} to 10^{-3}. It would be reasonable to assume that similar phenomena are occurring in plants too. Genome changes through unequal crossing over at multigene family sites and tandemly arranged rRNA gene (repetitious) loci were reported by Russell and Rodland (1986), Smith (1973) and Li and Graur (1991). Mitotic recombination readily occurs between genes on non-homologous chromosomes. Ten percent event is often associated with reciprocal recombination resulting in translocations (Scherer and Davis 1980).

TRANSPOSABLE ELEMENTS

Transposable elements were first discovered in maize but have not been found in many plants. However, newer elements are being reported from diverse plant species. The biology and molecular biology of these elements have been reviewed in depth (Shapiro 1983, Federoff 1983, Rubin 1983, Freeling 1984, Li and Graur 1991, Gillespie 1991). By analyzing the structure and fuction of these elements, it could be inferred that the key feature of these elements is a regulated disruption of chromosomes in cells in which they are active, and thus they influence the genome fluidity. McClintock (1950) first observed such changes in maize chromosomes, where a transposable element (*Ds*), sitting near the centromere of the short arm of chromosome 9 is responsible for chromosome breaks and chromosome rearrangements. Subsequent transpositions of *Ds* to a distal location along the same chromosome arm were found to chromosome 9 — thus leading to a chromosome instability. The insertion of a transposable element is accompanied by a few base pair duplications of host sequence; the extent of duplication is the characteristic of each cell type. With the duplication, complex rearrangements of chromosome arms could take place (Feeling 1984, Courage-Tebbe et al. 1983), as was observed in an allele of *alcohol dehydrogenase* gene site of maize containing *Mu*-I element (a transposon).

Transposable elements are of many kinds, but usually they carry genes or synthesizing enzymes for transposition. Another class of transposable element is retrotransposon. Retrotransposons are now found to be a common feature of all eukaryotic cells, irrespective of plants, animals or lower eukaryotes (Li and Graur 1991). The important feature of these elements are genes for *reverse transcriptase,* an enzyme which can synthesize DNA on a RNA template, and cDNA could be inserted on the host genomes by other gene products responsible for transposition. *Reverse transcriptase* is now considered a powerful element for addition of genetic material in eukaryotic organisms (Li and Graur 1991), and it might be the prime enzyme responsible for DNA duplication. It could also be mentioned that many repeats with active or relic sequences code for *reverse transcriptase,* and evolution of such repeats in eukaryote genomes could be mediated through retroelements operating at the ancestral source. This observation on relic (or active) sequences supports the "parasite hypothesis". Viral infection also leads transposon-like phenomenon, if the viral genome gets integrated into host genome and stays therein as silent parasites. So selfish DNA elements could be amplified horizontally through *reverse transcriptase* of retroelements. Any change through transposons or transposable-like elements could be permanent in the cell lineage provided the altered sequences become a part of normal chromosome complement, but the changes will be inherited only if they occur in germ line or could be transported to germ cells through mitosis. On the other hand, unstable genome changes are transient in the cell line if they are non-chromosomal and they are unlikely to be inherited, even if they occur in germ cells.

PLASTID GENOME

The plastid genome of higher plants is a circular chromosome of approximately 130 kilobase (Bedbrook and Kolodner 1980) where a 20 kb inverted repeat with rRNA gene sequences are found to be flanked by a large and small unique copy sequences (Palmer and Thompson 1982). There is a high degree of conservation of plastid gene sequences between highly divergent plant species, such as monocotyledons, dicotyledons and algae (Zurski et al. 1984). Only a few rearrangements are required to explain the changes, with one exception, in the plastid genome. The pea plastid genome lacks the inverted repeat structure and the gene order is scrambled compared to other legumes which have regular inverted sequences in their chloroplast DNA (Walbot and Cullis 1985). Palmer and Thompson (1984) hypothesized that the loss of inverted repeat structure in pea chloroplast DNA could be due to massive rearrangements within that genome. Such events must be extremely rare in plastid evolution.

Thus far there is no evidence of transfer of nuclear DNA sequences to the plastid genome (Walbot and Culis 1985). However, using spinach plastid probes Timmis and Scott (1983) discovered a plastid sequence in the nuclear genome of spinach. The exact mechanism of such transfer could not be explained, but it could be assumed that there must be some homologous sequences at the site where the transfer took place during evolution. It might be also possible that transfer of plastid genome fragments might be directed by a unknown vector to the nucleus. Alternatively, frequent organelle fusion might deliver plastid DNA to the nuclear compartment where some or all of its gets integrated (Walbot 1985). Therefore, there are possible mechanisms by which an increase and/or rearrangement of the plant genome could take place in nature, at homologous sequence sites.

MITOCHONDRIA GENOME

The mitochondria genome in higher plants is more complex and large than the plastid genome. They may be 218 kb in size in brassicas (Palmer and Shields 1984) to 2500 kb in cucurbits (Ward et al. 1981). Although the restriction pattern of the normal maize mitochondria genome is quite stable, over three generations studies in strain B37, many fragment differences appear after restriction enzyme digestions — indicating thereby the sequence divergence and rearrangements within mitochondria DNA, after the separation of normal and three male-sterile line of the same strain (Borck and Walbot 1982). It suggests that the genomes can be represented in the single "master" chromosome containing all the sequences. However, recombination apparently occurred between repetitive sequences on the master chromosome generating circular chromosomes at high frequency (Palmer and Shield 1984). Thus, the basic structure of the plant mitochondria genome is hypervariable and it could be highly amplified during evolution or environment pressure.

Mitochondria of higher plants also contains one or more small linear or circular plasmids. The size and copy number of those plasmids vary within a species (Palmer et al. 1983) — another source of genomic variation in the organelle. The presence of two linear plasmids (S1 and S2) in the cytoplasm of maize is found to be associated with male sterility. In the reverent progenies, which appear after several generations, there is less free plasmid and some S plasmid sequence appeared to integrate into the main mitochondria genome (Long and Dawid 1980). Perhaps this situation is not a unique feature for maize mitochondria alone, and this phenomenon may be happening in other plants also.

Mitochondria genomes also contain integrated segments of plastid DNA. Stern and Longdale (1982) discovered a 12 kb plastid-derived sequence in maize mitochondria genome. The sequence was termed as "promiscuous DNA" by some reviewers (Ellis 1982). The mode of this transfer is still undermined. However, a 252 bp sequence sharing 90 percent homology to the 3' end of plastid gene *psbA* was found in mitochondria genome (Walbot and Cullis 1985), and in both plastid and mitochondria DNAs, the sequence was found next to some active repetitive sequences responsible for recombination (Walbot and Cullis 1985).

CONCLUSION

The exchange of DNA sequences within a species or in an individual ultimately leads to genetic diversity. It is essential that at a population level, evolution and genetic phenomena be well understood. Genetic variation, life history traits, population dynamics and genetic population structure all shape and influence the way a species interacts with its environment and also with other species.

It is apparently from the discussion that plant genomes are variable, and that might be due to the presence of subset of repetitious sequences. It is also evident that plants are easily moulded by biotic and abiotic stress (Willis 1983, Walbot 1985) and sequences of repeat DNAs may be preserved and/or accumulated within a plant genome. A molecular drive is apparent therein for concerted evolution, where repetitive DNA elements accumulate mutations more rapidly than transcribed genes of known functions. Genetic mosaics within an individual plant may have some adaptive value. Inter- and intrachromosomal recombination occurs in plants and reports are available for information exchange from organelle genomes to nuclear genomes or within different organelle genomes. The role of repetitious sequences in such transfers is being speculated.

REFERENCES

Alberts, B., Bray, D., Lewis, J., Raff, M., Roberts, K. and Watson, J.D. 1983. *Molecular Biology of the Cell.* Garland, New York.

Arrighi, F.E., Hsu, T.C., Saunders, P. and Saunders, G.F. 1970. Localization of repetitive DNA in the chromosomes of *Microtus agretis* by means of in situ hybridization. *Chromosoma* 32: 224-236.

Bassi, P., Cremonini, P. and Cionini, P.G. 1982. Cytological localization of the inverted repeated DNA sequence in *Vicia faba. Chromosoma* 85: 453-459.

Bateson, W. and Pellew, C. 1920. The genetics of 'rogues' among culinary peas (*Pisum sativum*). *Proc. Roy. Soc. Ser. B.* 91:186-195.

Bedbrook, J.R. and Kolodner, R. 1980. The structure of the chloroplast DNA. *Ann. Rev. Plant. Physiol.* 30:593-620.

Bedbrook. J.R., Jones, J., O'Dell, M., Thompson, R. and Flavell, R.B.V. 1980. A molecular description of telomeric heterochromatin in *Secale* species. *Cell* 19:545-560.

Bennett, M.D. 1971. The duration of meiosis. *Phil. Trans. Roy. Soc. Ser. B.* 178:227-299.

Bennett, M.D. 1972. Nuclear DNA content and minimum generation time in herbaceous plants. *Proc. Roy. Soc. Ser. B.* 181:109-135.

Bennett, M.D. and Rees, H. 1969. Induced and developmental variation in chromosomes of meristematic cells. *Chromosoma* 27:226-244.

Bennett, M.D. and Smith, J.B. 1976. Nuclear DNA amounts in angiosperms. *Proc. Roy. Soc. Ser. B.* 274:227-274.

Bhattacharyya, N. 1991. Study of the structure and sequence of DNA in a member of Solanaceae. Ph.D. Thesis, Calcutta University, Calcutta, p. 110.

Bhattacharyya, N., Mukhopadhyay, D.K., Chaudhuri, I. and Chaudhuri, R.K. 1986. Repetitive DNA amount in *Solanum nigrum* genome. *Curr. Sci.* 55:569-571.

Borck, K.S. and Walbot, V. 1982. Comparisons of the restriction endonuclease digestion patterns of mitochondrial DNA from normal and male sterile cytoplasms of *Zea mays* L. *Genetics* 102:109-128.

Britten, R.J. and Khone, D.E. 1968. Repeated sequences in DNA. *Science* 161:529-540.

Britton, R.J. and Davidson, E.H. 1971. Repetitive and non-repetitive DNA sequences and a speculation on the origins of evolutionary novelty. *Quart. Rev. Biol.* 46:111-138.

Cavellier-Smith, T. 1989. The kingdom chromista. In: J.C. Green and B.S.C. Leadbeater (Ed.) *The Chromophyte Algae: problems and perspectives.* Oxford University Press, London.

Cave, M.D. 1972. Localization of ribosomal DNA within oocytes of house cricket, *Achete domestica. J. Cell. Biol.* 55:310-321.

Chaudhuri, R.K. 1975. Repeated DNA: an analysis. *Sci. Cult.* 41:4-10.

Commings, D.E. 1972. The structure and function of chromatin. pp. 237-431. In: H. Harri and K. Hirschhorn (Eds.) *Advances in Human Genetics.*

Chooi, W.Y. 1971. Comparison of the DNA of six *Vicia* species by the method of DNA-DNA hybridization. *Genetics* 68:213-230.

Courage-Tebbe, U., Doring, H.P., Federoff, N. and Starlinger, P. 1983. The controlling elements Ds at the *shrunken* locus in *Zea mays:* structure of the unstable *sh-m 5933* allele and several revertants. *Cell* 34:383-393.

Cullis, C.A. 1979. Quantitative variation of ribosomal RNA genes in flax genotrophs. *Heredity* 42:237-246.

Cullis, C.A. 1983. Environmentally induced DNA changes in plants. *CRC Crit. Rev. Ser. RI Sci.* 1:117-131.

Cullis, C.A. and Charlton, L.M. 1981. The induction of ribosomal DNA changes in plants. *Plant Sci. Lett.* 20:213-217.

Darlington, C.D. 1978. *The Little Universe Man.* George Allen and Unwin, London, P. 358.

Deumling, B. and Greilhuver, J. 1982. *Chromosoma* 84:535-543.

Dobzansky, T. 1969. Selection. *Encyclopedia Britinica* 20:181.

DuPraw, E.J. 1972. *Advances in Cell and Molecular Biology,* Vol. 2. Academic Press, New York, p. 786.

Dutta, S.K. and Chaudhuri, R.K. 1975. Transcription of repetitive DNA in *Neurospora crassa. Mol. Gener. Genet.* 136:227-232.

Ellis, J. 1982. Promiscus DNA - chloroplast genes inside plant mitochondria. *Nature* 299:678-679.

Falconer, D.S. 1981. *Introduction to Quantitative Genetics,* 2nd Edition. Longman, New York.

Federoff, N. 1983. Controlling elements in maize. pp. 1-63. In: J. Shapiro (Ed.) *Genetic Elements.* Academic Press, New York.

Flavell, R.B. 1980. The molecular characterization and organization of plant chromosomal DNA sequences. *Ann. Rev. Plant Physiol.* 31:569-596.

Flavell, R.B. 1982. Sequence amplification, deletion and rearrangement: major source of variation during species divergence. pp. 298-323. In: G.A. Grover and R.B. Flavell (Eds.) *Genome Evolution.* Academic Press, New York.

Flamm, W.G. 1972. Highly repetitive sequences of DNA in chromosomes. *Int. Rev. Cytol.* 32:1-52.

Freeling, M. 1984. Plant transposable elements and insertion sequences. *Ann. Rev. Plant. Physiol.* 35:277-298.

Gerste, D.U. and Burns, J.A. 1976. Enlarge euchromatic chromosomes ("megachromosomes") in hybrid between *Nicotiana tabacum* and *N. plumbaginifolia. Genetics* 46:139-156.

Gill, D.E. 1986. Individual plants as genetic mosaics: ecological organisms versus evolutionary individuals. pp. 321-334. In: M.J. Crawley (Ed.) *Plant Ecology.* Blackwell, Oxford.

Gillespie, J. 1991. *Causes of Molecular Evolution.* Oxford Univ. Press, U.K.

Goldsbrough, P.B. and Cullis, C.A. 1981. Characterization of the genes for ribosomal RNA in flax. *Nucl. Acid. Res.* 9:1301-1309.

Grant, V. 1975. *Plant Genetics.* Columbia University Press, New York.

Gupta, V.S. 1981. Structure and sequence organization of DNA in Eukaryotes with special reference to three Graminae Plant species. Ph.D. Thesis, Poona University, Pune, p. 245.

Hamlin, J.L., Milbrandt, J.D., Heintz, N.H. and Azizkhan, J.C. 1984. DNA sequence amplification in plant cells. *Int. Rev. Cytol.* 90:31-82.

Hess, O. 1971. Lampenbursten chromosomen, pp. 215-281. In: H.W. Altmann (Ed.) *Handbuch der Allgemeinen Pathologie* Vol. 2 Springer-Verlag, Berlin.

Hinegardner, R. 1976. Evolution of genome size, pp. 179-199. In: F.J. Ayala (Ed.) *Molecular Evolution.* Sinauer Associates, Massachusetts.

Jeffries, R.L. 1984. The phenotype: its development, physiological constraints and environmental signals, pp. 349-358. In: R. Dirzo and J. Sarukhan (Eds.) *Perspectives on Plant Population Biology.* Sunderland-Sinauer, Massachuetts, p. 648.

Jones, K.W. 1970. Chromosomal and nuclear location of mouse satellite DNA in individual cells. *Nature* 225:912-915.

Jones, K.W and Rees, H. 1982. *Chromosomes.* Academic Press, New York.

Kedes, L.H. 1979. Histone genes and histone messengers. *Ann. Rev. Biochem.* 48:837-870.

Kimura, M. 1968. Evolutionary rate at the molecular level. *Nature* 217:624-626.

Klekowski, E.J. Jr. 1988. Mutational loan on clonal plants: a study of two fern species. *Evolution* 38:417-426.

Knalmann, M. and Burger, E.C. 1977. Cytologische Localisation von 5S und 18/25S RNA Genorten in Mitose Chromosomen von *Vicia faba. Chromosoma* 61:177-192.

Krimer. D. and Van't Hof, J. 1983. Extrachromosomal DNA of pea (*Piscum sativum*) root-tip cells replicate by strand displacement. *Proc. Natl. Acad. Sci. USA.* 80:1933-1937.

Laird, C.D. 1971. Chromatid structure, relationship between DNA content and nucleotide sequence diversity. *Chromosoma* 32:306-378.

Laughman, J.R. amd Gabay-Laughman, S. 1983. Cytoplasmic male sterility in maize. *Ann. Rev. Genet.* 17:27-48.

Leutwiler, L.S., Hough-Laughman, B.R. and Meyerowitz 1984. The DNA of *Arabidiopsis thaliana. Mol. Gener. Genet.* 194:15-23.

Lever, B. and Hemeleben, V. 1979. Nucleosomal organization in active and inactive plant chromatin. *Plant Syst. Evol. Suppl.* 2:187-199.

Leweke, B. and Hameleben, V. 1982. Organization of rDNA in chromatin plants. *Cells Nucleus* 11:225-253.

Liebman, S. amd Picologlu, S. 1988. Recombination associated with yeast retrotransposons, pp. 193-227. In: Y. Koltin and M. Leibowitz (Eds.) *Viruses of Fungi and Simple Eukaryotes.* Dekker, New York.

Long, E.O. and Dawid, I.B. 1980. Repeated genes in eukaryotes. *Ann. Rev. Biochem.* 49:727-764.

Longley, A.E. 1952. Chromosome morphology in maize and its relatives. *Bot. Rev.* 18:399-412.

Li, W. and Graur, D. 1991. *Fundamentals of Molecular Evolution.* Sinauer Assoc., Sunderland, Massachusetts.

Lima-de-Faria, A. 1952. Chromatin analysis of chromosome complement of rye. *Chromosoma* 5:1-68.

Lima-de-Faria, A. 1954. Chromosome gradient and chromosome field in *Agapanthus. Chromosoma* 6:330-370.

Lima-de-Faria, A. 1959. Differential uptake of tritiated thymidine into hetero- and euchromatin in *Melanoplus* and *Secale. J. Biophys. Bichem. Cytol.* 6:457-466.

Lima-de-Faria, A. 1973. Evolutions defining the position of ribosomal cistrons in the eukaryotic chromosomes. *Nature New Biol.* 241:136-139.

Lima-de-Faria, A. 1974. The molecular organization of the chromomeres of *Achete* involved in ribosomal DNA amplification. *Cold Spring Harbor Symp. quant. Biol.* 38:559-571.

Lima-de-Faria, A. 1975. The relation between chromomeres, replicons, operons, transcription units, genes, viruses and palindromes. *Hereditas* 81:249-284.

Lima-de-Faria, A. 1980. Classification of genes, rearrangements and chromosomes according to the chromosome field. *Hereditas* 93:1-46.

Lima-de-Faria, A. 1983. *Molecular Evolution and Organization of Chromosomes.* Elsevier, New York, Oxford, p. 1186.

Lima-de-Faria, A. and Jaworsky, H. 1972. The reaction between the chromosome size gradient and the sequence of DNA replication in rye. *Hereditas* 70:39-58.

Maher, E.P. and Fox, D.R. 1973. Multiplicity of ribosomal RNA genes in *Vicia* species with different nuclear DNA contents. *Nature* 245:170-172.

Mallik, P. 1992. Cloning of a repeated DNA fragment and distribution of different repeat family in some memebrs of Solanaceae. Ph.D. Thesis, Calcutta University, Calcutta, p. 178.

McClintock, B. 1940. Mutable loci in maize. *Carnegia Inst. Washington Year Book* 48:142-154.

McClintock, B. 1950. The origin and behaviour of mutable loci in maize. *Proc. Natl. Acad. Sci.* 36:344-355.

McClintock, B. 1978. Mechanisms that rapidly reorganize the genome. *Stadler Genet. Symp.* 10:25-48.

McClintock, B. 1984. The significance of responses of the genome to challenges. *Science* 226:792-801.

McGhee, J.D. and Engel, J.D. 1975. Subunit structure of chromatin is same in plants and animals. *Nature* 254:449-450.

Mirksche, J.P. and Hotta, Y. 1973. DNA base composition and repetitious DNA in several cereals. *Chromosoma* 41:29-36.

Murray, M.G. and Kennard, W.C. 1984. Altered chromatin conformation of the higher plant gene phaseolin. *Biochemistry* 23:4225-4232.

Nagl, W. 1978. *Endopolyploidy and Polyteny in Differentiation and Evolution.* North Holland, Amsterdam, p. 415.

Ohno, S. 1974. Conservation of ancient linkage groups in evolution and some insight into the genetic regulatory mechanism of X-inactivation. *Cold Spring Harbor Symp. Quant. Biol.* 38:155-164.

Oschiem, Y.N. and Miller, O.L. 1983. Novel amplification and transcriptional activity of chorion genes in *Drosophila melanogaster* follicle cells. *Cell* 33:543-553.

Palmer, J.D. and Thompson, W.F. 1982. Chloroplast DNA rearrangements are more frequent when a large inverted repeat sequence is lost. *Cell* 29:537-550.

Palmer, J.D. and Schields, C.R. 1984. Tripartite structure of the *Brassica campestris* mitochondrial genome. *Nature* 306:437-440.

Palmer, J.D., Shields, C.R., Cohen, D.B. and Orton, T.J. 1983. An unusual mitochondrial DNA plasmid in the genus *Brassica. Nature* 301:725-728.

Pardue, M.L. and Gall, J.G. 1970. Chromosomal localization of mouse satellite DNA. *Science* 168:1356-1358.

Pardue, M.L. and Gall, J.G. 1972. Molecular Cytogenetics, pp. 65-99. In: M. Sussman (Ed.) *Molecular Genetics and Developmental Biology.* Prentice Hall, New Jersey.

Peacock, W.J., Appels, R., Dunsmuir, P., Lohe, A.R. and Gerlach, W.L. 1977. Highly repetitive DNA sequences, chromosomal localization and evolutionary conservation, pp. 494-506. In: B.R. Brinkley and K.R. Porter (Eds.) *International Cell Biology* 1976-77. The Rockefeller University Press, New York.

Pearson, O.H. 1968. Unstable gene systems in vegetable crops and implications for selection. *Hort. Sci.* 3:241-271.

Philips, R. 1978. Molecular cytogenetics of the nucleolar organizer region, pp. 711-741. In: D.B. Walden (Ed.) *Maize Breeding and Genetics.* Wiley Interscience, New York, p. 1015.

Price, H.J. 1976. Evolution of DNA content in higher plants. *Bot. Rev.* 42:27-52.

Price, H.J., Chambers, J.L., Bachman, K. and Riggs, J. 1983. Inheritance of nuclear 2C DNA content variation in interspecific and intraspecific hybrids of Microseries (Asteraceae). *Amer. J. Bot.* 70:1133-1138.

Primac, R.B. and King, H. 1989. Measuring fitness and natural selection in wild plant population. *Ann. Rev. Ecol. Syst.* 20:367-374.

Pruitt, R.E. and Meyerowitz, E.M. 1986. Characterization of the genome of *Arabidopsis thaliana. J. Mol. Biol.* 187:169-183.

Rae, P.M.M. 1972. The distribution of repetitive DNA sequences in chromosomes, pp. 109-149. In: E.J. DuPraw (Ed.) *Advances in Cell and Molecular Biology,* vol. 2. Academic Press, New York.

Raina, S.M. and Bishy, M.S. 1988. DNA amounts and chromosome compactness in *Vicia. Genetics* 77:65-77.

Rees, H. and Jones, G.H. 1972. The origin of the wild species variation in nuclear DNA content. *Int. Rev. Cytol.* 32:53-92.

Rivin, C.J. and Cullis, C.A. 1983. Modulation of repetitive DNA in the maize genome. *Genetics* 104:859-860.

Rogers, S.O. and Bendich, A.J. 1987. Hereditibility and variability in ribosomal RNA genes of *Vicia faba. Genetics* 117:285-295.

Rose, M.R. and Doolittle, W.F. 1983. Molecular biological mechanisms of speciation. *Science* 200:157-162.

Rothstein, R., Helms, C. and Rosenberg, N. 1987. Concerted deletions and inversions are caused by mitotic recombination between delta sequences in *Saccharomyces cerevisiae. Mol. Cell. Biol.* 7:1198-1207.

Rowland, R.E. 1980. Distribution of repetitive DNA sequences in chromosomes. *Exp. Cell. Res.* 126:444-448.

Rubin, G.M. 1983. Dispersed repetitive DNAs in *Drosophila,* pp. 329-361. In: J.P. Shapiro (Ed.) *Mobile Genetic Elements.* Academic Press, New York, p. 688.

Russell, J.P. and Rodland, K.D. 1986. Magnification of rRNA gene number in a *Neurospora crassa* strain with a partial deletion of nucleolus organizer. Chromosoma 93:337-340.

Scherer, S. and Davidson, R.W. 1980. Recombination of dispersed repeated DNA sequences in yeast. *Science* 209:380-384.

Scheiner, S.M. and Lyman, R.F. 1991. The genetics of phenotypic plasticity. II response to selection. *J. Evol. Biol.* 4:23-50.

Schimke, R.T., Kaufman, R.J., Alt, F.W. and Kellems, R.F. 1978. Gene amplificiation and drug resistance in cultured mammalian murine cells. *Science* 202:1051-1055.

Schimke, R.T., Beverly, S., Brown, P., Cassin, R. and Federspiel, N. 1984. Gene amplification and drug resistance in cultured mammalian cells. *Cancer Treat. Rev.*

Schlichting, C.D. 1986. The evolution of phenotypic plasticity in plants. *Ann. Rev. Ecol. Syst.* 17:667-693.

Sen, S. 1974. Differentiated nuclei and elucidation of chromosomal control of differentiation. *Nucleus* 17:42-45.

Shapiro, J.A. 1983. *Mobile Genetic Elements.* Academic Press, New York, p. 688.

Sharma, A.K. 1985. Chromosome structure. *Proc. Natl. Acad. Sci. Per. Rep. Ser.* 14:1-22.

Sharma, A.K., Chaudhuri, R.K. and Mukherjee, S. 1986. DNA content and repetitive DNA in *Lathyrus ativus* at interspecific level: Lathyrus and Lathyrism. *Proc. Collogue Lathyrus III:*85-90.

Shau, H., Chaudhuri, R.K and Sharma, A.K. 1980. DNA, RNA and protein content of isolated nuclei from different plant organs. *Ind. J. Exp. Biol.* 18:1519-1521.

Silander, J.A. 1985. Microevolution in clonal plants, pp. 107-152. In: *Population Biology and Evolution of Clonal Organisms.* JBC Jackson, L.W. Buss and R.E. Cook (Eds.), Yale Univ. Press, New Haven, CT.

Smith, D.W. 1973. DNA synthesis in prokaryotes: replication. *Prog. Biophys. Mol. Biol.* 26:321-408.

Sparrow, A.H., Price, H.J. and Underbrink, A.G. 1972. A survey of DNA content per cell and per chromosome or prokaryotic and eukaryotic organisms, pp. 451-494. In: H.H. Smith (Ed.) *Evolution of Genetic Systems.* Brookhaven Synmposia in Biology, Number 23. Gordon and Breach, New York, p. 694.

Spralding, A.C. and Maowald, A.P. 1980. Amplification of genes for chorion proteins during oogenis in *Drosophila melanogaster. Proc. Natl. Acad. Sci. USA* 77:1096-2002.

Spiker, S. 1975. An evolutionary comparison of plant histones. *Biochem. Biophys. Acta* 400:461-467.

Stern, D.B. and Longdale, D.N. 1982. Mitochondrial and chloroplast genomes of maize have a 12-kilobase DNA sequence in common. *Nature* 299:698-702.

Szostak, J.W. and Ingle, J. 1980. Unequal crossing over in the ribosomal DNA of *Saccharomyces cerevisiae. Nature*. 284:426-430.

Timmis, J.N. and Ingle, J. 1973. Environmentally induced changes in rRNA gene redundancy. *Nature New Biol.* 244:235-236.

Timmis, J.N. and Scott, N.S. 1983. Sequence homology between nuclear and chloroplast genes. *Nature* 305:65-67.

Thompson, J.D. 1991. Phenotypic elasticity as a component of evolutionary change. In: *Trends in Ecology and Evolution* 6:246-249.

Van't Hof, J., Mijirkneo, C.A. and Delihas, N.R. 1983. Excision and replication of extrachromosomal DNA of pea (*Pisum sativum*). *Mol. Cell. Biol.* 3:172-182.

Varley, J.M., MacGregor, H.C. and Erba, H.P. 1980. Satellite DNA is transcribed in lambrush chromosomes. *Nature* 283:686-688.

Walbot, V. and Cullis, C.A. 1985. Rapid genomic changes in higher plants. *Ann. Rev. Plant. Physiol.* 36:3637-396.

Walker, P.M.B. 1971. Repetitive DNA in higher organisms. *Prog. Biophys. Mol. Biol.* 23: 145-190.

Ward, B.L., Anderson, S.R. and Bendich, A.J. 1981. The mitochondrial genome is large and variable in a family (Cucurbitaceae). *Cell* 25:793-803.

Watson, J.D. 1976. *Molecular Biology of the Gene.* Third Edition, Metro Park, California, p. 739.

Weintraub, H. 1978. The nucleosome repeat length increases during erythropoiesis in the chick. *Nucl. Acid Res.* 5:1179-1188.

White, M.J.D. 1973. *Animal Cytology and Evolution,* Third Edition. Cambridge University Press, U.K., p. 9161.

Willis, C. 1983. The possibility of a stress-related evolution. *Heredity* 51:530-553.

Yakura, K., Kato, A. and Tanifuji, Sh. 1987. Cytological localization of the highly repeated DNA sequences, the Fok I sequence family and Bam H I sequence family in *Vicia faba* chromosomes. *Jap. J. Genet.* 62:325-332.

Zurwski, G., Clegg, M.T. and Brown, A.H.D. 1984. The nature of nucleotide sequence divergence between barley and maize chloroplast DNA. *Genetics* 106:735-746.

Biological Diversity: Problems and Challenges. Edited by S.K. Majumdar, F.J. Brenner, J.E. Lovich, J.F. Schalles, and E.W. Miller. © 1994, The Pennsylvania Academy of Science.

Chapter Twenty

THE ROLE OF BOTANIC GARDENS

JAMES P. FOLSOM
Huntington Botanical Gardens
1151 Oxford Road
San Marino, CA 91108

INTRODUCTION

Botanic Gardens generally have not been in the vanguard of organizations mounting effective programs to conserve rare, threatened, and endangered plants. The many possible explanations for this failure could involve discussions of purpose, management, staffing, funding, local pressures, politics, difficulties in networking with other gardens, problems in cooperating with national membership organizations, or even smug detachment. Rather than attempting to dissect the historical complacency of public gardens toward conservation of biodiversity, this chapter proposes to summarize the potential for excellent achievement in the current decade, as evidenced by the unique characteristics of botanical gardens and programs that have emerged within this community.

STRENGTHS OF NORTH AMERICAN BOTANICAL GARDENS

In relating mutual concerns and experiences, we discover that the modern botanic garden exists as one of various programs of public gardens, from traditional botanic collections to horticultural display gardens, commercial firms, educational museums, plant research programs, agricultural stations, teaching collections, zoological gardens, parks, and reserves that on this continent have come into being during the last century. Consequently, North America's botanic gardens are loosely connected through a range of organizations and programs, including the American Association of Botanical Gardens and Arboreta (AABGA), horticultural and trade groups, tourism and amusement gambits, landscape and design affiliates, museum and collections management professionals, conservation organizations and programs, historical associations, educators and promoters of volunteerism, plant collectors and their organizations, and elements of the plant science research community.

The current reality broadens considerably Victorian prototypes of the botanic garden (Bruinsma, 1976; Heywood, 1987), and brings an increased vitality and revised methods to the promotion, management, and achievements of today's public garden. The modern garden has been forced to learn business skills such as *marketing* and *strategic planning.* For financial survival these institutions have been pressed to push plants, education, and landscapes as public services. We analyze audiences, set income projections, and determine budgetary priorities. Programs that gardens are able to develop grow in the full light of realistic expectations for public support. We have learned how to fund projects in the absence of governmental largess through setting goals, interpreting results, and justifying costs.

From all aspects as public service organizations, public gardens have acquired a normal expectation for *volunteerism.* Not only does this interaction provide extra assistance for regular projects (such as propagation of important plants or tracking documentation), volunteers form a significant support constituency and channel of outreach for public gardens. Many of the regular chores associated with conservation programs can be organized and accomplished handily through volunteer assistance.

The new botanic gardens are those public gardens with a mission or purpose that supports maintaining, studying, and interpreting *collections* of living plants over long time periods (Raven, 1985). From experience both as museums and as scientific institutions, botanic gardens have focused on professional curation and evaluation of those collections. Over the last two decades most North American botanic gardens have formulated collections policies that seriously address issues of quality, focus, resources, documentation, and procedure. Leadership and funding from the Institute of Museum Services (IMS) have prompted gardens to bring professional standards to collections curation.

Clearly botanic gardens have knowledge and experience in *plant propagation* and *cultivation.* At other botanical institutions that lack horticulturally trained staff members, the handling of living plants acquired for research has always proven a limiting and wasteful exercise, whereas this type of expertise is considered a normal necessity at a garden. As gardens move to hire staff with better professional training and as micropropagation techniques become more standard, such capabilities should continue to improve.

Public institutions have become more clearly instruments of *display* and *education.* This aspect of botanic gardens is anticipated to grow as public service increases for a senior population increasingly dedicated to gardening and concerned about conservation, social causes, and quality of life. As universities fail to teach basic botany and this information consequently is excluded from grade and high school curricula, public gardens acquire a greater burden to inform youngsters as well.

Site specificity and *longevity* are forgotten attributes of botanic gardens. Basic to founding these institutions is the securing of property for collections and landscapes, as well as buildings and support structures. It is uncommon for a garden to change locations or eliminate important plantings.

In contrast with commercial establishments, agricultural research institutes, and governmental germplasm agencies, botanic gardens have *missions* that indicate long-standing interest in living plants and long-term dedication to growing collections (Given, 1987).

This interest and purpose includes *research*, but with the exception of a few gardens (Kew, New York, Missouri, Fairchild, etc.) the most important areas in advancement of plant science, descriptive studies and publication, have not depended on the activity of botanic gardens historically but have been supported largely at universities, museums, and other research institutes. Unfortunately, there is little indication that universities and natural history museums have a long-term commitment to the study of basic botany, evidenced by the gradual loss of botanical research and education from university programs since mid-century.

Even when bearing a research orientation, botanic gardens have dedicated themselves to general floristics or horticultural investigations. Therefore, at least in North America, gardens have had little impact on agricultural research, experimental science, habitat aquisition, political lobbying, and national conservation efforts, though the missions of most gardens would well-support cooperation with organizations dedicated to such programs.

EMERGING AND EXISTING PROGRAMS

Computerized access to collections. Promoted in North America by the AABGA, the Institute for Museum Services (IMS), the Center for Plant Conservation, and other organizations (Cullen and Lear, 1987), gardens have worked independently and in concert to computerize catalogs of plant holdings, mapping, and other aspects of collections management. There is every concern within this community that holdings of botanic gardens be computerized and verified with the greatest expediency. As other plant records (floras, herbaria, nomenclature) also yield to automation we discover that botanic gardens increasingly provide the stable centers for these long-term projects as well (Bisby et al. 1992), such as *Index Kewensis,* TROPICOS, etc.

The North American Plant Collections Consortium has evolved within the American Association of Botanical Gardens and Arboreta to formulate a network of botanic gardens "to coordinate and improve the collective living plant collections of North America, and to enhance the conservation and availability of this plant germplasm for current and future use." Eventually this consortium could bring new standards, coordination, and availability for living plants held in North American gardens—creating in essence a single networked collection as envisioned by others for seed banking (Gómez-Campo, 1987) and horticultural plants (Brickell and Sharman, 1986). A pilot program for a limited number of gardens began in 1992. Standards for verification, documentation, and replication of collections will be tested and refined during the first years of operation.

Center for Plant Conservation (CPC). Certainly the most significant program addressing the role of botanic gardens in the conservation of plants native to North

America has been the Center for Plant Conservation. This organization, created in 1984 and now based at Missouri Botanical Garden in St. Louis, MO, has as its primary objectives (CPC, 1990):

"1. to develop, maintain and coordinate the *National Collection of Endangered Plants* and its associated conservation, research, and education programs.
2. to develop and maintain comprehensive and broadly accessible information systems, national networks, and databases concerning the biology, horticulture, and conservation status of all the nationally endangered native plants of the United States.
3. to work with colleague organizations on collaborative integrated conservation projects combining species research, habitat management, and restoration of endangered plant populations in the wild.
4. to assist botanical gardens nationally in developing public awareness of plant endangerment and conservation efforts."

Currently, more than twenty participating botanic gardens manage conservation of over 300 native plant species under arrangements with CPC. The Center establishes standards and methods for determining the necessary program of acquisition and propagation a garden should attain to protect a given species of rare or endangered plant. CPC intends a fully integrated approach to conservation, from propagation and *ex situ* cultivation to reintroduction, habitat protection, research, computerized information access, plant economics, and public awareness. The goal of working with the continent's 3,000 plus threatened plants awaits increased funding for the Center's activities. The logical extension of this program is that as a model for similar activities in other regions of the world.

Regional Plant Conservation Programs. Many gardens associated with the CPC have their own programs that are more extensive than the activities coordinated through the Center. These gardens have learned to understand the relationships between their own objectives and federal/state regulations, such as the Endangered Species Act (Bartel, 1986) and the California Native Plant Protection Act (McMahan, 1986; Cochrane, 1986). In each case their projects relate to the most intrinsic conservation-worthy characteristics of botanic gardens.

Through interaction and cooperation with a range of conservation, governmental, and landowner groups, Berry Botanic Garden has built a seed banking program solidly integrated with all other aspects of collections management (McMahan, 1990). Conservation and land agencies begin to think of this garden as extending their own capabilities for germplasm storage, viability studies, and habitat restoration expertise, while the garden builds constituencies, successfully obtains outside funding, and formulates realistic strategies for collections and conservation policies. Like Berry, many gardens are learning that low-cost seed banking can enhance and help focus the conservation aspects of plant collections.

Concentrating on rare plants that have not proven good candidates for seed banking, Bok Tower Gardens organized a strong program that involves in depth understanding as to the cultivation of native plants as well as the potential and process for reintroduction. Lessons gained from the experiences at Bok Tower (Wallace, 1990) include a constant need for experimentation as to propagation, cultivation, and re-establishment techniques and an overwhelming need to expand the process of reintroduction in protected sites immediately. The scale at which reintroduction of these Florida upland scrub community natives must be accomplished appears to be much greater than is supportable through a simple and limited *ex situ* collection. The multi-dimensional work accomplished at Bok Tower Gardens indicates that botanic gardens should be gearing toward much larger and more dynamic reintroduction programs than may have been previously assumed would be necessary for success.

Bringing a fresh and rigorous approach to collections management, Desert Botanic Garden has not only applied museum standards to plant holdings, but has demanded that garden collections and research have applications to conservation of biodiversity. With focus on native systems, researchers at DBG have initiated projects to determine methods for habitat restoration of abandoned cotton farms. Working with the city of Mesa, AZ, randomized block trials address questions of timing and irrigation requirements for successful establishment.

San Antonio Botanical Garden continues to monitor reintroductions of *Styrax texana* (Cox, 1990), a native shrub that was known from only 40 extant wild specimens. Funded initially through the Institute of Museum Services, the project was continued through support by the U.S. Fish and Wildlife Service. Three sites were chosen and reintroduction efforts begun in 1987. The first flowering of these newly established populations occurred in spring 1992. Grazing pressure remains the significant problem to solve for a complete reintroduction program, but the work accomplished and the public awareness achieved are themselves great successes.

Conserving an entire native flora greatly imperiled by development and exotic introductions, is a big order almost anywhere. A program of the National Tropical Botanic Garden, the Hawaii Plant Conservation Center (HPCC), developed from a 1989 MacArthur Foundation grant is nearly that extensive. Of the native Hawaiian flora, 90% of plant species are endemic, 50% are threatened or endangered. HPCC has brought 70% of the 400 targeted plant species into protective cultivation. Beyond collection and cultivation, the program includes general floristic surveying, estimation as to plant status, studies in propagation techniques, collaboration with other gardens, and public education.

Outside the CPC network, many gardens conduct propagation, introduction, or management programs that concentrate on exotic rare plant species. Huntington Botanical Garden (Kimnach and Trager, 1992) manages the International Succulent Introductions (ISI) program that has operated for over thirty years. Of the approximately 2,000 different succulent plants that have been propagated and made available, most are species of rare plants that were distributed with original collection data. This program gives the Huntington Gardens strong entrée with individual collectors and commercial growers, allowing great opportunities for cooperative efforts in the propagation of succulent plants.

The University of California at Irvine Botanical Garden maintains a cryogenic seed and pollen bank integral with collection goals directed to study and conservation of irids, lilies, orchids, and other petaloid monocots. In addition to housing large representations of genera such as *Gladiolus, Moraea,* and *Babiana,* staff have experimented (Koopowitz and Kaye, 1983; Koopowitz, 1986) in maintaining one of the world's first orchid seed banks (dedicated to tropical orchids). The project as a whole invites participation not only of scientists, but also by commercial and lay growers in an effort to bring expanded support, concern, and utility to the garden conservation efforts.

Despite efforts at many gardens, zoos have outdistanced botanic gardens in cooperative efforts aimed at reproduction of endangered wildlife. In many instances zoos have bolstered their own botanical programs, which now may present some of the most innovative of conservation-wise activities at gardens. A recent mailing from the Cincinnati Zoo and Botanical Garden's Center for Reproduction of Endangered Wildlife (CREW) indicates a willingness to assist with tissue culture propagation of endangered plants held by gardens across the country. This organization even has a Plant Conservation Division, an aspect that most stand-alone botanic gardens have never featured.

International Trade and Botanic Gardens. The Convention on International Trade of Endangered Species (CITES) has modified considerably the operation of botanic gardens specializing in exotic plants (McGough, 1991). Positive effects have included a real change in professional and public attitudes concerning the acquisition and trade of wild-collected plants. Botanic gardens are able to set an example with private collectors concerning the dangers of collection pressures on limited populations. Moreover, the new regulations have forced proper behavior of researchers based in economically advantaged countries. We have all learned that it is appropriate and productive for North American botanic gardens to foster research and education in economically disadvantaged countries (Budowski, 1977).

Unfortunately, this international agreement has also politicized the acts of plant collecting and exchange. At the very time that living collections may actually receive proper curation and cultivation as well as find real utility in research and conservation, the political and bureaucratic obstacles to responsible field collecting have mushroomed. Procedures for application are often complex, sometimes irrational. Policy changes are continual, such that researchers are never certain as to the new circumstances. Enforcement is generally unidirectional, with importation to developing countries being unregulated as to CITES. Scientific exchange, though recently clarified and effectively expanded, remains murky due to varying dispensation of scientific authority from one country to the next. Rescue operations in the face of logging and homestead clearing are difficult to arrange and finance. Moreover, the clear disparity between limitations on collecting and lack of habitat protection in many countries disables arguments for right behavior and encourages lay collectors to disregard the law. This places botanic gardens, which court secular audiences, in the crucial position of explicator and trend setter in establishing new sensibilities regarding wild plant collections.

The legal residue of international agreements is large shipments of confiscated plant material. When not returned to the country of origin, these plants may be deposited at designated rescue centers. However, from the viewpoint of a rescue center that must accept an entire shipment, and agree to hold all plants under very strict limitations on distribution, the unpredictable and often unmanageable quantity of confiscated material swamps the capability of the limited numbers of designated rescue center gardens and forces rejection of shipments. From the viewpoint of governmental authorities, the problems in controlling this plant material once it has been "rescued" have disabled the entire concept and meant that for the last several years new rescue centers have been grudgingly designated.

International Coordination. North American botanic gardens that have been members of the Botanic Gardens Conservation Secretariat are familiar with the bewildering array of programs, materials, and structural changes generated by this organization relating to IUCN conservation strategies for gardens (Hamann, 1990; IUCN 1987; 1989a). The Secretariat arose as a small program within IUCN termed the Botanic Gardens Conservation Co-ordinating Body. The main efforts of that organization combined with those of the Threatened Plants Unit/Committee to yield listings of threatened plants cultivated in the world's botanic gardens.

Established in 1987, the Secretariat proposed (IUCN, 1987):

"1. to promote the implementation of the *Botanic Gardens Conservation Strategy.*
2. to monitor and co-ordinate *ex situ* collections of conservation-worthy plants.
3. to develop a program for liaison and training.
4. to arrange botanic gardens conservation congress every three year."

A complete document outlining the conservation strategy was published by the Conservation Secretariat in 1989 (IUCN, 1989a; Hamann, 1990). Sixty-eight points were made regarding gardens operations and the practice of conservation. From the bulletins and publications released, it is difficult to determine what programs the organization, now termed Botanic Gardens Conservation International, may actually manage on a permanent basis. Various pilots evolve, such as the monitoring effort termed the Botanic Gardens Plant Trade Commission (Jackson, 1991), but there is as yet no history of sustained accomplishment. North American gardens have barely participated in determining the directions and activities of the program.

The direct activities of the Threatened Plants Committee (TPC), now commonly termed the Threatened Plants Unit, continue. They have proven very important in coordinating the significant holdings of gardens internationally. The surveys and reports (Lucas and Synge, 1978; IUCN, 1989b) that have been issued attempt to bridge the gap between the current status of gardens documentation and automation and a future objective of complete computerized access and verification.

Perhaps the most venerable international system for exchange and contact between gardens has been the *index seminum.* Hundreds of gardens issue these seed lists under rules of conduct that are similar to the operation of herbaria. The entire cost of collection, publication, and mailing is born by the issuing institution with participation in the exchange open to all similarly-oriented institutions. Most gardens do not continue exchange with an institution that fails to issue its own *index seminum* on some regular basis. Reasonable guidelines have been suggested that encourage gardens to concentrate on field collected seed (with published collection data) native to their own regions (Heywood, 1976). Such collections should also be considered as prospects for adding samples to seed banks.

CLEAR ROLES OF BOTANIC GARDENS

A. Education. No organization bears greater responsibility for informing the public concerning the nature and importance of plants and their roles in natural systems than botanic gardens. The educational and outreach program of any botanic garden should make this basic message its greatest priority (Raven, 1976; 1985; 1992).

Education is also necessary for citizens and officials who, because of their dedication to plants, one would believe are already sympathetic and knowlegeable concerning plant conservation. Children of the Victorian collectors, some would say pillagers, the new botanic gardens must work to bridge gaps in communication and agreement that have developed between and among those who study plants (professional and lay scientists, nurserymen and horticulturists), those who collect plants (institutions, private collectors), and those who regulate plants (forestry, park, agricultural, and wildlife officials).

B. Managing Collections. Implicit in the definition of a botanic garden is creating and using collections of living plants for the public welfare. The constructive management of artificially maintained (a broader concept than *ex situ*) collections includes an obligation for professional curation - requiring focus, documentation, quality cultivation, tracking, and vouchering in accordance with clearly established policy. The only conservation-wise justifications for collections are:

1. interpretation and education,
2. investigation,
3. propagation and eventual reintroduction.

With greater purpose and rigor incorporated in the curation of conservation collections, it becomes more likely these collections will have real implications for protection of biodiversity (Given, 1987).

More cost-effective than extensive living collections, preserved collections have great implications for conservation. Ranging from herbarium tissue to seed banks to tissue culture samples, these population samples have compactness and appeal to future technological prowess. Borrowing a page from science fiction, one can envision the re-establishment of complex biotas if genetic stock is available. Because botanic gardens are in the long-term business of encountering large selections of plant species (through our own acquisitions as well as association with individuals and other collecting organizations) gardens are among the most logical sites for establishment of seed banks and other sampling/vouchering systems. Thorough integration of seed banking with institutional objectives and collections policy is essential (Gómez-Campo, 1987; McMahan, 1990).

C. Coordination. At a working level, botanic garden staff members have on-going contact with many individuals and organizations with interests in plant collections and biodiversity. Leadership could be taken by any number of botanic gardens to coordinate regional issues relating to collections and their roles in conservation, including acquisition, documentation standards, access, cataloging, automation, investigation, permanence (Poppendieck, 1976), conservation, and interpretation. These activities could be designed to coordinate the resources of private collectors, landowners, plant societies, commercial nurseries, regional and national organizations, governmental offices and authorities, and permitting authorities toward conservation and education activities in an infinite range of combinations. Such relationships are necessary to resolve problems of limited resources and space (Shaw, 1976) and will be absolutely essential for eventual reintroduction.

D. Propagation. Botanic gardens should be centers for information concerning the effective cultivation and propagation of threatened plants. This community must develop programs for training professional horticulturists in the management of collections, techniques needed for conservation-wise propagation of rare plants, and the preservation of natural resources.

E. Reintroduction may be viewed as a responsibility of land management organizations. However, it is impossible to believe that gardens should dedicate precious resources to maintaining, propagating, and studying living collections of rare plants as a mere exercise in scientific curiosity, with no implications for reintroduction and habitat restoration. Though not the initial task at hand, reintroduction is a logical, ultimate goal for all collections-oriented conservation programs.

F. In every way possible, gardens should foster basic descriptive research - by their own staffs, scientists at other institutions, and non-professional scientists (Givens, 1987). Botanic gardens can make available space for study, garden space and support for maintaining research collections, horticultural expertise, plant material, expedition support, and even in some cases grants and fellowships. Thus, researchers should expect assistance with the development of good handling techniques for monographic collections, rare plants, and plants with problematic survival potential. Moreover, botanic gardens should be integrated regularly with study programs of monographers, who continually build documented working collections of living plants that are regularly destroyed or lost at the conclusion of the investigation (Poppendieck, 1976; Raven, 1992). With the loss of botanical programs at many universities, the new botanic garden may remain the only institution dedicated to pure and applied plant research. Gardens could encourage local universities to maintain botanical programs through the potential avenues of support described above, as well as more direct intervention such as agreements for cooperative appointments. Moreover, given distinct objectives outlined by conservation policies, gardens research programs could encourage particularly the quest for invaluable data applicable to the preservation of biodiversity (Ayensu, 1981; Ashton, 1987; Raven, 1992).

G. Cooperation and Networking with other botanic gardens. Surveying major collections for confirmed presence of threatened plants has been an on-going program of the IUCN Threatened Plants Committee. It is critical that gardens survey, confirm identities, and voucher plants held in living collections - and that a system for general, automated access to that information be developed immediately.

Gardens must also support the establishment of new gardens, which are clearly under-represented in economically stressed regions of the world - often the very places of great biodiversity and uncontrollable forest destruction (Raven, 1976). As centers for research, collection, coordination, and education, botanic gardens are crucially needed where conservation problems seem most severe. Productive support mechanisms include assistance with staff training, technical support for development, help with building library resources, exchange of plant materials, and even financial backing. In final analysis, collections of rare plants are best maintained and better serve the purposes of conservation when housed, studied, propagated, and promoted in the areas and climates to which they are native (Heywood, 1987). Artificially supported cultivation of plants will always prove problematic (Poppendieck, 1976; Simmons, 1976; Ashton, 1987) and the appreciation of any rare plant necessary to foster its protection and reintroductions must arise in the land where the plant was born to prosper.

H. Establish Conservation Policies and Professionalize the Plant Conservation vocation. As gardens begin to establish formal policies, objectives, and goals related to plant conservation, there must be a movement to broaden the traditional monitoring role of curators and horticulturists to include extensive training in conservation science, techniques, and objectives (McMahan, 1990). This new cadre of professionals with distinct experience and long-term goals will be the driving force behind eventual global coordination of plant collections and plant conservation activities.

CONCLUSION

In an ideal world, gardens would be permanent centers of calculation, fostering long-term botanical study, conservation, and interpretation with individualized focus on given plant groups or ecosystems. Through non-exclusive coordination arrangements active gardens would assume the commitment to maintain contact with environmental groups, governmental authorities, individuals, ana other institutions concerning the future conservation status of the target plants. Gardens would stimulate and support symposia, habitat studies and population surveys, taxonomic clarification, thorough use of plant collections for study, distribution and conservation efforts, and programs for public awareness. Gardens might provide housing for a range of plant and conservation related, constituency-based societies and organizations, from national plant interest groups and professional groups to more politically active environmental programs.

In the gathering storms of overwhelming world cultural, social, economic, and environmental tragedy, botanic gardens must visualize beyond the chaos to a desired future of sustainability and restored natural balance. One hopes we can educate audiences so as to alleviate the potential full brunt of this crisis. Minimally, we must cooperate to provide time capsules and refugia that offer some promise in preserving global biodiversity through these new dark ages.

SELECTED BIBLIOGRAPHY

Ashton, P.S. 1987. Biological considerations in *in situ* vs *ex situ* plant conservation, in *Botanic Gardens and the World Conservation Strategy,* ed. Bramwell, et al. IUCN by Academic Press, London, pp. 117-130.

Ayensu, E.S. 1981. Assessment of threatened plant species in the United States, in *The Biological Aspects of Rare Plant Conservation,* ed. by H. Synge, John Wiley & Sons Ltd., Chichester, pp. 19-58.

Bartel, J.A. 1986. The Federal Listing of Rare and Endangered Plants: What is Involved and What Does It Mean, in *Conservation and Management of Rare and Endangered Plants* ed. T.S. Elias, California Native Plant Society, Sacramento, pp. 15-22.

Bisby, F.A., D.A. Sutton and G.F. Russell. 1992. Report of the Third Meeting of the Taxonomic Databases Working Group, Edinburgh, 19-21 October 1987, Huntia 8(2):93-109.

Brickell, C. and Fay Sharman. 1986. *The Vanishing Garden,* John Murray in association with The Royal Horticultural Society, London, 261 pp.

Bruinsma, B.F. 1976. Past: Aims and Objectives, in *Conservation of Threatened Plants,* ed. J.B. Simmons, et al. NATO Scientific Affairs Division by Plenum Press, New York, pp. 19-26.

Budowski, G. 1977. A Strategy for Saving Wild Plants: Experience from Central America, in *Extinction is Forever* ed. G.T. Prance and T.S. Elias, New York Botanical Garden, Bronx, pp. 283-300.

Center for Plant Conservation (CPC). 1990. Mission Statement, as approved by the Executive Committee on 10 August 1990, 6 pp.

Cochrane, S.A. 1986. Endangered Plants and California State Laws, in *Conservation and Management of Rare and Endangered Plants* ed. T.S. Elias, California Native Plant Society, Sacramento, pp.

Cox, P. 1990. Reintroduction of the Texas Snowbell, (*Styrax texana*), Endangered Species UPDATE, 8(1):64-65.

Cullen, J. and M. Lear. 1987. Principles and standards for the computerization of garden record schemes, as applied to conservation, with proposals for an International Transfer Format, in *Botanic Gardens and the World Conservation Strategy,* ed. Bramwell, et al. IUCN by Academic Press, London, pp. 301-337.

Davis, S.D. 1986. *Plants in Danger, What do we know?* IUCN, Kew, Richmond, 461 pp.

Given, D.R. 1987. What the conservationist requires of *ex situ* collections, in *Botanic Gardens and the World Conservation Strategy,* ed. Bramwell, et al. IUCN by Academic Press, London, pp. 106-116.

Gómez-Campo, C. A strategy for seed banking in botanic gardens; some policy considerations, in *Botanic Gardens and the World Conservation Strategy,* ed. Bramwell, et al. IUCN by Academic Press, London, pp. 151-160.

Hamann, O. 1987. The IUCN/WWF Plants Conservation Programme in action, in *Botanic Gardens and the World Conservation Strategy,* ed. Bramwell, et al. IUCN by Academic Press, London, pp. 31-43.

Heywood, V.H. 1976. The Role of Seed Lists in Botanic Gardens Today, in *Conservation of Threatened Plants,* ed. J.B. Simmons, et al., NATO Scientific Affairs Division by Plenum Press, New York, pp. 225-231.

Heywood, V.H. 1987. The changing role of the botanic garden, in *Botanic Gardens and the World Conservation Strategy,* ed. Bramwell, et al. IUCN by Academic Press, London, pp. 3-18.

IUCN. 1987. Botanic Gardens Conservation Secretariat - A Prospectus. Conservation Monitoring Centre, Kew, 11 pp.

IUCN Botanic Gardens Conservation Secretariat. 1989a. *The Botanic Gardens Conservation Strategy,* Kew, Richmond, 60 pp.

IUCN Botanic Gardens Conservation Secretariat. 1989b. *Extinct Plant Species of the World,* Survey, questionnaire and preliminary report on their occurrence in botanic gardens, 23 pp.

Jackson, P.S.W. 1991. BGCI Botanic Garden Plant Trade Commission to be established, *Botanic Gardens Conservation News* 1(9):48.

Kimnach, M. and J.N. Trager. 1992. The Huntington Botanical Gardens presents the 1992 offering of International Succulent Introductions, *Cactus and Succulent Journal* 64(2):82-89.

Koopowitz, H. and H. Kaye. 1983. *Plant Extinction: A Global Crisis,* Stone Wall Press, Washington, pp. 150-151.

Koopowitz, H. 1986. A Gene Bank to Conserve Orchids, AOSB, 55(3):247-250.

Lucas, G. and H. Synge. 1978. *The IUCN Plant Red Data Book,* IUCN, Kew, Richmond. 540 pp.

McGough, N. 1991. CITES and botanic gardens, Botanic Gardens *Conservation News* 1(9):44-47.

McMahan, L.R. 1986. Rare Plant Conservation by State Government: Case Studies From the Western United States, in *Conservation and Management of Rare and Endangered Plants,* ed. T.S. Elias, California Native Plant Society, Sacramento, pp. 23-31.

__________. 1990. Propagation and Reintroduction of Imperiled Plants, and the Role of Botanical Gardens and Arboreta, Endangered Species UPDATE, 8(1):4-7.

Poppendieck, H.H. 1976. Mesembryanthemums and the Problems of their Cultivation, in *Conservation of Threatened Plants,* ed. J.B. Simmons, *et al,* NATO Scientific Affairs Division by Plenum Press, New York, pp. 55-60.

Raven, P.H. 1976. Ethics and Attitudes, in *Conservation of Threatened Plants,* ed. J.B. Simmons, et al., NATO Scientific Affairs Division by Plenum Press, New York, pp. 155-179.

__________. 1985. Public Address, Symposium, 25 May, Rancho Santa Ana Botanic Garden, CA.

__________.1992. Public Address, American Association of Botanic Gardens and Arboreta Annual Meeting, 15 June, Columbus, OH.

Shaw, R.L. 1976. Future: Integrated International Policies, in *Conservation of Threatened Plants,* ed. J.B. Simmons, et al. NATO Scientific Affairs Division by Plenum Press, New York, pp. 39-47.

Simmons, J.B. 1976. Present: The Resource Potential of Existing Living Plant Collections, in *Conservation of Threatened Plants,* ed. J.B. Simmons, et al. NATO Scientific Affairs Division by Plenum Press, New York, pp. 27-38.

Wallace, S.B. 1990. Central Florida Scrub: Trying to Save the Pieces, Endangered Species UPDATE, 8(1):59-61.

Biological Diversity: Problems and Challenges. Edited by S.K. Majumdar, F.J. Brenner, J.E. Lovich, J.F. Schalles, and E.W. Miller.

Chapter Twenty-One

THE ROLE OF ZOOS AND AQUARIUMS IN THE CONSERVATION OF BIODIVERSITY

DON MOORE[A] and **ROBERT WIESE**[B]

[A]Thompson Park Conservancy
P.O. Box 8182
Watertown, NY 13601
and
[B]AAZPA Conservation Center
7970-D Old Georgetown Road
Bethesda, MD 20814

"In the preservation of biological diversity, the use of technology is a last resort. When the preservation of ecosystems falters, their fragments may have to be cared for piece by piece."[1]

ABSTRACT

Zoos and aquariums help conserve the world's biodiversity through captive breeding, research, public and professional education and *in situ* reserve support and conservation programs. Summarizing current financial support for field conservation by North American zoos, the authors recommend that zoos annually commit at least 1% of their institutional budgets to field conservation programs.

INTRODUCTION

Zoos and aquariums are like a modern "ark"[2] which can help stem the world's current flood of extinctions, this current loss of biological diversity.[3,4] The US Congress,[5] the World Bank,[6] and the world's leading conservation organizations (see [6,7]) have all formally recognized the role of zoos and aquariums in the conservation of biological diversity.

Zoological institutions began to have an influence on the conservation of biological diversity at the turn of the 20th century. Dr. William T. Hornaday, former Director of the New York Zoological Park, was one of the leading voices in the early conservation movement in the United States and envisioned the National Zoological Park as a center for conservation. His conservation efforts were critical in the establishment of the American Bison Society[8] and of several laws protecting wild birds from unregulated commercial slaughter.[9] The influence of zoo professionals such as Heini Hediger, Bernard Grzimek and others was also important in promoting early advances in captive animal management and the conservation of nature.[10]

Due to a modern "zoo Rennaisance" over the last twenty years, zoos and aquariums can be critical components in an integrated effort by governmental agencies and non-governmental organizations to conserve biodiversity through *ex situ* and *in situ* programs (see [11]).

Zoos and aquariums help conserve biodiversity in four ways.

1. **Captive breeding programs** provide a refuge for species declining in the wild and serve as a genetic reserve and source for reintroduction programs.
2. **Research** conducted by zoos and aquariums, both basic and applied, is important in gaining knowledge, expertise and new technologies critical to the management of both captive and wild populations.
3. **Conservation education,** both within the local community and around the world, is necessary to achieve a greater public conservation ethic. Zoos and aquariums educate both the lay public and wildlife professionals.
4. **Zoos and aquariums** provide support, both educational and financial, to *in situ* conservation efforts.

CAPTIVE BREEDING PROGRAMS

Most professional zoos and aquariums have made a major commitment to the conservation of biodiversity during the last 10-15 years through cooperative breeding programs for rare and endangered species. These programs offer a safe haven for species threatened with extinction until conditions in the wild can be improved. This is particularly important for species whose populations are greatly reduced or extirpated, e.g. the red wolf (*Canis rufus*), Arabian oryx (*Oryx leucoryx*) and the California condor (*Gymnogyps californianus*). For these species, the captive population acts as insurance against extinction while education and *in situ* programs are established. Once the original causes of population decline are neutralized, breeding of animals for reintroduction may be appropriate for many species.

The Species Survival Plan (SSP) of the American Association of Zoological Parks and Aquariums (AAZPA) was the first regional program to promote cooperative management of endangered species. Established in 1981, the SSP originally focused on the demographic and genetic problems that can occur within small populations.[12] Since 1981, a number of similar cooperative breeding programs have been started in other regions, such as the Australasian Species Management Program (ASMP), European Endangered Species Program (EEP), and the Species Survival Committee of Japan (SSCJ). As the regional programs become organized, they are beginning cooperative programs between regions as well, with the assistance of the IUCN/SSC Captive Breeding Specialist Group (CBSG).

Regional captive breeding programs coordinate the management of animals distributed across many institutions, in the best interest of a species. Through analysis of studbook data, specific animal-by-animal mating recommendations are made and management strategies are evaluated and improved. Participating institutions voluntarily implement the recommended protocols and transfer animals to facilitate desired pairings.

The ultimate goal of each captive breeding program must be to assist the survival of rare species in the wild. To this end the SSP attempts to preserve as much genetic diversity as possible within a limited "captive carrying capacity." Genetic diversity is important at both the population and individual levels for long-term survival and preservation of a species' ability to adapt to changing environmental conditions.[13,14,15,16]

The current management philosophy of the SSP program is to preserve allelic variation through the relative equalization of genetic contributions of founders to a descendant population. Within this context, the SSP also attempts to maximize individual heterozygosity through minimizing inbreeding. Of course, all breeding programs must also manage their populations within the context of the species' basic biology and social structure, if they are to succeed. The integration of genetic, demographic and biological criteria into one functional, cooperative and realistic plan is a challenging task for each captive breeding program.

The number of taxa involved in captive breeding programs continues to expand, as does the necessary philosophical and financial commitment of all participating institutions. There are currently 64 species, some with multiple subspecies, in the SSP programs in North America alone. The goal of the SSP is to have 200 programs by the year 2000.[17] Worldwide, over 200 taxa are already being managed or have been proposed in the five regions with existing programs (CBSG, unpubl. ms.). This number of taxa will expand greatly as new regions begin their own management programs.

Zoos and aquariums also maintain breeding populations of more common species. These programs provide animals for exhibit and public education without requiring removal of animals from the wild. Another advantage of this strategy is that husbandry concerns can be addressed before these species become threatened; hence, a viable captive population will exist if environmental or political conditions change quickly, putting the wild population at risk.[18]

The decision of which species are to be allowed on the zoo "ark" is not a trivial one. Space and resources for the operation of these cooperative programs are limited and not all species can be maintained. As responsible stewards of biological diversity, zoos and aquariums must prioritize species selection to preserve as much natural diversity as possible.[12,18] Strategic planning for the regional programs is carried out by Taxon Advisory Groups (TAGs). The TAG concept began in North America in conjunction with the SSP program, and TAGs are now being developed in other regions as well.

The primary responsibility of a TAG is to evaluate the present captive carrying capacity for a given taxonomic group (e.g. lizards, parrots, canids), and to recommend how available "ark" resources should be allocated to species at risk.[12] Through their membership of zoo professionals, university and field biologists, and governmental or NGO officials, TAGs prioritize species for SSP and studbook designation. They also address more general issues common to their taxon of concern, such as over-collection of fauna for laboratory use or for the pet trade, and global decline. TAGs are organized in a manner similar to the specialist groups of the IUCN/SSC and the International Council for Bird Preservation in the hope that similar organization will facilitate cooperation and communication. Broader representation in the species selection process should ensure that prioritization reflects real needs and is compatible with *in situ* conservation efforts. IUCN/SSC Taxon Specialist Groups' Captive Action Plans and IUCN/SSC Global Captive Action Plans are an important resource for the species selection process when they are available. Other factors considered in the TAG decision-making process include IUCN critical need criteria, availability of adequate founder stock, ability to achieve captive breeding success, taxonomic uniqueness, and the species' ability to serve as a flagship species for endangered ecosystems.

Many of the current threats to wildlife dictate that programs will need to manage populations in captivity for an extended period of time, i.e. 100-200 years.[19] However, some species experiencing local declines may benefit from short-term projects lasting only 5-25 years. Referred to as captive propagation rescue (CPR), these programs are designed to quickly provide individuals to reinforce natural populations that are so small and fragmented that they are not genetically or demographically viable. This must be in conjunction with the preservation, protection, or restoration of the native habitat. This level of conservation activity has received increased attention lately; many zoos and aquariums are involved in local projects. For example, the San Diego Zoo and Wild Animal Park is working with the San Clemente loggerhead shrike (*Lanius ludovicianus mearnsi*),[20] and Knoxville Zoo is breeding bog turtles (*Clemmys muhlenbergii*).[21]

While not a prerequisite for success, reintroduction programs are often associated with captive breeding programs in zoos. Red wolves howl again in South Carolina, condors soar in California, and Arabian oryx again roam the deserts of Saudi Arabia due to the reintroduction efforts of captive breeding programs.[17]

RESEARCH PROGRAMS

Successful conservation programs require detailed, accurate knowledge of species' basic biology, including physiology, nutrition, reproductive biology, animal behavior, and pathology. Research on captive populations allows zoo biologists to acquire this basic data, perfect wildlife management techniques, and develop veterinary protocols that can be applied to management of wildlife in natural habitats. Conversely, zoo biologists also study wildlife in the field to contribute to scientific knowledge and conservation (e.g. [22]).

Zoos and aquariums have been responsible for the development of good husbandry protocols for captive propagation and management of a wide variety of species. Zoos, in fact, are now net producers of wildlife, and in a few species "surplus" individuals have become a concern.[23] However, once husbandry and propagation techniques are developed, a species' chances for survival are increased greatly if its numbers decline precipitously in the wild, and a captive program is recommended.

The lesser apes (*Hylobates* spp.) are a good example of a species group for which an excellent captive husbandry and breeding record has been established over the last two decades.[24] Now that southeast Asian tropical forests are imperiled, critically endangered forms are being designated by the Gibbon SSP/TAG for captive propagation. The expertise previously gained in captive husbandry of this group will be critical for the survival of the declining populations.

Zoos and aquariums have access to a wide variety of species, and are in a unique position to perform basic research. Normal physiological values are more easily obtained from captive animals where urine, feces, blood and semen can be collected on a regular basis from readily identifiable individuals. Access to other animal products, such as mammalian placentas, is also virtually impossible in free-ranging wild populations. Zoo biologists have regular access to these resources and can use them efficiently for their own research or in collaboration with university researchers.[25]

Most zoos with endangered Asian and African elephants (*Elephas maximus* and *Loxodonta africana*), for instance, are able to collect blood regularly, and normal blood hormone cycles are now being established.[26] Normal values are used by researchers in a variety of disciplines, including nutrition, reproductive physiology, genetics and animal behavior to improve husbandry techniques. The knowledge of normal values is also important in monitoring wild populations for health and disease.[27]

Similarly, captive animals can be used for the development of veterinary medical procedures and treatments which may help both captive and wild populations.[28] A sound knowledge of infectious diseases is important to both treatment and prevention of those diseases in healthy captive populations. Sophisticated zoo research includes investigations in parasitology, microbiology and virology.[25] Infectious disease, vaccine and pathology research in captive wildlife contributes to the management of wild populations, especially as wild species are compressed into smaller protected areas and have more interspecific contact.[25] Zoo veterinarians assist with field wildlife procedures, especially with endangered species programs which have high public visibility. In New York State, for example, endangered massasagua rattlesnakes (*Sistrurus catenatus*) are captured in the field, but have radio transmitters inserted subcutaneously under sterile conditions at the Burnet Park Zoo veterinary clinic (Wallace and Johnson, pers. comm.); this reduces the risk of infection to the snakes, and guarantees the presence of a trained clinician if the snakes have a medical problem.

Zoos have refined techniques for the capture, restraint, identification, transfer and monitoring of wildlife (see [29]). Zoo veterinarians have refined the use of immobilizing chemicals, in terms of type, delivery, dosage, and combinations used.[29]

Wildlife identification techniques are often tested in zoos prior to field studies. Radio transmitters have been tested on captive animals before being used in the field; elephant infrasound transmitter collars, for example, were tested by Cornell researchers on elephants at Burnet Park Zoo before their use in Africa. Similarly, radio transmitters using satellite tracking technology were tested at the New York Zoological Park prior to use on elephants in Africa.

Detailed studies at several aquariums have allowed the collection of information that would have been impossible to collect under field conditions. Marine vertebrate science benefits greatly from these studies. Studies on the call and communication in orcas (*Orcinus orca*)[30] and growth rates of various shark species[31] are examples.

Many species are secretive in their feeding, reproductive and other behaviors, and are, therefore, difficult to observe under field conditions. Captive populations allow researchers to gather data on these subjects in a scientific manner. Observation of captive animal behavior before embarking on field research has also given many field biologists critical insight in developing necessary techniques for data collection, thus reducing the time and expense of refining techniques in the field.

Finally, extensive necropsy protocols allow researchers access to normal anatomical and pathological data. Many North American TAGs and SSPs have developed such protocols to allow access to materials that might otherwise be unavailable or denied for endangered species in the wild.

Cooperative breeding programs and population management have allowed zoo biologists to make a considerable contribution to small population management over the last ten years. This contribution includes analysis of small population genetics and demographics,[25,32] development of small population software for microcomputers, and of techniques for decision analysis for endangered species action plans.[33]

Many technologies developed or tested by zoos and aquariums are directly relevant to biodiversity conservation efforts in the field, especially in small, isolated islands of habitat in need of intensive management—large, semi-natural "zoos." To realize their full potential as actors on the conservation stage however, zoos and aquariums will have to increase their committment to research programs, and will have to increase their collaboration with the academic community.[12,18] Research priorities for single species conservation biology were recommended at a 1988 workshop sponsored by the National Zoo and the National Science Foundation.[34] Increased cooperation and communication among the academic and zoological community will reduce duplication of effort and will allow zoos and aquariums to make more efficient use of limited human and financial resources.

EDUCATION PROGRAMS

Zoos and aquariums have a tremendous potential to contribute to global and local conservation efforts through education of the public and wildlife professionals. The AAZPA-accredited zoos and aquariums in North America alone attract over

100 million visitors each year;[35] worldwide, the number is even greater.[6] Zoos have become a major focus for public conservation education: nearly 90 percent of North American zoos have an educational department, and environmental conservation has been the main focus of their programs for over a decade.[35]

For many people, in both industrialized and developing countries, a visit to the zoo is the only chance they will have to experience first-hand some of the world's most spectacular biological diversity. Whether viewing a herd of Pampas deer (*Ozotoceros bezoarticus*) in a natural habitat zoo in South America, a group of ruffed lemurs (*Varecia variegata*) in a zoo in Madagascar, or a pair of red wolves (*Canis rufus*) in a North American zoo, visitors may walk away with a greater aesthetic appreciation of their wildlife heritage, and perhaps a greater desire to be a participant in the conservation process. Wildlife education, which uses these "flagship" species, helps make zoos more exciting places where learning can occur (Aiello, pers. comm.).

Modern zoos do not rely solely on passive education of the zoo visitor. Zoos actively teach about conservation of biodiversity, ways to become active participants in the conservation process, and about backyard problems like local pollution and habitat loss.[36] This education is accomplished through excellent programming of exhibits, animal care staff and volunteers. Zoo education can be more: visitors should be taught about why habitat loss occurs, and what they can do about it, capitalizing on the connection between a zoo's animals and habitat conservation (Aiello, pers. comm.).

Zoos also provide on-site training opportunities for zoo personnel, wildlife managers and veterinarians working in many countries throughout the world.[37] Training of foreign nationals can be a critical step in helping local people and governments realize the importance of their commitment to and investment in conservation projects. Without local commitment, conservation projects will not succeed.

IN SITU CONSERVATION PROGRAMS

Historically, zoos and aquariums have invested relatively little in direct field conservation. However, there is a growing commitment to expand their role in this arena. During 1990-1991, AAZPA institutions supported over 400 conservation-related projects in 63 countries around the world.[38] Many zoo professionals serve on working committees of the IUCN's SSC,[12] or serve as advisors or consultants to governments and conservation organizations around the world (Rabb, pers. comm.). These activities constitute a major contribution on the part of zoos and aquariums to the management of diverse wild and captive species.

Funding is often limited for zoos in the industrialized countries, whether they are under the financial umbrella of municipal or state governments, or private societies. This funding level is generous, however, when compared to costs in many developing countries. A growing number of institutions are recognizing the need to use some of this funding to support biodiversity conservation in endangered animals' countries of origin. The authors recently completed a brief survey of financial support by North American zoos and aquariums for field conservation projects in 1990 and 1992, which produced some remarkable figures. Direct and indirect financial aid by the responding institutions to field conservation totalled over 8 million dollars. Some institutions were unable to contribute to field conservation due to local public trust statutes. Of the 32 institutions which did contribute to field conservation, financial support levels ranged from one thousand to 6.2 million dollars; this level of support represented on average 1.32% (range = .02 - 12%) of the total annual institutional budget, and did not include institutional expenditures of grants from government agencies. Although eight of the institutions decreased their support for *in situ* conservation between 1990 and 1992, three institutions began conservation funding during this time period, and field conservation fund growth averaged 128% (range = 3% - 500%) for the remaining institutions.

The largest contribution to *in situ* conservation was made by the New York Zoological Society, which operates the New York Aquarium, the Bronx Zoo, other metropolitan New York zoos and a rare animal breeding facility. Wildlife Conservation International, one of the world's largest and oldest conservation and field research organizations, is a prominent member of the New York Zoological Society's effort. Other outstanding North American institutions include the Smithsonian Institution ([38]; Rabb, pers. comm.) and the Chicago Zoological Society ([38]; Rabb, pers. comm.). These organizations commit large amounts of staff time and monetary resources for vehicles, employment of foreign nationals, and equipment and housing for researchers (Rabb, pers. comm.).

This growing involvement by zoos in field conservation has resulted in the development of regional and global Fauna Interest Groups (FIGs). FIGs coordinate the conservation activities of regional member institutions working in specific geographic regions, for instance Brazil, Madagascar and Zaire. FIG members may be actively involved in helping in-country wildlife agencies, NGOs and zoos establish protected areas, support existing parks, conduct field research, transfer technology by training foreign nationals, develop public education programs, and, when appropriate, develop cooperative captive propagation programs.[12,37] For example, members of the AAZPA regional Indonesia-Malaysia FIG, are supporting *in situ* efforts to preserve wildlife and habitats, and to develop field studies in the region. The Minnesota Zoological Gardens has "adopted" Ujung Kulon National Park on the island of Java; this "Adopt-A-Park" program finances about US$25,000 to buy equipment which will allow park staff to more effectively guard against poaching. The Smithsonian Institution, another member of this FIG, expended US$75,000 in 1991 in field training courses and development of an international center for conservation biology in Malaysia. The center is the outcome of six years of intensive conservation training programs between the Wildlife Department of Malaysia and the Smithsonian's National Zoological Park, and will award advanced degrees in conservation biology following academic training and field research in protected areas. The above monetary amounts do not include costs incurred by paid staff of the FIG-member institutions.[39]

The IUCN CBSG coordinates global equivalents of the regional FIGs. These groups are made up of professionals working in-country and regional FIG chairs. By pooling their resources and cooperating to minimize competition and duplication on various projects, the world's zoos and aquariums will have a greater impact on the conservation of global biological diversity.

CONCLUSION

Historically, wildlife biologists, ecologists and others have been skeptical of the motives behind zoo activities, especially when animal curators were only acquiring rare exhibit specimens to enhance their own prestige and egos. Twenty years ago, zoos procured animals from the wild for display. The zoo profession has changed dramatically since that era, and zoos are now committed to protecting endangered wildlife. Zoo staffs now have wildlife biologists and researchers who actively participate in field research, habitat protection, and other spheres beyond captive management and breeding,[40] and there is now more cooperation and collaboration with the wildlife conservation community. In 1980, the AAZPA designated conservation as its highest priority, which helped bring more credibility to the profession.[12] Zoos and aquariums have now become protectors rather than collectors of wildlife. Participation in the support and protection of national parks and reserves to sustain species in their natural habitat is increasing at a rapid rate, and will continue to increase. To achieve greater *in situ* conservation financing, we suggest that zoos commit a minimum of 1% of their total budget to cooperative, off-site conservation programs.

Conservation is a multi-faceted endeavor, involving numerous governmental and non-governmental organizations, and a wide variety of individual players around the world who must work cooperatively in order to achieve success.[11] Zoos and aquariums have an important role to play in maintaining the world's biological diversity, and should continue to work together with wildlife agencies and NGOs to conserve our remaining wildlife.

ACKNOWLEDGEMENTS

Our thanks to Dr. James Aiello, Burnet Park Zoo, and to Dr. B. Diane Chepko-Sade and numerous graduate students from her conservation biology seminar group at the SUNY College of Environmental Science and Forestry, Syracuse, for their constructive criticism of preliminary drafts of this paper. Kevin Willis also made numerous suggestions on the manuscript.

REFERENCES

1. Conway, W. 1988. Can technology aid species preservation? Pp. 263-268 In: Wilson, E.O. (ed.). 1988. *Biodiversity.* National Academy Press, Washington, D.C. 521 pp.

2. Conway, W. 1980. An overview of captive propagation. Pp. 199-208 In: M.E. Soule and B.A. Wilcox. *Conservation Biology: An Evolutionary-Ecological Perspective.* Sinauer Associates, Sunderland, Mass., 395 pp.
3. Wilson, E.O. (ed.). 1988. *Biodiversity.* National Academy of Sciences, Washington, D.C., 521 pp.
4. Myers, N. 1985. A look at present extinction spasms and what it means for the future evolution of species. Pp. 47-58 In: R.J. Hoage (ed.). *Animal Extinctions: What Everyone Should Know.* Smithsonian Institution Press, Washington, D.C., 192 pp.
5. OTA (U.S. Congress, Office of Technology Assessment). 1987. *Technologies To Maintain Biological Diversity.* U.S. Government Printing Office, Washington, D.C., 334 pp.
6. McNeely, J.A., K.R. Miller, W.V. Reid, R.A. Mittermeier and T.B. Werner. 1990. *Conserving The World's Biological Diversity.* International Union for the Conservation of Nature and Natural Resources, Gland, Switzerland; World Resources Institute/Conservation International/World Wildlife Fund-US/World Bank, Washington, D.C., 193 pp.
7. IUCN. 1980. *World Conservation Strategy.* International Union for the Conservation of Nature and Natural Resources/United Nations Environment Program/World Wide Fund for Nature. Gland, Switzerland.
8. Mergen, A. 1989. From bison to biopark: 100 years of the National Zoo. *Zoogoer* 18.
9. Matthiessen, P. 1987. *Wildlife in America.* Viking Penguin Inc., New York, 332 pp.
10. Grzimek, B. 1961. *Serengeti Shall Not Die.* E.P. Dutton, New York.
11. Moore, D. 1989. The red wolf: extinction, captive propagation and reintroduction. *Bull. New York State Museum* 471:136-139.
12. Hutchins, M. and R.J. Wiese. 1991. Beyond genetic and demographic management: the future of the Species Survival Plan and related AAZPA conservation efforts. *Zoo Biology* 10:285-292.
13. Soule, M.E. 1980. Thresholds for survival: maintaining fitness and evolutionary potential, pp. 151-169 In: M.E. Soule and B.A. Wilcox (eds.) *Conservation Biology,* Sinauer Assoc., Sunderland, MA.
14. Allendorf, F.W. and R.F. Leary. 1986. Heterozygosity and fitness in natural populations of animals, pp. 57-76 In: M.E. Soule and B.A. Wilcox (eds.) *Conservation Biology.* Sinauer Assoc., Sunderland, MA.
15. Ralls, K. and J. Ballou. 1986. Preface to the proceedings of the workshop on genetic management of captive populations. *Zoo Biology* 2:81-86.
16. Ralls, K., J. Ballou and A. Templeton. 1988. Estimates of lethal equivalents and the cost of inbreeding in mammals. *Conservation Biology* 2:185-193.
17. Wiese, R.J., B. Zebrowski, L. Willbanks, K. Asis, M. Hutchins. In press. Species Survival Plans: Strategies for wildlife conservation. Legacy Publishing, Santa Barbara, CA. 64 pp.
18. Wiese, R.J. and M. Hutchins. In press. The role of zoos in amphibian and reptile conservation. In: J.B. Murphy, J.T. Collins and K. Adler (eds.). *Captive Management and Conservation of Amphibians and Reptiles,* SSAR Publication.

19. Foose, T.J. 1983. The relevance of captive populations to the conservation of biotic diversity. Pp. 374-401 In: C.M. Schonewald-Cox, S. Chambers, B. MacBryde and W.L. Thomas (Eds.) *Genetics And Conservation: A Reference For Managing Wild Animal And Plant Populations.* Benjamin Cummings Publishing Co., Menlo Park, CA, 722 pp.
20. Kuehler, C. 1991. Here today, gone to Maui: efforts to save endemic island species. *Zoonooz* 64(10):10-14.
21. Tryon, B.W. 1990. Bog turtles (*Clemmys muhlenbergii*) in the south—a question of survival. *Bull. Chic. Herp. Soc.* 25:57-66.
22. Schaller, G.B. 1972. *The Serengeti Lion.* Univ. Chic. Press, Chicago, 480 pp.
23. Lacy, R.C. 1991. Zoos and the surplus problem: An alternative solution. *Zoo Biology* 10:293-297.
24. Dahl, J.F. 1989. An inexpensive, climate-controlled enclosure for gibbons utilizing appropriate technology, pp. 323-335 In: E.F. Segal (ed.) *Housing, Care and Psychological Well-being of Captive and Laboratory Primates.* Noyes Publ., Park Ridge, NY, 544 pp.
25. Benirschke, K. 1983. The impact of research on the propagation of endangered species in zoos. Pp. 402-413 In: C.M. Schonewald-Cox, S. Chambers, B. MacBryde and W.L. Thomas (Eds.) *Genetics And Conservation: A Reference For Managing Wild Animal And Plant Populations.* Benjamin Cummings Publishing Co., Menlo Park, CA, 722 pp.
26. Cooper, K.A., J.D. Harder, D.H. Clawson, D.L. Frederick, G.A. Lodge, H.C. Peachley, T.J. Spellmire and D.P. Winstel. 1990. Serum testosterone and musth in captive male African and Asian elephants. *Zoo Biology* 9:297-306.
27. Gage, L.J., L. Amaya-Sherman, J. Roletto and S. Bentley. 1990. Clinical signs of San Miguel sea lion virus in debilitated California sea lions. *J. Zoo and Wildl. Med.* 21:79-83.
28. Hutchins, M., T. Foose and U.S. Seal. 1991. The role of veterinary medicine in endangered species conservation. J. Zoo and Wildl. Med. 22:277-281.
29. Fowler, M.E. 1986. *Zoo And Wild Animal Medicine.* W.B. Saunders, Philadelphia, PA, 1127 pp.
30. Bowles, A.E., W.G. Young and E.D. Asper. 1988. Ontogeny of stereotyped calling of a killer whale calf, *Orcinus orca* during her first year. *Rot. Fiskideildar* 11:251-275.
31. Schmid, T.H., F.L. Murru, F. MacDonald. 1990. Feeding habits and growth rates of bull (*Carcharinus leucas*), sandbar (*Carcharinus plumbeus*), sandtiger (*Eugomphodus taurus*) and nurse (*Ginglymostoma cirratum*) sharks maintained in captivity. *J. Aquaculture and Aquatic Science* 5:100-105.
32. Foose, T.J. and J. Ballou. 1988. Population management: theory and practice. *International Zoo Yearbook* 27:26-41.
33. Seal, U.S. 1988. Intensive technology in the care of *Ex Situ* populations of vanishing species. Pp. 289-295 In: E.O. Wilson (ed.) *Biodiversity.* National Academy of Science, Washington, D.C.
34. Wildt, D.E. and U.S. Seal. 1988. Research Priorities For Single Species Conservation Biology. (Summary ms, published).

35. Prescott, J. and M. Hutchins. 1991. Joining efforts for the preservation of biodiversity. *Trans. 56th N.A. Wildl. & Nat. Res. Conf.,* pp. 227-232.
36. Lenox, S. 1987. Our new exhibit stinks—"Boston Harbor: The Place, The Problem, The Plan." *AAZPA 1987 Reg Proc:* 66-72.
37. Wemmer, C., C. Pickett, J.A. Teare. 1990. Training zoo biology in tropical countries: a report on method and progress. *Zoo Biology* 9:461-470.
38. Hutchins, M., R.J. Wiese, K. Willis and S. Becker. 1991. AAZPA annual report on conservation and science. AAZPA Conservation Center, Bethesda, MD. 263 pp.
39. Wemmer, C. and R. Tilson (Ch). 1991. Report of the Indonesia-Malaysia Fauna Interest Group, Singapore, 26-29 September 1991.
40. Seal, U.S. 1991. Life after extinction. *Symp. Zool. Soc. London* 62:39-55.

Biological Diversity: Problems and Challenges. Edited by S.K. Majumdar, F.J. Brenner, J.E. Lovich, J.F. Schalles, and E.W. Miller.

Chapter Twenty-Two

BIODIVERSITY AND ENDANGERED SPECIES RECOVERY PROGRAMS

CRAIG JOHNSON

Office of the Assistant Secretary for Fish and Wildlife and Parks
Department of the Interior
1849 C Street, NW
Mail Stop 3156
Washington, D.C. 20240

INTRODUCTION

In terms of biodiversity, the Endangered Species Act of 1973 (ESA) is both savior and pariah; it is both conservation biology in practice and a shining example of how not to conserve biotic diversity. With the ESA, the United States made an unprecedented commitment to end centuries of environmental neglect, abuse, and destruction that had resulted in the extinction of the passenger pigeon and the heath hen and the endangerment of the bald eagle. It converted an ethical commitment to other species of plants and animals in legislation. To achieve this goal, the ESA created programs, concepts, and tools intended to return threatened and endangered species and the ecosystems that support them from the brink of extinction. It supported these programs, concepts, and tools with strong legislative authority and mandates.

At the same time, many authors point to the ESA as a failed model for conserving biodiversity by noting that it reacts to ecological damage and destruction instead of preventing it; that it cannot protect landscapes and ecosystems from the perturbations and stresses that result in insidious population declines and community collapse until after the damage has been done (Salwasser 1991, Scott et al. 1988). Others look no further than the species that have gone extinct while protected by the ESA to demonstrate that the ESA has failed in its purpose (Noss 1991, Reffault 1991).

Ultimately, the ESA will probably never succeed or fail to conserve elements of biodiversity. On a planet experiencing such dramatic ecological change induced by a single species, each successful recovery from the brink of extinction is only temporary or tentative. Every failure to recover species from extinction only highlights the tentativeness of our successes. When a species went extinct it was because our biological knowledge failed, our ecological understanding failed, or our commitment to make difficult social choices failed. As long as it exists, the ESA will be a continual experiment in which the people of the United States test our biological and ecological tools, our ethical and social commitment, and our willingness to integrate other species into our social and economic structure.

When viewed as an experiment, the ESA has created a multitude of small laboratories in which to develop and test tools and techniques to conserve biodiversity. These laboratories consist of the more than 600 species listed as threatened or endangered in the United States, the more than 3,000 species being considered as candidates for federal protection, and the ecosystems that support each of those species. Within each of these laboratories, federal, state, and private organizations and individuals try to recover species from the brink of extinction by experimenting with genetics, populations, habitats, ecosystems, and human societies.

In this paper I examine the relationship between endangered species recovery and conservation of biodiversity. I will examine that relationship by showing that many of the approaches developed to conserve biodiversity originated or were tested in the countless experiments to recover species from the brink of extinction.

THE GOALS AND TOOLS OF CONSERVING BIODIVERSITY

There seem to be as many definitions of biodiversity as there are authors writing about the subject. Even the scientific community has not agreed on a single definition of the concept, which explains much of the confusion and disagreement that surrounds discussions of biodiversity. Biodiversity is defined in this paper as a concept that represents the degree of nature's variety, including the number and frequency of ecosystems, species, populations, and genes in a geographic area and the ecological processes that interconnect them (after MacNeeley et al. 1990).

The ultimate objective of efforts to conserve biodiversity is to avert and reverse the rate of species extinction that scientists expect to occur in the next twenty to fifty years. Achieving this goal will depend on much more than endangered species programs, which are intended to recover species from the brink of extinction. The success or failure of efforts to conserve biodiversity will depend on programs that conserve native habitats, species, and populations from the adverse effects of human activity before endangerment occurs. These efforts will be centered around four primary goals (after Keystone 1991):

- Maintain ecosystem integrity by protecting ecosystems from further declines in structure, composition, and function. Achieving this goal will require an emphasis on restoring damaged and degraded ecosystems and eliminating or mitigating the source of the damage or degradation.

- Maintain species diversity by protecting the variety of each region's native flora and fauna. This includes preserving the structure and composition of biotic communities by preventing species and other taxa from the population declines that lead to endangerment.
- Maintain genetic diversity by protecting the variety of the Earth's genetic resources. This includes maintaining representative populations, races, and varieties by protecting species across their entire range and in all of the habitats to which they have become adapted.
- Maintain the balance between ecological and socio-economic needs by developing a planning approach that identifies, *beforehand,* the ecological consequences of human social and economic change.

Achieving these goals will require a comprehensive and integrated strategy composed of a wide range of elements that involve public and private organizations. Several authors have outlined comprehensive strategies to conserve biodiversity (Blockstein 1989, McNeely et al. 1990, Reid and Miller 1989, Soulé 1991). I have consolidated those strategies into eight elements that should be taken by individuals, organizations, and governments interested in conserving biodiversity:

1. *Inventory.* E.O. Wilson (1985) identified one of the most formidable problems facing efforts to conserve biodiversity: we do not know how many species of organisms exist on Earth. Estimates have ranged from a low of 5 million to a high of more than 100 million (Wilson 1985, 1988). Although we have formally described about 1.7 million species, we cannot begin to grasp the consequences of lost biotic diversity without a better understanding of how much diversity there is.
2. *Assessment.* A comprehensive strategy to conserve biodiversity must include methods that will determine the health of ecosystems and biotic communities. This element is often overlooked in most strategies to conserve biodiversity, but it is essential because so many ecosystems and biotic communities have been degraded, modified, fragmented, or simplified even though they have not been destroyed. Assessment helps establish a baseline for future monitoring and will identify prescriptions for populations and ecosystems that need rehabilitation or restoration (Schaeffer et al. 1988).
3. *Planning.* Biodiversity is threatened by habitat destruction and modification resulting from the socio-economic demands of an ever-expanding human population. Reversing the trend will require changes in government policies, integrated land-use management, and a better understanding of the social and ecological consequences of expanding the human domain (Reid and Miller 1989). These changes can only occur through planning at the local, regional, and national levels that integrates all other elements of this comprehensive approach within a larger concept of land-use planning and management.
4. *In Situ Conservation.* Natural and semi-natural ecosystems and landscapes are the primary refuges of biotic diversity. Consequently, *in situ* conservation of biodiversity has emphasized establishing protected natural and semi-natural areas within specially designated boundaries. More comprehensive approaches to conserving biodiversity will also include ecological restoration and rehabilitation, translocation, and reintroduction. Most of these activities are what Soulé (1991)

calls *inter situ* conservation and extractive reserves, or activities on lands outside the boundaries of established protected areas. This broader concept of *in situ* conservation, habitat protection includes land acquisition (fee-title or easements), protective land management, pre-development coordination, and regulatory protection.

5. *Ex Situ Conservation.* This consists of conservation in zoos, aquaria, botanic gardens, and germplasm banks. Although far more artificial than *in situ* forms of conservation, many organisms will endure only in these conditions because their native habitats have been destroyed or their populations have become nonviable (Soulé 1991). *Ex situ* conservation can also be a temporary measure that allows captive propagation followed by reintroduction into the wild.
6. *Monitoring.* This element is often overlooked as an element of biodiversity strategies. However, monitoring is essential to measure the effectiveness of the other elements of this comprehensive conservation strategy. Monitoring makes it possible to apply remedial or corrective measures if a management action fails and makes it possible to improve management treatments.
7. *Changing Attitudes and Values.* This element is traditionally called "education," but the purpose is not to simply educate, but to change the value systems and belief structures that determine the outcome of human decision making. Many individuals who are educated in the benefits of conserving biodiversity still engage in lifestyles that reflect traditional economic beliefs and are either unwilling or unable to make socioeconomic decisions that integrate the value of biodiversity. Education will have to do more than increase the amount of information if it will have lasting benefit to biodiversity: it will have to expand the underlying philosophies, ethical assumptions, and economic assumptions to make it clear that the human economy depends on the function of Earth's biosphere (Daly 1991, Daly and Cobb 1989). Changing how those value systems and belief structures are expressed in national accounting systems is at least as important as changing individual attitudes since those systems can often influence institutions and land use policies more than individual beliefs (Daly 1991, Reid and Miller 1989).
8. *Improve the Information Applied to Human Actions and Decisions.* This element is traditionally called "research," but biodiversity will not be conserved simply by more research, but by dramatic improvements in the amount and quality of information that is *applied* to human action and decisions. To conserve biodiversity, we not only need research to improve our ecological and biological knowledge, we need research to improve information for environmental restoration, environmental assessment, land use planning, population management, ecosystem management, and economic analysis. Consequently, this element includes further development of integrated information systems—including field sampling procedures, electronic data systems, museum collections, geographic information systems, computer modeling, and mechanisms for data exchange—that will improve the effectiveness and efficiency of efforts to conserve biodiversity. It also includes improving the information that determines the sociological, economic, and philosophical premises underlying our society's actions and decisions.

THE GOALS AND TOOLS OF ENDANGERED SPECIES RECOVERY

Endangered species recovery is the singular goal of all programs associated with the ESA. Tobin (1990) compared the endangered species program with a medical intensive care unit: the purpose of both is to return the patients to a level of health and security. The U.S. Congress provided federal resource management agencies with a wide variety of tools to achieve this goal, but they have at least one prerequisite: the species, subspecies, population, or variation must be on the federal list of threatened or endangered species.[1] Once a taxon is listed as a threatened or endangered species, it receives the full benefits of the ESA and all tools available to recover the species can be applied.

The key elements of a comprehensive strategy to recover threatened and endangered species from the brink of extinction are almost identical to those used to conserve biotic diversity. The ESA established recovery plans, critical habitat designation, land acquisition, cooperation from state resource management agencies, interagency consultations, and law enforcement as the primary tools for recovering threatened and endangered species. However, the USFWS and National Marine Fisheries Service (NMFS) apply a greater set of tools and techniques to the task of recovering threatened and endangered species, to clarify the relationship between endangered species recovery and biodiversity. I will present this comprehensive strategy using the same headings contained in the biodiversity conservation strategy outlined above.

Inventory

In the past inventories have not been as important for endangered species recovery because most federally listed species were vertebrates whose distribution and abundance had been fairly well studied. This changed in the early 1980s with the increased number of threatened and endangered plant and invertebrate species. Recovery plans for these plants and animals increasingly recommended surveys throughout the historic ranges of these species to identify new sites and locations (USFWS 1983a, 1990a). Systematic inventories as an element of species recovery reached a new plateau as species from less familiar ecosystems, like cavedwelling and aquatic invertebrate species, became listed in the mid-1980s. For example, as scientists search for additional locations of the Ozark cavefish (*Amblyopsis rosae*) in the karst cave structures of Arkansas, Missouri, and Oklahoma they are discovering and describing new taxa from these subterranean ecosystems. Surveys for the Iowa pleistocene snail (*Discus macclintocki*) on the moist cliff slopes of Iowa and Minnesota identified several new species of vertigo snails. In time, surveys associated with endangered species recovery will substantially increase our awareness of native flora and fauna by focusing attention on taxa that are not as charismatic as vertebrate species.

---

[1]The ESA defines the terms species to include any subspecies of fish or wildlife or plant, and any distinct population segment of any species of vertebrate or wildlife which interbreeds when mature. Consequently, several populations of vertebrates (e.g., interior population of least tern, *Sterna antillarum,* interior and Atlantic coast populations of piping plover, *Charadrius melodus*) are listed as "species" for the purposes of the ESA.

Assessment

As stated earlier, there are no standard methods or techniques for assessing the health of ecosystems, even though several approaches have been proposed (Kay 191, Schaeffer et al. 1988). Similarly, there are no standard methods or techniques for assessing the health or viability of populations and species although some have been proposed (Auerbach 1976, Barrett and Rosenberg 1981). Population viability analysis has now replaced most other assessment technologies that had been used for endangered species recovery. Population viability analysis has been applied to a wide variety of threatened and endangered mammals and birds and has started to see use with arthropods and plants (CBSG 1990, 1992a). In many areas of the country, any problem encountered with threatened and endangered species looks to population viability analysis for its solution.

In the absence of alternatives, few people are addressing the limits and limitations of population viability analysis. Nevertheless, the assessment needs of endangered species recovery will require more tools than population viability analysis. Additional tools are necessary to more effectively assess the ecological causes and consequences of species decline, particularly for invertebrate and plant species. Assessment tools that can be used in conjunction with monitoring programs for threatened and endangered species are also necessary. Finally, tools are necessary for assessing the effectiveness of management prescriptions and the effects of public policy and land use decisions on threatened and endangered species (Bean et al. 1991, Reid and Miller 1989).

Planning

The ESA provides several tools for planning the recovery of threatened and endangered species. The foremost, or most well-known tool, is the recovery plan that is developed for each species on the threatened and endangered species list. The planning process required to receive a permit to incidentally take listed animal species, which is commonly called habitat conservation planning, is becoming much more familiar to private individuals and local governments as the numbers of habitat conservation plans increase. The final process, which is most familiar to federal agencies, results from the requirement for federal agencies to consult with the FWS or the NMFS on actions that may affect federally listed species.

Recovery Planning

Once a species is listed as federally threatened or endangered, the FWS or the NMFS is required to develop a recovery plan for the species. These plans serve as a template for the actions necessary to restore and secure a threatened or endangered species as a viable, self-sustaining member of its host ecosystem (USFWS 1983b). Recovery plans are intended to provide step-by-step outlines of all actions necessary to counter threats to a species' continued survival. These plans identify any habitat protection, habitat restoration, captive propagation, translocation, and research that is necessary to recover a threatened or endangered species.

The current process for developing and implementing recovery plans has serious problems and has received much criticism. The most common criticism is that recovery plans take an inordinate amount of time to prepare and revise and, consequently, too few species have complete recovery plans (GAO 1992, Reffault 1991, Tobin 1990). The second most common criticism is that FWS and NMFS are given too few resources to implement the recovery plans (Campbell 1991, Tobin 1990). Clark and Harvey (1991) accurately characterized endangered species recovery as a complex and difficult process which has neither precedent nor prescription; all of us are learning as we go.

In addition to these problems, recovery planning has been hampered by limited scientific understanding of extinction, population and community ecology, and restoration ecology. The earliest recovery plans were based on traditional approaches to population management and did not incorporate concepts of minimum viable population size, small population dynamics, genetic considerations, environmental restoration, biosphere preserves, and preserve design (USFWS 1976, 1978a, 1978b). They represented the best knowledge and thinking available when they were written.

As the tools of population management and conservation biology have improved, those improvements have been incorporated into the recovery planning process. The most significant advance resulted from the advent of inexpensive and powerful computers that have increased the volume of information that can be integrated in recovery planning and the speed at which it that information can be analyzed. The availability of computer technology combined with recent advances in community ecology, stress ecology, conservation biology, genetic analysis, epidemiology, and environmental chemistry to greatly improve the recovery planning process. Population viability analyses now often initiate recovery plan development or revision (CBSG 1990, 1992a, 1992b). Although these analyses do not assure threatened and endangered species recovery, they provide heuristic procedures for developing and evaluating management presciptions. The effectiveness of future recovery planning will depend on more reliable tools for evaluating community ecology, population viability and vulnerability, determining the effect of generic bottlenecks on species viability, and the ecology of metapopulations.

Other Planning

One of the fundamental protections provided by the ESA is protection from killing, harming, or otherwise "taking" threatened and endangered species. Under the ESA, "taking" includes destruction or significant alteration of the habitat on which threatened and endangered species depend. This prohibition against habitat alteration is often difficult to enforce and has created considerable conflict between economic interests and endangered species conservation.

The ESA contains a provision that allows private landowners to incidentally "take" individual threatened or endangered organisms. In exchange, the landowner must develop and implement a plan that minimizes and mitigates the impacts of the taking and promotes conservation of the species being taken. These plans, which are usually called habitat conservation plans (HCPs), have become significant tools for effecting land use planning to benefit threatened and endangered species. The planning process surrounding HCPs is still evolving and has a wide number of problems; however, the HCP planning process has benefited from recent improvements in reserve design, population viability analysis, resource economics, and biodiversity conservation (Bean et al. 1991). Although these technologies have not yet been fully integrated into the HCP planning process, their availability greatly enhances the potential contribution of the HCP planning process to threatened and endangered species conservation.

Thus far, only seven HCPs have been developed and approved and most of these have been in California. The first HCP, which was for San Bruno Mountain near San Francisco, California, has become the model of an ideal HCP even though it appears to be the exception rather than the rule. This HCP, which protects the habitat for the endangered mission blue (*Plebejus icatrioides missionensis*) and the San Bruno elfin butterflies (*Callophrys mossi bayensis*), permanently protects about ninety percent of the habitat for these two species. Limited development is allowed on the remaining ten percent of the habitat. The San Bruno HCP also protects the diversity of habitats found on the mountain and has conveyed about 790 hectares of the mountain into public ownership (Bean et al. 1991).

Twenty other HCPs are now being developed throughout the country and for a wide variety of species. As these develop, each HCP could provide living laboratories in which to test and improve preserve design population viability models, ecological retoration methods, sustainable development models, and resource economics. In fact, improvements in habitat conservation planning will depend on effective development and integration of these tools.

In Situ Conservation

Habitat protection, or *in situ* conservation, has been the centerpiece of all efforts to conserve threatened and endangered species. Until recently, virtually all of these efforts have tried to create exclusive preserves that exclude human use of the habitats occupied by threatened or endangered species. This approach to endangered species conservation relies on the principle that "nature knows best." More than 40 national wildlife refuges have been established specifically for endangered and threatened species and federally listed species occur on more than 260 other national wildlife refuges. More than 200 threatened and endangered species occur on lands managed by the U.S. Forest Service.

Unfortunately, this approach to endangered species conservation may not have been as successful for many species as resource managers hoped. As Forman (1987) pointed out "no boundary in a landscape is impermeable and no action is isolated." Many of the lands that have been protected for federally listed species are either too small to assure the species' continued survival, are themselves fragmented by structures like highways, or are subjected to external phenomena like atmospheric or waterborne contaminants that do not recognize lines placed on land surface maps (GAO 1987, NPS 1980, Shafer 1990). A survey of the National Wildlife Refuge System conducted by the USFWS (1982) determined that threatened and endangered species were threatened on 154 refuges. The Wilderness Society (1988) cited key deer in National Key Deer Wildlife Refuge as an example of a federally listed species that is threatened by human encroachment and pollution on a preserve established to protect it. These problems have forced biologists associated with endangered species recovery to broaden the scope of *in situ* endangered species conservation.

Because few areas in the conterminous United States can be considered either "pristine" or free from contaminants or exotic species, habitat restoration will increasingly become a tool for conserving endangered species and biodiversity (Jordan et al. 1988). Recovery of the endangered Kirtland's warbler in Michigan requires restoring seral stages of native jack pine (*Pinus banksiana*) that have been allowed to mature into older forest stages because of fire suppression. This restoration has not only stopped further declines in the Kirtland's warbler population, it has allowed the population of this endangered species to increase to levels not recorded since 1961. As techniques for restoring wetlands, grasslands, forests, and desert ecosystems improve, habitat restoration will become increasingly important for *in situ* conservation of threatened and endangered species.

The increasing number of federally listed species on private lands has also led resource managers to explore ways to protect endangered and threatened species *in situ*, while allowing continued human use of those private lands. This approach, which is consistent with the concept of "biosphere reserves" developed by UNESCO (1974) is now used increasingly throughout the world (Batisse 1986m, Gregg and McGean 1985, Hart 1966, McNeely et al. 1990, Tangley 1988). The concept results from an understanding that exclusive preserves, by themselves, will not be able conserve all species, genetic resources, and ecological processes (Harris 1984, Soulé 1986). Biosphere reserves represent a planned network of exclusive preserves within a matrix of restored lands and lands with compatible economic and social uses.

To help recover threatened and endangered species in the United States, the USFWS, U.S., Forest Service, and several state resource management agencies are now using protective easements, conservation agreements, and cooperative management agreements to protect threatened and endangered species in place. The states of Minnesota and Wisconsin have developed conservation agreements with private landowners that protect federally listed plant species from habitat loss and prevent habitat degradation resulting from herbicide and pesticide applications. At the same time, these programs work with landowners to develop agricultural uses that are compatible with the continued survival of threatened and endangered species.

The Wisconsin program, which was modeled after Minnesota's program, has effectively protected colonies of the prairie bush clover (*Lespedeza leptostachya*), prairie white-fringed orchid (*Platanthera leucophaea*), northern monkshood (*Aconitum noveboracense*), Fassett's locoweed (*Oxytropis campestris*), dwarf lake iris (*Iris lacustris*) and dune thistle (*Cirsium pitcheri*) on private lands throughout Wisconsin. The states of Illinois, Iowa, Michigan, and Missouri have developed similar programs to protect federally listed plant and animal species within their jurisdictions.

In Indiana, a combined effort between federal and state resource management agencies, conservation organizations, and private individuals to benefit the endangered white cat's paw pearly mussel (*Epioblasma sulcata delicata*), will also benefit a diverse assemblage of species in the Fish Creek watershed. That effort emphasizes wetland restorations and conservation agreements on private lands to reduce erosion and improve water quality within the watershed. It also includes cooperative agreements and management to protect the main channel of the creek from efforts to remove snags and other natural stream material that regulate water velocity, provide nutrient sources, and help maintain the water quality on which the endangered mussel depends for its survival.

The approaches to *in situ* conservation of threatened and endangered species still have problems. Restorations methods are still evolving and often fail to produce self-sustaining ecosystems that remain free of exotic species. Preserve and reserve designs are often poorly planned, creating small habitat islands that are incapable of withstanding external stressors. In other instances, they rely on serendipity rather than planning for their creation. These problems, once recognized, can be addressed and resolved. They will have to be addressed because the people of the United States are increasingly unwilling to set more land out of production; in the future, in situ conservation of threatened and endangered species will see increasingly rely on a matrix of lands patterned after the concept of biosphere reserves.

Ex Situ Conservation

Several well publicized efforts to return endangered species to the wild were preceded by *ex situ* conservation efforts. The California condor, the black-footed ferret, the red wolf, the American peregrin falcon, the whooping crane, and the bald eagle provide examples of ex situ conservation of federally threatened and endangered species. Each of these *ex situ* conservation efforts has been applauded as a success. California condors bred in captivity have been reintroduced into the wilds of California, although the final results of that experiment remain to be seen. The red wolf has been reintroduced into the coastal swamps and forests of North Carolina and American peregrine falcons now successfully rear young in cities throughout the eastern United States.

Unfortunately, each of these ex situ conservation efforts was preceded by failure to protect the species in its native habitat. *Ex situ* conservation of threatened and endangered species has been, and will continue to be, a tool of last resort. Nevertheless, new threats to threatened and endangered species will result in a dramatic increase in *ex situ* conservation efforts in the near future. A massive effort to translocate and propagate threatened and endangered mussel species will be required to protect the richest mussel fauna in the world from the zebra mussel (*Dreissena polymorpha*) invasion. The entire native unionid mussel fauna in the Great Lakes, Mississippi, Ohio, Cumberland, and Tennessee River drainages may have to be translocated if they are to survive the zebra mussel. The entire native mussel fauna on the Canadian side of the Detroit River, which forms the border of Michigan; a large colony of the northern riffleshell mussel, which has been proposed as an endangered species may be extirpated by the zebra mussel if it is not translocated. Some of the species may be removed from the wild entirely and exist only in captive situations. These efforts pose new challenges for *ex situ* conservation of threatened and endangered species and will require new tools, techniques, and approaches.

Monitoring

Monitoring is integral to threatened and endangered species recovery. The ESA requires the FWS and NMFS to periodically monitor federally listed species to determine their status. The ESA also mandates a five-year monitoring period for any species that is removed from the list because its recovery goals have been achieved. In practice, however, monitoring has been a problem with many threatened and endangered species because of limited resources. The FWS (1990b) reported that the status was unknown for 19.4 percent (113) of the federally listed species, largely because monitoring had not happened. Although this is encouraging for the four-fifths of the listed species whose status has been monitored, it is not as encouraging for the species that have not been monitored. Without being noticed or detected species like the neosho madtom (*Noturus placidus*) and Curtis' pearly mussel (*Epioblasma floretina curtisi*) have declined dramatically.

Much of the monitoring for threatened and endangered species is conducted by other federal agencies and private citizens to comply with incidental take permits. These monitoring programs are usually designed and overseen by the various offices of the FWS, even though that agency does not undertake them. They are also closely associated with activities that may adversely affect the survival of threatened or endangered species. For example, the U.S. Army Corps of Engineers monitors the status of federally listed mussels in the mainstem of the Ohio and Mississippi Rivers as part of navigation studies. Similarly, permit holders for the San Bruno Mountain, Coachella Valley, and Riverside County HCPs monitor the status of federally listed species covered by the HCPs (Bean et al. 1991).

Changing Attitudes and Values

Public attitudes about threatened and endangered species determines the level of the United States' commitment to endangered species conservation. Ultimately, the continued success of endangered species conservation depends on public attitudes and values. Even species that have been recovered from the brink of extinction can become endangered again very quickly if public attitudes toward conservation and environmental health change dramatically.

Unfortunately a dramatic conflict—either real or perceived—between people and an endangered species can undo a lot of progress. The conflict between Tellico Dam and the Tennessee snail darter resulted in a "major assault on the Endangered Species Act" in 1979 (Dingell 1991). In the past few years, the conflict between the northern spotted owl and logging interests in the Pacific northwest states has had a considerable effect on public opinions about endangered species conservation.

This element of a comprehensive strategy to conserve threatened and endangered species remains underused and unreliable. In the long term, it may rely more on changes in economic models and assumptions, like those proposed by Daly (1991) and Boulding (1991), than on a dramatic increase in biotic egalitarianism by the public.

Improving the Information Used

This element of threatened and endangered species recovery has never been fully explored much less exploited. For the most part, those involved in federal and state recovery programs still emphasize traditional biological and ecological research as the principle mechanism for improving the amount of information used to recover threatened and endangered species. Although these research activities will continue to be essential to endangered species recovery efforts, the scope of the questions asked and the tools developed will have to expand significantly if most of these species will be conserved.

Recovering threatened and endangered species will continue to generate questions that will fuel biological and ecological research activities for years to come. Every listed species provides a stimulus for new inquiry in evolution, genetics, biogeography, population biology, community ecology, stress ecology, land management, landscape architecture, epidemiology, and economic ecology. Every failure to stabilize or recover a species provides motivation to generate an answer to those questions. Yet those answers will have little effect on threatened and endangered species if they are not integrated into recovery programs and activities.

Failure to integrate and synthesize new and existing information into recovery efforts and recovery decision making is perhaps the greatest problem facing endangered species conservation. Clark and Harvey (1991) called this problem "intelligence failure" and outlined its effect on black-footed ferret recovery efforts. Despite considerable progress in recent years, integration of existing information on the biology, ecology, biogeography, status, and trends of federally listed species is ineffective or non-existent in the FWS and NMFS (Bean et al. 1991, Clark and Harvey 1991, Weinberg 1986).

Many biologists and resource managers involved in endangered species recovery either do not have access to the most current information on a species or they have not integrated it into their day to day activities and decision making. Consequently, advances in reserve design, information systems, population dynamics, genetics, ecological modeling, and natural resource economics have not been followed by a proportional increase in the effectiveness of recovery activities. In the future, more effective recovery efforts will rely on advances in research not traditionally used by life scientists and resource managers. In particular, future recovery efforts will rely on integrating new approaches to collecting, exchanging, synthesizing, and distributing technical and social information. It will also rely on improvements in resource economics, organization and management structures, problem solving and conflict resolution, and training to make certain that existing and new information is incorporated into threatened and endangered species recovery efforts (Clark and Harvey 1991).

CONCLUSIONS

Conservation of biodiversity and endangered species recovery are irrevocably tied in an endless feedback loop. Each is the teacher and student of the other. Conservation biologists study the successes and failures of endangered species conservation to develop new approaches and tools to conserve biodiversity. Practitioners of endangered species recovery study new approaches developed by conservation biologists and apply them to recovery efforts. Conservation biology develops many of the tools, techniques, and approaches; endangered species conservation provides most of the experience. In the United States, conservation biology provides the hypothesis and endangered species recovery provides the experiment that tests the hypothesis.

Despite the relationship between the means of conserving biodiversity and endangered species recovery, the goals are very different. Biodiversity is concerned with preventing the ecosystem declines that endanger native flora and fauna while endangered species recovery is concerned with curing the problem once it arises. Thus far, conserving biodiversity has largely been a theoretical exercise with limited empirical testing. Conservation biology is just beginning to develop the paradigms, theoretical foundations, and technologies that will make its prescriptions more predictable and reliable. The results it has produced have had mixed success.

Similarly, endangered species recovery has been a mixed success. Few species have been removed from the list of threatened and endangered species because their populations are no longer threatened with extinction. At the same time, the numbers of species added to the list continue to grow at an alarming rate; the result is further dilution of the resources available to recover federally listed species. Because species and ecosystems do not come with an instruction booklet that outlines what to do in an emergency, resource managers involved in endangered species recovery must learn as they go along. In some instances, experience with one species can be transferred to another species; in other instances, attempts to transfer information have not helped.

Conservation biologists are developing tools for conserving earth's genetic, species, ecosystem, and landscape diversity. When they mature, these tools will represent that missing booklet of instructions on what to do when species or ecosystems find themselves in an emergency situation. Until then, efforts to recover species from the brink of extinction will represent experiments testing the theories and approaches developed by conservation biologists. We can only hope that the subjects survive the experiments and last long enough to find a cure for their predicament.

BIBLIOGRAPHY

Auerbach, S.J. 1976. Current perceptions and applicability of ecosystems analysis to impact assessment. *Ohio J. Sci.* 78:163-174.

Barrett, G.W. and R. Rosenberg. 1981. Stress effects on natural ecosystems. John Wiley and Sons; New York, New York.

Batisse, M. 1986. Developing and focusing the biosphere reserve concept. UNESCO *Nature and Resources* 22:1-10.

Bean, M.J., S.G. Fitzgerald, and M.A. O'Connell. 1991. Reconciling conflicts under the Endangered Species Act. The habitat conservation planning experience. World Wildlife Fund; Washington, D.C.

Blockstein, D.E. 1989. Toward a federal plan for biological diversity. *Issues in Science and Technology* 5:63-67.

Boulding, K.E. 1991. What do we want to sustain? Environmentalism and human evaluation, pp. 22-31. In: R. Constanza (ed.) *Ecological economics. The science and management of sustainability.* Columbia University Press; New York, New York.

Campbell, F. 1991. The appropriations history, pp. 134-146. In: K.A. Kohm (ed.) *Balancing on the brink of extinction. The Endangered Species Act and lessons for the future.* Island Press; Washington, D.C.

Clark, T. and A. Harvey. 1991. Implementing recovery policy: learning as we go? pp. 147-163. In: K.A. Kohm (ed.) *Balancing on the brink of extinction. The Endangered Species Act and lessons for the future.* Island Press; Washington, D.C.

Daly, H.E. 1991. Steady-state economics. Second Edition. Island Press; Washington, D.C.

Daly, H.E. and Cobb, J.B., Jr. 1989. For the common good: redirecting the economy toward community, the environment, and a sustainable future. Beacon Press; Boston, Massachusetts.

Dingell, J.D. 1991. The Endangered Species Act: legislative perspectives on a living law pp. 25-30. In: K.A. Kohm (ed.) *Balancing on the brink of extinction. The Endangered Species Act and lessons for the future.* Island Press; Washington, D.C.

Forman, R.T.T. 1987. Emerging directions in landscape ecology and applications in natural resources management, pp. 59-88. In: R. Herrmann and T.B. Craig (eds.) *Conference on science in National Parks: the fourth triennial conference on research in the National Parks and equivalent reserves.* The George Wright Society and the U.S. National Park Service; Washington, D.C.

Gregg, W.P. and B.A. McGean. 1985. Biosphere reserves: their history and promise. *Orion* 4:40-51.

Harris, L.D. 1984. The fragmented forest: island biogeography theory and the preservation of biotic diversity. The University of Chicago Press; Chicago, Illinois.

Hart, W.J. 1966. A systems approach to park planning. IUCN; Morges, Switzerland.

International Union for the Conservation of Nature, Captive Breeding Specialist Group (CBSG). 1990. Florida key deer population viability assessment. Captive Breeding Specialist Group; Minneapolis, Minnesota.

International Union for the Conservation of Nature, Captive Breeding Specialist Group (CBSG). 1992a. Karner blue butterfly population, habitat, and viability analysis. Captive Breeding Specialist Group; Minneapolis, Minnesota.

International Union for the Conservation of Nature, Captive Breeding Specialist Group (CBSG). 1992b. Kirtland's warbler population and habitat viability assessment. Captive Breeding Specialist Group; Minneapolis, Minnesota.

Jordan, W.R., R.L. Peters and E.B. Allen. 1988. Ecological restoration as a strategy for conserving biological diversity. *Environmental Management* 12:55-72.

The Keystone Center. 1991. Biological diversity on federal lands. Report of a Keystone Policy Dialogue. The Keystone Center; Keystone, Colorado.

McNeely, J.A., K.R. Miller, W.V. Reid, R.A. Mittermeier and T.B. Werner. 1990. Conserving the world's biological diversity. IUCN; Gland, Switzerland; World

Resources Institute, The World Bank, World Wildlife Fund, and Conservation International; Washington, D.C.

Noss, R.F. 1991. From endangered species to biodiversity, pp. 227-236. In: K.A. Kohm (ed.) *Balancing on the brink of extinction. The Endangered Species Act and lessons for the future.* Island Press; Washington, D.C.

Reffault, W. 1991. The endangered species lists: chronicles of extinction? pp. 77-85. In: K.A. Kohm (ed.) *Balancing on the brink of extinction. The Endangered Species Act and lessons for the future.* Island Press; Washington, D.C.

Reid, W.V. and K.R. Miller. 1989. Keeping options alive: the scientifc basis for conserving biodiversity. World Resources Institute; Washington, D.C.

Salwasser, H. 1991. In search of an ecosystem approach to endangered species conservation, pp. 237-255. In: K.A. Kohm (ed.) *Balancing on the brink of extinction. The Endangered Species Act and lessons for the future.* Island Press; Washington, D.C.

Scott, J.M., B. Csuti, K. Smith, J.E. Estes and S. Caicco. 1988. Beyond endangered species: an integrated conservation strategy for the preservation of biological diversity. *Endangered Species Update* 5:43-48.

Schaeffer, D.J., E.E. Herricks and H.W. Kerster. 1988. Ecosystem health: I. measuring ecosystem health. *Environ. Manage.* 12:445-455.

Shafer, C.L. 1990. Nature preserves. Smithsonian Institution Press; Washington, D.C.

Soulé, M.E. (ed.) 1991. Conservation: tactics for a constant crisis. *Science* 253:744-750.

Soulé, M.E. 1986. Conservation biology: the science of scarcity and diversity. Sinauer Associates Inc.; Sunderland, Massachusetts.

Tangley, L. 1985. A national biological survey. *BioScience* 35:686-690.

Tangley, L. 1988. A new era for biosphere reserves. *BioScience* 38:148-155.

Tobin, R. 1990. The expendable future. U.S. politics and the protection of biological diversity. Duke University Press; Durham, North Carolina.

United Nations Education, Scientific, and Cultural Organization (UNESCO). 1974. Task force on: criteria and guidelines for the choice and establishment of biosphere reserves. Final Report, Man and the Biosphere Report Series Number 22, UNESCO; Paris, France.

U.S. Fish and Wildlife Service. 1976. California condor recovery plan. U.S. Department of the Interior, Fish and Wildlife Service; Washington, D.C.

U.S. Fish and Wildlife Service. 1978a. Black-footed ferret recovery plan. U.S. Department of the Interior, Fish and Wildlife Service; Washington, D.C.

U.S. Fish and Wildlife Service. 1978b. Colorado squawfish recovery plan. U.S. Department of the Interior, Fish and Wildlife Service; Washington, D.C.

U.S. Fish and Wildlife Service. 1982. Field station threats and conflicts: National Wildlife Refuges, National Fish Hatcheries, research centers. U.S. Department of the Interior, Fish and Wildlife Service; Washington, D.C.

U.S. Fish and Wildlife Service. 1983. National recovery plan for northern monkshood (*Aconitum noveboracense*). U.S. Department of the Interior, Fish and Wildlife Service; Twin Cities, Minnesota.

U.S. Fish and Wildlife Service. 1990a. Decurrent false aster, *Boltonia decurrens,* recovery plan. U.S. Department of the Interior, Fish and Wildlife Service; Twin Cities, Minnesota.

U.S. Fish and Wildlife Service. 1990b. Report to Congress: endangered and threatened species recovery program. U.S. Department of the Interior, Fish and Wildlife Service; Washington, D.C.

U.S. General Accounting Office. 1987. Parks and recreation: limited progress made in documenting and mitigating threats to the parks. U.S. General Accounting Office; Washington, D.C.

U.S. General Accounting Office. 1992. Endangered Species Act: types and numbers of implementing actions. U.S. General Accounting Office; Washington, D.C.

U.S. National Park Service. 1980. State of the parks-1980: a report to the Congress, U.S. Department of the Interior, National Park Service; Washington, D.C.

The Wilderness Society. 1988. Ten most endangered National Wildlife Refuges. The Wilderness Society; Washington, D.C.

Wilson, E.O. 1985. The biological diversity crisis. BioScience 35:700-706.

Wilson, E.O. 1988. The current state of biological diversity crisis, pp. 3-18. In: E.O. Wilson (ed.) *Biodiversity.* National Academy Press; Washingon, D.C.

Biological Diversity: Problems and Challenges. Edited by S.K. Majumdar, F.J. Brenner, J.E. Lovich, J.F. Schalles, and E.W. Miller.

Chapter Twenty-Three

INCREASING BIOLOGICAL DIVERSITY THROUGH RESTORATION OF DISTURBED ECOSYSTEMS

FRED J. BRENNER
Biology Department
Grove City College
Grove City, PA 16127

INTRODUCTION

The alteration of natural ecosystems by man will often either eliminate or change the functional values of these systems. Man has been altering natural ecosystems since before recorded history and evidence of these alterations are still being documented by archeologists around the world. In many cases, the restoration or rehabilitation of the disturbed ecosystem to their original function and value may not be possible or in some cases even desirable. This chapter focuses on the traceable problems of the restoration or rehabilitation of ecosystems function and value while at the same time incorporating regional diversity concepts.

The relatively new discipline of restoration ecology provides an opportunity to establish a new frontier in both theoretical and applied ecology (Cairns 1988). The restoration ecologist may be involved in biotechnology by genetically altering organisms to metabolize toxic wastes thereby removing these materials from polluted ecosystems or by developing a community of plants that would have limited growth but at the same time provide a diversity of food and cover for wildlife. Man has been altering natural ecosystems for food and fiber production for centuries and these activities have often resulted in pollution or destruction of natural ecosystems. For example, the introduction of European cattle into Africa has not only resulted in the alteration of the landscape but has produced management problems as well.

European cattle characteristically only feed on a small fraction of the native vegetation, are vulnerable to disease and their compact fecal material relatively resistant to re-cycling (Cairns 1988). Hopcraft (1975) demonstrated that by using native species for meat production not only were the pounds of meat/hectare increased but also the management costs were reduced as well since these species feed on a wider variety of native plant species. Moreover, their fecal material is more readily recycled into the natural ecosystems and in addition, it can be used as fuel thereby reducing the destruction of native vegetation for firewood. Overall, the use of native species in place of European cattle was found to be far less damaging to natural ecosystems since they evolved in these systems.

The restoration ecologist is forced by necessity to study the entire system rather than the individual components of the system in isolation from one another. If the objective is to restore diversity to these systems, the restoration ecologist must also rely on historic data, but unfortunately, these data are often lacking or at best fragmental in their occurance. It is not possible to restore the diversity and function of an ecosystem one species at a time, the restoration ecologist must consider among others the spatial relationship, species interaction, and physical and chemical aspects of the ecosystem. Restoration ecology therefore involves the replacement of structural and functional characteristics that have been altered or destroyed within the system and in many cases, they may have more social, economic or ecological value than currently or in some cases, previously existed on the site prior to disturbance. For example, if a corn field is destroyed by mining, the reclaimed ecosystem may have more diversity and wildlife habitat than existed prior to mining.

INCREASING BIODIVERSITY THROUGH STREAM RESTORATION

Streams and rivers are among the most enduring geomorphic features on the surface of the earth (Gore and Bryant 1988). These lotic environments provide a diversity of habitats for a variety of aquatic organisms including both invertebrate and vertebrate species. Though rivers and streams represent a small fraction of the available freshwater ($<0.2\%$), the natural replenishment of this component of the hydrologic cycle has resulted in humans as perceiving this resource more available and less destructable than other freshwater systems (Gore and Bryant 1988). Unfortunately, the misuse of these streams and rivers often results in severe damage to these ecosystems.

Stream and river systems have been degraded by a variety of land use practices including mining, highway construction, railroads, channel modification and impoundments for flood control or irrigation (Gore and Bryant 1988, Chapman and Khudson 1980, Brenner 1980, Brenner et al. 1986, 1991a). In addition, these ecosystems may also be adversely impacted by drainage from agricultural lands (Brenner et al. 1991) and urban areas. With the worldwide decline in biodiversity (Wilson 1988), the restoration and maintenance of stream and river ecosystems is imperative (Gore and Bryant 1988).

In contrast to terrestrial restoration, stream and river systems must take into account the systems that interact to produce the lotic ecosystem. For example, the source of point and non-point pollution must be modified and connected and it is imperative that the riparian habitat be restored. The restoration of riparian vegetation will not only reduce pollution loading (Brenner et al. 1991b), but will also provide habitat for a variety of species that either use or are affected by these habitats (Brenner 1980, Brenner et al. 1984, 1991b). If it is necessary to restore the channel, a number of hydrologic features must be examined. In most streams, the discharge pattern is fairly predictable; hence, the channel configuration should be designed to stimulate natural meanders and decompositional patterns. In addition, the bank and stream bed must be stabilized either by restoring vegetation patterns or by the installation of erosion control structures. It may also be necessary to install in-stream structures to improve habitat for both fish and macro-invertebrates. These structures should be designed to enhance the existing fauna as well as, attracting colonization from upstream and downstream areas. Once the stream is restored, colonization will usually occur within a short time frame (months rather than years) (Bradt 1975, Brenner 1980, Ney and Mauncy 1980, Woods and Griswold 1980, Griswold et al. 1982, Brenner et al. 1991b, Brenner and Helm 1991). The restoration of stream ecosystems therefore is a process of pollution abadment, stream bank and bed stabilization and habitat and colonization enhancement.

HYDROLOGY

If the stream restoration is to succeed, the stream channel must be stabilized. It must be recognized that streams and rivers are dynamic systems subject to a variety of hydraulic forces acting on the channel. Variation in stream flow will result in changes in the erosion and deposition patterns thereby changing geometry of the channel.

Gore and Bryant (1988) indicated that four major factors: geology, hydrologic, hydraulic, and geometric interact to establish equilibrium conditions in streams. The geologic features include soil types and topography, both of which influence the amount and type of sediments entering the stream thereby affecting meander patterns. Hydrologic factors also affect channel morphology due to changes in stream flow and run-off patterns. Although, long term climatic factors will affect stream flow, changing land use and alternation of vegetation within the watershed will have immediate effects on channel morphology due to changes in run-off characteristics. Hydraulic factors including channel geometry, depth, velocity and slope directly affect sediment transport, bank erosion and pool and ripple formation because of the differential transport in the particle size and water velocities (Gore and Bryant 1988) (Figure 1).

Meandering is a common feature of streams and must be taken into account in designing a stream restoration project. Bloom (1978) suggested that meanders are randomly generated since curves function to minimize variance. The development of meanders is an important aspect in stream restoration since they not only provide channel stability but habitat diversity as well (Figure 2).

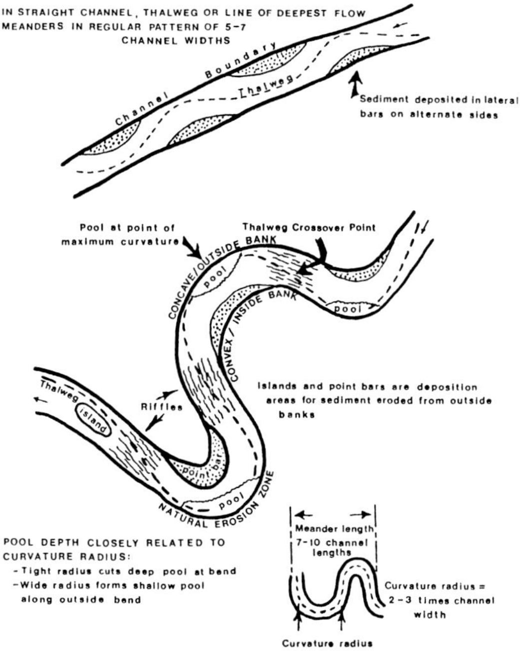

FIGURE 1. Meander and sediment characteristics of a typical stream system.

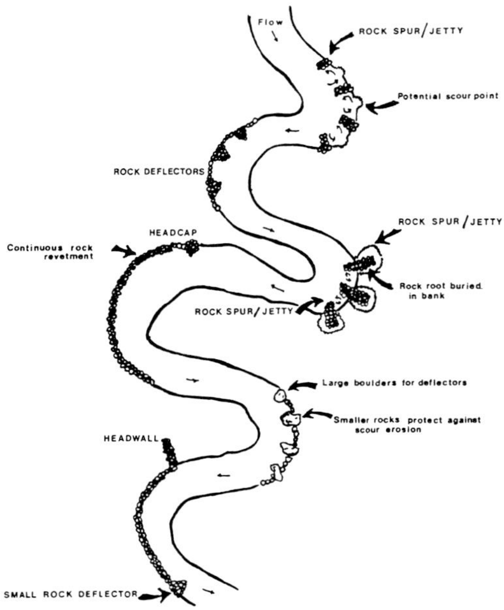

FIGURE 2. Common practices used for stream stabilization and habitat improvement.

In order to incorporate meandering into a stream restoration project, four basic hydrologic and hydraulic parameters: wavelength, sinuosity radius of curvature and peak-to-amplitude (meander belt width) should be used in the design (Gore and Bryant 1988). The linear wavelength is twice the linear distance between successive inflection points and sinuosity is the ratio of thalwey distance to arc distance (Figure 3) (Gore and Bryant 1988). The radius of curvature is calculated as the radius of a circle that passes through one and two points nearby and the meander belt width is the distance across the meander from successive convex peaks (Figure 3). According to Gore and Bryant (1988), in order for a stream to remain stable the quantity of sediment (Q_s) and the sediment particles (d) must be proportional to the discharge of water (Q_w) and the slope of the stream (S) which may be as expressed as Q_sd or Q_wS. If a portion of a stream is disturbed, these values may have to be adjusted. For example, if the discharge (Q_w) is reduced then the quantity of sediment (Q_s) would decline providing that the slope (S) and particle size (d) remained unchanged.

Brenner et al. (1991b) indicated that nutrient and fecal coliform concentrations within a stream system are directly related to sediment loading which in turn is a function of land use and rainfall within the watershed. The potential sediment yield within a watershed may be estimated by the Universal Soil Lost Equation (USDA 1983 and Brenner and Steiner 1987):

$$A = RK\,LSCP$$

where A = Soil Loss per unit area in metric tons/hectare (tha^{-1})
R = Rainfall and run-off ($tmhc^{-1}$ cm hr^{-1} x 10^{-1})
K = Soil erodability factor (thc^{-1} $cm\ hr^{-1}$ x $1010^{-2})^{-1}$
L = Slope length
S = Slope steepness
C = Cover and management factor
P = Support practice

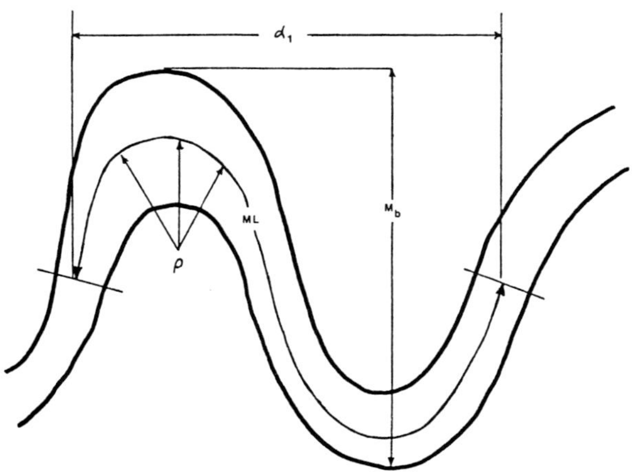

FIGURE 3. Meander diagrams for use in the design of stream systems. Aı linear wavelength, ML meander length; M_b meander belt width; pı radius of curvatures (modified from Gore and Bryant 1988).

The values for R, K, C, and P may be obtained from the U.S. Soil Conservation Service Technical Guide (USDA 1983).

Since land use and soil characteristics are factors in determining sediment yield and run-off to a given stream, they will affect channel size and the flow characteristic of that stream. Within a given watershed, the effective rainfall is that portion of the precipitation that produces direct run-off that enters the stream channel. The amount of run-off in turn depends largely on detention, infiltration, evapotransporation, etc. and is related to soil type, type of vegetation, and the amount of impervious cover. The potential run-off (Pe) from accumulated rainfall P, run-off Q and infiltration plus initial abstraction (F + la) may be calculated assuming that

$$\frac{F}{S} = \frac{Q}{Pe} \quad \text{Eq. 1}$$

where F is the infiltration in cm occurring after run-off begins, S is the potential obstruction in cm, Q is the actual run-off in cm, and Pe is the potential run-off or effective storm run-off. (Rainfall minus the control absorption in cm)

With P = Pe - Q Equation 1 can be written as:

$$Q = \frac{Pe^2}{Pe + S} \quad \text{Eq. 2}$$

The initial obstruction (l_a) in cm can be estimated as $l_a = 0.25$ Eq. 3.

Therefore, $Pe = P - l_a = P - 0.25$ Eq. 4, where P is the total run-off in cm. Substituting Equation 4 in Equation 2,

$$Q = \frac{(P - 0.25)^2}{P + 0.85} \quad \text{Eq. 5}$$

For further information on the application of these equations, refer to the U.S. Soil Conservation Service Technical Manual (USDA 1983). For further information on the application of these equations in the development of run-off models, refer to the U.S. Soil Conservation Technical Manual (USDA 1983).

The integration of the hydrologic, hydraulic, geometric, and potential runoff models along with land use characteristics is essential in the stream restoration in terms of both maintaining channel integrity as well as habitat diversity.

RESTORATION OF WATER QUALITY

An improvement in water quality is an essential aspect of stream restoration. A water quality base for stream ecosystems was the focus of The Water Quality Act inacted by Congress in 1965 which required each state to establish water quality standards for every stream within their boundaries (Table 1). Based on these criteria, it is possible to establish goals for water quality improvement within any given stream ecosystem. Prior to the establishment of these goals, studies should be undertaken to establish a water quality base line currently existing in the stream and to identify all the potential sources of both point and non-point pollution. These data are necessary to be sure that the water quality improvement goals are achievable within the finances available for the project. If finances are limited, then goals have to be adjusted to identify and categorize the pollution sources as to the area that will achieve the greatest improvement in water quality with the technology and monies available for the project.

The condition of the watershed has a major impact on water quality in the stream being considered for restoration. In agricultural watersheds, a substantial portion of the pollution load results from cropland drainage, livestock being pasture within or adjacent to stream systems and septic drainages (Brenner et al. 1991b). Streams in drainage basins with little or no vegetation will carry a large suspended load and therefore, the water chemistry will reflect the chemical characteristics of the soils and land use within the drainage basin (Gore and Bryant 1988). Whereas in a vegetated drainage basin, the water chemistry is reflected in the dissolved load rather than the suspended load.

In 1988, Gore and Bryant discussed four methods of restoring water quality to a disturbed stream: isolation, removal, transfer, and dilution.

The isolation of contaminants has a limited application for stream ecosystems but in lakes, the contaminant will precipitate and eventually be covered by sediments thereby isolating it from the water column. In streams, the sediments are continually being transported by the stream system often with severe impacts downstream (Gore and Bryant 1988). It may be possible; however, to isolate contaminates by the construction of sediment basins that will increase retention, and function as a lake

TABLE 1

Examples of discharge water quality standards intended to protect aquatic life. In stream concentration should be equal to or exceed these discharge standards. (Pennsylvania Code Title 25 - Chapter 93)

Parameter	Criteria
Aluminum	Maximum 0.1 mg/l of the 96 hour LD_{50} for representive species.
Alkalinity	Minimum 20 mg/l 25 $CaCO_3$ except where natural conditions are less.
Coliform Bacteria	Geometric mean of 200/100 ml based on five consecutive samples (May 30 - Sept. 30) 200/100 ml remainder of the year.
Dissolved Oxygen	Warm water fishery - Minimum daily mean 5.0 mg/l. Cold water fishery - Minimum daily mean 6.0 mg/l.
Iron	Total daily mean 1.5/l dissolved mean 0.3 mg/l.
Total Nitrogen	Maximum 10 mg/l.
pH	6.0 - 9.0
Sulfate	Maximum 250 mg/l.
Temperate F	Warm water fishery 40-87. Cold water fishery 38-66. Discharge may not change temperature more than 0.2 F/hr.
Total Dissolved Sollas	Mean 500 mg/l Maximum 750 mg/l
Turbidity	30 NTU - 30 May - 15 September Remainder of the year: Mean 40 NTU Maximum - 150 NTU.

ecosystem thereby restoring the potential problems associated with sediment transport.

The removal of pollutants from stream systems should, if possible, be accomplished at the source rather than after it enters the stream. Eventually, the removal of pollutants at the source not only is cost effective, but will also eliminate the potential mechanical damage to the stream system during the removal process. Moreover, because of the generally low concentrations of these substances, the removal of these pollutants is both ineffective and costly (Gore and Bryant 1988). If in-stream removal is necessary, precautions should be taken to protect the integrity of the channel.

The transfer of substances is merely the removal of a contaminate by stream flow, but this may only transfer the problem to the downstream areas. Dilution, as the term implies, dilutes the pollutant by increased water volume. These contaminants may be removed as physical characteristics of stream changes. For example, as the stream becomes more alkaline heavy metals tend to precipitate from the water column.

It should be emphasized that a hoistic approach is necessary in order to insure an improvement in water quality. Moreover, as is discussed in the next section, the establishment and maintenance of riparian buffer zones is a cost effective way of improving and maintaining water quality.

RIPARIAN VEGETATION

The re-establishment and maintenance of riparian zones including wetlands has been shown to be an effective and cost effective method of reducing nutrient, sediment and bacteria loading from non-point sources (Brenner et al. 1991a). In addition, these riparian zones provide shading thereby reducing water temperatures and providing detritus which is an important aspect of community dynamics (Vannote et al. 1980) as well as providing a diversity of wildlife habitats and bank stabilization.

For a successful restoration of a riparian area, livestock must be removed from the area and provided a water supply elsewhere in the pasture. If it is necessary for livestock to cross over a stream, they should be restricted to one region via fencing and a constructed cattle crossing (Figure 4). Once the watering devices, cattle

FIGURE 4. Design for cattle crossing that provides access to water but at the same time protect the stream. Stream damage can be minimized by constructing access ramps from railroad ties.

crossings, and fencing have been installed, a riparian buffer zone of at least 100 m should be developed. These zones should be planted, if possible, with a variety of wildlife food and cover species native to the area. In the northeastern United States, these species could include alder (*Alnus rugosa, A. serrulata*), dogwood (*Cornus amomum, C. stonifera*), Green ash (*Fraxinus pennsylvanica*) Pin oak (*Quercus palustris*), buttonbush (*Cephalanthus occidentalis*), among others. If native alder species are unavailable, European alder (*Alnus glutinosa*) have been used successfully in developing riparian zones on mined land (Brenner 1990). This species grows rapidly on moist soils thereby providing cover in a relatively short period of time. In addition, European alder along with native alders are legume species which enhance soil fertility for other species.

As early as 1981, Brenner and Artuso recommended the use of natural and/or man-made wetlands for pollution abadment from agricultural lands. These authors suggested that all agricultural drain tiles be discharged into a wetland for the removal of nutrients and other pollutions prior to their discharge into receiving streams. This concept was later incorporated into a restoration stream model by

TABLE 2

Habitat Structures to Enhance the Biodiversity in Stream Ecosystems.

Structure	Utility	Stiting Criteria
Deflectors	Redirecting current	Streams of various size
	Stabilizing thalweg	Gradients <3%
	Scouring pools	Bank stability opposite deflector
	Silt removal	Alternating banks, 5-7 channel widths for senicous flow
	Erosion abatement	
	Consolidating low flows	Anchorer into bank 1.5 m
	Increased pool/riffle ratios	Efficient with natural materials
Dams	Pool formation/control	Low end of steep break in gradient
	Holding spawning gravels in up-stream areas	Stable substrate and banks
		Anchor into bank 2 m
	Fish passage	Successive structures, 5-7 channel widths
	Sediment control	
	Collection of organic debris	Heights <0.3m
		Spawning gravels between if no passage
Boulder Placement	Added rearing habitat	During low flow
	Cover	0.6-1.5 m diameter
	Restore meanders	Granitic types preferred
		Embedded a short distance
		Greatest effect in reaches with <20% pool area
		Natural materials are most economical
Trash Catchers	See Dams (above)	Small, headwater stream
		High gradient
		1/3 cost of log dams

Petersen et al. (1992). Wetlands have also been recommended as treatment systems for acid mine drainage (Hedin 1989), municipal sewage (Kadlic 1989), and for agricultural storm water management (Hammer 1989). These reviews provide extensive documentation on the effectiveness of wetlands for both point and non-point source pollution abadment. Whenever possible, wetlands should be incorporated into the restoration of riparian zones both from the standpoint pollution abadment as well as increasing the habitat diversity of the region.

AQUATIC LIFE

Macro-invertebrates

Benthic macro-invertebrate communities are an essential aspect of the biological diversity of a stream system since they provide a critical pathway for the transportation and utilization of energy within stream ecosystems (Gore and Bryant 1988). The distribution and zonation of macro-invertebrates is a continuum from the headwaters to the mouth of the stream. This continuum is a factor not only of the allochthmous input (particulates) and autochthomous production (Vannote et al. 1980), but also of stream hydrology and channel hydraulics (Gore and Bryant 1988). In addition, macro-invertebrates are an important component of the types of stream fishes including brook trout (*Salvelinus fontinalis*) and sculpins (*Cottus* spp.) (Gore and Bryant 1988). Hence, stream restoration must be designed to provide habitat diversity in order to encourage the colonization by macro-invertebrates, therefore increasing the biological diversity of the stream system.

Following a disturbance, streams generally recolonize from four sources (Gore and Bryant 1988): (1) Drift from upstream communities, (2) upstream migration, (3) movement from within the substrate, and (4) colonization from aerial sources (Oviposition by adults). Drift is a continuous process in stream systems that is influenced by a variety of internal and external factors (Walton 1980). The recolonization by drift organisms requires that an undisturbed upstream source of colonizers be maintained. Upstream migration is an important aspect of colonization since as much as 20 percent of colonizing macro-invertebrates may be due to upstream migration. The recolonization of intermittant streams or those with porous substrates and flood plains rely on substrate movement. Regardless of the method of recolonization, streams have been shown to recolonize within a few months following restoration due to either natural disturbances or the activity of man (Brandt 1975, Woods and Griswold 1980, Brenner 1980, Ney and Muncy 1980, Griswold et al. 1982, Brenner et al. 1986, Brenner and Helm 1991). Recolonization may be enhanced by establishing meanders and adequate pool/ripple ratios along with the placement of fish habitat improvement devices (Figure 5) (Gore and Bryant 1988).

The establishment of diverse macro-invertebrate communities requires little capital investment beyond that required for channel stabilization, establishment of riparian areas and general habitat improvement, but it is essential in maintaining a stable ecosystem.

Fish Habitat Improvement

The ultimate objective of any stream restoration project is to enhance the biological diversity of the stream communities. Once the stream channel has been stabilized and the riparian zones established, water quality should improve and the aquatic community will become more diverse. A properly designed stream restoration project will include adequately designed ripple and pool areas (3) for food productions (Macro-invertebrates) and spawning sites. The importance of re-establishing riparian zones for shading and additional habitat diversity cannot be over emphasized. If the ultimate objective is to re-establish a salmonid fishery, a cobbled and gravel substrate (0.6 - 7.6 cm diameter) along with water depths > 15 cm and ripple areas with velocities between 15 and 90 cm/sec. are preferred by most cold water fish (Wesche 1985). The dissolved oxygen content should be at least 5.0 mg/l for successful embryonic development.

A variety of fishery habitat improvements can also be employed to increase habitat diversity within the stream system (Figure 5). The most common structures used for habitat enhancement include: current deflectors, jack dams, bank covers and boulder placement (Table 2, Figure 5). These structures have been reported to increase production and recruitment within a year. Duff (1982) concluded that instream dams have the greatest potential as fish habitat enhancement devices. Pools

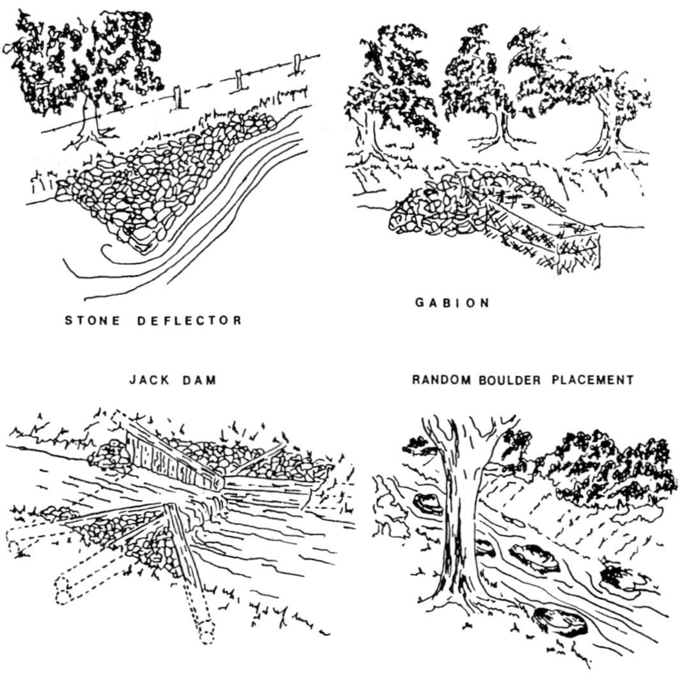

FIGURE 5. Four common fish habitat improvement devices.

created by dams increase available cover, resting and feeding opportunities for most fish species in addition to the procedures mentioned in Table 2, rock riprap (0.1 to 0.8 D) and log overhangs (snags) have also been used to enhance fishery habitats.

CONCLUSION AND RECOMMENDATIONS

The design and restoration of streams and river ecosystem requires an interdisciplinary approach which will involve civil engineers, hydrologists, and ecologists. Once the engineers and hydrologist have designed the stream channel, then ecologists should become involved in the replacement of riparian zones and habitat improvement devices. Ecologists should also be involved in channel design to insure adequate meanders and ripple/pool ratios are incorporated into channel design.

The various components of the ecosystem should be surveyed prior to restoration to establish base land data and then the stream should be monitored during and after restoration to determine if futher amelioration procedures are required to produce the desired diversity and productive ecosystem.

INCREASING BIODIVERSITY THROUGH RESTORATION OF TERRESTRIAL AND WETLAND SYSTEMS

Reclamation of disturbed lands provides an opportunity not only to increase regional biodiversity but also to develop a mosaic of wildlife habitats. In the past, many of these lands have been reclaimed, using species exotic to the region while not incorporating native flora into the reclamation plan. For example, Brenner (1984, 1990) indicates that of the 66 species of grasses, legumes, and forbs currently recommended for surface mine reclamation in the United States, only six are native to the northeast. Likewise, of the 52 species of shrubs, 14 are native to the northeast United States. While on the other hand, 41 of the 52 tree species recommended for the establishment of forest communities and/or wildlife habitat are native to the northeastern United States.

In actual practice, the majority of these disturbed sites are being reclaimed as grasslands either to abide by current regulations or as a matter of convenience since grasses and legumes will establish cover in a shorter time frame than trees or shrubs. As a result, these areas often lack not only the biological diversity that existed on the site prior to disturbance but these reclamation practices also fail to take into account regional biodiversity and wildlife habitats as well. It is possible, however, that by providing habitat heterogeneity during reclamation that the values and functions existing on the site prior to disturbance may also be re-established during reclamation.

It is well known that wetlands are a declining resource in the eastern United States (Tiner 1987) as well as elsewhere in the world but through the reclamation of mine lands it is possible to increase these diverse and productive ecosystems. Brenner (1987, 1990) advocated increasing wetlands and other critical habitats, including those of endangered species. (Brenner 1986) during the reclamation of mined lands. In addition to wetlands, the reclamation of these lands also provides an opportunity to establish a variety of diverse terrestrial communities without a substantial increase in costs (Brenner 1990).

For over four decades, wetlands have been developed on coal mine lands, gravel pits and other disturbed sites in the United States and elsewhere in the world. Numerous studies have documented the fishery (Brenner 1973, Jones et al. 1985, Brenner et al. 1987) and wildlife potential of these habitats (Amphibians, Fowler et al. 1985, Brooks et al. 1985, Brenner and Hofius 1990; Reptiles, Brooks et al. 1985, Brenner and Hofius 1990; Birds, Brenner 1973, Brenner and Mondok 1979, Brenner and Kelly 1981, Brooks et al. 1985, Brenner and Sterner 1987, Sandusky 1978, Brenner and Hofius 1990; Wray et al. 1978, Mammals, 1982. Brenner et al. 1981, Brooks et al. 1985, Brenner and Hofius 1990, Lacki et al. 1991).

Although these wetlands are used extensively by wildlife, there is often little if any planning for the integration of these habitats into the overall reclamation of the site. It is possible through reclamation to reverse the loss of wetlands and other critical habitats not only for the benefit of wildlife but also increasing regional biodiversity as well thereby benefiting all segments of society.

WETLAND DEVELOPMENTS

Prior to the actual construction of wetlands and associated upland habitats, a plan must be developed which not only considers the previous land use but future use and regional biodiversity as well. The construction of wetlands should be given top priority in the reclamation and diversions should be constructed in a manner that will channel surface water into the developing wetlands (Figure 6). This will not only allow for the management of surface water but will also insure an adequate water supply to the wetlands (Brenner 1988). Brenner (1986, 1987, 1990) describes how basic methods for wetland construction on disturbed lands involving either the removal, transportation, and placement of wetland soils and associated vegetation into constructed wetlands and/or the conversion of sedimentation and treatment ponds or the construction of wetlands during the recontouring of the site. In the first instance, a shallow impoundment (maximum depth 30 cm) is prepared in an area with sufficient drainage to insure an adequate water supply. Once a site has been constructed, wetland materials are removed from disturbed wetlands on adjacent sites, transported and incorporated into the constructed wetland. Once the transplanted material is in place, water should be released immediately into the constructed wetland. This procedure has been used successfully on three mine sites in western Pennsylvania (Brenner 1988, 1990; Brenner and Sterner 1988; Brenner and Hofius 1990) and the use of these materials as mitigation for wetland disturbance has resulted in a three fold increase in regional wetland resources.

The second procedure also involves the creation of shallow impoundments and the channelization of surface water from the mine site into the constructed wetland. Once the wetland has been constructed, the area is planted with wetland species according to the hydrologic regime existing on the site. Both of these procedures have been used successfully in establishing wetlands on mine lands and after two growing seasons there was not a significant difference in the number of species or plants/m^2 between the two types of wetlands (Brenner and Sterner 1988, Brenner 1990). If sedimentation or treatment ponds are being converted into wetlands, shallow water areas and irregular shorelines should be incorporated either into the original design or installed once mining has been completed. Once this has been completed, the ponds can be planted with wetlands species as described previously. All wetlands may be further enhanced by the placement of waterfowl nesting structure for both cavity and open nesting species (Bishop and Barratt 1970, Doty and Lee 1974, 1975, Doty 1979, Brenner and Mondok 1979, Brenner and Sterner 1988, Brenner 1990, Brenner and Hofius 1990, and Brenner and McDonald 1986).

BUFFER ZONES AND UPLAND HABITATS

Once the wetlands have been constructed and planted, a 100 m buffer zone should be created using a native tree and shrub species which are beneficial as food and cover for wildlife.

In addition, riparian habitats should be created by planting a 50 m tree and shrub zone along both sides of all water courses (Figure 6). Additional upland habitats can be developed by planting hedgerows along the major contour intervals thereby providing food and cover for a variety of wildlife species. Brenner and Clark (1987) found that passerine birds used these habitats extensively for nesting especially when crabapple (*Pyrus coronaria*), hawthorn (*Crateagus* spp), dogwoods (*Cornus* spp) are incorporated into the planting regime. If the site is terraced, these terraces can

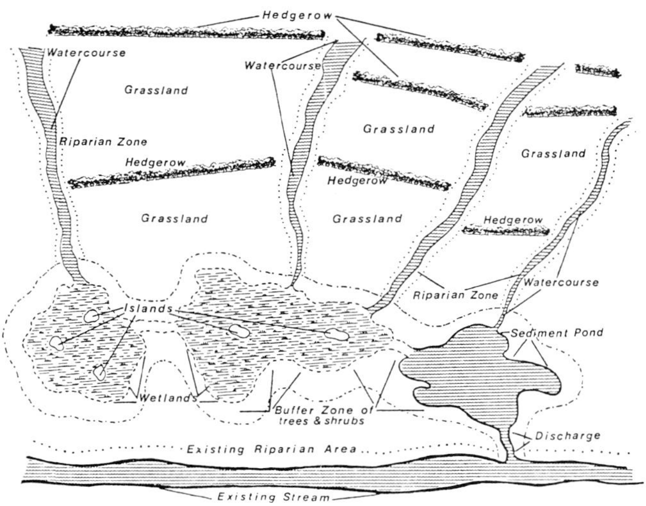

FIGURE 6. Integration of wetland and upland habitats to enhance species diversity on restored lands.

be planted with species similar to those used to establish hedgerows and riparian zones around wetlands and water courses. These areas will provide habitats for a variety of forest wildlife including ruffed grouse (*Bonasa umbellus*), (Kimmel and Samuel 1978, Kimmel 1978, Brenner and Michaliski 1984, Brenner 1989), wild turkey (*Meleagris gallopavo*), (Brenner and Brown 1989) and white-tailed deer (*Odocoileus virgininanus*) (Brenner et al 1977).

SPECIES SELECTION

A total of 20 different herbaceous wetland species have been planted successfully in western Pennsylvania (Brenner 1990) and similar results have been achieved in Indiana (Thompson 1984). Whenever possible, it is advantageous to use species native to the area, rather than introducing exotics which may replace native species and/or become a pest species in later years. For example, reed canary grass (*Phalaris arundinacea*) has been found to be a pest species in some areas (Grese 1991) but on sites previously described, reed canary grass was planted along with switch grass (*Panicum virgatum*) and other native species and at present, it has not become a problem since reed canary grass was replaced by switch grass and the other native species. In upland areas, species such as crown vetch (*Lotus corniculatus*), and K31 fescue (*Festuca elatior*) should be avoided since except for the first one or two years after planting, they have little if any value as food and cover for wildlife (Brenner et al 1975, Brenner 1978) as well as tending to discourage the establishment of native flora (Brenner and Goughler 1983). It is recommended that if the ultimate objective is to increase the diversity of wildlife habitats in the region species such as big blue stem (*Andropogon gerardii*), little blue stem (*Schizachyrium scoparium*), timothy (*Phleum pratense*), and clovers (*Trifolium* spp) be used in conjunction with or in place of the standard fescue, bluegrass (*Poa compressa*), and timothy legume mixture since they provide excellent food and cover for wildlife as well as not inhibiting the invasion of native plant species (Brenner and Goughler 1983).

In regard to trees and shrubs, European black alder (*Alnus glutinosa*) is the only non-native species currently being used extensively for the development of wildlife habitats in western Pennsylvania. Based on previous studies (Brenner et al 1975, 1990) it has not become a nuisance species. It is recommended that species such as autumn olive (*Elaeagnus umbellata*) be avoided since they spread rapidly beyond the planting site often crowding out more beneficial native species. On the other hand, black locust (*Robinia pseudoacacia*) does well on disturbed sites providing not only an abundant food source for wildlife but its nitrogen fixing ability makes it a beneficial species especially in conjunction with black alder in lowland areas. Other species recommended for the establishment of riparian and buffer zones as well as upland shrub and forested habitats include dogwoods, oaks (*Quercus* spp), crabapple, hawthornes, aspens (*Populus tremuloides, P. grandidentata*) and ashes (*Fraxinus* spp) along with other species native to the region. Most of the species may be obtained as seedlings either from governmental agencies or from commercial nurseries.

WILDLIFE USE

The use of avian fauna has been suggested as an index of habitat quality for natural wetlands systems (Glooschanko et al 1988, Cable et al. 1989) and hence, avian fauna diversity should be applicable for the assessment of man-made wetlands as well. To date on the wetlands constructed on mine lands in western Pennsylvania, over 60 species of birds representing 23 families and 18 different feeding guilds have been observed using these wetlands and adjacent uplands habitats for some phase of their life history. Included within this avian fauna were seven species of special concern in Pennsylvania (Genoways and Brenner 1985). These species inclue the kind rail (*Rallus elegans*), osprey (*Pandion haliaetus*), short-eared owl (*Asio flammeus*), bobolink (*Dolichonyx oryzivorus*), and the northern harrier (*Circus cyaneus*). Although Gill (1985) considers the osprey to be extirpated in western Pennsylvania, Leberman (1988) suggests that it might be an occassional breeder but no confirmed nesting has been documented in western Pennsylvania. Ospreys were observed feeding on one of the constructed wetlands in 1989 and 1990. The short-eared owl was confirmed to be nesting on one of the reclaimed areas. In addition to these species, uplands sandpipers (*Bertramia longicauda*) and vesper swallows (*Pooecetes gramineus*), both species of concern in Pennsylvania, have been observed nesting in grasslands adjacent to the wetlands areas. The great egret (*Casmerodius albus*) was observed visting a sediment pond that was converted into a wetland in 1991.

These constructed wetlands are also used extensively by migratory waterfowl. Over the last five years, an average of five broods of Canada geese (*Branta candensis*) and seven mallard (*Anas platyrhynchos*) broods have been observed using these wetlands each year. In addition, of the 10 wood duck (*Aix-sponsa*) nesting boxes placed on these wetlands, four were used the first year increasing to eight after five years and two of the boxes were used by nesting American kestrels (*Falco sparverius*). 20 of the 24 blue bird (*Sialia sialis*) nest boxes were used by nesting tree swallows (*Iridoprocne bicolor*) producing an average of 100 fledglings/year. Within three years of their establishment, American woodcock (*Scolopax minor*) were observed using the riparian and buffer zone areas around these wetlands during their spring and fall migrations and at least one female nested on one of the sites. In addition to the species using the wetlands and associated habitats, ringneck pheasant (*Phasianus colchicus*) and bobwhite quail (*Colinus veirginianus*) broods were observed in the grasslands adjacent to the buffer zones.

The avian fauna using these areas were assigned points depending on the population and breeding status of the different species using the wetlands and adjacent upland habitat (common = 10 points, vulnerable = 100 points, threatened = 150 points, and endangered = 200 points). The species using wildlife habitats developed on the three mine sites totaled 2,190 points compared to 250 points accumulated by species using five natural wetlands of similar size located within five km of the mine sites.

In addition to the avian fauna, these wetlands were also used by 17 different mammalian species and eight species of amphibians and reptiles. It is of interest to note that pickerel frog (*Rana palustris*) which is declining in abundance in the region was found in large numbers using these constructed wetlands.

CONCLUSIONS AND RECOMMENDATIONS

Based on these studies, it is apparent that a mosaic of wetland and upland habitats will provide wildlife habitat functions and values similar to those occurring on undisturbed sites. The similarity in the number and type of species along with the densities of vegetation on wetlands receiving supplemental planting compared to those receiving transplated wetland materials indicated that both procedures will produce functional wetlands within two years of their construction. But if the objective is to develop a wetland with identical species composition as those being disturbed, then the removal, transportation, and placement into the constructed wetlands would be the preferred procedure. However, this usually involves a considerable amount of time and expense in the permitting process and these values and functions may be duplicated by the development and planting of wetland species in the constructed wetlands. The development of buffer zones and riparian areas along water courses will provide additional nesting habitats along with food and cover for wildlife. The rapid establishment of the functions and value of these habitats along with their multiple use by wildlife and the possible increase in regional biodiversity, suggest that the construction of wetlands and associated habitat are being given serious consideration by the mining industry and regulatory agencies alike.

LITERATURE CITED

Binns, N.A. 1986. Stabilizing eroding stream banks in Wyoming. Wyoming Game and Fish Department.

Bishop, R.A. and R. Barratt. 1970. Use of artificial nest baskets by mallards. J. Wildl. Mgmt. 34:734-738.

Bloom, P.L. 1978. Geomorphology: A systemic analysis of late Cenozoic land forms. Prentice Hall, Englewood Cliffs, NJ.

Bradt, P.T. 1975. The impact of flooding on the water quality in a trout stream. Proceed. Penn. Acad. Sci. 49:47-50.

Brenner, F.J. 1973. Evaluation of abandoned strip mines as fish and wildlife habitat. Trans. N.E. Wildl. Conf. 30:205-229.

Brenner, F.J. 1978. Food and cover evaluation of strip mine plants as related to wildlife management. pp. 294-305. In D.E. Samual, J.R. Stauffer C.H. Hocutt and W.T. Mason Jr. Surface Mining and Fish/Wildlife Needs in the Eastern United States: U.S. Fish and Wildlife Service.

Brenner, F.J. 1980. Impacts of impoundments on six small watersheds in Pennsylvania. pp. 291-302. In L.A. Krumholz (ed). The Warm Water Stream Symposium. Amer. Fish. Soc. Washington, D.C.

Brenner, F.J. and F. Artuso, 1981. Chemical and biological characteristics of agricultural drainage tile outflows. Water Pollution Control Association of Pennsylvania Magazine. Sept.-Oct. p. 46-52.

Brenner, F.J. 1984. Restoration of natural ecosystems on surface coal mines in the northeastern United States. pp. 211-255. In T.N. Vezlroglu (ed.), The Biosphere: Problems and Solutions: Amsterdam: Elsevier.

Brenner, F.J. 1986. Evaluation and mitigation of wetland habitats on mined lands. National Symposium on Mining, Hydrology, Sedimentology, and Reclamation: pp. 181-184.

Brenner, F.J. 1986. Habitat preservation and development for rare and endangered species management. pp. 37-52. S.K. Majumdar, F.J. Brenner and A.F. Rhoads (eds.) Endangered and Threatened Species Programs in Pennsylvania and Other States: Causes, Issues, and Management: Pennsylvania Academy of Science.

Brenner, F.J. 1987. Surface coal mine reclamation: An opportunity for wetland development. pp. 338-344. Increasing Our Wetland Resources: National Wildlife Federation.

Brenner, F.J. 1988. The role of wetlands in water management, mine drainage on surface mined lands. Proceed. Australian Mining Industry Council Environmental Workshop 1:14-27.

Brenner, F.J. 1989. New strategies to encourage ruffed grouse. pp. 305-353. S. Atwater and J. Schnell (eds.). Ruffed grouse: Harrisburg: Stackpole Books.

Brenner, F.J. 1990. Mine reclamation: Opportunity for critical habitat development. Ecosystem Management: Rare Species and Significant Habitats: pp. 235-238. New York State Museum Bulletin 471.

Brenner, F.J., R.H. Crowley, J.J. Musaus, J.H. Gott III. 1975. Evaluation and recommendation of strip mine reclamation procedures for maximum sediment erosion control and wildlife potential. Third Symposium Surface Mining and Reclamation. vol. 1:2-24.

Brenner, F.J., M.J. Masaus and W. Granger. 1977. Selectivity of species by white-tailed deer on strip mine lands in Mercer County, Pennsylvania. Proceed. Penn. Acad. Science. 51:105-108.

Brenner, F.J. and J.J. Mondok. 1979. Waterfowl nesting rafts designed for fluctuating water levels. J. Wildl. Mgmt. 43:978-982.

Brenner, F.J. and J. Kelly. 1981. Characteristics of bird communities on mined lands in Pennsylvania. Environ. Manage. 5:441-449.

Brenner, F.J., R.B. Kelly and J. Kelly. 1982. Mammalian community characteristics on surface mine lands in Pennsylvania. Environ. Manage. 5:214-249.

Brenner, F.J. and C. Goughler. 1983. Role of nurse, cover crops in encouraging establishment of natives on reclaimed lands: An evaluation (Pennsylvania). Restoration and Management Notes. 1:31-32.

Brenner, F.J. and S. Michalski III. 1984. Evaluation of surface coal mines as ruffed grouse habitat. pp. 87-100. Better Reclamation with Trees.

Brenner, F.J., W. Kanour, B. Weston, G. Valerio, and K.P. Grayburn. 1986. Impact of flood control reservoirs and pollution in flux on the Sandy Creek Watershed, Mercer County, Pennsylvania Environ. Mgmt. 10:214-253.

Brenner, F.J and R. McDonald. 1986. Evaluation of waterfowl production on private lands. Trans NE Wildl. Soc. 43:8-14.

Brenner, F.J. and R.P. Steiner. 1987a. Alternative reclamation strategies for mined lands. pp. 115-130. In S.K. Majumdar, F.J. Brenner and E.W. Miller (eds.). Environmental Consequences of Energy Production: Problems and Prospects. Pennsylvania Academy Science.

Brenner, F.J. and R.J. Clark. 1987. Plant species used as nest sites by passerine birds on reclaimed surface coal mines (Pennsylvania). Restoration and management Notes 5:46-47.

Brenner, F.J., J. Edmundson, M. Werner and T. McGrath. 1987b. Plankton, chlorophyll characteristics and fishery potential of surface coal mine lakes in western Pennsylvania. Proceed. Penn. Acad. Science 61:147-152.

Brenner, F.J. and B.A. Sterner. 1988. First year evaluation of mitigated wetlands on two mine sites in western Pennsylvania. Mine Drainage and Surface Mine Reclamation. 2:133-139.

Brenner, F.J. and C. Brown. 1990. Habitat characteristics and possible impact of fall hunting on the establishment of turkey populations in western Mercer and Crawford Counties, Pennsylvania. J. Penn. Acad. Science. 64:121-126.

Brenner, F.J. and D.L. Hofius. 1990. Wildlife use of mitigated wetlands on surface mined lands in western Pennsylvania. Proceed. Mining and Reclamation Conference. 2:373-382.

Brenner, F.J. and J. Helm. 1991. Macro-invertebrate recolonization and water quality characteristics of a reconstructed stream after surface mining in northwestern Pennsylvania, USA. Intern. J. Surface Mining and Reclamation 5:11-15.

Brenner, F.J., V. Raup and W. Stiffler. 1991a. Characteristics of fish communities above and below six flood control impoundments in Mercer County, Pennsylvania. J. Penn. Acad. Sci. 65:24-30.

Brenner, F.J., J.J. Mondok and R.J. McDonald Jr. 1991b. Impact of riparian areas and land use on four non-point source pollution parameters in Pennsylvania. J. Penn. Aad. Sci. 65:65-70.

Brooks, R.P., J.B. Hill, F.J. Brenner and S. Capets. 1985. Widlife use of coal mines in western Pennsylvania. pp. 327-352. In R.P. Brook, D.E. Samuel, J.B. Hill (eds.). Wetlands and Water Management on Mined Lands.

Cable, T.T., U. Black and V.R. Holmes. 1989. Simplified method for wetland habitat assessment. Environ. Manage. 13:207-215.

Cairns, J. Jr. 1988. Restoration ecology: The new frontier. pp. 2-11. In J. Cairns Jr. (ed.) Rehabilitating Damaged Ecosystems. CRC Press: Boca Raton, Florida.

Chapman, D.W. and E. Knudsen. 1980. Channelization and livestock impacts on salmonid habitat and biomass in western Washington. Trans. Amer. Fish. Soc. 109:357-363.

Commonwealth of Pennsylvania. Title 25. Environmental Resources Chapter 93. Water Quality Standards.

Doty, H.A., F.B. Lee and A.D. Krouse. 1985. Use of artificial nest baskets by mallards. Wildl. Soc. Bull. 3:66-73.

Doty, J.A. and G.B. Lee. 1974. Homing to nest baskets by wild female mallards. J. Wildl. Mgmt. 38:714-720.

Doty, H.A. 1979. Duck structures in prairie wetlands. J. Wild. Mgmt. 43:976-979.

Duff, P.A. 1982. Historical perspective of stream habitat improvement in the Rocky Mountain area. Rocky Mtn. Stream. Management Workshop. Wyoming Game Fish Department. Laramic.

Fowler, D.K., D.M. Hill and L.J. Fowler. 1985. Colonization of coal surface mine sediment ponds in southern Appalachia by aquatic organisms and breeding amphibians. pp. 261-287. In R.P. Broods, D.E. Samual, J.B. Hill (eds.). Wetlands and Water Management on Mined Lands.

Genoways, H.H. and F.J. Brenner. 1985. Species of special concern in Pennsylvania. Spec. Pulb. No. 14. Carnegie Museum of Natural History.

Gill, F.B. 1985. Birds. pp. 292-354. In H.H. Genoways and F.J. Brenner (eds). Species of Special concern in Pennsylvania: 299-301. Spec. Publ. No. 14. Carnegie Museum of Natural History.

Glooschenko, V., J.H. Archbold and D. Herman. 1988. The Ontario wetland evaluation system: replacability and bird habitat selection. pp. 115-127. In D.D. Hook (ed.). The Ecology and Management of Wetlands. Portland; Timber Press.

Gore, J.A. and F.L. Bryant. 1988. River and stream restoration. pp. 23-38. In J. Cairns Jr. (ed.). Rehabilitating Damaged Ecosystems. CRC Press. Boca Raton, Florida.

Grese, R.E. 1991. The landscape architect and problem exotic plants. In J.B. Burley (ed.). Reclamation Vision and Realities: 7-16.

Griswold, B.L., C.J. Edwards and L.C. Woods III. 1982. Recolonization of macro-invertebrates and fish in a channelized stream after a drought. Ohio J. Sci. 82:96-102.

Hammer, D.A. 1988. Constructed wetlands for treatment of industrial and non-point source pollutions. pp. 333-349. Wetlands Ecology and Conservation: Emphasis in Pennsylvania Academy Science. 394 pp.

Hedin, R.S. 1989. Treatment of coal mine drainage with constructed wetland, pp. 349-362. In S.K. Majumdar, R.P. Brooks, F.J. Brenner, and R.W. Tiner Jr. (eds.). Wetlands Ecology and Conservation: Emphasis in Pennsylvania. Pennsylvania Academy Science. 394 pp.

Hopcraft, D. 1975. Productivity comparison between Thomson's Gazelle and cattle, and their relationship to the ecosystem in Kenya. Unpub. Ph.D. Thesis. Cornell Univ. Ithaca, NY.

Kadlec, R.H. 1989. Wetlands for treatment of municipal wastewater. pp. 300-314. In S.K. Majumdar, R.P. Brooks, F.J. Brenner, and R.W. Tiner Jr. (eds.). Wetlands Ecology and conservation: Emphasis in Pennsylvania. Pennsylvania Academy Science. 394 pp.

Kimmel, R.O. 1978. Ruffed grouse brood habitat on reclaimed surface mine in West Virginia. Unpub. Ph.D. Thesis. West Virginia University, 118 pp.

Jones, D.W., H.J. McElliogott and R.H. Manoz. 1985. Biological, chemical and morphological characterization of 33 surface mine lakes in Illinois and Missouri. St. Louis: Peabody Coal Company.

Lacki, M.J., J.W. Hummer and H.J. Webster. 1991. Effect of reclamation technique on mammal communities inhabiting wetlands on mine lands in east central Ohio. J. Ohio Acad. Sci. 91:154-159.

Leberman, R.C. 1988. A field checklist of the birds of western Pennsylvania and adjacent regions. Spec. Publ. 13. Carnegie Museum of Natural History.

Miller, J.G. and R. Tibbott. No Date. Fish habitat improvement for streams. Penn. Fish. Commission.

Ney, J.J. and M. Mauncy. 1980. Impact of a small impoundment on benthic macro-invertebrate and fish communities in a head water stream in the Virginia piedmont. pp. 102-112. In L.A. Krumholz (ed.). The Warm Water Stream Symposium. Amer. Fish. Soc. Washington, D.C.

Peterson, R.C., L.B.M. Peterson and J.J. LaCoursiere. 1992. A building block model for stream restoration. pp. 293-311. In P.J. Boon, P. Calow, and G.E. Petts (eds.). River Conservation and Management. John Wiley & Sons, Chickester. 47 pp.

Sandusky, J.E. 1978. The potential for management of water fowl nesting habitats on reclaimed lands. pp. 325-327. In D.E. Samuel, J.R. Stauffer, C.H. Hocutt, and W.H. Mason (eds.). Surface Mining and Fish/Wildlife Needs in the Eastern United States.

Thompson, C.B. 1984. Experimental practices in surface coal mining-creating wetlands habitats. Wetlands Newsletter April-May 1984:15-16.

United States Department of Agriculture. 1983. Soil conservation technical guide. U.S. Government Printing Office. Washington, D.C.

Vannote, R.L., G.W. Minshall, K.W. Cummins, J.R. Sedell and C.E. Cushing. 1980. The river continuum concept. Can. J. Fish. Aquat. Sci. 37:130-177.

Walton, O.E. Jr. 1980. Invertebrate drift from predator-prey associations. Ecology 61:1486-1497.

Wesche, T.A. 1980. The WRRI trout cover rating method, development, and application. Water Resource Ser. No. 78. Univ. Wyoming, Laramie.

Wilson, E.O. (ed.). 1988. Biodiversity. National Acad. Press. Washington, D.C.

Woods, L.C. III and B.L. Griswold. 1980. Channelization and mitigation: their effects on macro-invertebrate communities of the Oletangy River. Columbus, OH. pp. 113-118. In L.A. Krumholz (ed.). Warm Water Stream Symposium. Am. Fish Soc. Washington, D.C.

Wray, T. III, P.B. Wachenhut and R.C. Whitmore. 1978. The reproductive biology of passerine bird breeding on reclaimed surface mines in West Virginia, pp. 333-344. In D.E. Samuel, J.R. Stauffer, C.H. Hocutt, and W.F. Mason, Jr. (eds.). Surface Mining and Fish/Wildlife Needs in Eastern United States.

Wray, T. III, K.A. Strait and R.C. Whitmore. 1982. Reproductive success of grassland sparrows on reclaimed surface mines in West Virginia. Auk. 99:157-168.

Biological Diversity: Problems and Challenges. Edited by S.K. Majumdar, F.J. Brenner, J.E. Lovich, J.F. Schalles, and E.W. Miller. © 1994, The Pennsylvania Academy of Science.

Chapter Twenty-Four

SINGLE SPECIES VERSUS COMMUNITY AND ECOSYSTEM MANAGEMENT: Their Impact on Biological Diversity

WILLIAM C. HUNTER

U.S. Fish and Wildlife Service
75 Spring Street, SW
Atlanta, GA 30303

INTRODUCTION

The main focus of wildlife management during the 20th Century has been on the conservation and management of game species. Management for waterfowl, deer, turkey, trout, bass, grouse, woodcock, elk, and quail, for examples, traditionally has been (and still is) assumed by many wildlife managers to be important also in conserving all other species and for stemming the tide of loss of many American wildlife species. However, this assumption is increasingly being challenged, both within the profession and from outside environmental groups.

Recognition among many wildlife conservationists that an ever-increasing number of species were extinct, facing extinction, or becoming very rare, despite traditional approaches to wildlife and fisheries management, culminated in the passage of the Endangered Species Act of 1973, as amended. However, this law is now widely perceived as resulting in management and conservation of only a few high profile species. Meanwhile, most other species and many threatened ecosystems, that may or may not have listed species to represent them, are unprotected in the face of ever-increasing human disturbance.

Most recently, the level of continuing losses to entire ecosystems and associated communities of organisms has led to much discussion on whether there is a need for additional laws that mandate protection of biological diversity. Paramount in this discussion is a clear definition of what biological diversity is and at what level does a manager need to know about whether an action is beneficial or detrimental to the maintenance of biological diversity. At the center of this debate is how to best approach the protection of biological diversity within the confines of traditional species-specific approaches versus an ecosystem approach.

The first question that must be asked in this debate is whether there is truly a dichotomy in the two approaches. The second question is, if there is a difference, how does a manager decide what the priority actions would be at the local land management level that would allow for maintaining or enhancing biological diversity at the global level. Finally, how can the maintenance of biological diversity using an ecosystem approach be best integrated on both public and private lands when there are many other uses of these lands, some of which may not be compatible with the goal of "maximizing" biological diversity at the local level for the maintenance of global biological diversity.

In this chapter I provide a working definition of the components involved in understanding the concept of biological diversity. Next, I provide a brief history of wildlife management in North America to place the present discussion of biological diversity in perspective. This is followed by a discussion of strategies involving game and endangered species and how these can address larger groups of species and their ecosystems, which contributes at least indirectly to the maintenance of biological diversity.

WHAT IS BIOLOGICAL DIVERSITY?

A concise definition of biological diversity is the full complement of the native plant and animal species occurring in their natural or normal patterns of abundance. A more detailed definition of biological diversity is the full range of variety and variability within and among living organisms and the interactions through ecological processes under which they all occur, including ecosystems or community diversity, species diversity, and genetic diversity (modified from Office of Technical Assessment 1987). Various other definitions exist, but the proliferation of interpretations for what biological diversity means has done little to definitively explain what this concept means to the operations of wildlife managers at the local level. Nevertheless, a discussion of the four generally recognized components of biological diversity (genetic, species, community-ecosystem, landscape-region) is important for developing management strategies to address this concept. Each of these components must have compositional, structural, and functional attributes in order for the concept of biological diversity to be fully understood (Franklin 1988, Noss 1990).

Genetic diversity.—Maintenance of the full variety of genetic material occurring within a population is essential for withstanding environmental perturbations and disease. Genetic diversity also can be measured among separated populations or subspecies to help understand the full variation inherent within a widespread species occurring in a variety of environments. Thus, at both the population and species levels, genetic diversity is essential for allowing the species as a whole to evolve within a variable environment.

Maintenance of a genetically viable population refers to the number of reproducing individuals considered necessary to ensure a population's persistence for a determined number of years. In theory, an effective population size of 500 reproducing individuals is considered necessary to avoid gene loss from a population due to genetic drift for 200 years (Franklin 1980, Frankel and Soule 1981, Frankel 1983). A population size of 50 reproducing individuals is considered necessary to minimize proliferation of deleterious genetic material through inbreeding.

The demographics of a species (i.e., overlapping generations, sex ratios, differential reproductive and survival qualities among individuals) may require substantially higher numbers of reproductive individuals than 500 for an effective population size to have viability (Reed et al. 1988). The minimum viable population concept is most important for maintaining populations that experience little interchange with other populations of the same species. Where there is frequent interchange among adjacent but different populations, the concept of "metapopulation" (i.e., all the populations within a landscape necessary to maintain the species in the area) becomes most important for measuring genetic diversity.

Loss of genetic diversity can occur when entire populations are lost from the species that may take with them genes or combinations of genes not found in other populations. Such losses would also constitute loss to global genetic diversity. Genes also can be lost from populations due to genetic drift when populations fall below minimum viable size and go through "bottle-necks." Natural or artificial reestablishment of gene flow from other populations may be necessary to stabilize the original population. In these latter situations, artificial reestablishment of gene flow should be conducted with caution so as to return diversification of the gene pool without eliminating ("swamping out") any remnant genes or combinations of genes that are of survival value to the original population in its environment.

Species diversity.—The component of biological diversity most familiar to the average wildlife manager is the number of species (species richness) and their abundances relative to each other (evenness or species diversity index) within a habitat or a land management unit. However, neither high species richness nor a high degree of evenness in the abundances of all component species necessarily equates to maximizing biological diversity in any one local area of interest. Among other factors that must be considered when judging biological diversity are the relative status of each species globally, the importance of the area under consideration for each species, and the taxonomic uniqueness of each species (e.g., sole surviving representative of a family of species).

Net increases in species richness and evenness indeed may occur by diversifying habitats locally (e.g., creating extensive edge). However, such actions may actually be detrimental to biological diversity when species requiring large contiguous patches of relatively uniform habitat are lost. As a related problem, local areas of high species richness may be composed of mostly generalist "weedy" species and incorrectly identified as more important than areas with low species richness, but with high uniqueness among the few species present (e.g., a cave system).

Community-ecosystem diversity.—The maintenance of all species in their normal patterns of abundance within a local area requires stability among biological communities (including all vegetative successional stages and species assemblages) within and among all ecosystems occurring within that area. The existence of a healthy ecosystem assumes that all the necessary abiotic (e.g., hydrology, geology, soils, climate) and biotic (e.g., intra and interspecific) interactions within that ecosystem are functioning properly. The relative importance of each ecosystem within an area should be dependent upon the number of specialist species dependent upon each ecosystem and how important each ecosystem is for maintaining the full variety of organisms found using the area under consideration.

Many generalist species occur in more than one ecosystem or habitat type. Many specialist species occur principally in one ecosystem or habitat type. Other species may require more than one ecosystem or habitat type for completing complex life cycles and these species may be either generalists or specialists depending on how many habitat types can fulfill their requirements for survival at the local level. Thus, it is important to maintain the variety of habitats within and among ecosystems as much as possible, while not unintentionally eliminating habitat conditions for sensitive species in the process.

Communities of organisms are thought by some as highly interactive "superorganisms" evolving together, while others think of communities as assemblages of species with independent evolutionary histories whose ecological tolerances randomly overlap (see James et al. 1984). For the purposes of conserving biological diversity, biological communities are best thought of as groups of species that may or may not be unique to particular ecosystems but are at least loosely associated with each other. With this in mind, the loss or modification of a particular ecosystem may adversely affect some species more than others, but with conservation of both species groups considered equally important locally. However, the latter groups of species may be stable in other ecosystems, with the former group most important for the purposes of conserving biological diversity at the landscape level.

Landscape-regional diversity.—The persistence of viable populations of all species occurring within a landscape requires that all ecosystems are functional within or among enough local areas to support those populations or metapopulations within that landscape. Loss and maintenance of biological diversity is best tracked at the landscape level. The landscape level is also where judgements are best made on needed management adjustments for the local land management level to maintain balance among habitats or among successional types of a single habitat.

Rarely is it possible on a single land management unit to provide a large variety of habitats in all successional stages (supporting a high number of species) without losing sensitive species in the process. Thus, the landscape level allows for a detailed assessment of where it would be necessary to recommend management adjustments to maintain or avoid the loss of local populations of species, with unique genetic material, or entire communities within a region. The likelihood of experiencing loss of biological diversity at a regional scale is greatly diminished when all local land managers cooperate with each other within a landscape. With loss of biological diversity within all regions minimized, the loss of global biological diversity is also minimized.

MAINTAINING BIOLOGICAL DIVERSITY IN NORTH AMERICA — PAST AND PRESENT

Modern wildlife conservation has its origins during the Teddy Roosevelt era of the late 1800's and early 1900's, with President Roosevelt himself a well-known outdoor sportsman. Roosevelt and his administration were instrumental in the protection of all migratory birds against exploitative market hunting, especially symbolized with the protection of colonial wading birds nesting on Pelican Island in Florida. The establishment of Yellowstone National Park during this same time period highlighted the need for protecting large areas with all its component game and nongame species. Society now realizes that setting aside a place like Yellowstone is not adequate for the conservation of a "Greater Yellowstone Ecosystem" (Brussard 1991).

Actually, the prophecies of environmental degradation and loss to biological diversity have been in existence for quite sometime, with Henry David Thoreau and John Muir among those credited with raising the environmental consciousness of their fellow Americans. Perhaps the most eloquent of the Twentieth Century prophets is no less a figure than Aldo Leopold. Leopold (1949) in his *A Sand County Almanac* proclaimed "last call" for many ecosystems despite our best efforts to preserve what was left of "natural areas."

Our society's emphasis on commodity wildlife and recreation in concert with the expansion of industrialism and transportation spelled the doom to many an ecosystem and countless undescribed species, not to mention the better-known ones. Indeed, there are few, if any, remaining ecosystems in North America outside of Alaska and northern Canada that have not been detrimentally modified since Leopold's last call. Presently, even many areas in Alaska and Canada are under new and expanding threats.

As a revered game biologist, Leopold (1949:191) writes with disgust over predator control programs conducted in "wilderness" in the name of big-game management, of which he himself contributed to as a young man. Leopold's view of how society has approached land management as a commodity rather than with a "Land Ethic" is as pertinent today as it was over 40 years ago, but now the problems associated with a management focus on game species are added to the continuing expansion of poor forest and range management, mineral extraction activities, agriculture, and urbanization. These activities expand at the expense of all wildlife and far exceed the problems associated with a focus to proliferate a few game species.

Nevertheless, credit must be given to the outdoor sportsmen without whom very little would have been done for wildlife at all during the last half-century. Most actions undertaken to conserve nongame as well as game species since the 1930's have been spearheaded and largely funded by revenues generated from hunters and fishermen. Private and public land managers can best justify their involvement and financial support for actions such as wetland restoration and protection within rural communities because of widespread recognition of waterfowl as a harvestable natural resource. The same interest is not as apparent for preserving wetlands just because it is right to do so, without the emphasis on waterfowl, especially when economic benefits (albeit short-sighted) are seen for converting wetlands to other uses.

We should not think that commodity wildlife is only synonymous with game management, as it is important to look closely at what drives nongame and watchable wildlife programs. "Non-consumptive" use of wildlife now exceeds hunting in popularity and is about equal with fishing (U.S. Fish and Wildlife Service 1988). Yet the success of such programs is gauged by the number of people watching and how much money they spend locally, how much bird seed is sold, and how many binoculars and field guides are sold. There is presently no good way to accentuate the positive of how many species and ecosystems are being maintained in healthy condition through these programs.

Unfortunately, little or no money spent in the pursuit of observing nature finds its way back to conserving the natural resource responsible for supporting these activities. In addition, many wild areas are under new threats of being loved to death (e.g., most major national parks and wild areas near major metropolitan areas) as ecotourism increases exponentially while protective measures for these wildlands remain desperately underfunded. Thus, there is little effort to expand management activities over and above actions taken to produce game species, which would be necessary if ecosystem-level management is to be achieved.

We are left with an imperfect approach to wildlife management today with society and the wildlife profession not yet mature enough to fully embrace Leopold's Land Ethic approach to conservation of natural resources. However, it is the focus on a few commodity species that allows for the protection of many wildlife habitats from being lost completely to other economic interests. The challenge for wildlife managers today is to develop strategies that expand efforts to conserve ecosystems while still managing for the game, endangered, and watchable wildlife resources that receive widespread support, at least until society has a longer-term vision of natural resource management.

STRATEGIES FOR ADDRESSING BIOLOGICAL DIVERSITY CONCERNS

Game species approaches.—Many game species are generalists in that each species may occur in a variety of habitats (e.g., different types of forested or wetland landscapes) as long as they each provide essential food, water, and cover. In addition, game species can be thought of as specialists in the sense that a variety of habitats (e.g., within a specific forested or wetland landscape) or an area of transition between habitats (i.e., habitat edge or ecotone) may be necessary to provide those essential elements. Thus, providing habitat conditions that satisfied the needs of many game species requiring substantial edge habitat was considered sufficient for providing the variety of habitat conditions necessary to support all species within a land management unit.

Habitat edges essentially allow for a synergistic effect between the basic elements of the two habitats that come together with additional features arising that attract wildlife not frequently found on either side of the edge. This so-called "edge-effect" is well illustrated when mature forest and open habitats (e.g., grasslands, early-successional forest, farmlands) come together. Greater sunlight on the trees lining the edge allows for extensive vegetative development at all vertical layers and therefore provide substantially greater cover and food resources in the understory and midstory than what might be expected in the interior of a close-canopied forest. Species found along forest edges include those using both open-areas and mature forests, as well as those preferring to feed in open areas near to forest cover and those requiring the understory or midstory vegetation available along forest edges.

Thus, it is understandable why wildlife managers have worked to create habitat-edges with vigor and conviction, especially for managing deer, turkey, grouse and woodcock. The edge-effect concept itself is considered one of the most important ecological tenets to be applied in managing natural resources (Leopold 1933, Odum 1959, Yoakum and Dasmann 1971). However, there is a rising chorus of opposition to the continuous creation of edge within forested landscapes. This opposition is especially prevalent within the ornithological community since birds are the best studied group of organisms, besides game species, using forested landscapes (Harris 1988, Robinson 1988).

Many species of forest-dwelling birds, particularly long-distance or neotropical migrants, have been shown to decline when forest patch size falls below a certain threshold (Ambuel and Temple 1983, Askins and Philbrick 1987, Robbins et al. 1989a). However, what is more alarming is that what has been observed at the local level is now being shown at the landscape and global levels (Robbins et al. 1989b). Why these declines are occuring for such a diverse groups of species is a subject undergoing debate and there is clearly more than one cause (Askins et al. 1990, Finch 1991, Terborgh 1989).

Failure to successfully raise young along an ever-increasing amount of forest-edge habitat on North American breeding grounds is one of the leading hypotheses. Predators (e.g., squirrels, raccoons, jays, crows) and brood parasites (cowbirds) are more efficient as the habitat edge increases relative to interior habitat. Many species initially attracted to forest-edges eventually disappear (Wilcove 1985, Robinson 1988, Temple and Cary 1988, Yahner and DeLong 1992).

The issue of forest fragmentation and its effects on forest-dwelling species is particularly relevant in the edge-effect controversy as forests in many regions of eastern North America are taking on the characteristics of habitat islands (some large and some small) within a landscape dominated by farmlands and residential development. There are more forested acres now in eastern North American than in the 1920's, but the new forests in many areas are of a highly manipulated nature with many stands in relatively unproductive short-rotation pine plantation. Much of the remaining more productive stands nevertheless occur in a wide variety of age classes, sizes, and shapes, with variable species composition.

Since many landscapes are already heavily fragmented, many scientists and environmentalists have argued, based on an ecosystem view, that it is no longer appropriate to further open forests within local land management units under public domain in the name of a few game species (Harris 1988). Yet many bird species that require early successional habitats, again many of which are neotropical migrants, are also undergoing dramatic declines (Robbins et al. 1989b). In addition, evidence exists that creation of forest openings through even-aged management in larger forested tracts may actually result in benefits to both forest-dwelling and early-successional species (Thompson et al. 1992).

One hypothesis for the detected declines of early-successional neotropical migrants may be because of the overall increase in forest cover in eastern North America. Also, the reduction of farmlands that were allowed to go fallow, the trend for smaller and widely-spaced clearcuts on public lands, and the very short-term suitability (2-3 years) of such habitats have placed a premium on large patches of early-successional habitat. Many early-successional neotropical migrants are as susceptible to depredation and brood parasitism when nesting in small patches as their forest-interior counterparts nesting along forest edges (Nolan 1978). Predators and brood parasites do not occur in overwhelming numbers within the larger forested landscapes, allowing for edges and early-successional habitats of suitable size to be of maximum benefit without many of the detrimental side-effects (Wilcove 1985, Thompson 1992).

Thus, we are left with a conservation strategy that emphasizes long rotations (>100 years) and relatively large clearcuts (15-40 ha) for managing bird species associated with extensive mature forested tracts, early-successional stages, and habitat edges that would be compatible with the needs of most game species (Harris 1984, DeGraaf 1987, Robbins 1988). Another option would be to conclude that effective management of all forest seral stages within a local land management unit is not achievable and that the best approach would be to coordinate and cooperate with other public and private land managers for best maintenance of landscape biological diversity. Because of differing management objectives, some managers favor early-successional species while others have greater opportunities to benefit and maintain area-sensitive forest-dwelling species.

Obviously, these recommendations are most appropriate for broadleaf dominated forests in eastern North America and may be of little relevance to other forest types and for other regions. In addition, management for birds still is not ecosystem management, but it is a stepping stone towards an ecosystem approach and away from a single-species approach. Thus, the needs of other species groups requiring forested landscapes may or may not be addressed by this approach in all situations. Another group of vertebrates suitable for investigation of forestry management and completely interactive with the local environment (as opposed to species that migrate) are the amphibians, especially salamanders (Hairston 1987).

In addition, the plants that make up the forest ecosystems to begin with, other than the harvestable tree species and the species that "compete" with maximum commodity production, are all too rarely considered in today's approach to large-scale forestry. Avoiding undue disturbance of the underground ecosystem of a forest is essential for rejuvenating that forest. Yet, widespread use of intensive mechanical site preparation and indiscriminate application of herbicides for the short-term encourages opportunistic weedy plants and works against the reestablishment of woody species, to include eventually the trees that are to be perpetuated as a "renewable" resource (Maser 1988). Nevertheless, the idea that the bigger the better for habitat patches of all successional stages (but especially mature forest) in an otherwise fragmented landscape should indeed be consistent with maintenance of biological diversity at the landscape as well as the global spatial scales (Harris 1984).

In addition to terrestrial game species, waterfowl have played an extremely important role in the conservation of wetlands during this century, especially recently with the implementation of the North American Waterfowl Management Plan (U.S. Fish and Wildlife Service 1987). This Plan has been instrumental in the formation of cooperative joint ventures among governmental agencies and private conservation and hunting groups. Results have been impressive with increasing reforestation efforts of bottomland hardwoods, increasing the acreage and management of emergent wetlands, and encouraging use of agricultural areas to provide seasonal benefits to waterfowl.

However, modifications to management practices for waterfowl are needed to benefit the full range of species using these restored wetlands and seasonally managed agricultural areas. First, reforested areas should be placed in such a way as to increase the size of already existing forests. Second, water should be made available both earlier and later than is normally provided for migrating and overwintering waterfowl in order to overlap peak periods of shorebird migration, marshbird nesting, and post-breeding dispersal of long-legged wading birds (Fredrickson and Taylor 1982, Fredrickson and Reid 1986). Other wetland values and groups of organisms (especially, aquatic plants and invertebrates, amphibians, and fishes) are likely to be benefitted overall with greater attention to all water-dependent birds, but again more documentation is needed to support this view.

As with management targeting terrestrial game and waterfowl species, management of game fishes can be beneficial by allowing for the conservation of ecosystems supporting other aquatic species, but can be harmful if careful attention is not given to the needs of nongame species. Aquatic systems throughout North American have been altered by damming, channelization, contaminants, and through changes in erosion runoff (for examples see Karr et al. 1985, Minckley and Deacon 1991). An active sports fishery can be maintained among the most altered systems through "natural" spawning or by artificial stocking. However, native nongame fishes and aquatic invertebrates suffer great decreases in most manipulated systems. These losses are reflected in the increasing number of extinct, Federally threatened and endangered species, and species that are candidates for listing associated with aquatic systems throughout North America.

Endangered species approaches.—The widespread perception that protection of endangered species through the Endangered Species Act ignores ecosystem values is untrue both in concept and in practice. The Endangered Species Act is best thought of as a safety net when society and the actions of Federal agencies have failed in their Congressional mandate to maintain the ecosystems upon which listed species depend. Congress has mandated that Federal agencies shall not conduct actions or have policies that would likely jeopardize the continued existence of listed species. The best way for Federal agencies to avoid being confronted with a potential jeopardy situation is by fully complying with the Congressional mandate that all Federal agencies shall use their authorities to conserve (recover) listed species. The best way to reach recovery for endangered species is to conserve their ecosystems.

The Endangered Species Act was never intended to stop projects that may harm listed species, but was intended to do everything possible to resolve potential conflicts in such a manner that the species and its ecosystem were not lost. The fact that some species have become extinct since the passage of Endangered Species Act, that few species have been fully recovered, and that many candidate species are awaiting listing is not a failure of the Act itself as much as a failure of society to force full compliance and funding of the Act. What is difficult to measure is how many species with their ecosystems intact are still in existence due to the Act.

There is no question that recovery efforts have concentrated on a few high-profile glamour species. However, recovery efforts for many of these species allow for recovering the ecosystems they occur in and all the associated species they occur with. Remnants of many South Florida ecosystems still exist (albeit under grave threat) because of endangered species protection and recovery efforts, as for the Florida panther (*Felis concolor coryi*), sea turtles, and Lake Wales Ridge scrub plants (U.S. Fish and Wildlife Service 1991, 1992; National Marine Fisheries Service and U.S. Fish and Wildlife Service 1991). The same can be said for remnants of historical aquatic ecosystems in the western United States as presented in great detail by Minckley and Deacon (1991).

Other examples include the need for conserving the San Marcos cave fauna in central Texas as essential for maintaining all ecosystems in the area (including those inhabited by people) dependent upon the Edward's Plateau aquifer. The highly publicized situation involving the northern spotted owl (*Strix occidentalis caurina*) clearly is an issue involving the conservation of entire ecosystems supporting a biologically diverse community of plants and animals. Unfortunately, this issue is often touted in the mainstream media as placing the importance of one species over the economic stability of a region, which is a false dichotomy. A final example comes from the Southeast where the red-cockaded woodpecker (*Picoides borealis*) will need to be recovered by the return of the almost eliminated fire-maintained longleaf pine ecosystem, which supports one of the most biologically diverse communities of plants and animals in the Southeast (Noss 1988). At least some declining early-successional neotropical species now associated with clearcuts (e.g., the prairie warbler [*Dendroica discolor*]) and several game species (e.g., northern bobwhite [*Colinus virginanus]*) were once most closely associated with the fire-maintained longleaf pine ecosystem (Nolan 1978, Engstrom et al. 1984).

CONCLUSION

Even the best legislation intended for maintenance of biological diversity through ecosystem approaches to management or to amend the Endangered Species Act into being more responsive to ecosystems are not likely to be solutions without effective development and implementation of regulations based on concise and unambiguous definitions (Karr 1990). There are unlikely to be any real changes to day-to-day activities without a basic change on how already strong legislation presently is complied with nation-wide. In reality, maintenance of biological diversity is a social issue that is best addressed through changes in society's attitudes followed by significant financial support from the private sector. Without a moral and financial commitment, land managers are likely to find themselves attempting to address maintenance of biological diversity on the few acres they control. These efforts are likely to end in futility, because of little coordination to balance differing land management objectives with all other managed lands in the areas.

Our society continues to fail to think of wildlife beyond their short-term economic importance and the wildlife profession continues to fail in leading society into understanding Aldo Leopold's Land Ethic. Neither production of game species nor increasing wildlife observation opportunities are adequate approaches for achieving ecosystem-level management. However, management emphasis for a few game or watchable wildlife species is still better than the conversion of lands to a state where they are unsuitable for most species of wildlife.

While our knowledge base continues to expand on how single-species and ecosystem approaches are to be integrated, it must be realized that management of natural resources will continue in one form or another. As a society, we must be careful not to advocate either of the two extreme reactions likely to occur as this process evolves. Preservation (i.e., "let nature take its course") of all natural resources for the stated reason that we do not have all the answers is unacceptable and will not be tolerated by a commodity driven society such as ours. With the same thought, we must resist the tendencies to react over the confusion of what biological diversity means to management by continuing to focus on a few game species; it is always easier to focus on a few species than groups of species. Avoiding these two extremes will require identifying local priorities that take into account the global status of organisms and stepping this information down to the local land manager (e.g., Millsap et al. 1990, Noss 1990). Such a process should take into account the relative importance of the land management unit to each species and the likely actions of adjacent land managers within the landscape. Species and groups (guilds) of species, along with the ecosystems that support them, can then be managed in such a way as to make sure regional biological diversity is maintained.

There is hope in that many Federal and State agencies and the public at large is moving towards an appreciation of biological diversity and the need to manage for species and groups of species at the ecosystem level. One such initiative is the National Fish and Wildlife Foundation sponsored Partners in Flight program that is intended to address management of a fairly large group of species with widely differing habitat needs (see Finch 1991). Enthusiasm for this program is gaining and it is unique in that it is the first such initiative to tie the problems these species face in temperate habitats with the problems they face in tropical habitats. In essence, neotropical migrants service as ambassadors for all species experiencing declines in both temperate and tropical zones with the realization that all human residents in the Western Hemisphere share both in the blame and in finding the solution to the biological diversity crisis we now face.

LITERATURE CITED

Ambuel, B. and S.A. Temple. 1983. Area-dependent changes in the bird communities and vegetation of southern Wisconsin forests. Ecology 64:1057-1068.

Askins, R.A., J.F. Lynch and R. Greenberg. 1990. Population declines in eastern North America. Current Onithol. 7:1-57.

__________ and M.J. Philbrick. 1987. Effects of changes in regional forest abundance on the decline and recovery of a forest bird community Wilson Bull. 99:7-21.

Brussard, P.F. 1991. The Greater Yellowstone Ecosystem. Conserv. Biol. 5:336-337.

DeGraaf, R.M. 1987. Managing northern hardwoods for breeding birds. Pp. 348-362 *in* Nyland, R.D., ed. Managing northern hardwoods. Soc. Amer. Foresters Pub. No. 87-03.

Engstrom, R.T., R.L. Crawford and W.W. Baker. 1984. Breeding bird populations in relation to changing forest structure following fire exclusion: a 15-year study. Wilson Bull. 96:437-450.

Finch, D.M. 1991. Population ecology, habitat requirements, and conservation of neotropical migratory birds. USDA For. Ser. Gen. Tech. Rept. RM-205.

Frankel, O.H. 1983. The place of management in conservation. Pp. 1-14 *in* Schonewald-Cox, C.M., S.M. Chambers, B. MacBryde and L. Thomas, eds. Genetics and conservation. Benjamin/Cummings Publ. Co., Inc., Menlo Park, Calif.

__________ and M.E. Soule. 1981. Conservation and evolution. Cambridge Univ. Press, Cambridge, Mass.

Franklin, I.R. 1980. Evolutionary change in small populations. Pp. 135-150 *in* Soule, M.E. and B.A. Wilcox, eds. Conservation biology. Sinauer Assoc., Inc., Sunderland, Mass.

Franklin, J.F. 1988. Structural and functional diversity in temperate forests. Pp. 166-175 *in* Wilson, E.O., ed. Biodiversity. National Academy Press, Washington, D.C.

Fredrickson, L.H. and F.A. Reid. 1986. Wetland and riparian habitats: a nongame management overview. Pp. 59-96 *in* Hale, J.B., L.B. Best and R.L. Clawson, eds. Management of nongame wildlife in the Midwest: a developing art. Proc. Symp. 47th Midwest Fish and Wildl. Conf.

__________ and T.S. Taylor. 1982. Management of seasonally flooded impoundments for wildlife. U.S. Fish and Wildlife Resour. Publ. 148:1-29.

Harris, L.D. 1984. The fragmented forest: island biogeography theory and the preservation of biotic diversity. University of Chicago Press, Chicago.

Harris, L.D. 1988. Edge effects and conservation of biotic diversity. Conserv. Biol. 2:330-332.

Hairston, N.G., Sr. 1987. Community ecology and salamander guilds. Cambridge Univer. Press, Cambridge, Mass.

James, F.C., R.F. Johnson, N.O. Wamer, G.J. Niemi and W.J. Boecklin. 1984. The Grinnellian niche of the Wood Thrush. Am. Nat. 124:17-30.

Karr, J.R. 1990. Biological integrity and the goal of environmental legislation: lessons for conservation biology. Conserv. Biol. 4:244-250.

__________, L.A. Toth and D.R. Dudley. 1985. Fish communities of midwestern rivers: a history of degradation. BioScience 35:90-95.

Leopold, A. 1933. Game Management. Charles Scribner and Sons. New York.

__________. 1949. A sand county almanac. Oxford Univer. Press, London.

Maser, C. 1988. The redesigned forest. R. & E. Miles, San Pedro.

Millsap, B.A., J.A. Gore, D.E. Runde and S.I. Cerulean. 1990. Setting priorities for the conservation of fish and wildlife species in Florida. Wildl. Monogr. 111,1-57.

Minckley, W.L. and J.E. Deacon, eds. 1991. Battle against extinction: native fish management in the American West. University of Arizona Press, Tucson, Arizona.

Nolan, V., Jr. 1978. The ecology and behavior of the Prairie Warbler *Dendroica discolor.* Ornithol. Monogr. 26:1-595.

National Marine Fisheries Service and U.S. Fish and Wildlife Service. 1991. Recovery plan for U.S. population of loggerhead turtle. National Marine Fisheries Service, Washington, D.C.

Noss, R.F. 1988. The longleaf pine landscape of the Southeast: almost gone and almost forgotten. Endangered Species Update 5:1-8.

Noss, R.F. 1990. Indicators for monitoring biodiversity: a hierarchical approach. Conserv. Biol. 4:355-364.

Odum, E.P. 1959. Fundamentals of ecology. 2nd ed. W.B. Saunders Company, Philadelphia.

Office of Technical Assessment. 1987. Technologies to maintain biological diversity. U.S. Government Printing Office, Washington, D.C.

Reed, J.M., P.D. Doerr, J.R. Walters. 1988. Minimum viable population size of the red-cockaded woodpecker. J. Wildl. Manage. 52:385-391.

Robinson, S.K. 1988. Reappraisal of the costs and benefits of habitat heterogeneity for nongame wildlife. Pp. 145-155 *in* Transactions of the Fifty-third North American Wildlife and Natural Resources Conference.

Robbins, C.S. 1988. Forest fragmentation and its effects on birds. Pp. 61-65 *in* Johnson, J.E., ed. Managing north central forests for non-timber values. Soc. Amer. Foresters Publ. No. 88-04.

Robbins, C.S., D.K. Dawson and B.A. Dowell. 1989a. Habitat area requirements of breeding forest birds of the middle Atlantic states. Wildl. Monogr. 103,1-34.

Robbins, C.S., J.R. Sauer, R.S. Greenberg and S. Droege. 1989b. Population declines in North American birds that migrate to the neotropics. Proc. Natl. Acad. Sci. 86:7658-7662.

Temple, S.A. and J.R. Cary. 1988. Modelling dynamics of habitat-interior bird populations in fragmented landscapes. Conserv. Biol. 2:340-347.

Terborgh, J. 1989. Where have all the birds gone? Essays on the biology and conservation of birds that migrate to the American tropics. Princeton Univ. Press.

Thompson, F.R., W.D. Dijak, T.G. Kulowiec, D.A. Hamilton. 1992. Breeding bird populations in Missouri Ozark forests with and without clearcutting. J. Wildl. Manage. 56:23-30.

U.S. Fish and Wildlife Service. 1987. North American Waterfowl Management Plan. Washington, D.C.

__________. 1988. 1985 National Survey for Fishing, Hunting, and Wildlife Associated Recreation. Washington, D.C.

__________. 1992. Draft Florida panther habitat preservation proposal. Atlanta, GA.

__________. 1991. Project Proposal for the Lake Wales Ridge National Wildlife Refuge. Atlanta, GA.

Wilcove, D. 1985. Nest predation in forest tracts and the decline of migratory songbirds. Ecology 66:1211-1214.

Yahner, R.H. and C.A. Delong. 1992. Avian predation and parasitism on artificial nests and eggs in two forested landscapes. Wilson Bull. 104:162-168.

Yoakum, J. and W. Dasmann. 1971. Habitat management practices. Pp. 173-231 *in* Giles, R.H., ed. Wildlife Management Techniques. 3rd ed. The Wildlife Society, Washington, D.C.

Biological Diversity: Problems and Challenges. Edited by S.K. Majumdar, F.J. Brenner, J.E. Lovich, J.F. Schalles, and E.W. Miller.

Chapter Twenty-Five

DIVERSITY OF TERRESTRIAL VERTEBRATES IN PENNSYLVANIA

C.J. McCOY[1], KENNETH C. PARKES[2] and JANIS C. SACCO[3]

[1]Curator, Section of Amphibians and Reptiles
[2]Curator, Section of Birds
[3]Postdoctoral Fellow, Section of Mammals
Carnegie Museum of Natural History
Pittsburgh, Pennsylvania 15213

INTRODUCTION

The term biodiversity is currently in vogue, and appears frequently in both scientific literature and the popular press. The simplest definition of biodiversity, which is the number of recognizable biological entities living in a defined area, is widely understood. As a consequence of that understanding, usage of the term biodiversity is generally appropriate in the context in which it appears.

However, for the scientist who requires precise and quantifiable definitions for terms applied to natural systems, the objective definition and measurement of biodiversity are frustrating. There are two reasons for this. First, natural biological systems are not static constructs. In these systems dynamic processes such as succession, competition, emigration, immigration, speciation, and extinction continuously affect and change diversity. Thus any quantitative measurement of biodiversity is a temporal artifact. Although such measurements are useful for comparisons, in a practical sense they are unrepeatable.

The second problem in measurement of biodiversity lies in delimitation of the area in which the measurement is made. Commonly, diversity estimates or measurements are related to poorly defined ecological units such as "The Amazon

Basin," or broadly inclusive geographic areas, "North America" for example. In either case the delimited area, its flora, and it fauna grade imperceptably into adjacent ecosystems or geographic areas. Precise delimitation of ecosystems or floral or faunal regions, which would be the only natural units for biodiversity measurement, is impossible.

The most common "solution" to the problem of delimiting natural areas for diversity measurement is to ignore it, and to base diversity measurements not on natural areas but on political units. As political boundaries are rarely coincident with the limits of natural areas, biodiversity measurements of political units are biologically meaningless. However, we recognize the usefulness of such measurements to government agencies which are mandated to function within political units.

Here we provide an estimate of the diversity of terrestrial vertebrate animals in Pennsylvania, based on the number of species of amphibians, reptiles, birds, and mammals currently or recently resident within the boundaries of the Commonwealth. Our estimate includes data from the historical record, which possibly results in an overestimate of the diversity of the extant fauna. We have noted major faunal changes such as extirpations, immigrations, and major range changes where they have been documented. The diversity estimate for Pennsylvania is compared with the number of species of each group in the world, in North America, and in adjacent states. Finally, faunal diversity in several physiographic (and roughly ecological) regions of Pennsylvania is compared.

COMPARATIVE DIVERSITY

The species diversities of the classes Amphibia, Reptilia, Aves, and Mammalia in Pennsylvania, adjacent states, North America, and the world are compared in Table 1. State diversities are compared both as the number of species in each Class and the proportion of the North American fauna represented. Extinct species are not included in the lists. Of recently extirpated species those for which suitable

TABLE 1

Comparative Diversity of Amphibians, Reptiles, Birds and Mammals in Pennsylvania and Surrounding States. Percentages are the Proportion of the North American Fauna Represented in each Category. See text for sources of data.

	Amphibians	Reptiles	Birds	Mammals
Number of species				
World	3,260	5,954	8,590-9,672	4,629
North America	221	275	535	364
Pennsylvania	36 (16%)	37 (14%)	197 (37%)	68 (19%)
Maryland	36 (16%)	38 (14%)	—	62 (17%)
New Jersey	26 (12%)	21 (12%)	—	58 (16%)
New York	26 (12%)	29 (11%)	—	66 (18%)
Ohio	32 (15%)	35 (13%)	193 (36%)	54 (15%)
West Virginia	33 (15%)	33 (12%)	180 (34%)	66 (18%)

habitat and the potential for re-establishment of breeding populations exist are included. Some extirpated species have been re-established through introductions, and these have been included (Table 2). Feral species were not counted, but long-established breeding exotics such as Ring-necked Pheasant (*Phasianus colchicus*), House Sparrow (*Passer domesticus*), European Starling (*Sturnus vulgaris*), House Mouse (*Mus musculus*), and Norway Rat (*Rattus norvegicus*) are included.

Sources of data for Table 1 are as follows: *amphibians and reptiles*, world;[1] North America,[2] Pennsylvania,[3] adjacent states;[4] *birds*, world,[5,6] North America,[7] Pennsylvania,[8,9] Ohio;[10,11] West Virginia;[12] *mammals*, world;[13] North America;[14] Pennsylvania;[15] Maryland;[16] New York;[18] Ohio;[19] West Virginia.[20]

The amphibian and reptile diversity of Pennsylvania and adjacent states is similar—11 to 16 percent of North American species in each group. There is a latitudinal trend in both amphibians and reptiles, with more southerly states having more species (New York versus Maryland). However, in Pennsylvania the rich salamander fauna of the Appalachian uplands compensates for the lower diversity of northern frogs.

There are documented breeding records of 197 species of birds in Pennsylvania, which is 37 percent of the breeding species of North America. This substantial diversity probably is a result of the geographic position of Pennsylvania, which lies in a transition zone between northern and southern avifaunas. Breeding bird diversities in adjacent Ohio (36 percent) and West Virginia (34 percent) are comparable to that in Pennsylvania. Bird species lists for Maryland, New Jersey, and New York are not comparable to that for Pennsylvania, as there are major breeding bird habitats in those states (seacoasts, high elevations) which are not present in the Commonwealth.

There are 68 species of resident mammals recorded in Pennsylvania, or 19 percent of the species known from North America. This diversity is comparable to that

TABLE 2

Extirpated Mammals of Pennsylvania;ffiffl (ADE is approximate date of extirpation).

Species	Original distribution	ADE
Beaver (*Castor canadensis*)	Statewide	about 1900[2]
Wolf (*Canis lupus*)[1]	Statewide	1892
Wolverine (*Gulo gulo*)[1]	Northcentral PA	1870's
Marten (*Martes americana*)	Northern PA	1901
Fisher (*Martes pennanti*)	Central & northern PA	1921 or 1923
Mountain Lion (*Felis concolor*)[1]	Mountainous areas	1871[3]
Lynx (*Lynx canadensis*)[1]	Northern and central PA	1903
Moose (*Alces alces*)[1]	Northeastern PA	1790-1805
Wapiti (*Cervus elaphus*)	Statewide	1870's[4]
Bison (*Bison bison*)[1]	Western Pennsylvania	see note[5]

Notes: [1] permanently extirpated, not included in faunal list;
[2]reintroduced about 1917, now established and widespread;
[3]numerous reported sightings after this date are considered unconfirmed;
[4]reintroduced beginning 1913, small herd established;
[5]occurrence in Pennsylvania questionable.[24]

of mammals in New York and West Virginia (both 18 percent) and slightly higher than in other adjacent states. Again, the greater diversity of mammals in Pennsylvania is related to the mixture of boreal and austral species in a faunal transition zone.

HISTORICAL CHANGES IN DIVERSITY

Twenty-six species of terrestrial vertebrates have been extirpated from the Pennsylvania fauna, and two species that formerly were part of the fauna of the Commonwealth are extinct (Table 3). All of these extirpations and extinctions are the direct result of the European colonization of North America. Although the heavily wooded mountains of north-central Pennsylvania may appear pristine to the casual observer, even there three centuries of intense human activity have left a mark on the fauna. Some of the extirpated species were the preferred targets of subsistence or market hunters, and others were predators incompatible with a human population that depended on domestic animals for food. Many more were victims of either the pervasive habitat modification that accompanied the spread of European colonists, or the radical changes that resulted from the growth of the industrial economy.

The only extirpated amphibian species is the Tiger Salamander (*Ambystoma tigrinum*), which was last collected in Chester County around 1850. Breeding populations persist nearby in adjacent Delaware, Maryland, and New Jersey. Three turtle species no longer occur in Pennsylvania. The Smooth Softshell (*Apalone mutica*) is a riverine species that requires muddy substrate. It was adversely affected by changes in drainage patterns and water quality that accompanied navigation improvements, such as damming and dredging, on the Allegheny and Ohio rivers. Populations exist farther down the Ohio Drainage, so there is potential for immigration. Blanding's Turtle (*Emydoidea blandingi*), which occurred in Pymatuning Swamp and the Lake Erie shoreline, likely suffered from loss of marsh habitats in both areas. Currently, waif individuals regularly appear on Presque Isle, and the species could possibly become established there. The Common Mud Turtle (*Kinosternon subrubrum*), which occurred until recently in Tinicum Marsh on the edge of Philadelphia, has not been found in several years and may be extirpated.

The bird population of the Commonwealth has suffered the heaviest toll. The Passenger Pigeon (*Ectopistes migratorius*) and Carolina Parakeet (*Conuropsis carolinensis*), both of which formerly occurred in Pennsylvania, are extinct. An additional 12 species that still breed in good numbers elsewhere in North America, but no longer in Pennsylvania, are listed in Table 4, with the areas where these species formerly occurred, and the approximate date of extirpation.

TABLE 3

Extirpated or Extinct Species of Pennsylvania Amphibians, Reptiles, Birds and Mammals.

	Amphibians	Reptiles	Birds	Mammals	Total
Extirpated	1	3	12	10	26
Extinct	—	—	2	—	2

Direct persecution by humans was undoubtedly the major if not the only factor in the loss of the Passenger Pigeon and Carolina Parakeet. Human disturbance of nesting pairs was responsible for the cessation of breeding at Presque Isle of the Piping Plover and the two terns. The three duck species nested at Pymatuning for a few years immediately after closure of the dam that formed Pymatuning Lake, when abundant submerged aquatic vegetation was available.[21] There is no definitely known explanation for loss of the four songbirg species.[22]

Nine or possibly ten species of native mammals have been extirpated in Pennsylvania (Table 4), most of them disappearing before 1900 and one as early as 1805.[23] Of these, the Wolf (*Canis lupus*) and Mountain Lion (*Felis concolor*) were actively hunted and bountied to reduce depredations on domestic livestock. Moose (*Alces alces* probably were near the southern limit of their range in Pennsylvania, and may have occurred only sporadically in the Commonwealth. Wolverine (*Gulo gulo*), Fisher (*Martes pennanti*), Marten (*Martes americana*), and Lynx (*Lynx canadensis*) were also at southern range limits in Pennsylvania, and were probably extirpated when the mature forests of northern Pennsylvania were cutover and extensively burned in the mid- to late Ninteenth Century. The Marten has appeared sporadically in Pennsylvania since 1957, probably the result of immigration of individuals from New York. The Fisher has been reintroduced in both New York and West Virginia, and may eventually become re-established in Pennsylvania. The last native Beaver (*Castor canadensis*) in Pennsylvania were killed around 1900, after extensive trapping and logging continuing from earliest colonial times contributed to decades of decline. Wapiti (*Cervus elaphus*) were hunted throughout Pennsylvania, and the last native animals were eliminated in the 1870's. Both Beaver and Wapiti were reintroduced from non-Pennsylvania stocks, and have become established (Table 4). There is controversy over whether Bison (*Bison bison*) occurred naturally in Pennsylvania.[24] Perhaps the few historical records, which date

TABLE 4

Extirpated Breeding Birds of Pennsylvania. (ADE is approximate date of extirpation).

Species	Area	ADE	References
Common Loon (*Gavia immer*)	Monroe Co.	1955	(9)
Redhead (*Aythya americana*)	Pymatuning, Crawford Co.	1937[1]	(21)
Ring-neck Duck (*Aythya collaris*)	Pymatuning, Crawford Co.	1939[2]	(21)
Common Merganser (*Mergus merganser*)	Presque Isle, Erie Co.	?	
Ruddy Duck (*Oxyura jamaicensis*)	Pymatuning, Crawford Co.	1936[3]	(21)
Piping Plover (*Charadrius melodus*)	Presque Isle, Erie Co.	mid to late 1950's	(29)
Caspian Tern (*Sterna caspia*)	Presque Isle, Erie Co.	1924[4]	
Common Tern (*Sterna hirundo*)	Presque Isle, Erie Co.	1966	
Bewick's Wren (*Thryomanes beweickii*)	Southwestern PA	1977	
Loggerhead Shrike (*Lanius ludovicianus*)	Several counties	recent	(22)
Lark Sparrow (*Chondestes grammacus*)	Southwestern PA	1902[5]	(22)
Bachman's Sparrow (*Aimophila aestivalis*)	Southwestern PA	1930's	

Notes: [1]nested 1936 and 1937; [2]nested 1936-1939; [3]nested 1935 and 1936; [4]last breeding year; [5]breeding records 1897 and 1902.

to as late as about 1810, were based on wandering individuals or small groups rather than resident herds.

The habitat degradation and human interference that affected the 26 extirpated and two extinct species of Pennsylvania terrestrial vertebrates continue to affect remaining native species, albeit more subtly than during the initial waves of colonization when massive habitat destruction occurred. The results are most easily discernible in the case of species that are habitat specialists, and thus are normally patchily distributed. When a patch of habitat is altered or destroyed, the specialist species is faced with migration to another patch of suitable habitat, or extirpation. As patches of habitat become fewer and farther apart, extirpation is increasingly the result. There are many habitat specialist species in the Pennsylvania fauna that have suffered serious population declines because of habitat destruction. Among species that could be cited as examples are the Bog Turtle (*Clemmys muhlenbergi*),[25] and Massasauga (*Sistrurus catenatus*),[26] each of which requires wetlands of a specific type, and the Indian Bat (*Myotis sodalis*),[27] which requires undisturbed limestone caves for hibernating sites.

For terrestrial vertebrate species that have greater tolerance for habitat disturbance or broader ecological valences, populations may be steady or even increasing. For example, habitat loss for birds in Pennsylvania in the past half century has been minimal. On the contrary, forests once cleared for agriculture or clear-cut for timber have regenerated providing increased habitat for woodland species. There was not much natural open habitat in pre-colonial Pennsylvania. Therefore, the restoration of strip-mined habitats and extensive plantings of grasses have resulted in great local increases in grassland species such as Savannah (*Passerculus sandwichensis*), Grasshopper (*Ammodramus savannarum*), and Henslow's (*Passerherbulus henslowii*) sparrows, the Bobolink (*Dolichonyx oryzivorus*), and the Dickcissel (*Spiza americana*). An additional 16 species of birds have either extended their ranges into Pennsylvania, expanded their distribution in the Commonwealth, or become much more abundant over the past 30 to 50 years (Table 5). In some cases the reason for the increase in abundance is clearly related to human activities. For example, all raptors have increased in numbers since organochlorine pesticides and hawk hunting were prohibited, and Wild Turkey (*Meleagris gallopavo*) and Mourning Dove (*Zenaida macroura*) populations have responded to regulation of sport hunting. In other cases the relationship is not clear. Both the Carolina Chickadee (*Parus carolinensis*) and Blue-winged Warbler (*Vermivora pinus*) are expanding their ranges in Pennsylvania at the expense of related species. Such successional change could result from a competitive advantage conferred by habitat modification that favored one species over the other, or it could be entirely due to natural processes. Although water pollution remains a problem in some drainages, continued attention to pollution control should improve water quality throughout the Commonwealth to the benefit of bird species that depend on aquatic organisms for food.

REGIONAL DIVERSITY

Pennsylvania is a complex mosaic of physiographic diversity. However, we identified eight broadly-defined regions or provinces which were used to compare the physiographic distribution of terrestrial vertebrates: Southwestern Pennsylvania; Glaciated and Unglaciated Allegheny Plateau; Valley and Ridge Province; Pocono Plateau; Piedmont; Coastal Plain; Lake Erie Plain. These regions have greater relevance to the diversity of less vagile (non-bird) terrestrial vertebrates.

The modern distribution of amphibians and reptiles in Pennsylvania results from post-glacial northward dispersal from more southerly Pleistocene refugia. Seventeen species of amphibians and 11 species of reptiles are statewide in distribution, and occur in all of the physiographic regions. Seven species (one amphibian, six reptiles) reach their northern range limits within the boundaries of the Commonwealth,

TABLE 5

Species of Birds with Major Range or Abundance Increases in Pennsylvania.

Species	Change	Reason
Cooper's Hawk (*Accipiter cooperii*)	increase	protection of raptors
Red-tailed Hawk (*Buteo jamaicensis*)	increase	protection of raptors
Bald Eagle (*Haliaeetus leucocephalus*)	increase since 1950's	pesticide restrictions and protection of raptors
Osprey (*Pandion haliaetus*)	increase	pesticide restrictions and protection of raptors
Peregrine Falcon (*Falco peregrinus*)	increase	pesticide restrictions and protection of raptors
Wild Turkey *Meleagris gallopavo*	increase and range expansion	regulation of hunting
Mourning Dove (*Zenaida macroura*)	increase	regulation of hunting
Chuck-will's-widow (*Caprimulgus carolinensis*)	expansion into PA	?
Red-bellied woodpecker (*Melanerpes carolinus*)	northward expansion, increase	?
Blue Jay (*Cyanocitta cristata*)	increase	?
Carolina Chicadee (*parus carolinensis*)	eastward expansion[1]	?
Carolina Wren (*Thryothorus ludovicianus*)	increase	mild winters
Blue-winged Warbler (*Vermivora pinus*)	increase[2]	?
Yellow-throated Warbler (*Dendroica dominica*	range expansion and increase[3]	?
Summer Tanager (*Piranga rubra*)	range expansion and increase	?
Northern Cardinal (*Cardinalis cardinalis*)	range expansion and increase	?

Notes: [1]displacing Black-capped Chicadee (*P atricapillus*) in Pittsburgh region;
[2]displacing Golden-winged Warbler (*V. chrysoptera*) in Eastern Pennsylvania;
[3]now breeds in Southwest and Southeast Pennsylvania.

mostly at or south of glacial boundaries.[28] The uplands of central Pennsylvania (Valley and Ridge, Allegheny Plateaus, Pocono Plateau) form a significant barrier to dispersal of lowland amphibian and reptile species. Twenty-nine species have distributions that are influenced by this feature: five amphibians and seven reptiles are found both east and west of the uplands, but not at higher elevations. Seven amphibians and four reptiles occur only west of the central uplands, and four amphibians and two reptiles only to the east. Of the latter one amphibian and one reptile are found only in the Coastal Plain Province. Four species of amphibians occur in Pennsylvania only in the Southwestern Pennsylvania Province. One amphibian and three reptiles are restricted to upland regions. Four species of reptiles with prairie or midwestern affinities occur in the westernmost part of the Commonwealth, from the Lake Erie Plain southward. The Hellbender (*Cryptobranchus alleganiensis*), a completely aquatic salamander, occurs in rivers and large tributaries in both the Ohio and Susquehanna drainages.

Considered by physiographic region (Table 6), the amphibian fauna of Southwestern Pennsylvania is the most diverse (31 species), followed by the Valley and Redge Province (27) and the Piedmont (25). The least diverse amphibian faunas are found on the Pocono Plateau and the Lake Erie Plain (19 each).

The reptile fauna of the Glaciated Allegheny Plateau is the most diverse (28 species), followed by the Piedmont (25) and both Southwestern Pennsylvania and the Valley and Ridge Province (24 each). The unglaciated section of the Allegheny Plateau (15 species), and the Lake Erie Plain (14) have the fewest reptile species.

Except in a few places, Pennsylvania suffers from a near lack of natural lakes, swamps, and marshes, thus eliminating much avian diversity from a large part of the state. Water birds are found chiefly in the northwestern glaciated area, including the Lake Erie Plain, and what is left of the marshes in the southeast (such as Tinicum National Wildlife Refuge). Elsewhere water birds can be found along the state's principal rivers and on artificial lakes, but usually as migrants, not breeders. Pennsylvania does not have the range of altitudes found in New York and West Virginia, but there is nevertheless some altitudinal and habitat stratification of the avifauna. A few species are confined to higher areas or conifer-dominated woodlands, and the same species may be found in cold hemlock gorges. Pennsylvania definitely is in a zone of transition between northern and southern avifaunas, as illustrated by the wood warblers (Parulidae). Species for which Pennsylvania is at or near the

TABLE 6

Diversity of Pennsylvania Amphibians, Reptiles, and Mammals by Physiographic Region (1 = Southwestern Pennsylvania; 2 = Glaciated Allegheny Plateau; 3 = Unglaciated Allegheny Plateau; 4 = Valley and Ridge Province; 5 = Pocono Platuea; 6 = Piedmont; 7 = Coastal Plain; 8 = Lake Erie Plain).

Regions	1	2	3	4	5	6	7	8
Amphibians	31	23	22	27	19	25	20	19
Reptiles	24	28	15	24	18	25	20	14
Mammals	56	47	56	60	45	41	26	27

southern edge of the breeding range include the Nashville (*Vermivora ruficapilla*), Magnolia (*Dendroica magnolia*), Yellow-rumped ("Myrtle", *D. coronata*), and Mourning (*Oporornis philadelphia*) warblers, and the Northern Waterthrush (*Seiurus noveboracensis*). Southern species extending north to Pennsylvania and not much beyond include the Yellow-throated (*D. dominica*), Worm-eating (*Helmitheros vermivorus*), and Kentucky (*O. formosus*) warblers.

The greatest diversity of mammals is found in the Valley and Ridge Province (60 species), followed by Southwestern Pennsylvania (56) and the Unglaciated Allegheny Plateau (56). The lowest diversity is exhibited by the mammal faunas of the Coastal Plain (26) and Lake Erie Plain (27). There are a number of taxa with affinities for particular physiographic provinces. For example, the Maryland Shrew (*Sorex cinereus fontinalis*) is limited to the Piedmont and Valley and Ridge provinces. The Indiana Bat (*Myotis sodalis*) is limited to Southwestern Pennsylvania and the Valley and Ridge Province. The distribution of the Eastern Spotted Skunk (*Spilogale putorius*) in Pennsylvania is restricted to the southern border of the Valley and Ridge Province, which represents the northern limit of its range. Finally, the Marsh Rice Rat (*Oryzomys palustris*), which reaches the northern limit of geographical distribution in Pennsylvania, is restricted to the Coastal Plain.

In contrast to these a number of species are, in appropriate habitats, widespread and occur in all physiographic regions of the Commonwealth. These include: Short-tailed shrew (*Blarina brevicauda*), White-footed Mouse (*Peromyscus leucopus*), Meadow Vole (*Microtus pennsylvanicus*), Eastern Chipmunk (*Tamias striatus*), Gray Squirrel (*Sciurus carolinensis*), and Striped Skunk (*Mephitis mephitis*). Finally, two ubiquitous and opportunistic species, Virginia Opossum (*Didelphis virginiana*) and Raccoon (*Procyon lotor*), merit special mention for their extraordinary versatility and adaptive success throughout Pennsylvania despite the disturbance created by human activity.

CONCLUSIONS

1. The terrestrial vertebrate fauna of Pennsylvania includes 36 species of amphibians, 37 reptiles, 197 breeding birds, and 68 mammals. The diversity of amphibians and reptiles is similar to that of adjacent states. The diversities of upland birds and mammals are relatively greater, as Pennsylvania is in a transition zone between southern and northern faunas and has elements of each.

2. Twenty-six species of native terrestrial vertebrates (1 amphibian, 3 reptiles, 12 birds, 10 mammals) have been either temporarily or permanently extirpated from the Pennsylvania fauna since colonial times, and two bird species that occurred in the Commonwealth are extinct. Most of these extirpations and extinctions are the direct result of human activities, or habitat modifications that accompanied European colonization of North America. Habitat destruction continues to reduce populations of some species, in particular habitat specialists. Some habitat modifications, such as increased acreage of grasslands, have favored some species and their populations have increased.

3. The distribution of terrestrial vertebrates in Pennsylvania is generally correlated with broad physiographic and ecological regions. A number of species are restricted in their distributions within the Commonwealth to one or a few such regions. The greatest diversity is found in the physiographic regions with the greatest variety of habitat types.

LITERATURE CITED

1. Duellman, W.E. 1979. The numbers of amphibians and reptiles. Herp. Review, 10:83-84.
2. Collins, J.T. 1990. Standard common and current scientific names for North American amphibians and reptiles, 3rd ed. SSAR Herp. Circular (19):1-41.
3. McCoy, C.J. 1982. Amphibians and reptiles in Pennsylvania: Checklist, bibliography, and atlas of distribution. Carnegie Mus. Nat. Hist., Spec. Publ. 6, 91 pp.
4. Conant, R. and J.T. Collins. 1991. A field guide to amphibians and reptiles, eastern and central North America, 3rd. ed. Houghton Mifflin, Boston, xx + 450 pp.
5. Mayr, E. and D. Amadon. 1951. A classification of Recent birds. Amer. Mus. Novitates, (1496):1-42.
6. Sibley, C.G. and B.L. Monroe, Jr. 1990. Distribution and taxonomy of the birds of the world. Yale University Press, New Haven, xxiv + 1, 111 pp.
7. American Ornithologists' Union. 1957. Check-list of North American birds, 5th ed. A.O.U., Washington, D.C., xiv + 691 pp.
8. Peterson, R.T. 1963. The birds. Life Nature Library. Time, Inc., New York, 192 pp.
9. Wood, M. 1979. Birds of Pennsylvania. Pennsylvania State Univ., University Park, xvi + 140 pp.
10. Peterjohn, B.G. 1989. The birds of Ohio. Indiana Univ. Press, Bloomington, xviii + 237 pp.
11. Peterjohn, B.G. and D.L. Rice. 1991. The Ohio breeding bird atlas. Ohio Dept. Natural Res., Columbus, xiv + 415 pp.
12. Hall, G.A. 1983. West Virginia birds. Carnegie Mus. Nat. Hist., Spec. Publ. 7, viii + 180 pp.
13. Hilson, D.E. and D.M. Reeder. 1993. Mammal species of the world. Smithsonian Institution Press, Washington and London, xviii + 1206 pp.
14. Jones, J.K., D.C. Carter, H.H. Genoways, R.S. Hoffman, D.W. Rice and C. Jones. 1986. Revised checklist of North American mammals north of Mexico. Occas. Pap. Mus. Texas Tech Univ., (107):1-22.
15. Kirkland, G.L. and P.M. Krim. 1990. Survey of the statuses of the mammals of Pennsylvania. J. PA Acad. Sci., 64:33-45.
16. Paradiso, J.L. 1969. Mammals of Maryland. N. Amer. Fauna, (66):i-iv + 1-193.
17. Van Gelder, R.G. 1984. The mammals of the state of New Jersey: A preliminary annotated list. New Jersey Audubon Soc. Occas. Pap., (143):1-20.

18. Hall, R.E. 1981. The mammals of North America. John Wiley and Sons, New York, xv + 1, 181 pp.
19. Gottschang, J.L. 1981. A guide to the mammals of Ohio. Ohio State Univ. Press, Columbus, xi + 176 pp.
20. Schlitter, D.A., personal communication.
21. Grimm, W.C. 1952. Birds of the Pymatuning region. Pennsylvania Game Comm., Harrisburg, vi + 226 pp.
22. Todd, W.E.C. 1940. Birds of western Pennsylvania. Univ. Pittsburgh Press, Pittsburgh, xvi + 710 pp.
23. Williams, S.L., S.B. McLaren and M.A. Burgwin. 1985. Paleoarchaeological and historical records of selected Pennsyvlania mammals. Ann. Carnegie Mus., 54:77-188.
24. Guilday, J.E. 1963. Evidence for buffalo in prehistoric Pennsylvania. PA Archaeol., 33:135-139.
25. Bury, R.B. 1979. Review of the ecology and conservation of the Bog turtle, *Clemmys muhlenbergii.* U.S. Fish and Wildlife Serv., Resource Publ., (219):1-9.
26. Reinert, H.K. 1985. Eastern massasauga, *Sistrurus c. catenatus,* pp. 273-275. In, H.H. Genoways and F.J. Brenner (Eds.), *Species of Special Concern in Pennsylvania.* Carnegie Mus. Nat. Hist., Spec. Publ. 11, 430 pp.
27. Hall, J.S. 1985. Indiana Bat, *Myotis sodalis,* pp. 357-357. In, H.H. Genoways and F.J. Brenner (Eds.), *Species of Special Concern in Pennsylvania.* Carnegie Mus. Nat. Hist., Spec. Publ. 11, 430 pp.
28. Smith, A.G. 1945. *Agkistrodon mokeson mokeson* (Daudin) in Pennsylvania. Proc. PA Acad. Sci., 19:69-72.
29. Stull, J., J.A. Stull and G.M. McWilliams. 1985. Birds of Erie County, Pennsylvania. Allegheny Press, Elgin, PA, 173 pp.

Biological Diversity: Problems and Challenges. Edited by S.K. Majumdar, F.J. Brenner, J.E. Lovich, J.F. Schalles, and E.W. Miller. © 1994, The Pennsylvania Academy of Science.

Chapter Twenty-Six

DIVERSITY STATUS OF TERRESTRIAL INSECTS

THEODORE BURK
Department of Biology
Creighton University
Omaha, Nebraska 68178-0103

INTRODUCTION

Most animals are insects. Although the exact number of insect species is still very uncertain, and the number of individual insects is probably impossible to calculate, the information we do have leads to the invevitable conclusion that insects are the dominant animals of the earth's terrestrial ecosystems. Over 50% of all species of organisms, over 70% of animal species, and over 90% of terrestrial animal species are insect species (May 1978, Wilson 1988). E.O. Wilson, the ant specialist, estimates that all of the individuals in just the 10,000 species of ants comprise 10^{15} living individuals at any given time (Wilson 1988). Insects comprise over 40% of the animal biomass in the Amazonian rain forest, with ants and termites accounting for 75% of the insect biomass; there can be as many as eight million ants and one million termites per hectare. Wilson once identified 43 species of ants, from 26 genera, from a *single tree* in Peru, roughly equalling the species diversity of all of Britain or Iowa (Holldobler & Wilson 1990). Thus is it obvious that any consideration of biological diversity must examine the incredible variety and abundance of terrestrial insects. In this chapter, I begin by considering just how many insect species there are. Next, various suggestions as to why insects are so incredibly diverse are examined. Threats to insect diversity are considered in the next section. Finally, after quickly restating the reasons that biodiversity, particularly insect biodiversity, should be conserved and treasured, I present a number of ideas for steps that could be taken to preserve insect biodiversity.

THE EXTENT OF INSECT DIVERSITY

How many insect species are there? To be brief, no one really knows; in fact, until the past decade few systematic attempts have been made to find out. Since a stimulating paper by Erwin in 1982, however, there has been a lively discussion of just how many species of organisms, especially insects, are currently alive. This discussion has been recently and clearly summarized by May (1990, 1992).

In his 10th edition of *Systema Naturae* that established the binomial nomenclature system, Linnaeus described 2,102 species of insects (Metcalf 1940). With the work of the great Victorian naturalists and others, the known number of insect species grew tremendously, and by the middle of the 20th century, a rough consensus as to the number of species had developed. In 1940, Metcalf estimated there were about 1.5 million insect species; Sabrosky (1952) estimated 686,000 insect species were named, and the final total seemed "sure to be somewhere in the millions." Southwood (1978) estimated there were just over a million insect species described, while Arnett (1985) arrived at the number of 750,000 described species of insects. While clearly the numbers varied, most estimates came up with figures of 750,000 to 1.5 million known species, and 3-5 million in all - described and undescribed.

Erwin (1982) published a brief paper that suggested those estimates might be off by an order of magnitude or more. Using what has become a standard technique, insecticidal "knock-down" fogs applied to whole trees, Erwin collected about 1200 species of beetles from the canopies of 19 individuals of one species of tree in Brazil. He then used a variety of assumptions to extrapolate from this number to obtain an estimate of the total number of insect species. He estimated that 160 of the beetles were host-specific to that species of tree; that beetles comprise 40% of all known arthropod species; that there are two insect species in the tropical forest canopy for every species elsewhere in the forest; that there are about 50,000 species of tropical forest trees with similar specific endowments of insects. This produced an estimate of about 30 million tropical forest arthropods. Although Erwin stopped at this point, one could continue the extrapolation - the tropics probably contain 1/2 to 2/3 of all species and insects comprise a conservative 80% of arthropods. This would give an estimate for global insect biodiversity of about 36-48 million species.

Many papers in the last decade have examined Erwin's assumptions (Stork 1988, Thomas 1990, Adis 1990, Gaston 1991, May 1986, 1988, 1990, 1992). He may have overestimated the percentage of specialist beetle species and the number of tree "species groups" (his 50,000 number) - lowering these would lower his final total. However, he may have underestimated the proportion of species that live elsewhere than in the forest canopy, and overestimated the percentage of insects that are beetles - adjusting these would raise his total, possibly to as much as 80 million!

May (1990, 1992) has recently provided excellent review of Erwin's estimate and how it is affected by changing the various assumptions. He has also reviewed a number of other techniques developed by others for independently arriving at an estimate of insect species diversity. One technique is to extrapolate from past and current rates of discovery of new species. Most birds and mammals are known by now (only a handful of new species of each is described each year), and half of currently known species had already been described by the middle of the 19th century. By contrast, half of the currently recognized species of insects had not been discovered until 1925, and 5-10,000 new species are still being named every year. Using standard statistical techniques, but plugging in the variable estimates of known species, extrapolation from these numbers gives estimates of 4-7 million total insect species (however, there may be a sharp increase in number of newly described species soon, given the increased recent emphasis on exploration of the mostly unknown tropical insect diversity).

Another technique, used by Raven (1985), assumes that birds and mammals are essentially all known, and works from the fact that for those groups there are about two tropical species for every temperate one. We can further assume that temperate insects are pretty well known (some scientists might dispute that). For insects, 2/3 of known species are from the temperate zone; extrapolating from the known number of temperate species, based on the premise that these are really only 1/3 of the total, gives an estimate of 3-5 million total insect species.

Another method discussed in May's papers is that of Hodkinson & Casson (1991), who have worked on bug (hemipteran) species on the tropical island of Sulawesi in Indonesia. Using very intensive sampling methods, they have collected 1690 species of bugs, of which 63% were previously undescribed (over 90% of Erwin's Brazilian beetles were undescribed). If the Sulawesi bugs are typical of tropical insects generally, then we could say that the approximately 1 million currently described insect species represent 37% of the real total, giving an estimated real total of 2.7 million species. Another way to work from the bug data is as follows: they found 1700 bugs on 500 species of trees; if there are 50,000 tropical tree species, and if bugs are 7.5% of all insect species, then we can arrive at an estimate of 2.3 million total insect species.

May himself has developed a method for estimating species numbers based on species-size relationships. As one goes down in animal linear body dimensions from a few meters to about a cm, an empirical rule becomes apparent that for each tenfold reduction in length (or 1000-fold reduction in weight), there are 100 times the species. The curve holds very well for the size ranges where animal groups are well known. May suggests that if the observed pattern is extrapolated down to animals with characteristic lengths of as little as 0.2 mm, the total number of terrestrial animal species would be about 10 million, with insect species accounting for perhaps about 7 million. As May (1990) frankly admits, however, one would have more confidence in these estimates based on species-size relationships "if we had a better understanding of the physiological, ecological or evolutionary factors generating species-size distributions," which at the moment are not much more than poorly understood empirical generalizations.

Unfortunately, I cannot conclude this section by telling you with any confidence how many species of insects there are. However, since the renewal of interest in the issue after the paper by Erwin, most "moderate" estimates have increased from a previous estimated total in the range of 2-3 million to current estimates of around 10 million (Stork 1988, Gaston 1991). What is perhaps most remarkable is that we are so ignorant, not only of how many insects there are, but even of the number that have already been named and described. As May (1988) points out, it is certainly not beyond our capabilities to have a complete, centralized archive and data base - the Library of Congress has no trouble keeping track of millions of books!

CAUSES OF INSECT DIVERSITY

The spectacular diversity of insects provokes speculation on its causes from almost everyone who writes about them. A general biology book in front of me (Curtis and Barnes 1989) attributes the success of insects to their exoskeleton (which made possible the many specialized appendages of insects), small size, highly specialized ecological niches, association with a great diversity of microenvironments provided by vascular plants, complete metamorphosis, and well developed nervous system. Interestingly, many of these rather general explanations ("just so stories"?) are also represented among the reasons for insect diversity provided by ecologists and entomologists who have considered the question more deeply.

Recent ecological theory on species diversity can in a sense be said to date to the seminal paper "Homage to Santa Rosalia, or why are there so many kinds of animals?," by G.E. Hutchinson (1959). Hutchinson concluded that the extremely high diversity of insects resulted from the interaction of three factors: (1) their small size, which given their association with plants, allowed them to take advantage of the many ecological niches provided by (2) the complex architectural diversity of higher plants, with (3) interspecific competition acting to prevent niche overlap and lead to many specialized phytophagous insects occupying small specialized niches.

A very influential paper bearing on the question of insect diversification was Ehrlich and Raven's (1964) "Butterflies and plants: a study in coevolution." As pointed out by Strong et al. (1984), up to half of all species of organisms are higher plants or phytophagous insects, and although less than one-third of insect orders can be described as "significantly phytophagous," all of the four large insect orders fit that description (Coleoptera, Diptera, Lepidoptera, and Hymenoptera). Ehrlich and Raven suggested that these facts were not coincidence, but reflected the reciprocal evolutionary influence of the two groups on each other. In fact, the now common term "coevolution" was coined by Ehrlich and Raven to designate the reciprocal evolutionary interaction between specific groups of organisms such as butterflies and dicotyledonous plants. They noted that phytophagy by insects has promoted the evolution of secondary plant defence chemicals to deter herbivory. A plant that evolves new chemical defenses is able to "enter a new adaptive zone," free of generalist insect herbivore pressure, which allows subsequent adaptive radiation by that plant group. However, should an insect species evolve metabolic/physiological or other mechanisms for overcoming the new plant chemical defenses, it breaks through into a new adaptive zone itself. It and its descendent species can adaptively radiate onto the different species of plants whose chemical barrier it has pierced. Important to the Ehrlich/Raven (hereafter E/R) hypothesis is the presumption that "metabolic adjustments" made by insects to specialize on one group of plants and one set of plant chemicals render those insects inefficient at utilizing other plant groups.

Ehrlich and Raven also pointed out that mechanical defenses of plants may sometimes act in an analogous manner in setting up new adaptive zones, and that incorporation of plant defensive chemicals into insect bodies for defense against the insects' natural enemies promotes species diversity by opening up "Batesian mimicry niches" for edible insects that mimic the appearance of chemically protected insects.

E/R suggest that co-evolution may promote increased species diversity in contexts other than plants and phytophagous insects. For example, there are at least several hundred thousand insects that are parasitoids of other insect species. Perhaps piercing host defenses opens up new adaptive radiation possibilities for insect parasitoid groups: "food chains comprising green plants, insect herbivores, and insect parasitoids include over half of all known species of [non-microbe] organisms" (Hawkins & Lawton 1987). Even if parasitoid groups are not as tightly linked to host species groups as are phytophagous insect groups to plant groups, the proliferation of phytophagous insects nevertheless makes available a large number of new host niches for specialist insect parasitoids to evolve to fill. As Vane-Wright (1978) has argued, even without highly specific coevolutionary linkages, "diversity promotes diversity" by creating new ecological niches generally. Finally, E/R note that many insects have "invented new niches" by becoming symbiotic associates with other insects, as for example in the case of lycaenid ("blue") butterfly larvae living as partners or predators inside ant nests. Recent studies based on modern cladistic taxonomic methods have provided generally strong support for the E/R model (Mitter et al. 1988, Farrell et al. 1992), which has come to be called the "escape and radiation model."

The E/R plant-insect coevolution model has been very influential, but some critics have supported other explanations for insect species diversity, either to supplement the coevolution model or to replace it. Gilbert and Smiley (1978) suggested that species diversity in phytophagous insects would be promoted not just by plant chemistry differences, but also by the evolution of morphological differences among plants to make them less similar in appearance to each other and therefore harder to recognize by host-seeking insects. This concept is known as the "limiting similarity hypothesis." Similarly, Futuyma and Moreno (1988) suggested that speciation may have been promoted in plants and insects via phytophagous insect chemical-based host-location behavior. Many phytophagous insect species locate their hosts via their typical olfactory profile, and plants may "escape" from insects not so much by evolving new poisons as by evolving different "bouquets." However, if insects happen by chance to discover the new, differently smelling hosts, new specialized adaptive zones again become available. Since many phytophagous insects "imprint" onto the odors of the hosts they were reared on, races of an insect species that come to attack new hosts can become isolated on that host and eventually become new species distinct from their ancestral populations on the traditional hosts. The "limiting similarity hypothesis" mentioned above can be applied one step higher in food chains, too: species diversity in phytophagous insects may be favored as a way to limit the success of their natural enemies' attacks.

Bernays (1991), citing Southwood (1978), pointed out that phytophagous insects have to deal with both a plant's chemistry and the problem of remaining "attached" to the host plant. She suggested that, in response, different species have evolved different morphological specializations that allow them to live on, under, or in plants. These, analogous to adaptations for overcoming plant chemical defenses, could act as entryways into new adaptive zones, resulting in adaptive radiation and increased number of species.

Otte (1989), focusing specifically on Hawaiian insect diversity, stressed not plant diversity but general habitat diversity as a key factor in insect speciation and diversification. In Hawaii, water barriers between islands, lava fields, gulches, isolated plateau and valley habitats, lava tubes, and other factors have split apart or isolated insect populations and promoted their diversification into new species. Factors such as these help explain why, for example, Hawaii has twice as many cricket species as the continental United States. Other factors that may promote insect diversity in Hawaii may apply more generally. One of these is sexual selection, in which isolated populations of species develop unique sexual displays which ultimately cause each population to become a reproductively isolated species (Ringo 1977). Another is the founder effect, in which newly established peripheral populations of a species become genetically distinct due to the non-random gene pool carried by their founders, combined with random genetic drift experienced by such isolated, small populations (Carson 1971).

Several studies have been published that examine the number of different insect species associated with various species of plants (for example, Lawton 1978). The two major factors that seem to be important are geographic range and plant form. Plant species with large ranges are host to more phytophagous insects (although this is over the whole range - they may not be attacked by more insects in a given location), while large and long-lived plants (trees) are hosts to more insect species than other plants. Lawton explains the first finding by suggesting that as plants increase their range, they are colonized by insects present in the local "species pool" of the newly occupied area. These insect colonizers then adapt to and become specialists on the newly-appearing plant host, at the same time becoming new species distinct from the ancestral insect species in the origin species pool. The plant form effect is basically due to the increased "architectural complexity" of a large plant such as a tree. Insects find a greater variety of resources and a larger number of specialized locations to live on, under, or in from a large tree. "Most insects exploit very restricted parts of their hosts" (Lawton 1978). Also, the seasonal changes in resources (buds, young leaves, older leaves, flowers, fruit, seeds, etc.) of large, perennial plants provide new specialized niches that promote insect diversity. In addition, natural enemies may promote the evolution of small, localized, specialized niches, as insects attempt to move out of exposed locations in search of "enemy free spaces" in plants. Studies of insect parasitoids show that the number of parasitoids attacking an insect species is strongly affected by where the insect lives - for example, species that live inside roots and stems have fewer parasitoid species than ones that live on or in leaves (Hawkins & Lawton 1987). It occurs to me that the chemical coevolution and plant architecture ideas may be "fruitfully" combined - overcoming a class of plant chemical defenses opens up new adaptive zones for an insect group, but the plant architecture provided by the new plant "victim" may ultimately determine how many insect species will be able to radiate inside that adaptive zone.

The two most comprehensive attempts to explain insect diversity are those of Southwood (1978) and May (1978). Southwood argued that insects, due to their small size, are able to survive on specialized resources that would not support larger species of animals; their environment is "patchier" than that of other organisms, and such factors as interspecific resource competition and the evolutionary search for "enemy free space" promote speciation and specialization on small niche "patches." Physiological evolution of diapause allows different species of insects to specialize on seasonally different resources. Diel specialization of different species may also occur, although much less commonly. Southwood suggests that, even when a particular resource niche isn't sufficient for a species to specialize on it alone, insects are adept at utilizing "fractional niches," via polymorphism within species and - most importantly - because of the evolution of complete metamorphosis. Eighty-eight percent of insects are endopterygotes (insects with complete metamorphosis) and Southwood argues that one of the major reasons for the success of endopterygote insects is the capacity that metamorphosis gives an insect species to take advantage of different niches in different locations at different times over the course of a single life history. From two inadequate niches, an insect with complete metamorphosis can patch together a successful single life history. Evolution of flight, unique among the invertebrates, allows insects to rapidly colonize new habitats and establishes insects in isolated peripheral situations where genetic drift, natural selection, and sexual selection bring about rapid speciation. According to Southwood, insects have high "spread potential" and are "continually probing the environment, 'throwing together' new combinations of insect and habitat." Such phenomena help explain, for example, why Hawaii has 6500 endemic species of insects (99% of the insects native to Hawaii), coming from only an estimated 250 original immigrant species. Another factor that is important in some groups, especially the Hymenoptera and Isoptera (ants, bees, wasps, and termites) is the evolution of advanced social colonies, which allows some insect species to "live like animals many times the size of insects"; in other words to fill ecological roles equivalent to large vertebrate herbivores or carnivores. Southwood succinctly summarizes what he feels are the major keys to insect species diversity: "size, metamorphosis, and wings."

May (1978), like Southwood, stresses small size and dispersal ability as keys to the insects' success. Small size, with concomitant short generation times and high reproductive potential, allows them to be highly successful "fugitive species" that can "escape" from superior competitors into new and different niches. Short generation times, combined with high mortality rates and large populations, accelerate insect evolution, compared to many of the plants and "higher animals" with whom they are coevolving. Dispersal abilities lead to insect species often being composed of many scattered subpopulations, each with the potential to evolve into distinct species.

In sum, there is no single explanation for the amazing biodiversity of insects. Undoubtedly important, however, are their flight capability, which gets them into new habitats and breaks them into isolated subpopulations; small size, which gives them the ability to occupy small specialized niches; metamorphosis and diapause, which allow them to live on fractional or seasonal niche bases; and high rates of evolution, which allow them to coevolve successfully with higher plants, with each other in the form of interspecific competitors and parasitoids, and with a wide range of non-insect natural enemies. Their segmented, exoskeletal body with its paired appendages and their highly developed nervous system provide malleable material upon which evolution can act in conjunction with the factors listed above, producing a marvelous range of species with elaborate and specialized morphology, physiology, and behavior.

THREATS TO INSECT DIVERSITY

The threats to insect diversity are in essence the same as the threats to biodiversity of other organisms, and ultimately trace themselves to the large and ever-increasing human population. The consequences for insects of the world's present human population of over 5 billion, and its projected human population of 12 billion a century from now, take a variety of specific manifestations.

As with so much else about insect biodiversity, no one knows how many insect species have already gone extinct in the recent past due to human activities, although a number of species known to be extinct can be mentioned, such as the Xerces blue butterfly and the Rocky Mountain locust (Pyle et al. 1981). Likewise, because of the prevailing underappreciation of the significance of insects to natural ecosystems - among biological professionals as well as the public at large, definite information as to the number and identity of seriously threatened or endangered species of insects is sparse.

The U.S. Fish and Wildlife Service recently (1989) published a list of extinct, endangered, and threatened U.S. invertebrates; that list named 83 U.S. species of insects as extinct and another 92 as not having been recorded in the past 25 years. Currently, 17 insect species (12 of which are lepidopterans) are listed under the U.S. Endangered Species Act, with another 504 listed as possibly qualifying for listing (Opler 1991). The total of listed and candidate species amounts to less than 1% of U.S. insect species (88,000), but that low percentage probably reflects nothing more than general ignorance about the status of most insect populations. In Europe, 96 of 380 species of butterflies are considered to be threatened with extinction (Boussienguet 1991). In Britain, about 14% of insect species are listed as threatened in the British Red Data Book for Insects (Foster 1991). In Brazil, only about 1000 of the 3130 butterfly species are considered to be well enough studied that their population status can be evaluated. Of those, about 10% are considered threatened with extinction, with 23 present on Brazil's endangered species list (Brown 1991). As the above numbers indicate, only in places such as Europe and North America or only for conspicuous groups of insects such as butterflies, is even barely adequate information available. For other groups and in other places, it is very difficult to say how many insect species have recently gone extinct or are threatened with extinction in the near future. Overall, the International Union for the Conservation of Nature (UCN) *1990 Red List of Threatened Animals*, which lists 1047 birds (about 1/8 of all bird species), lists 1220 insects (if there are 10 million insect species, that amounts to about 0.01% of insect species). Over 1/3 of the insects listed are lepidopterans, another 30% are beetles; by contrast, less than 2% of the insect species listed are flies, although Diptera approximately equals Lepidoptera in number of species and there is no reason to believe that dipteran species are less threatened than lepidopteran species.

Many different estimates of the rate of species extinction can be found; these are listed and evaluated by Lugo (1988). Since adequate data to arrive at reasonable estimates of current rate of species losses are unavailable, it may be better to lay out some long-term prospects. One of the more robust generalizations to emerge from the field of island biogeography is the "90-50 rule" (Wilson 1985), which states that for a 90% reduction in habitat area, there is usually a 50% loss of species. If we assume that 2/3 of 9 million insect species live in the tropics, and that ultimately about 90% of tropical rain forests will be lost in the next half century (Sutton and Collins 1991), then we might predict 3 million insect species will go extinct. Lugo (1988) points out that this scenario may be overly pessimistic. In Puerto Rico, 99% of the original forest was removed, but coffee plantations and second growth forests maintained at least a 15% forest cover and many species were able to persist in these alternative habitat types. About 11% of native birds went extinct. If that percentage is used, then perhaps as "few" as 700,000 insect species extinctions will occur. Whatever the number, a biological tragedy of immense proportions is possible if not likely.

The threats to insect diversity are many and varied. Most significant is undoubtedly habitat loss and degradation. Loss of primary forests - tropical and temperate - by logging or conversion to agricultural land, spreading urbanization, mining, water impoundment, draining of wetlands, groundwater extraction, movement away from traditional agricultural practices (such as maintenance of hedgerows) to intensive modern agriculture, construction of roads and "cleaning" of roadside verges, and fragmentation of large natural areas into a scattered mosaic of small natural patches all have significant effects on insect populations (Hafernik 1992, Pyle et al. 1981, New 1984).

Pollution of various kinds threatens insect species. Loss of temperate forests to acid rain, other forms of air pollution, and effects of water pollution on terrestrial insects that have aquatic larvae are all significant. Especially worrisome are the possible results of global warming due to an enhanced greenhouse effect over the next century (Greenslade and New 1991). Significant geographic shifts in the earth's climatic regions are expected, with changes occurring at such high rates that vegetational communities may be unable to migrate quickly enough to keep up. The result is likely to be significant reduction in area of many habitat types and possibly extinction of many plant species. This will cause "extinction chains" spreading down to insect species dependent on those habitats and plants for survival.

Introduction of alien species around the world presents another serious threat to native insect species; according to Howarth and Ramsey (1991), "the world's biota is being homogenized." When chestnut blight almost eliminated American chestnuts, five species of moths whose larvae were chestnut specialists were driven to extinction (Howarth and Ramsey 1991). Much of the threat to Hawaii's endemic species of insects comes from the 2500 now - established alien invertebrate species (Howarth and Ramsey 1991). Aliens compete with native species, are predators upon them, eat native plants that native species are dependent upon, are sometimes toxic to native species, spread diseases, alter the litter and soil community and disrupt nutrient cycling, and provoke control measures that could drastically affect native species. Aerial spraying to eradicate Mediterranean fruit flies threatened a number of local endemic species in the San Francisco Bay area in the early 1980's, and contemplated control projects against Hawaii's introduced tephritid fruit fly pests could have a devastating impact on native insect species. In some cases, introduction of an alien host leads a native insect to become an agricultural pest; Colorado potato beetles were an innocuous consumer of native nightshade species until potatoes were introduced to them - now they are a significant worldwide pest. One-fourth of Hawaii's established alien insects were deliberately introduced, and a point worth keeping in mind is that biocontrol introductions can have devastating unforeseen effects on non-target species. In Hawaii, extinction of 15 species of native moths can be attributed to introduced biocontrol agents; 5 of these species were the intended victims, but 10 others were innocent third parties. Despite the widespread impression to the contrary, biological control programs can indeed pose ecological hazards. Of course, the prospect of the widespread release of genetically engineered biological control agents, from microbes to insects, poses another potential threat to native species, and the possible effects of such releases should be extremely carefully researched before any are conducted.

Other threats to insect species exist. Recreational vehicles can destroy sensitive habitats such as sand dunes, host to many highly localized endemic species (New 1984). Profusion of electric lights may endanger some species that are attracted to them.

Two factors that, perhaps rather surprisingly, cannot be unquestionably blamed for the extinction of any insect species are insecticides and overcollection (Hafernik 1992, Pyle et al. 1981). It is perhaps wise to be guarded in the first case, however, as failure to wipe out a carefully monitored pest population during a control program does not necessarily mean that some unmonitored or perhaps even scientifically unknown species was not eliminated in ignorance. Also, as severe stressors, pesticides could alter competitive outcomes and predator/prey relationships, or reduce populations to such a small size that random extinction becomes more likely. In the case of collecting, concern about overcollecting of certain colorful and conspicuous species, such as the birdwing butterflies of Papua New Guinea and the Apollo butterfly of Europe, has led to their listing among the banned species of the Convention on the International Trade in Endangered Species (CITES). A number of conservation and scientific organizations have developed guidelines for ecologically responsible collecting, a leading example being the guidelines of the Joint Committee for the Conservation of British Insects (JCCBI) (New 1984). To repeat, however, it is not believed that current collecting practices are a serious threat to any insect species.

VALUE OF INSECT BIODIVERSITY

The average citizen of the world may feel that the loss of a few, or even many, insect species poses no serious problem; in fact, it may even be desirable. A major task of ecologists and educators is to convey to the general public (and to non-entomologist professionals!) the importance of preserving insect biodiversity.

Insects provide us with a number of crucial "goods and services" (Morris et al. 1991), and should be preserved for their ecological, economic, intellectual, and aesthetic value (Pyle et al. 1981).

As already mentioned, in most of the world's natural terrestrial ecosystems, insects are significant if not dominant herbivores, carnivores, and macro-decomposers. They are the most significant group of pollinators of flowering plants, both wild and domesticated. They quite probably are much more important for the maintenance of ecosystems than are vertebrates; they are "the little things that run the world" (Wilson 1987). They can serve us as sensitive monitors of ecological deterioration, they provide extensive natural pest control, they constitute a still largely untapped source of species for biological control of pest weeds and insects - especially introduced alien pests, they may possess countless varieties of useful pharmaceutical chemicals, and they act as the key natural force driving the evolution in plants of the natural biologically-active chemicals and pest-resistance genes that are the key resources for the modern biotechnology industry (Morris et al. 1991).

Insects are sometimes economically important "livestock," as when mantids, ladybugs, and exotic butterflies are collected and sold commercially. "Butterfly ranching" is now a significant agricultural practice in Papua New Guinea (New 1984). They are important "deadstock" as preserved specimens in a growing trade. Wild insects provide a few commercial products (although most insect products - honey, silk, lac, cochineal, beeswax - come from domesticated species). They are an important food source in many parts of the world, and perhaps in a world with an ever-increasing protein-short human population, should become a much more important one.

Insects have provided and should continue to provide paradigm examples of natural phenomena for study by biologists. Where would genetics, embryology, or modern eukaryotic molecular biology be without *Drosophila*; the study of evolution without Hawaiian *Drosophila* or Mullerian and Batesian mimic butterflies; or the study of animal behavior without the social insects? In all parts of modern biology, from molecular biology to community ecology, natural communities of insects provide key material for the intellectual pursuits that distinguish humans as a species.

Finally, insects, both in their own right and as a part of the natural whole, have great aesthetic and ethical value. Insects provide intricate, miniature beauty, of a type that is perhaps better appreciated in Oriental cultures than in Western ones, but whose appreciation adds immensely to the lives of those attuned to it. And it will be a crucial test of the moral worth of the human species whether or not it can develop the ethical attitude that all parts of the natural world are worthwhile for their own sake, and not just for their economic value to greedy, self-centered and short-sighted humans.

STEPS TO PRESERVE INSECT BIODIVERSITY

The threats to insect biodiversity are many; however, so are the steps that are available for the preservation of insect biodiversity. In this section, just a few of those possible steps are listed, from ones requiring the mobilization of large-scale movements to others than can be carried out by anyone in their own back yard.

Undoubtedly a highest priority is ensurance of suitable habitats. Additional reserves need to be set aside. A feeling of some conservationists is that if large reserves are established in areas which will preserve particular large vertebrate species, those reserves will automatically preserve plants, insects and other invertebrates, too. Unfortunately, such an "umbrella approach" (Sutton and Collins 1991) fails to take into account the fact that the distribution of endangered insects does not necessarily overlap that of endangered birds and mammals (Hafernik 1992). Critical habitats need to be identified and set aside with insects in mind. In some ways, preservation of insects may be more cost efficient than preservation of larger organisms. Insect habitats often consist of small patches, and while it may take many square kilometers of reserve to save a vertebrate, a few hectares may be sufficient for some endangered insects. Quality of small habitat patches may be more crucial than quantity of reserve (Ehrlich 1992). However, as always, the larger the individual reserve, the better. Ehrlich and Murphy (1987) suggest that the amount of necessary habitat should be estimated, then a reserve twice as big be set aside. Ehrlich (1992) suggests that vertebrate-sized reserves be created, but that they be located in areas with the diversity of habitat patches that will preserve the greatest number of invertebrates within them. Reserves also need to be managed in such a way as to ensure a continued diversity of habitat patches; as Hafernik (1992) says, reserve management shouldn't be "strictly for the birds."

Preservation of natural habitats on islands, with their large number of vulnerable endemics, needs to be a high priority (Howarth and Ramsey 1991). Also, given that the world of the future is, at best, going to consist of a mosaic of primary and secondary vegetational communities, intermingled with agricultural and urban lands, ways to promote species diversity in such mosaic landscapes need to be developed. Small scale habitat patches should be encouraged - roadsides, field edges, small ponds and wetlands, flower-rich grasslands, suburban gardens and parks (Usher and Jefferson 1991). Preservation of species on private lands, via such incentives as conservation easements, should be encouraged. Rehabilitation or restoration of severely degraded habitats, focus of the dynamic new field of restoration ecology, should be carried out with maximum insect biodiversity as a prime objective (Jordan et al. 1987).

More interventionist steps also can be taken. Captive breeding programs for insects should be more widespread and, when carefully done, reintroductions of locally-extirpated species can be carried out, as in the example of the reintroduction of the large copper butterfly from Holland to Britain (Pyle et al. 1981).

A second major priority in the preservation of insect diversity is an increase in our knowledge base. We need to overcome the "taxonomic impediment" of not knowing how many and what species are out there (Howarth and Ramsey 1991). Top priority needs to be given to faunal surveys; however, if there are as many as 9 million unidentified insect species to be found and catalogued, things cannot be done at the traditional pace. As Wilson (1985) has put it, to just describe and name that many species using the traditional methods would take the complete lifetimes of 25,000 taxonomists. Therefore, we need to carry out "quick and dirty" surveys (Ehrlich 1992) that allow us to discover species and identify crucial and endangered habitats quickly. Simple, standard methods, to be carried out by basically-trained "barefoot taxonomists" need to be developed (Sutton and Collins 1992); regional "sorting centers" linked by computer databases should be established (Yen and Butcher 1992). Regional atlases, based on a grid system as used in Europe, need to be developed for the rest of the world (Opler 1991). Selected insect groups should be identified as "indicator groups" and intensively surveyed to identify key habitats for preservation; Sutton and Collins (1991) suggest butterflies, odonates, sphinx moths, katydids, and dung beetles, and perhaps also termites, leafhoppers, ground beetles, and pyralid moths.

Important, but less urgent than the quick and dirty studies just mentioned, is the continuing need for rigorous, long-term research on specific populations and demographic factors affecting them, on the role of insects in ecosystem functioning, and so on (Ehrlich 1992).

A third major priority area for preservation of insect biodiversity is education. This has a number of aspects. First, the general public's attitudes toward insects need to be reshaped; they need to be informed about the crucial role of insects in natural ecosystems and the enjoyment that can be derived from the study and appreciation of the insect diversity around them. The ignorance of insects and the cultural bias in Western societies for large conspicuous species (what Ehrlich (1992) calls "charismatic megafauna") needs to be replaced by knowledge about and interest in other creatures. Second, private and governmental conservation agencies need to be better informed about the key role of insects, so as to take insects into account more fully in agency activities. The disparity mentioned earlier between our knowledge about endangered birds and mammals versus our ignorance about endangered plants and invertebrates reflects not so much a problem in the general public as a problem among professional biologists. Third, the critical shortage of taxonomists needs to be remedied. A recent "taxonomy of taxonomists" (Gaston and May 1992) showed that the average insect species receives about 1% as much attention by professional biologists as the average vertebrate, that there are proportionately about 1/50 as many insect taxonomists as vertebrate taxonomists, that the taxonomists who exist are overwhelmingly located in the temperate zones far away from the center of insect diversity, and that the taxonomic community has the top-heavy age structure of a waning profession. In fact, Gaston and May say that if a species were discovered with the age structure of the taxonomic profession, it would immediately be designated as endangered!) Fryxell (1985) has called for a rejuvenation of the teaching of botanical and zoological groups in schools and colleges, not from the old perspective of memorizing the names of lots of different taxa, but from the perspective that such knowledge is critical for the future generation of biologists and educated laypersons striving to preserve the earth's biodiversity. Finally, students and amateur naturalists need to be educated to carry out their vitally necessary part in the accumulation of information about faunal identity, distribution, and abundance.

Government and private agencies have lots to do, too. Although many useful steps are under way (as reviewed by Pyle et al. 1981), much more could be done. New Zealand has a Department of Conservation (Howarth and Ramsey 1991), and so should other countries. In the United States, proposals exist for establishment of a National Institutes of the Environment, a National Biodiversity Center, and a National Biological Survey (Pennisi 1991, Tangley 1985). Yen and Butcher (1982) suggest that every state or province should have a person in charge of invertebrate conservation. Agencies such as the U.S. Fish and Wildlife Service should be more adequately staffed, so that endangered species can be identified more quickly and added to endangered species lists in less than the many years that process now takes (Opler 1991). As many countries as posible should enact comprehensive legislation similar to the U.S. Endangered Species Act of 1973. This act, for all its limitations, is still probably the most effective piece of conservation legislation anywhere in the world. Much better coordination is needed among the government and private organizations involved in the environment, including the alphabet soup of international agencies (IUCN, UNEP, WRI, WWF, FAO, UNESCO, etc.). Scientific and conservation organizations should run joint programs, and there should be involvement of farmers' and growers' organizations as well. Programs designed to maintain agricultural land and keep farmers on the land should be encouraged, and governments should alter agricultural subsidy programs to encourage preservation of traditional agricultural practices that were more favorable to wildlife than intensive fencerow to fencerow practices (Erhardt and Thomas 1991).

Governments and private conservation organizations should encourage ventures aimed at utilizing insects commercially. "Chemical prospecting" (Eisner 1983) should be encouraged, as well as creation of more enterprises like the "butterfly ranches" of New Guinea that encourage the reproduction of saturniid moths and birdwing butterflies for collection and sale. Butterfly ranchers even have their own government marketing board (Morris et al. 1991).

Governments should also screen more closely the introduction of alien species, as well as exercise close regulation of releases of genetically-engineered species into the general environment. Traditional biocontrol programs should take more account of possible effects on non-target species.

Finally, following the dictum of "thinking globally, but acting locally," individuals need to get increasingly involved in preservation of insect biodiversity. Concerned citizens should join local, national, and international conservation organizations, including especially those devoted to insect conservation (a good list can be found in Xerces Society/Smithsonian Institution [1990]). They should keep in constant communication with their local and national government representatives, encouraging positive legislation and executive action. They should get involved in the collection of survey material - obtaining the kind of faunal survey and distributional data needed will only be possible if an army of amateurs gets involved, as has traditionally been the situation in ornithology. They should patronize insect zoos, butterfly houses, and the like, both to help promote a friendlier attitude toward insects and to encourage the growth of these insect-utilizing commercial enterprises. They should create insect-attracting natural patches in their own backyards and encourage their employers and local governments to do so. New (1984) relates the story of an English woman who, over the course of 11 years, recorded 1226 insect species in her urban garden. Above all, they should be willing to get involved in a steady, persistent way over the long haul, because although failure at biodiversity protection could come quickly, success at biodiversity protection will be a long-term effort.

CONCLUSION

Although the exact number of living insect species is not known - it may be around ten million - insects probably comprise 90% of terrestrial animal species. Their ecological importance matches their diversity. The major causes of this tremendous insect diversity are probably their small size and concomitant small specialized niches, their coevolution with higher plants, their evolution of flight, and their evolution of complete metamorphosis. Insect diversity is threatened by many factors, including especially loss and degradation of habitat, introduction of alien species, extinction chains in which insects become extinct as a consequence of disappearance of their host species, and pollution - especially in the form of greenhouse gases and acid precipitation. Nevertheless, insects have enormous ecological, economic, intellectual, and ethical value to humans, and insect biodiversity should be maintained. Preservation and restoration of habitat, increase in our knowledge of insect diversity and the threats to it, positive actions by governmental and private organizations, and increased activity by concerned citizens at local, national, and international levels can all help to preserve the diversity of the species that, more than any other, are the earth's representative organisms.

REFERENCES

Adis, J. 1990. Thirty million arthropod species - too many or too few? *J. Tropical Ecol.* 6:115-118.

Arnett, R.H., Jr. 1985. *American Insects: a Handbook of the Insects of America North of Mexico.* Van Nostrand Reinhold Company, New York, pp. 850.

Bernays, E.A. 1991. Evolution of insect morphology in relation to plants. *Phil. Trans. R. Soc. Lond.* B 333:257-264.

Boussienguet, J. 1991. Problems in assessment of biodiversity, pp. 31-35. In: D.L. Hawksworth (Ed.) *The Biodiversity of Microorganisms and Invertebrates: Its Role in Sustainable Agriculture.* CAB International, Wallingford, England, pp. 302.

Brown, K.S., Jr. 1991. Conservation of neotropical environments: insects as indicators, pp. 349-404. In: N.M. Collins and J.A. Thomas (Ed.) *The Conservation of Insects and Their Habitats.* Academic Press, London, pp. 388.

Carson, H.L. 1971. Speciation and the founder principle. *Stadler Symp.* 3:51-70.

Curtis, H. and N.S. Barnes. 1989. *Biology.* Worth Publishers, Inc., New York, pp. 1192.

Eisner, T. 1983. Chemicals, genes, and the loss of species. *Nat. Cons. News* 33(6):23-24.

Erhardt, A. and J.A. Thomas. 1991. Lepidoptera as indicators of change in the semi-natural grasslands of lowland and upland Europe, pp. 213-236. In: N.M. Collins and J.A. Thomas (Ed.) *The Conservation of Insects and Their Habitats.* Academic Press, London, pp. 388.

Ehrlich, P.R. 1992. Population biology of checkerspot butterflies and the preservation of global biodiversity. *Oikos* 63:6-12.
Ehrlich, P.R. and D.D. Murphy. 1987. Conservation lessons from long-term studies of checkerspot butterflies. *Conserv. Biol.* 1:122-131.
Ehrlich, P.R. and P.H. Raven. 1964. Butterflies and plants: a study in coevolution. *Evolution* 18:586-608.
Erwin, T.L. 1982. Tropical forests: their richness in Coleoptera and other arthropod species. *Coleopt. Bull.* 36:74-75.
Farrell, B.D., C. Mitter and D.J. Futuyma. 1992. Diversification at the insect-plant interface: insights from phylogenetics. *Bioscience* 42:34-42.
Foster, G.N. 1991. Conserving insects of aquatic and wetland habitats, with special reference to beetles, pp. 237-262. In: N.M. Collins and J.A. Thomas (Ed.) *The Conservation of Insects and Their Habitats.* Academic Press, London, pp. 388.
Fryxell, P.A. 1985. Up with diversity. *Bioscience* 35:735-737.
Futuyma, D.J. and G. Moreno. 1988. The evolution of ecological specialization. *Ann. Rev. Ecol. Syst.* 19:207-233.
Gaston, K.J. 1991. The magnitude of global insect species richness. *Conserv. Biol.* 5:283-296.
Gaston, K.J. and R.M. May. 1992. Taxonomy of taxonomists. *Nature.* 356:281-282.
Gilbert, L.E. and J.T. Smiley. 1978. Determinants of local diversity in phytophagous insects: host specialists in tropical environments, pp. 89-104. In: L.A. Mound and N. Waloff (Ed.) *Diversity of Insect Faunas.* Blackwell Scientific Publications, Oxford, England, pp. 204.
Greenslade, P. and T.R. New. 1991. Australia: conservation of a continental island fauna, pp. 33-70. In: N.M. Collins and J.A. Thomas (Ed.) *The Conservation of Insects and Their Habitats.* Academic Press, London, pp. 388.
Hafernik, J.E., Jr. 1992. Threats to invertebrate biodiversity: implications for conservation strategies, pp. 171-195. In: P.L. Fielder and S.K. Jain (Ed.) *Conservation Biology: the Theory and Practice of Nature Conservation, Preservation and Management.* Routledge, Chapman, and Hall, Chapman, New York, pp. 507.
Hawkins, B.A. and J.H. Lawton. 1987. Species richness for parasitoids of British phytophagous insects. *Nature* 326:788-790.
Hodkinson, I.D. and D. Casson. 1991. A lesser predilection for bugs: Hemiptera (Insecta) diversity in tropical rain forests. *Biol. J. Linn. Soc. Lond.* 43:101-110.
Holldobler, B. and E.O. Wilson. 1990. *The Ants.* Harvard University Press, Cambridge, Massachusetts, pp. 732.
Howarth, F.G. and G.W. Ramsay. 1991. The conservation of island insects and their habitats, pp. 71-107. In: N.M. Collins and J.A. Thomas (Ed.) *The Conservation of Insects and Their Habitats.* Academic Press, London, pp. 388.
Hutchinson, G.E. 1959. Homage to Santa Rosalia, or why are there so many kinds of animals? *Am. Nat.* 93:145-159.
International Union for the Conservation of Nature. 1990. *The 1990 IUCN Red List of Threatened Animals.* IUCN, Gland, Switzerland, pp. 192.
Jordan, W.R. III, M.E. Gilpin and J.D. Aber (Ed.) 1987. *Restoration Ecology: A Synthetic Approach to Ecological Restoration.* Cambridge University Press, Cambridge, England, pp. 342.

Lawton, J.H. 1978. Host-plant influences on insect diversity: the effects of space and time, pp. 105-125. In: L.A. Mound and N. Waloff (Ed.) *Diversity of Insect Faunas.* Blackwell Scientific Publications, Oxford, England, pp. 204.

Lugo, A.E. 1988. Estimating reductions in the diversity of tropical forest species, p. 58-70. In: E.O. Wilson and F.M. Peter (Ed.) *Biodiversity.* National Academy Press, Washington, D.C., pp. 538.

May, R.M. 1978. The dynamics and diversity of insect faunas, pp. 188-204. In: L.A. Mound and N. Waloff (Ed.) *Diversity of Insect Faunas.* Blackwell Scientific Publications, Oxford, England, pp. 204.

May, R. 1986. How many species are there? *Nature* 324:514-515.

May, R. 1988. How many species are there on Earth? *Science* 241:1441-1449.

May, R.M. 1990. How many species? *Phil. Trans. R. Soc. Lond.* B 330:293-304.

May, R.M. 1992. How many species inhabit the earth? *Sci. Am.* 267(4):42-48.

Metcalf, Z.P. 1940. How many insects in the world? *Entomol. News* 51:219-222.

Mitter, C., B. Farrell and B. Wiegmann. 1988. The phylogenetic study of adaptive zones: has phytophagy promoted insect diversification? *Am. Nat.* 132:107-128.

Morris, M.G., N.M. Collins, R.I. Vane-Wright and J. Waage. 1991. The utilization and value of non-domesticated insects, pp. 319-347. In: N.M. Collins and J.A. Thomas (Ed.) *The Conservation of Insects and Their Habitats.* Academic Press, London, pp. 388.

New, T.R. 1984. *Insect Conservation: An Australian Perspective.* Dr. W. Junk Publishers, Dordrecht, Netherlands, pp. 184.

Opler, P.A. 1991. North American problems and perspectives in insect conservation, pp. 9-32. In: N.M. Collins and J.A. Thomas (Ed.) *Conservation of Insects and Their Habitats.* Academic Press, London, pp. 388.

Otte, D. 1989. Speciation in Hawaiian crickets, pp. 482-526. In: D. Otte and J.A. Endler (Ed.) *Speciation and its Consequences.* Sinauer Associates, Inc., Sunderland, Massachusetts, pp. 769.

Pennisi, E. 1991. Biodiversity rides a popular wave. *Scientist* 5(8):1,8-10.

Pyle, R., M. Bentzien and P. Opler. 1981. Insect conservation. *Ann. Rev. Entomol.* 26:233-258.

Raven, P.H. 1985. Disappearing species: a global tragedy. *Futurist* 19:8-14.

Ringo, J.M. 1977. Why 300 species of Hawaiian *Drosophila*? The sexual selection hypothesis. *Evolution* 31:694-696.

Sabrosky, C.W. 1952. How many insects are there? pp. 1-7. In: United States Department of Agriculture. *Insects: the Yearbook of Agriculture. 1952.* United States Government Printing Office, Washington, D.C., pp. 780.

Southwood, T.R. 1978. The components of diversity, pp. 19-40. In: L.A. Mound and N. Waloff (Ed.) *Diversity of Insect Faunas.* Blackwell Scientific Publications, Oxford, England, pp. 204.

Stork, N.E. 1988. Insect diversity: facts, fiction and speculation. *Biol. J. Linn. Soc. Lond.* 35:321-337.

Strong, D.R., J.H. Lawton and R. Southwood. 1984. *Insects on Plants: Community Patterns and Mechanisms.* Harvard University Press, Cambridge, Massachusetts, pp. 313.

Sutton, S.L. and N.M. Collins. 1991. Insects and tropical forest conservation, pp. 405-424. In: N.M. Collins and J.A. Thomas (Ed.) *The Conservation of Insects and Their Habitats.* Academic Press, London, pp. 388.

Tangley, L. 1985. A national biological survey. *Bioscience* 35:686-690.

Thomas, C.D. 1990. Fewer species. *Nature* 347: 237.

U.S. Fish and Wildlife Service. 1989. Endangered and threatened wildlife and plants: annual notice of review. *Fed. Reg.* 54:554-579.

Usher, M.B. and R.G. Jefferson. 1991. Creating new and successional habitats for arthropods, pp. 263-291. In: N.M. Collins and J.A. Thomas (Ed.) *The Conservation of Insects and Their Habitats.* Academic Press, London, pp. 388.

Vane-Wright, R.I. 1978. Ecological and behavioural origins of diversity in butterflies, pp. 58-70. In: L.A. Mound and N. Waloff (Ed.) *Diversity of Insect Faunas.* Blackwell Scientific Publications, Oxford, England, pp. 204.

Wilson, E.O. 1985. The biological diversity crisis. *Bioscience* 35:700-706.

Wilson, E.O. 1987. The little things that run the world (the importance and conservation of invertebrates). *Conserv. Biol.* 1:344-346.

Wilson, E.O. 1988. The current state of biological diversity, pp. 3-18. In: E.O. Wilson and F.M. Peter (Ed.) *Biodiversity.* National Academy Press, Washington, D.C., pp. 538.

Xerces Society/Smithsonian Institution. 1990. *Butterfly Gardening: Creating Summer Magic in Your Garden.* Whitecap Books, Vancouver, Canada, pp. 192.

Yen, A.L. and R.J. Butcher. 1992. Practical conservation of non-marine invertebrates. *Search* 23:103-105.

Biological Diversity: Problems and Challenges. Edited by S.K. Majumdar, F.J. Brenner, J.E. Lovich, J.F. Schalles, and E.W. Miller.

Chapter Twenty-Seven

DIVERSITY STATUS OF AQUATIC INSECTS

MEL ZIMMERMAN
Department of Biology
Lycoming College
Williamsport, PA 17701

INTRODUCTION

Biological diversity is being rapidly lost through extinction of species or whole populations and human alteration of ecosystems, in all parts of the world, especially in the tropics (Wilson, 1988). Global interdependence, both environmental and economic, makes loss of biodiversity anywhere in the world a matter of concern for everyone. This critique will examine the biodiversity status of aquatic insects. Estimates of the total number of insect species in the world vary widely and range from 750,000 (Wilson, 1988) to 30 million (Wolf, 1987). Burk (see chapter 26) reviews the various hypotheses which has lead to this wide range of estimates. Even less exact figures are known on the aquatic insect orders (see Table 1). Cheng (1976) estimated that the number of aquatic insects is about 25,000 - 30,000 world wide (based on 3% of the total estimated species of insects at that time). Lehmkuhl (1979) projected 7% of North American insects as aquatic or semiaquatic (based on a total of 91,000 species of insects in North America). Not only is information for an accurate count unavailable, the definition of an aquatic insect also is debated by authors (Lehmkuhl, 1979; Hilsenhoff, 1991; Resh and Rosenberg, 1984).

STATUS OF AQUATIC INSECT TAXONOMY

Insects that live on or in water come from a remarkably diverse assemblage of 13 orders of the arthropod class Insecta (see Table 1). One of these orders, Collembola, may no longer be considered an insect and is placed by some in the class Entognatha (see Hilsenhoff, 1991). Nine of the thirteen orders have terrestrial adults with aquatic immature stages. The remaining three orders can have all life history stages in the water. Only five of the orders are considered completely aquatic with almost all species having aquatic larvae (Ephemeroptera, Odonata, Plecoptera, Trichoptera and Megaloptera). Of this group, the Trichoptera (caddisflies) show the greatest number of species. MacKay and Wiggins (1979) view this diversity as an expression of ecological opportunities made possible by the secretion of silk. By using silk to fashion variety of shelters and other devices, larval Trichoptera have evolved adaptive systems to exploit a wide range of resources. Seven of the remaining orders (Heteroptera, Coleoptera, Diptera, Lepidoptera, Neroptera, Hymenoptera and

TABLE 1

Orders Containing Aquatic Insects.
Adopted from Hilsenhoff (1991); Lehmkuhl (1979); Rosmer (1981).

		North America North of Mexico		World
Order	Aquatic Stage (excluding egg)	Approximate Number of Aquatic Species in North America	Non-aquatic Species * (exceptions occur)	Approximate Total Number Species (all habitats)
Ephemeroptera (Mayflies)	Nymphs	575	None	2,100
Odonata (Damselflies & Dragonflies)	Nyphs	415	None	5,000
Plecoptera (Stoneflies)	Nymphs	550	None	1,900
Megaloptera (Dobsonflies & Alderflies)	Larvae	40	None	750
Trichoptera (Caddisflies)	Larvae	1,340	* None	10,000
Hemiptera (True Bugs)	All	400	4,600	60,000
Neuroptera (Lacewings)	Larvae	5	300	4,000
Coleoptera (Beetles)	All	1,100	30,000 total	300,000
Lepidoptera (Moths)	Larvae	50	10,000 total	175,000
Collembola (Springtails)	All	96	700 total	3,500
Orthoptera (Grasshoppers & Crickets)	All	5	1,000	12,000
Diptera (True Flies)	Larvae	10% of total?	17,000 total	90,000
Hymenoptera (Wasps)	Larvae, parasites	0.5# of total?	17,000 total	120,000
		approx. 7,500 total		

Orthoptera) are partially aquatic, in which most species are terrestrial but there are species or entire families that have one or more life stages adopted for living in the aquatic environment. In this latter group, the aquatic Diptera, especially the Chironomidae, have evolved mechanisms to exploit environments with low dissolved oxygen. The remaining order, Collembola, is semiaquatic with some mostly moist terrestrial species occupying the surfaces of aquatic environments. The aquatic members of these orders have been extremely successful in occupying all types of freshwater environments ranging from shallow or deep lentic systems, slow or fast lotic environments, as well as surface films, tree holes, hyporheic zones, temporary or seasonal aquatic habitats, and seep areas (see review by Resh and Rosenberg, 1984). A few occupy salt lakes, thermal springs or brackish water. Still fewer are considered truly marine (mainly a few species of Hemiptera, Diptera, Coleoptera and Tricoptera). Discussions of why insects failed to dominate the seas can be found in Hinton (1976), Hynes (1984) and Lehmkuhl (1979).

Hynes (1984) explains that taxonomic analysis of aquatic insects began in Western Europe about a century ago. In fact, work is so complete that it is possible to map out the broad distributional patterns of most species, especially in well studied areas at the British Isles, France, Scandinavia and Germany. However, a continuous trickle of new species continues to be described particularly in such difficult groups as Baetidae (Ephemeroptera) and Chironomidae (Diptera). Hynes (1984) further explains that the study of aquatic insects in Europe has not only been favored by a concentrated population of interested entomologists, but also by the fact that it has a more restricted species diversity compared to other areas. This is in part due to the severe change associated with the Pleistocene glaciations. This resulted in a much more extensive area of severe climate than occurred, for example, in North America or Northern Asia (Hynes, 1984).

Ward and Whipple (1918) attempted the first comprehensive work on aquatic insects in North America. This was later followed by Pennak (1953, 1978). However, only in the last fifteen years have several major works been developed (Merritt and Cummins 1978, 1984; McCafferty, 1981; Hilsenhoff, 1991). Some excellent regional keys exist for northeastern North America (Peckarsky et al., 1990) southeastern United States (Brigham et al., 1982) and a classic work on California by Usinger (1956). The most recent comprehensive descriptions of North American genera of mayflies, stoneflies and caddisflies are by Edmunds et al. (1976), Stewart et al. (1988) and Wiggins (1977), respectively. Even with all this work, species identification is limited and new genera and species are constantly being found.

Central and South America continue to receive attention on certain groups (Ephemeroptera, Plecoptera, Trichoptera and Coleoptera) (see MacKay and Wiggins, 1979; Brittain, 1982; Brown, 1987; and Flowers, 1991). However, to date, no large scale systematic approaches have occurred. The same can be said for parts of Asia, Africa and Australia (see Obeng, 1981; Awachie, 1981 and Williams, 1981). Winterboum and Gregson (1989) recently published a guide to the aquatic insects of New Zealand. As with many other groups of organisms, species richness appears greatest for aquatic insects in the tropics than in many temperate zone areas of the world (see Stout and Vandermeer, 1975). However, Illies (1989) found tropical stonefly fauna to be surprisingly depauperate (i.e., 68 species in South America versus 340 in Europe) even though rhithron streams can be found throughout the middle latitudes. Furthermore, McElravy et al. (1981) observed that although the caddisfly fauna of Panama was more diverse than most comparative sites in the Holarctic region, species richness was similar to streams in Pennsylvania and South Carolina. After reviewing the literature, Stanford and Ward (1983) conclude that: 1) species diversity in tropical stream insects may well be largely influenced by patterns of environmental variability and periodic, unpredictable disturbance in the same manner as for temperature regions, and 2) that diversity of all or most lotic forms may actually be lower in the tropics compared to optimal habitats in more northerly latitudes. Clearly, although we have come a long way in our understanding of the taxonomy of aquatic insects, an enormous amount of work remains to be done, especially in the tropics. Rapid development in our understanding of the ecology of aquatic insects has also occurred.

THE ECOLOGY OF AQUATIC INSECTS

Aquatic insects are ubiquitous in freshwater and function as major converters of energy (Benke, 1984). They do so as primary aquatic consumers/herbivores (Lamberti and Moore, 1984), detritus processors (Anderson and Sedell, 1979), and predators (Peckarsky, 1984). They are important in the cycling of nutrients and nutrient spiraling (Merritt and Cummins, 1984; Meyer, 1990), as fish food organisms (Healey, 1984) and as food for diving birds while aquatic larva and terrestrial birds when transformed to winged adults (see review in Hynes, 1984). The development (see outline in Merritt and Cummins, 1978) of functional feeding groups for aquatic insects (shredders, collectors, scrapers, predators) has lead to a better understanding of the aquatic food web. It is not uncommon for shredders, which break apart and consume coarse particulate organic matter (CPOM $>$ 1 mm), to comprise more than 30% of the invertebrate density in forested streams. Food web changes occur in different habitats (see Meyer, 1990) and following perturbations (see Cummins et al., 1989). The trophic status of a lake will drastically affect the diversity, distribution, and abundance of its insect fauna (see Wiederholm, 1984). Insects in streams are influenced by allochthonous inputs of energy associated with stream width, discharge, gradient, type and density of riparian vegetation. These factors change in a predictable sequence as stream order increases, and the aquatic insect fauna generally reflect changes in these factors (see Merrit et al., 1984). The role of aquatic insects as the principle benthic converters of the photosynthetic production of plants into food for fish is undoubtedly among the major economic benefits to mankind that inland waters provide.

One of the most fascinating characteristics of the aquatic insects as a group is their diverse and often intriguing colonization behavior (see review by Wiley and Kohler, 1984). As larva, many aquatic insects are transported down stream by current as "drift." In this way new areas can be colonized. Waters (1972) classified drift into three categories: 1) catastrophic (pulsed), 2) behavioral (periodic), and

3) constant (continuous). As adults, most aquatic insects have winged forms which allow them to fly and colonize new areas. This is a distinct advantage in dispersal that the majority of other aquatic invertebrates lack (Sweeney, 1984). Insects exhibit a wide diversity of life history patterns in aquatic habitats (see Butler, 1984; Sweeney, 1984).

The diversity of aquatic insects and their often close adaptation to fairly limited ecological niches makes them excellent indicators of water quality (Wiederholm, 1984). Several biotic indicies utilizing aquatic insects as monitors of water quality have been developed (see Chutter, 1972; Hilsenhoff, 1982, 1988). The ecological respones to many human activities that impact on surface waters are generally related to three basic physico-chemical and biological changes:

(i) increase in area, depth and volume of a water body (for example, dam construction),
(ii) decrease in the area, depth and volume (for example, drainage or channelization projects), and
(iii) changes in water quality which may be traceable to a variety of pollution inputs (oxygen demanding, organic wastes, heavy metal, synthetic chemicals including biocides).

Numerous articles and books have dealt with the subject of aquatic pollution and its biological effects. However, as the field of pollution ecology has broadened, it has become increasingly more difficult to find the necessary width and depth of coverage in treatments of this subject in a single volume. Hynes (1960) is probably still the best single work on the biology of polluted waters. More recent reviews are by Welch (1980), Widerholm (1984) and Ford (1989). Organic pollution generally reduces the number of insect species; however, since some species benefit from the increased food supply or have adaptations to survive in low oxygen, the total density of organisms may increase (Hilsenhoff, 1987). While a number of studies document reduced species diversity of aquatic insects as a result of acidification (Hall et al., 1980; Otto and Svensson, 1983; Weiderholm, 1984 and Ford, 1989), others show increases in biomass (Nilssen, 1980, Henrikson and Oscarson, 1981). MacKay and Kersey (1985) found that total abundance of benthic invertebrates increases with decreasing pH in a number of Canadian streams because of increased representation of acid tolerant chironomid species. A variable in the understanding of the effects of heavy metals on aquatic insects appears to be whether the stress occurred continuously or intermittently. Sheehan and Winner (1984) studied two streams receiving continuous additions of copper and found good correlations between mean macroinvertebrate abundance and copper concentrations. However, in a third stream receiving periodic inputs of copper and other metals, no correlation between concentration and abundance was shown. The authors attributed the lack of correlation to insect avoidance of pollution events by burrowing into the sediments and avoiding the stress.

While research on the input of these perturbations on aquatic insects has and will continue to be important, perspectives based on isolated components (i.e. pH affects on . . . , temperature effect on . . . , substrate affects on . . . , pesticide effects on . . ., etc.) will remain ecologically incomplete. Holistic views of the importance of aquatic insects in the dynamics of freshwater systems are being developed. The river continuum concept first proposed by Vannote et al., (1980) has led to a broader view of streams and their food webs. Most recently Gregory et al., (1991) advocated that an expanded focus of research investigating the links between land and water should occur. The riparian zones are the interfaces (ecotones) between terrestrial and aquatic ecosystems. These zones are not easily delineated but are composed of mosaics of landforms, communities and environments within the larger landscape. What humans do to the landscape (urbanization, deforestation, etc.) will have a direct affect on the aquatic system's web of life.

Additional Impacts of Aquatic Insects and Man

Hynes (1984) outlines a number of ways aquatic insects have impinged upon humans. This ranges from severe nuisance conditions and/or allergic reactions caused by mass emergences of midges, caddisflies and/or mayflies in many parts of the world (see Beadle, 1974; Fremling, 1960 a,b) to the human suffering caused by biting flies. Three of the important families of blood-sucking Diptera, the Culicidae (mosquitoes), the Simuliidae (blackflies) and the Tabanidae (horseflies) have aquatic larvae and are of great medical and veterinary importance (see Schmidt and Roberts, 1989). Mosquitoes are known transmitters or vectors for a wide range of diseases or infections, such as malaria, yellow fever, dengue, encephalitis, and the filarial worms that cause elephantiasis, to name a few. Similarly, blackflies and horseflies also transmit filarial diseases (such as onchocerciasis or river blindness caused by the nematode *Onchocerca* and carried by the blackfly *Simulium* in Africa and South America). It is economically, and also environmentally, important to know just where a species breeds so that control measures can be applied precisely, and with minimal effects on other organisms.

CONCLUDING REMARKS

It is impossible to consider the use of freshwater for any purpose involving quality, fisheries, recreation, aesthetics, agriculture, aquaculture, and tropical diseases without soon confronting aquatic insect topics. The biodiversity dilemma for aquatic insects is the same as for other species - their biological importance is often a mystery and their potential value to society an open question. Since distinct populations, especially invertebrates, have not even been delimitated, especialy in the tropics, it is impossible to protect each individual species. Conservation of biological diversity has long been seen as a matter of creating parks and reserves free from human interference. However, Wolf (1987) explains, park designation is a static solution to a dynamic problem and is not itself enough to avert mass extinction. Clearly, it is difficult, and maybe impossible, to protect biological communities from assaults carried by air, water or from the effects of rapid climatic change. Furthermore, perhaps as much as 1.3 billion hectares (Wilson, 1988) would have to be set aside to conserve representative samples of the earth's ecosystems. Recently, Ehrlich and

Ehrlich (1992) suggest that the United States lead the world in formulating an "Endangered Community" or "Endangered Humanity" Act. This act would deem as law the maintenance of the maximum number of relatively undisturbed natural ecosystems in as pristine a condition as possible. A world version of this act would also have to address human population control.

LITERATURE CITED

Anderson, N.H. and J.R. Sedell. 1979. Detritus processing by macroinvertebrates in stream ecosystems. *Annu. Rev. Entomol.* 24:351-377.

Awachie, J.B.B. 1981. Running water ecology in Africa. pp. 339-366. In: Lock, M.A., and D.D. Williams. (Ed.). *Perspectives in Running Water Ecology.* Plenum Press, New York, pp. 430.

Beadle, L.C. 1974. *The Inland Waters of Tropical Africa.* Longman Group LTD., London, pp. 365.

Benke, A.C. 1984. Secondary production of aquatic insects. pp. 289-322. In: Resh, V.H., and D.M. Rosenberg (Ed.). *The Ecology of Aquatic Insects.* Praeger, New York, pp. 625.

Brigham, A.R., W.U. Brigham, A. Gnilka, (Ed.). 1981. *The Aquatic Insects and Oligochaetes of North and South Carolina.* Midwest Aquatic Enterprises. Mahomet, Illinois, pp. 837.

Brittain, J.E. 1982. Biology of mayflies. *Annu. Rev. Entomol.* 27:119-147.

Brown, H.P. 1987. Biology of riffle beetles. *Annu. Rev. Entomol.* 32:253-273.

Butler, M.G. 1984. Life histories of aquatic insects. pp. 24-55. In: Resh, V.H., and B.M. Rosenberg (Ed.). *The Ecology of Aquatic Insects.* New York, pp. 625.

Cheng, L. (Ed.). 1976. *Marine Insects.* American Elsevier Pub. Co., New York, pp. 581.

Chutter, F.M. 1972. An empirical biotic index of the quality of water in south African streams and rivers. *Water Research.* 6:19-30.

Cummins, K.W., M.A. Wilzbach, D.M. Gates, J.B. Perry and W.B. Taliaferro. 1989. Shredders and riparian vegetation. *BioScience.* 39:24-30.

Edmunds, G.F., S.L. Jensen and L. Berner. 1976. *The Mayflies of North and Central America.* University of Minnesota Press, Minneapolis, pp. 330.

Ehrlich, A. and P. Ehrlich. 1992. Needed-an endangered humanity act? *The Amicus Journal.* 14:26-27.

Flowers, R.W. 1991. Diversity of stream-living insects in northwestern Panama. *J.N. Am. Benthol. Soc.* 10:322-324.

Ford, J. 1989. The effects of chemical stress on aquatic species composition and community structure. pp. 99-144. In: Levin, S.A., M.A. Harwell, J.R. Kelley, K.D. Kimball (Ed.). *Ecotoxicology: Problems and Approaches.* Springer-Verlag. New York. pp. 547.

Fremling, C.R. 1960a. Biology of large mayfly *Hexagenia Bilineata* (Say), of the upper Mississippi River. *Iowa State University Agricultural and Home Economics Experimental Station Research Bulletin.* 482:841-852.

Fremling, C.R. 1960b. Biology and possible control of nuisance caddisflies of the upper Mississippi River. *Iowa State University Agricultural and Home Economics Experimental Station Research Bulletin.* 483:856-879.

Gregory, S.V., F.J. Swanson, W.A. McKee and K.W. Cummins. 1991. An ecosystem perspective of riparian zone. *BioScience.* 41:540-551.

Hall, R., G. Likens, S. Fiance and G. Hendrey. 1980. Experimental acidification of a stream in the Hubbard Brook Experimental Forest, New Hampshire. *Ecology* 61:976-989.

Healey, M. 1984. Fish predation on aquatic insects. pp. 255-288. In: Resh, V.H., and D.M. Rosenberg (Ed.). *The Ecology of Aquatic Insects.* Praeger, New York, pp. 625.

Henrikson, L. and H.G. Oscarson. 1981. Corixids Hemiptera-Heteroptera, the new top predators in acidified lakes. Verh. Internat. Verein. Limnol. 21:1616-1620.

Hilsenhoff, W.L. 1982. *Using a Biotic Index to Evaluate Water Quality in Streams.* Technical Bulletin No. 132. Department of Natural Resources, Madison, Wisconsin, pp. 93.

Hilsenhoff, W.L. 1991. Diversity and classification of aquatic insects and Colembola, pp. 593-663. In: J.H. Thorp and A.P. Covich. *Ecology and Classification of North American Freshwater Invertebrates.* Academic Press, Inc. New York, pp. 991.

Hilsenhoff, W.L. 1987. An improved biotic index of organic stream pollution. *Great Lakes Entomol.* 20:31-39.

Hinton, H.E. 1976. Enabling mechanisms, pp. 71-83. In: *Proceedings of the Fifteenth International Congress of Entomology.* International Congress of Entomology. Washington, DC, pp. 824.

Hynes, H.B.N. 1984. The relationship between the taxonomy and ecology of aquatic insects, pp. 9-23. In: Resh, V.H., and D.M. Rosenbert (Ed.). *The Ecology of Aquatic Insects.* Praeger, New York, pp. 625.

Hynes, H.B.N. 1960. *The Biology of Polluted Waters.* Liverpool University Press. Liverpool, England. pp. 202.

Illies, J. 1969. Biogeography and ecology of neotropical freshwater insects, especially those from running water. pp. 685-708. In: Fittkaw, E.J. (Ed.). *Biogeography and Ecology in South America.* Dr. W. Junk. The Hague. pp. 956.

Lamberti, G.A. and J.M. Moore. 1984. Aquatic insects as primary consumers. pp. 164-195. In: Resh, V.H., and D.M. Rosenberg (Ed.). *The Ecology of Aquatic Insects.* Praeger, New York, pp. 625.

Lehmkuhl, D.M. 1979. *How to Know the Aquatic Insects.* Wm. C. Brown, Co., Dubuque, IO, pp. 168.

MacKay, R.J. and G.B. Wiggins. 1979. Ecological diversity in Trichoptera. *Annu. Rev. Entomol.* 24:185-208.

MacKay, R.J. and K.E. Kersey. 1985. A preliminary study of aquatic insect communities and leaf decomposition in acid streams near Dorset, Ontario. *Hydrobiologia* 122:3-11.

McCafferty, W.P. 1981. *Aquatic entomology. The Fisherman's and Ecologist's Illustrated Guide to Insects and Their Relatives.* Science Books International, Boston, Massachusetts, pp. 448.

McElravy, E.P., V.H. Resh, H. Wolda and O.S. Flint, Jr. 1981. Diversity, seasonality and annual variability of caddisflies from a "non-seasonal" tropical environment. pp. 149-156. In: Moretti, G.P. (Ed.). *Proceedings of Third International Symposia on Trichoptera.* Series Entomolgica, Vol. 20. Dr. W. Junk. The Hague. pp. 845.

Merritt, R.W. and K.W. Cummins, (Ed.). 1978. *An Introduction to the Aquatic Insects of North America.* 1st Edition. Kendall/Hunt, Dubuque, Iowa, pp. 722.

Merritt, R.W. and K.W. Cummins, (Ed.). 1984. *An Introduction to the Aquatic Insects of North America.* 2nd Edition. Kendall/Hunt, Dubuque, Iowa, pp. 722.

Merritt, R.W., K.W. Cummins and T.M. Burton. 1984. The role of aquatic insects in the processing and cycling of nutrients. pp. 134-163. In: Resh, V.H., and B.M. Rosenberg (Ed.). *The Ecology of Aquatic Insects.* New York, pp. 625.

Meyer, J.L. 1990. A blackwater perspective on riverine ecosystems. *BioScience.* 40:643-651.

Nilssen, J.P. 1980. Acidification of a emale watershed in southern Norway and some characteristics of acidic aquatic environments. *Hydrobiologie* 65:177-207.

Obeng, L.E. 1981. Man's impact on tropical rivers. pp. 265-288. In: Lock, M.A., and D.D. Williams (Ed.). *Perspectives in Running Water Ecology.* Plenum Press, New York, pp. 430.

Pekarsky, B.L. 1984. Predator-prey interactions among aquatic insects. pp. 196-254. In: Resh, V.H., and D.M. Rosenberg (Ed.). *The Ecology of Aquatic Insects.* Praeger, New York, pp. 625.

Peckarsky, B.L., P. Fraissinet, M.A. Penton and D.J. Coklin, Jr. 1990. *Freshwater Macroinvertebrates of Northeastern North America.* Cornell University Press, Ithaca, NY, pp. 422.

Pennak, R.W. 1953. *Freshwater Invertebrates of the United States.* 1st Ed. John Wiley and Sons. New York, pp. 803.

Pennak, R.W. 1978. *Freshwater Invertebrates of the United States.* 2nd Ed. John Wiley and Sons. New York, pp. 803.

Resh, V.H. and D.M. Rosenberg, (Ed.). 1984. *The Ecology of Aquatic Insects.* Praeges, New York, pp. 625.

Romoser, W.S. 1981. *The Science of Entomology.* 2nd Ed., MacMillion Pub. Co., Inc. London, pp. 575.

Schmidt, G.D. and L.S. Roberts. 1989. *Foundations of Parasitology.* Mosby Co., Saint Louis, MO, pp. 750.

Sheehan, P.J. and R.W. Winner. 1984. Comparison of gradient studies in heavy-metal-polluted streams. pp. 255-271. In: Sheehan, P.J., D.R. Miller, G.C. Butler, and P. Boudeau (Ed.). *Effects of Pollutants at the Ecosystem Level.* Scope 22. Wiley. New York. pp. 581.

Stanford, J.A. and J.V. Ward. 1983. Insect species diversity as a function of environmental variability and disturbance in stream systems. pp. 265-277. In: Barnes, J.R. and G.W. Minshall (Ed.). *Stream Ecology: Application and Testing of General Ecological Theory.* Plenum Press. New York. pp. 399.

Stewart, K.W. and B.P. Stark. 1988. *Nymphs of North American Stonefly Genera (Plecoptera).* Vol. 12. Thomas Say Foundation, pp. 460.

Stout, J. and J. Vandermeer. 1975. Comparison of species richness for stream-inhabiting insects in tropical and mid latitude streams. *Amer. Nat.* 109:263-280.

Sweeney, B.W. 1984. Factors influencing life-history patterns of aquatic insects. pp. 56-100. In: Resh, V.H., and B.M. Rosenberg (Ed.). *The Ecology of Aquatic Insects.* New York, pp. 625.

Usinger, R.L. (Ed.). 1956. *Aquatic Insects of California with Keys to North America genera and California species.* University of California Press, Berkely and Los Angeles, pp. 508.

Vannote, R.L., G.W. Minshall, K.W. Cummins, J.R. Sedell and C. E. Cushing. 1980. The river continuum concept. *Can. J. Fish. Aquat. Sci.* 37:130-137.

Ward, H.B. and G.C. Whipple (Ed.). 1918. *Freshwater Biology.* John Wiley and Sons, Inc. New York, pp. 1111.

Waters, T.F. 1972. The drift of stream insects. Annu. Rev. Entomol. 17:253-272.

Welch, E.B. 1980. *Ecological Effects of Waste Water.* Cambridge University Press, Cambridge. pp. 337.

Wiederholm, T. 1984. Responses of aquatic insects to environmental pollution. pp. 508-557. In: Resh, V.H., and D.M. Rosenberg (Ed.). *The Ecology of Aquatic Insects.* Praeger, New York, pp. 625.

Wiggins, G.B. 1977. *Larvae of the North American caddisfly genera (Trichoptera).* University of Toronto Press, Toronto, pp. 401.

Wiley, M. and S.L. Kohler. 1984. Behavioral adaptations of aquatic insects. pp. 101--127. In: Resh, V.H., and B.M. Rosenberg (Ed.). *The Ecology of Aquatic Insects.* New York, pp. 625.

Williams, W.D. Running water ecology in Australia. pp. 367-392. In: Lock, M.A., and D.D. Williams (Ed.). *Perspectives in Running Water Ecology.* Plenum Press, New York, pp. 430.

Wilson, E.O. (Ed.). 1988. *Biodiversity.* National Academy Press, Washington, D.C. pp. 521.

Winterboum, M.J. and K.L.D. Gregson. 1989. Guide to the Aquatic Insects of New Zealand. *Entomological Society of New Zealand, pp. 95.*

Wolf, E.C. 1987. On the Brink of Extinction: Conserving the Diversity of Life. World Watch Paper 78, pp. 54.

Biological Diversity: Problems and Challenges. Edited by S.K. Majumdar, F.J. Brenner, J.E. Lovich, J.F. Schalles, and E.W. Miller.

Chapter Twenty-Eight

CHANGES IN AMPHIBIAN BIODIVERSITY ASSOCIATED WITH 25 YEARS OF PINE FOREST REGENERATION: Implications for Biodiversity Management

BRUCE W. GRANT[1,2], KENT L. BROWN[3], GARY W. FERGUSON[3], and J. WHITFIELD GIBBONS[1]

[1]University of Georgia
Savannah River Ecology Laboratory
Drawer E
Aiken, SC, 29802
[2]Department of Biology
Widener University
Chester, PA 19013
[3]Department of Biology
Texas Christian University
Fort Worth, Texas 76129

ABSTRACT

Global biodiversity decline, especially for amphibians, is presently of international concern. Data are critically lacking on the effects of human impacts on biodiversity, especially regarding silvicultural management of public and private land. In response to this need, we present data on variation in amphibian biodiversity in replicated stands of one-, three-, eight- and twenty-six-year old loblolly pine trees managed by the USFS on the Savannah River Site in South Carolina. According to censusing using terrestrial drift fences and pitfall traps, from March to November, 1979, intermediate-aged stages of regeneration had higher amphibian species diversity than did either the youngest or oldest stand. Neither the total number of amphibians nor an estimate of total amphibian biomass were correlated with regeneration age.

During the process of even-aged pine forest regeneration, forest structural diversity is initially low after clearcutting and replanting, followed by a decade or more of dense foliage and high habitat structural diversity. As the pine monoculture matures, periodic burning and thinning clears understory vegetation and reduces structural diversity at ground level where most amphibians occur. Thus, we suggest that in our study plots amphibian biodiversity might have been greater in more structurally diverse stages of regeneration, especially near ground level, although the precise mechanisms remain unknown.

Studies are need to elucidate determinants of diversity within managed landscapes. In particular, field studies of the mechanisms underlying amphibian population dynamics in response to forest structural diversity would enable land managers to refine existing and develop new management techniques that minimize biodiversity loss while maximizing economic yield.

keywords: biodiversity, amphibian decline, forest regeneration

INTRODUCTION

The problem of global biodiversity decline, especially for amphibians, is of international concern (*e.g.,* Pechmann et al. 1991, Wake 1991). In response, resource managers are turning to the scientific community to develop methods to meet the joint objectives of biological conservation and sustainability of economic yield. Biodiversity is a critical ecosystem character to preserve not only because of the potential economic or medical benefits yet to uncover in existing life forms, but because maintaining naturally occurring levels of biodiversity is synonymous with maintaining an intact ecosystem within which we and our agricultural products must live (Wilson 1985). Additionally, interest is increasing in the role of biodiversity in buffering ecological food chains against environmental variation due either to natural or anthropogenic causes (Lubchenco et al. 1991).

Competent stewardship and use of our environment demands an understanding of the structure and function of managed ecosystems and the realization that methods to conserve biodiversity and attain sustainable economic yield are interrelated (*e.g.,* Norse et al. 1986). Such stewardship involves developing models which can accurately predict the effects of specific economically necessitated impacts on ecosystem stability and resiliency as well as on the persistence of rare life forms (Soulé and Wilcox 1980, Akçakaya 1992, Duffy and Meier 1992). Predictive models demand intensive interdisciplinary research efforts of population and community ecology to elucidate paths of energy and material flow through "functional groups"

(Chapin et al. 1992) in an ecosystem, since it is precisely these paths that are affected by economic land use.

It is well known that reptiles and amphibians can play critical roles in temperate ecosystems (Gibbons and Bennett 1974, Burton and Likens 1975, Gibbons 1988). These animals have low metabolic rates, high ecological conversion efficiences, explosive reproductive potentials, and provide important links in ecological food chains, the loss of which may lead to ecosystem destabilization and collapse (Burton and Likens 1975, Bury and Corn 1988, Vitt et al. 1990).

Recent techniques in wood pulp production often cause a greater reduction in understory vegetation and more even-aged regenerating stands than did methods used in the past (Bury and Corn 1988). The effects of these environmental differences on the animal component of the managed ecosystem are poorly known, even though a large fraction of North American forest wildlife undoubtedly depends on managed sites. Due to the extreme sensitivity of amphibians to environmental change (Wilbur 1980, Grant et al. 1992), we suggest that these organisms might provide excellent indication of the severity of environmental perturbation to forest ecosystem function due to silvicultural practice. In this paper, we present data on patterns of amphibian biodiversity within the managed landscape of the southeastern Coastal Plain on the Savannah River Site (SRS) in South Carolina. Our ultimate aim is to provide resource managers with ecological guidelines to develop management techniques that minimize biodiversity loss subject to economic constraints.

MATERIALS AND METHODS

Study Sites. The study was conducted from March to November, 1979, on the Savannah River Site (SRS) in the Upper Coastal Plain near Aiken, South Carolina, USA. This is a 780 km^2 National Environmental Research Park devoted, in part, to long-term ecological research and to the production of wood for pulp and lumber by the United States Forest Service (USFS) often using loblolly pine *(Pinus taeda)*. The management activities of the USFS on the SRS have been extensive and well-documented, thus ample replicates of loblolly pine stands of known age were available on the SRS as was the management history of the various stands. Study sites were selected in pine stands that were either one, three, eight, or twenty-six years old in 1979, based upon records on file at the USFS station on the SRS. Within each age class there were three replicate study sites, each >20 ha, all >1 km apart, and as closely matched in plant species cover and composition, soil type, and management history as possible. Also, no study site was adjacent to permanent surface water, such as reservoirs, streams or large Carolina bays, in order to minimize the edge effects of the highly diverse spectrum of amphibians associated with aquatic habitats in the southeastern USA (Bennett et al., 1980).

Pine stands of one year in age contained large numbers of short trees 0.5 m or less in height, dominated by loblolly, but also included scrub oak (*Quercus* spp.), cherry (*Prunus* spp.), and sweetgum (*Liquidambar styraciflua*). Common understory plants included dog fennel (*Eupatorium* spp)., goldenrod (*Solidago*), rabbit tobacco (*Grapaleum* spp.), broomsedge (*Andropogon* spp.), witchgrass (*Leptiloma* spp.), and *Lespedeza* spp. Also, the physical effects of the previous episode of harvesting and clearing were still residual as exposed unvegetated soil which occupied about a quarter to a third of the ground area of one-year-old stands.

Three-year-old pine stands had pine trees about 2 m in height and the spectrum of other plant species was about the same as in one-year-old stands. However, three-year-old stands had much higher plant density and unvegetated soil was greatly reduced relative to one-year-old stands.

Eight-year-old stands had pine trees about 6 m in height, the canopy approached closure, and the understory was denser than in younger-aged stands in this study. In addition to all of the previously listed plant species, eight-year-old stands included hickory (*Carya* spp.) and hawthorn (*Crataegus* spp.). Pine straw and witchgrass covered most of the forest floor.

Twenty-six-year-old pine stands were almost pure loblolly pine (only a few scattered oaks and cherry trees remained) with a dense seemingly unbroken upper canopy of about 20 m in height. Periodic burning by the USFS had cleared much understory vegetation; however, since the most recent prescribed burn was at least several years before this study began (exact burn dates for each 26-year replicate are unknown), a thick mat of pine straw, typically about 20 cm deep, completely blanketed the forest floor.

Amphibian Censusing. Within each of the three study sites per age class, a pair of permanent sampling stations was constructed about 100 m apart. Each sampling station consisted of a drift fence and array of pitfall traps (Gibbons and Bennett 1974, Gibbons and Semlitsch 1981). Drift fences were made of four 15 m sections of aluminum flashing, 60 cm high, arranged in an X-shaped pattern with two 37.7 liter pitfall traps on opposite sides of the fence at both ends of each section with a total of 16 traps per station. Pitfall traps were checked nearly every day from mid-March to mid-November, 1979, except during periods of low amphibian activity in mid-summer when traps were checked every other day. All animals collected were identified to species and sex, measured to the nearest mm in snout-vent length, toe-clipped according to the area in which they were found, and released on the opposite side of the fence.

Diversity and Evenness Estimation. Techniques of estimating community diversity are as many and varied as are ecological communities, and diversity indices differ greatly in their implicit assumptions and biological interpretability. Since our study deals with relative differences in diversity among study plots, we do not feel the need to argue the case for any particular method and instead we selected the more commonly used indices: the number of species, the Shannon index, and the Simpson Index. We refer interested readers to excellent discussions of diversity indices in Pielou (1977), or Engen (1978).

The equation for the Shannon index, H′, is

$$H' = -\sum_{i}^{S} p_i \ln(p_i)$$

where S is the total number of species and p_i is the proportion of the ith species in the sample. A biological interpretation for this index can best be drawn from the exponent of H′ which represents the equivalent number of equally common species in the community (John W. MacArthur, personal communication). The Shannon index simultaneously estimates two components of community structure: the total number of species and the relative abundance of each (e.g., if S = 1, then H′ = —ln(1) = 0 and the community is of zero diversity and has e° = 1 equally common species).

The equation for the Simpson index, H_s, is

$$H_s = 1 - \sum_{i}^{S} \frac{n_i * (n_i - 1)}{N * (N - 1)}$$

where S is the total number of species, N is the total number of individuals observed, and n_i is the number of observations of the ith species in the sample. The Simpson index estimates the probability that any two individuals drawn at random from the community will be of different species (e.g. if S = 1, then $H_s = 1 - 1 = 0$).

In addition, we also estimated amphibian community evenness using the equation

$$\text{Evenness} = H' / \ln(S)$$

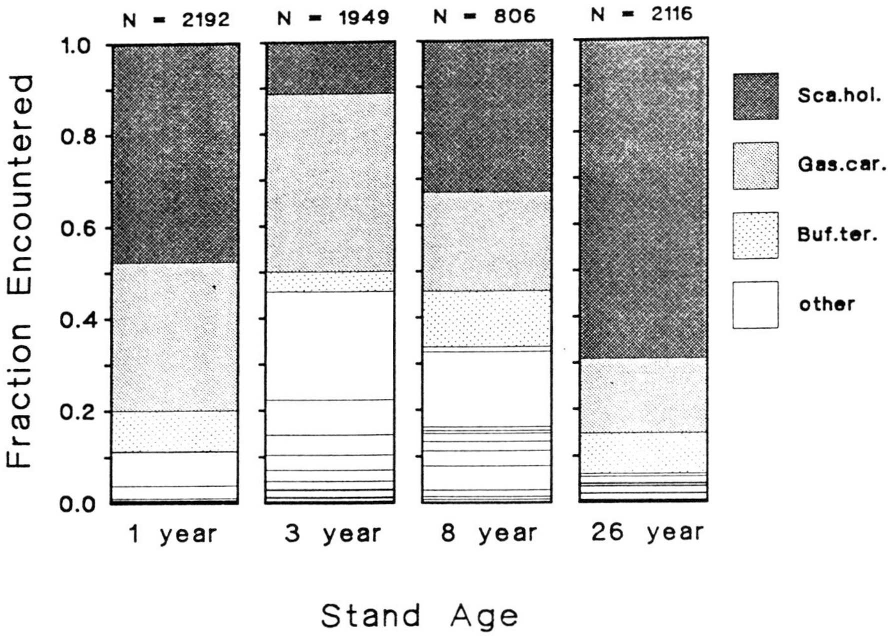

FIGURE 1. Encounter frequencies of the three most common amphibian species *(Scaophiopus holbrooki, Gastrophryne carolinensis,* and *Bufo terrestris)* as well as the relative representation of other taxa, among regenerating pine stands of different ages on the Savannah River Site in South Carolina. Sample sizes for the pooled totals among replicates appear across the figure top.

where S is the total number of species in the entire community (Pielou 1977) within each age class, and H′ is the Shannon diversity index. The Evenness index varies between 0 for communities dominated by a single species to 1 for communities of equally occurring species.

RESULTS

Over the nine month census period, a total of 7063 amphibians from 21 species were encountered among all censuses at each of 6 drift fence/pitfall trap arrays within each age class of regenerating pine trees. Table 1 shows the complete species list for these samples and Figure 1 depicts the fractions encountered of the three most common amphibian species (*Scaphiopus holbrooki, Gastrophryne carolinensis,* and *Bufo terrestris*) as well as the relative representation of other taxa.

The total number of captures per array did not differ among classes of regenerating pine (Table 2, ANOVA $F_{3, 20} = 1.51$, $P > 0.05$), despite that about half as many captures occurred on plots in the 8-year-old stand relative to all other stands. The total amphibian biomass, as estimated by the total snout-vent lengths of all censused amphibians, was significantly different (Table 2, ANOVA $F_{3, 20} = 3.13$, $P = 0.048$); however, the difference was due to the highest and lowest values being in the intermediate-aged stands (3- and 8-yr-old stands, respectively).

According to either the number of species observed per array, the Shannon diversity index, or the Simpson diversity index, the intermediate-aged pine stands (3 and 8 yrs) exhibited the highest amphibian species diversity (Figure 2 [upper row and lower left], MANOVA Wilks's Lambda $F_{9, 44} = 3.30$, $P < 0.005$). Typically, diversity was highest in the 3-year-old stand and lowest in the 1 year old stand (rankings and significant differences among least squared means appear in Table 3).

Similar to the pattern among stands in amphibian diversity, the evenness of the amphibian community was also highest in the intermediate-aged stands and lowest in the 1- and 26-yr-old stands (Figure 2 [lower right], ANOVA $F_{3, 20} = 7.10$, $P < 0.005$, rankings and significant differences among least squared means appear in Table 3).

DISCUSSION

Pine stands represent about a third of the land area on the 720 km^2 Savannah River Site (SRS). Despite the low plant species diversity necessarily associated with pine monoculture, the number of amphibian species found among all age classes of regenerating pines is quite high ahd represents about half of all 45 amphibian species recorded from the SRS (Gibbons and Semlitsch 1991). One factor potentially contributing to the high diversity of amphibians within SRS pine stands is transient immigration from regional, albeit non-adjacent, sources. The SRS is a National Environmental Research Park, a large fraction of which is relatively pristine habitat. Since SRS amphibians must develop and metamorphose in standing water, immigration must be the cause of all incidence of amphibians on our study plots. Indeed, as several authors have observed, in systems for which "source" pools provide colonists for "sink" localities

TABLE 1

Complete species list of amphibians encountered at 24 drift fence/pitfall trap array in 4 differently aged pine stand (March-November 1979).

stand age ——>	1 year old						3 year old						8 year old						26 year old					
species / station ID ——>	A	B	C	D	E	F	A	B	C	D	E	F	A	B	C	D	E	F	A	B	C	D	E	F
Ambystoma talpoideum (mole salamander)	0	2	0	0	0	3	134	141	84	60	34	4	4	0	0	0	4	1	4	0	0	0	2	6
Ambystoma opacum (marbled salamander)	0	0	0	0	0	0	23	35	1	3	0	1	0	0	0	0	1	4	5	3	0	0	1	0
Ambystoma tigrinum (tiger salamander)	0	0	1	0	1	0	4	4	1	1	1	0	0	0	0	0	0	0	0	0	0	0	1	0
Notophthalmus viridescens (red-spotted newt)	1	0	0	0	0	0	6	13	24	2	1	1	2	0	0	0	3	9	2	0	0	0	0	0
Plethodon glutinosus (slimy salamander)	3	0	0	0	0	0	2	0	0	0	0	0	10	3	5	7	1	0	4	3	2	2	3	14
Pseudotriton ruber (red salamander)	1	0	1	1	0	0	0	0	0	0	0	1	4	0	0	0	0	0	0	0	0	0	2	1
Eurycea quadridigitata (dwarf salamander)	0	0	0	0	0	0	0	0	0	0	0	0	0	0	0	0	0	0	0	0	0	0	1	0
Scaphiopus holbrooki (eastern spadefoot toad)	202	140	243	442	8	12	45	47	50	22	21	35	12	5	53	33	85	77	999	228	115	64	39	18
Bufo terrestris (southern toad)	17	12	23	17	87	34	4	14	31	23	6	7	18	5	16	10	31	17	10	12	38	53	35	35
Bufo quercicus (oak toad)	0	0	0	0	0	1	0	0	1	3	0	1	0	0	1	1	0	3	2	1	10	12	0	0
Pseudacris crucifer (spring peeper)	2	0	0	0	0	2	5	13	5	5	0	0	0	0	0	0	7	4	0	1	0	0	0	0
Hyla gratiosa (barking treefrog)	0	0	0	0	0	0	0	0	1	2	0	0	0	0	0	0	0	0	0	0	0	0	0	1
Hyla chrysoscelis (Cope's gray treefrog)	0	0	0	0	0	0	0	0	0	0	0	0	0	0	0	0	0	0	0	0	0	6	1	0
Hyla avivoca (bird-voiced treefrog)	0	0	0	0	0	0	0	0	0	0	0	0	0	0	2	0	0	0	0	0	0	0	0	0
Pseudacris triseriata (striped chorus frog)	0	0	0	0	2	0	0	0	0	0	0	0	0	0	0	0	0	0	0	0	0	0	0	0
Pseudacris nigrita (southern chorus frog)	0	0	0	0	0	0	3	1	0	2	0	0	0	0	0	0	0	0	0	0	0	0	0	0
Pseudacris ornata (ornate chorus frog)	72	32	14	21	16	5	52	41	27	26	2	0	5	2	2	3	58	61	18	11	0	0	0	0
Gastrophryne carolinensis (eastern narrow-mouthed toad)	54	114	32	17	171	319	65	122	341	143	33	49	4	4	35	24	57	49	38	35	115	59	44	50
Rana catesbeiana (bullfrog)	0	0	0	0	0	0	1	0	0	0	1	1	4	32	0	1	2	3	2	1	0	0	0	0
Rana clamitans (green frog (bronze frog)	4	2	1	0	0	1	29	4	0	0	0	0	4	7	0	0	3	2	2	1	0	0	0	0
Rana utricularia (southern leopard frog)	5	8	2	1	40	3	37	30	7	4	1	5	0	0	1	0	2	3	1	2	0	0	1	0

TABLE 1 cont.

ranked in order of decreasing total occurrence species	TOTALS	STAND AGE 1	3	8	26
Scaphiopus holbrooki (eastern spadefoot toad)	2995	1047	220	265	1463
Gastrophryne carolinensis (eastern narrow-mouthed toad))	1974	707	753	173	341
Bufo terrestris (southern toad)	555	190	85	97	183
Ambystoma talpoideum (mole salamander)	483	5	457	9	12
Pseudacris ornata (ornate chorus frog)	468	160	148	131	29
Rana utricularia (southern leopard frog)	153	59	84	6	4
Ambystoma opacum (marbled salamander)	77	0	63	5	9
Notophthalmus viridescens (red-spotted newt)	64	1	47	14	2
Rana clamitans (green frog (bronze frog)	60	8	33	16	3
Plethodon glutinosus (slimy salamander)	59	3	2	26	28
Rana catesbeiana (bullfrog)	48	0	3	42	3
Pseudacris crucifer (spring peeper)	44	4	28	11	1
Bufo quercicus (oak toad)	36	1	5	5	25
Ambystoma tigrinum (tiger salamander)	14	2	11	0	1
Pseudotriton ruber (red salamander)	11	3	1	4	3
Hyla chrysoscelis (Cope's gray treefrog)	7	0	0	0	7
Pseudacris nigrita (southern chorus frog)	6	0	6	0	0
Hyla gratiosa (barking treefrog)	4	0	3	0	1
Hyla avivoca (bird-voiced treefrog)	2	0	0	2	0
Pseudacris triseriata (striped chorus frog)	2	2	0	0	0
Eurycea quadridigitata (dwarf salamander)	1	0	0	0	1

(which in our case are aquatic breeding grounds and upland pine stands, respectively) regional biogeographic patterns may greatly complicate the interpretation of local species diversity as well as community structure and function (Pulliam 1988, Cornell and Lawton 1992). However, if it were true that our study plots happen to have been selected in localities with equal access from source pools, which was a stated but untested objective, these considerations do not explain why differentially diverse communities

TABLE 2. Total number and total snout-vent lengths (an estimate of biomass) of amphibians censused among pine stands of different ages.

Stand Age	1	3	8	26
Number of Individuals	365.3	324.8	134.3	352.7
(±1 S.E.)	±28.9	±79.2	±35.5	±149.8
Total Snout-Vent Length (m)	10.64	12.55	4.55	8.69
(±1 S.E.)	±0.67	±2.71	±0.80	±2.55

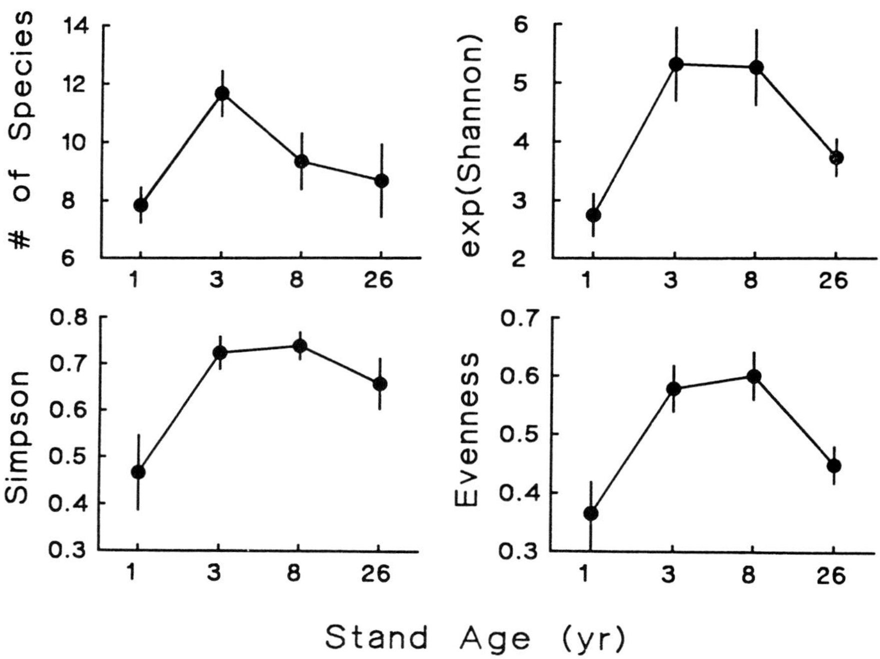

FIGURE 2. Estimates of amphibian community diversity and evenness among regenerating pine stands of different ages on the Savannah River Site. Error bars are ±1 S.E. among the six replicates per age class. The upper left plot is the number of species observed. The upper right plot is the exponent of the Shannon index which represents the equivalent number of equally common species (John W. MacArthur, personal communication). The lower left is the Simpson index. The lower right is the community evenness. (See MATERIALS AND METHODS for equations used).

TABLE 3. Rankings and significant differences in amphibian diversity and evenness among pine stands of different ages from separate ANOVA's of each diversity estimate (horizontal lines indicate overlapping, $p > 0.05$, least-squared means).

Diversity Rankings and Overlap							
Number of species	3 yr	>	8 yr	>	26 yr	>	1 yr
Shannon, H′	3 yr	>	8 yr	>	26 yr	>	1 yr
Simpson, H_S	8 yr	>	3 yr	>	26 yr	>	1 yr
Evenness Ranking and Overlap							
Evenness	8 yr	>	3 yr	>	26 yr	>	1 yr

assembled from source pools on replicate pine stands of different ages. The question is — were local, stand-age dependent processes at work?

Our results suggest that amphibian biodiversity was highest in pine stands that were intermediate in age. Reduced diversity in the youngest and oldest pine stands was proximally due to the presence of large numbers of amphibians from two species (*Scaphiopus holbrooki* and *Gastrophryne carolinensis*) which accounted for over 80% of all observations in these stands (Figure 1). In contrast, these taxa only accounted for about 50% of the observations in intermediate-aged stands. In 3-year-old stands, more species of salamanders were common (especially *Ambystoma*), whereas in 8-year-old stands, the increased diversity was proximally due to relatively greater representation by more species of anurans (*e.g., Bufo terrestris* and *Pseudacris ornata*). However, these proximate relations provide poor explanation of the overall pattern of amphibian community biodiversity in Figure 2 for which a wider perspective is demanded.

Pioneering work by MacArthur and MacArthur (1961) demonstrated the importance of vegetational structural diversity to avian communities. These authors observed that plant species diversity did not affect bird community diversity present at each sampling locality. Instead, the foliage height diversity, estimated as the equivalent number of equally vegetated height classes of foliage, was positively correlated with the diversity of bird species (both comparisons based on Shannon indices). In other words, in multi-layered forests with dense understories, more bird species were resident, which suggested to the authors that structural diversity may provide a better indicator of "niche" diversity for birds.

Although quantitative studies such as this comparing data on forest structural diversity with amphibian species diversity are presently lacking (the present study included), anecdotal evidence abounds. For example, Raphael (1988) found that structural components of regenerating forests, and not stand age per se, were important determinants of amphibian occurrence. In particular, the presence of downed woody debris and/or an intact hardwood understory was correlated with the presence of several species of salamanders in northwest California forests. Similar results were reported in Aubry et al. (1988), who found that the occurrence of salamanders in Washington state forests was more related to the presence of an understory of coarse woody debris, such as from downed logs and branches, than to the age of the overstory. Bury and Corn (1988) also emphasized the importance of forest structural diversity, in the form of downed woody debris, the absence of which was associated with decreased amphibian biodiversity.

Although we do not have quantitative data on habitat structural diversity from our study plots, 3- and 8-year-old pine stands are densely foliated as well as vegetationally and structurally diverse habitats. During the process of even-aged pine forest regeneration as is typical on the SRS, forest structural diversity is initially low after clearcutting and replanting, followed by a decade or more of dense foliage and high habitat structural diversity. As the pine monoculture matures, periodic burning and thinning clears understory vegetation and reduces structural diversity at ground level where most amphibians occur. Thus, we suggest that in our study plots amphibian biodiversity might have been greater in more structurally diverse stages of regeneration, especially near ground level, although the precise mechanisms remain unknown.

A clear implication from our and from previous studies is that the management techniques that retain rather than eliminate natural components of regenerating stands would provide for the habitat needs of many amphibians, and mitigate against biodiversity decline. An heuristic model of the relationship between forest structural diversity, amphibian biodiversity and sustainable pulp yield appears in Figure 3. Management techniques such as prescribed burning would tend to decrease structural diversity and push the system to the left, thereby increasing pulp yield while decreasing amphibian biodiversity. The optimal management strategy (indicated by the '*') is the one for which the lines relating yield and biodiversity cross. The model in Figure 3 underscores three areas of critically needed research: (1) how much control can feasible management practices exert on habitat structural diversity?, (2) what is the precise relation between understory foliage and stand yield?, and (3) how do amphibian populations respond to specific habitat features which can be affected by management?

Of the three, amphibian ecologists can contribute greatly to the third through intensive field studies to assess mechanisms linking habitat vegetational or structural diversity and amphibian population and community structure. Are less common taxa drawn to less common microhabitats found only in plots of high structural diversity? Is the diversity of amphibian food items highest in plots of high structural diversity? Is there more cover and concomitant protection from predation or physical stress in intermediate-aged stands with dense foliage? Studies of

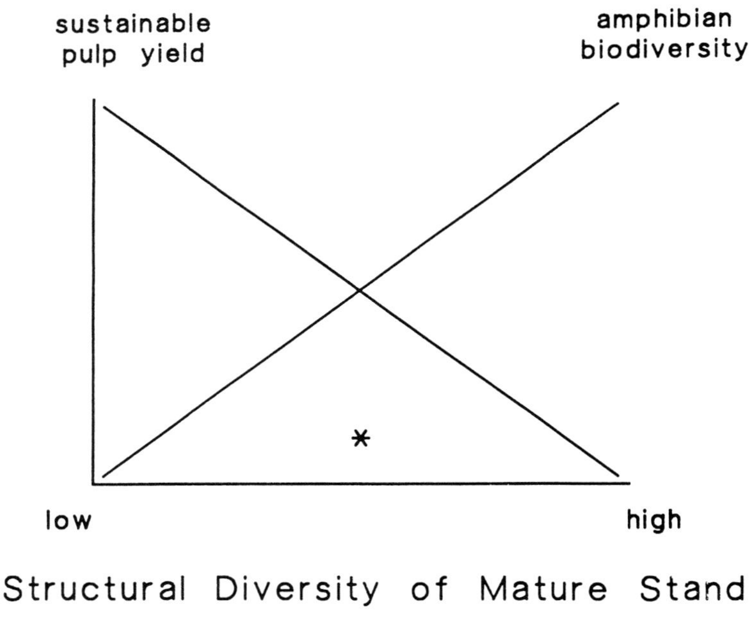

FIGURE 3. An heuristic model of the relationship between forest structural diversity, amphibian biodiversity and sustainable pulp yield. The optimal management strategy (indicated by the '*') is the one for which the lines relating yield and biodiversity cross (see text for explanation).

amphibian autecology designed to address these questions would lead to models to estimate the relationship between forest management and biodiversity.

Putting it all together to determine where the yield and biodiversity lines cross in Figure 3 will require a comprehensive method of ecological cost accounting (Westman 1990). A common currency must be found for the equations relating habitat diversity on the x-axis to both yield and biodiversity on the y-axis. All too frequently, cost accounting weights too heavily on short term economic benefit, and should instead be based on rigorous analyses over the long term, weighing appropriate scientific and economic criteria (Westman 1990). Established ecological principles must be part of the equation, and ecologists are the only ones in a position to advise on ecosystem-level effects of various policy decisions. For amphibians as well as any other taxa, models of population dynamics would identify key sensitivities and lead to development of new management techniques to minimize biodiversity loss and mitigate impact on pulp production as part of an integrated ecological management scheme.

There is an urgent need for an ecosystems approach to the problem of the preservation of biodiversity (Scott et al. 1987). But, rather than abandoning the species approach which has proven extremely useful in the past, research efforts should be devoted to identifying functional ecological groups and determine the precise effects of various scenarios of land use on these "keystone" taxa (Scott et al 1987). The highest priorities for protection are those individual species that play critical roles in ecosystem structure and function, regardless of their taxonomic standing and concomitant public appeal. The alternative, which is occurring ever more frequently, is a mode of management reactivity in which emergency steps are taken, often involving high profile media coverage, when particular species become endangered following unconstrained, and often governmentally encouraged, exploitation of their critical habitat. Resource managers and policy makers must move in the direction of proactivity for which biological conservation and sustainability of economic yield are viewed as important joint objectives of land management. Clearly, ecologists must be involved in this process because only ecologists can identify critical links in ecosystems and guide management policy along an ecologically sound course.

ACKNOWLEDGEMENTS

Experimental design and data collection for this manuscript were in partial fulfillment of the degree of Master of Science by Kent L. Brown at Texas Christian University, 1980. Discussion and assistance of Ray Semlitsch, John Pinder, Gary Moran, Wayne Hendricks, Vince DeMarco, and Garfield Keaton of the Savannah River Ecology Lab and David Samuel of the U.S. Forest Service proved invaluable during early stages of this project.

More recent contributors of critical insight and catalytic inspiration include Joe Pechmann, Ruth Estes, Joe Mitchell, Tom Jenssen, and Trip Lamb. We especially thank Jeff Lovich for his patience and flexibility during the production process.

This ms has benefitted from illuminating comments of Tracy Lynch, Jeff Lovich, and an anonymous reviewer. Support for this project was possible through contract DE-AC09-76SR00819 between the University of Georgia and the U.S. Department of Energy.

LITERATURE CITED

Akc̊akaya, H.R. 1992. Population viability analysis and risk assessment. Pages 148-157 *in* D.R. McCullough and R.J. Barrett (eds.) *Wildlife 2001: Populations.* Elsevier Science Publ., Inc. London, England.

Aubry, K.B., L.L.C. Jones, and P.A. Hall. 1988. Use of woody debris by plethodontid salamanders in douglas-fir forests in Washington. Proc. Symp.: Manag. of Amphib., Rept., and Small Mammals in N. Am., 19-21 July, 1988, Flagstaff, Arizona. p. 32-37 US Dept. of Agric., Forest Service, Rocky Mountain For. and Range Exper. Stat., Fort Collins, CO 80526, General Technical Report RM-166.

Bennett, S.H., J.W. Gibbons, and J. Glanville. 1980. Terrestrial activity, abundance and diversity of amphibians in differently managed forest types. *American Midland Naturalist* 103:412-416.

Burton, T.M., and G.E. Likens. 1975. Energy flow and nutrient cycling in salamander populations in the Hubbard Brook experimental forest, New Hampshire. *Ecology* 56:1068-1080.

Bury, R.B., and P.S. Corn. 1988. Douglas-fir forests in the Oregon and Washington Cascades: relations of the herpetofauna to stand age and moisture. Proc. Symp.: Manag. of Amphib., Rept., and Small Mammals in N. Am., 19-21 July, 1988, Flagstaff, Arizona. p. 11-22 US Dept. of Agric., Forest Service, Rocky Mountain For. and Range Exper. Stat., Fort Collins, CO 80526, General Technical Report RM-166.

Chapin, F.S., E.D. Schulze, and H.A. Moore. 1992. Biodiversity and ecosystem processes. *Trends Ecol. Evol.* 7:107-108.

Cornell, H.V., and J.H. Lawton. 1992. Species interactions, local and regional processes, and limits to the richness of ecological communities: a theoretical perspective. *J. Animal Ecology* 61:1-12.

Duffy, D.C. and A.J. Meier. 1992. Do Applalacian herbaceous understories ever recover from clearcutting? *Conservation Biology* 6:196-201.

Engen, S. 1978. *Stochastic Abundance Models.* John Wiley and Sons, New York. 126 pages.

Gibbons, J.W. 1988. The management of amphibians, reptiles and small mammals in North America: the need for an environmental attitude adjustment. Proc. Symp.: Manag. of Amphib., Rept., and Small Mammals in N. Am., 19-21 July, 1988., Flagstaff, Arizona. p. 4-10 US Dept. of Agric., Forest Service, Rocky Mountain For. and Range Exper. Stat., Fort Collins, CO 80526, General Technical Report RM-166.

Gibbons, J.W., and D.H. Bennett. 1974. Determination of anuran terestrial activity patterns by a drift fence method. *Copeia* 1974:236-243.

Gibbons, J.W., and R.D. Semlitsch. 1981. Terrestrial drift fences with pitfall traps: An effective technique for quantitative sampling of animal populations. *Brimleyana* 7:1-16.

Gibbons, J.W., and R.D. Semlitsch. 1991. *Guide to the Reptiles and Amphibians of the Savannah River Site.* University of Georgia Press, Athens.

Grant, B.W., A.D. Tucker, J.E. Lovich, A.M. Mills, P.M. Dixon, and J.W. Gibbons. 1992. The use of coverboards in estimating patterns of reptil and amphibian biodiversity. Pages 379-403 *in* D.R. McCullough and R.H. Barrett (eds.). *Wildlife 2001: Populations.* Elsevier Science Publ., Inc. London, England.

Lubchenco, J., A.M. Olson, L.B. Brubaker, S.R. Carpenter, M.M. Holland, S.P. Hubbell, S.A. Levin., J.A. MacMahon, P.A. Matson, J.M. Melillo, H.A. Mooney, C.H. Peterson, H.R. Pulliam, L.A. Real, P.J. Regal, and P.G. Risser. 1991. The sustainable biosphere initiative: an ecological research agenda. *Ecology* 72:371-412.

MacArthur, R.H., and J.W. MacArthur. 1961. On bird species diversity. *Ecology* 42:594-598.

Norse, E.A., K.L. Rosenbaum, D.S. Wilcove, B.A. Wilcox, W.H. Romme, D.W. Johnston, and M.L. Stout. 1986. *Conserving Biological Diversity in Our National Forests.* Prepared by the Ecological Society of America for The Wilderness Society. 116 pages.

Pechmann, J.H.K., D.E. Scott, R.D. Semlitsch, J.P. Caldwell, L.J. Vitt, and J.W. Gibbons. 1991. Declining amphibian populations: the problem of separating human impacts from natural fluctations. *Science* 253:892-895.

Pielou, E.C. 1977. *Mathematical Ecology.* John Wiley and Sons, Inc. New York, NY. 385 pages.

Pulliam, H.R. 1988. Sources, sinks and population regulation. *American Naturalist* 132:652-661.

Raphael, M.G. 1988. Long-term trends in abundance of amphibians, reptiles, and mammals in douglas-fir forests of northwestern California. Proc. Symp.: Manag. of Amphib., Rept., and Small Mammals in N. Am., 23-31 July, 1988, Flagstaff, Arizona. p. 4-10 US Dept. of Agric., Forest Service, Rocky Mountain For. and Range Exper. Stat., Fort Collins, CO 80526, General Technical Report RM-166.

Scott, J.M., B. Csuti, J.D. Jacobi, and J.E. Estes. 1987. Species richness. *Bioscience* 37:782-788.

Soulé, M.E., and B.A. Wilcox, (eds.). 1980. *Conservation Biology: An Evolutionary-Ecological Perspective.* Sinauer, Sunderland, Massachusetts, USA.

Vitt, L.J., J.P. Caldwell, H.M. Wilbur, and D.C. Smith. 1990. Amphibians as harbingers of decay. *Bioscience* 40:418.

Wake, D.M. 1991. Declining amphibian populations. *Science* 253:860.

Westman, W.E. 1990. Managing for biodiversity. *Bioscience* 40:26-32.

Wilbur, H.M. 1980. Complex life cycles. *Ann. Rev. Ecol. Syst.* 11:67-93.

Wilson, E.O. 1985. The biological diversity crisis. *Bioscience* 35:700-706.

Biological Diversity: Problems and Challenges. Edited by S.K. Majumdar, F.J. Brenner, J.E. Lovich, J.F. Schalles, and E.W. Miller. © 1994, The Pennsylvania Academy of Science.

Chapter Twenty-Nine

FACTORS INFLUENCING FISH SPECIES DIVERSITY IN THE GREAT LAKES AND EFFECT OF RECENT INTRODUCTIONS ON LAKE ERIE

JOHN M. CAMPBELL[1] and **ROGER B. KENYON**[2]

[1]Department of Biology
Mercyhurst College
Erie, PA 16546
and
[2]Lake Erie Research Unit
Pennsylvania Fish and Boat Commission
P.O. Box 531
Fairview, PA 16415

INTRODUCTION

The North American Great Lakes (Figure 1) collectively contain the largest continuous body of freshwater in the world (Wetzel 1983). Erosion by glacial ice formed the basins of the lakes during the repeated advance and retreat of the Wisconsin ice sheet approximately 15,000 — 5,000 years ago (Hough 1958). The 177 species of fish known to have occurred in the Great Lakes and their tributaries (Underhill 1986) represent nearly one-fifth of approximately 950 species of freshwater fish indigenous to North America (Briggs 1986) and about 2% of approximately 8,000 freshwater fishes known worldwide (Wooten 1990).

There has probably been more written about the fish communities of the Great Lakes than of any other freshwater system in the world. Hubbs and Lagler (1964) and Scott and Crossman (1973) address the taxonomy of Great Lake fishes. Bailey and Smith (1981) and Underhill (1986) describe the late Pleistocene and Holocene history of the lakes and origins of their fish fauna. Reviews on the effects of over-fishing, introductions and eutrophication on Great Lakes fisheries are found in Smith (1972), Christie (1974) and Emery (1985). In this chapter, we will not attempt to summarize the entire body of published literature pertaining to the biological diversity of Great Lakes fishes. Rather, our purpose in the first part of the chapter will be to compare species richness among the lakes, and evaluate the importance of geographical, historical and biological factors that Wootton (1990) has suggested may influence fish species diversity. In the second part of the chapter, we discuss the effects of recent introductions on the fish community of Lake Erie.

COMPARISON OF FISH SPECIES DIVERSITY AMONG THE LAKES

Geographical Factors Affecting Species Richness

Summaries of the physical features of the Great Lakes and the number of fish species known to have occurred in individual lakes and their tributaries are summarized in Tables 1 and 2. The most northern of the five lakes (Superior) contains the lowest number of fish species, while the most southern lake (Erie) has the highest number (considering lake species alone). This conforms to Wootton's (1990) generalization that fish species richness increases with decreasing latitude. Lake Superior is colder than the other lakes due to its high latitude location and thermal stability imparted by great depth and volume (Table 1). Warm-water fish families Ictaluridae and Centrarchidae are collectively represented by a single species in Lake

FIGURE 1. Map of the Great Lakes of North America, showing major drainage connections.

Superior, and cool-water families Esocidae and Percidae are collectively represented by only 5 species (Table 3). The shallower, warmer waters of Lake Erie have been known to support 19 and 16 species of those warm- and cool-water groups. The lakes at latitudes and depths intermediate between Superior and Erie contain intermediate numbers of species in warm- and cool-water families. More species of the cold-water family Salmonidae occur in the deeper, higher latitude lakes (13-19 species) than in Lake Erie (8 species). Christie (1974) gives more details regarding the relationship between species richness and north-south differences in climate and lake morphology.

The eastern lakes (Ontario, Erie and Huron) have greater species richness than the western ones (Superior and Michigan), although if tributary species are included

TABLE 1

Locations, physical characteristics, and ranges of rates of primary production of the North American Great Lakes. Data from Hutchinson (1957), Ryder (1972), Wetzel (1983), Quinn (1992) and Lesht and Brandner (1992). Morphoedaphic Index calculated using Ryder's (1972) values for mean depth.

	Lake Superior	*Lake Michigan*	*Lake Huron*	*Lake Erie*	*Lake Ontario*
Latitude North	46°30′ -49°00′	41°35′ -46°00′	43°00′ -46′00°	41°25′ -42°50°	43°15′ -44°15′
Longitude West	84°30′ -92°05′	84°45′ -88°00′	80°00′ -84°45′	78°50′ -83°30′	76°10′ -79′50′
Surface Area (km²)	82,100	57,800	59,600	25,700	18,960
Drainage Area (km²)	227,376	117,741	128,490	58,530	70,655
Max. Depth (m)					
Hutchinson (1957)	307	265	223	64	225
Ryder (1972)	397	285	229	64	237
Mean Depth (m)					
Hutchinson (1957)	145	99	76	21	91
Ryder (1972)	149	84	60	18	80
Total Volume (km³)	12,100	4,920	3,540	484	1,640
Hydr. Residence Time (yr)	173	62	21	2.7	7.5
Max. Avg. Daily Surf. Temp. (°C)	15	23	20	25	—
Total Diss. Solids (ppm)	60	118	108	133	134
Morphoedaphic Index TDS/Mean Depth	0.41	1.40	1.80	7.39	1.68
Primary Production (mg C m^{-2} day^{-1})	50 -260	70 -1030	150 -700	120 -1690	60 -1400

in the totals, Lake Michigan would be considered most diverse. Wootton (1990) made no global generalizations regarding the influence of longitude on fish species diversity, except to indicate that the Pacific coast of North America has more (marine) species than the Atlantic coast. The eastern half of the North American continent contains about four times as many freshwater species as the western half, mostly due to the major adaptive radiations that occurred in the Mississippi River Basin (Briggs 1986). Since the fish fauna of the western Great Lakes originated mostly from glacial refugia in the Mississippi drainage (Bailey and Smith 1981, Underhill 1986), it is unlikely that the trend (among the Great Lakes) for increasing diversity from west to east is related to the continental trend described by Briggs (1986). Latitudinal and morphological variations among the lakes mentioned previously, and the fact that the western lakes drain through Huron, Erie and Ontario into the St. Lawrence River (Figure 1) are probably major determinants of east-west variations in species richness among the lakes.

Higher species richness may occur in the eastern lakes, especially Ontario, due to the St. Lawrence River connection to the Atlantic Ocean and proximity of the Susquehanna and Hudson river drainages. Apparently many fish species colonized the Great Lakes from the Atlantic drainage during the postglacial period, before the separation of Lakes Ontario and Erie by Niagara Falls (about 10,000 years ago) (Underhill 1986). From that time until the Welland Canal was constructed in 1825 (Emery 1985), fish in Lake Ontario were unable to bypass Niagara Falls to enter the western lakes (Figure 1). Consequently, Lake Ontario and its tributaries contain more species from the Atlantic drainage than the other Great Lakes (Underhill 1986).

Based upon island biogeography theory (MacArthur and Wilson 1967, Simberloff 1974), large areas are expected to contain more species than small areas. The correlation between species richness and area is good for rivers of the world and African Lakes, but poor for North American lakes (Barbour and Brown in Wootton 1990). The data for the Great Lakes indicate an inverse correlation between lake area and number of fishes (exclusing tributaries) (Tables 1 and 2). This is no surprise, in view of latitudinal

TABLE 2

Number of different species of fish known to have occurred in the Great Lakes and their tributaries, including introduced and extirpated forms. Summarized from data in Bailey and Smith (1981) and Underhill (1986). Totals for introduced and extirpated forms are for individual lakes and their tributaries combined.

	Number of Forms					
	Lake Superior	Lake Michigan	Lake Huron	Lake Erie	Lake Ontario	Total Basin
Total in Lake(s)	44	76	93	101	95	128
Total in Tributaries	70	128	107	120	115	163
Lakes and Tributaries	80	140	115	131	127	177
Introduced/ Colonized	11	18	15	18	9	—
Extirpated	1	9	7	10	8	—

and morphological differences previously mentioned and the drainage interconnections among the lakes

Historical Factors Affecting Species Richness

In addition to colonization phenomena previously mentioned (Bailey and Smith 1981, Underhill 1986), historical factors affecting the process of speciation (isolation, time and speciation rate) also influence fish species richness (Wootton 1990). Endemic species of the Great Lakes include the blue pike (*Stizostedion vitreum glaucum*) of Erie and Ontario and five coregonids (*Coregonus hoyi, johannae, kiyi, nigripinnis* and *reighardi*) known from several of the lakes (except Erie) (Underhill 1986).

TABLE 3

Number of species of fish in families known to have occurred in the Great Lakes, including introduced and extirpated forms. Numbers in parentheses are totals including species known to have occurred in the lakes' tributaries. Summarized from data in Bailey and Smith (1981) and Underhill (1986).

Family	Lake Superior	Lake Michigan	Lake Huron	Lake Erie	Lake Ontario	Total Basin
Petromyzontidae	2(4)	3(5)	2(4)	2(4)	2(3)	3(5)
Polyodontidae	0(1)	0(1)	1(1)	1(1)	0(0)	1(1)
Acipenseridae	1(1)	1(1)	1(1)	1(1)	1(1)	1(1)
Lepisosteidae	1(1)	2(3)	1(1)	2(2)	1(1)	3(3)
Amiidae	1(1)	1(1)	1(1)	1(1)	1(1)	1(1)
Anguillidae	1(1)	1(1)	1(1)	1(1)	1(1)	1(1)
Hiodontidae	0(0)	1(1)	1(1)	1(1)	1(1)	1(1)
Clupeidae	2(2)	2(2)	2(2)	2(2)	3(3)	3(3)
Salmonidae	15(16)	18(19)	19(20)	8(9)	13(14)	20(21)
Osmeridae	1(1)	1(1)	1(1)	1(1)	1(1)	1(1)
Umbridae	0(1)	1(1)	1(1)	1(1)	1(1)	1(1)
Esocidae	2(2)	3(3)	2(3)	3(4)	4(4)	4(4)
Cyprinidae	6(21)	11(39)	21(31)	20(38)	22(39)	27(53)
Catostomidae	2(5)	5(14)	9(12)	13(16)	7(10)	13(16)
Cobitidae	0(0)	0(0)	0(1)	0(0)	0(0)	0(1)
Ictaluridae	0(3)	4(7)	6(6)	9(11)	5(8)	9(11)
Aphredoderidae	0(0)	0(1)	0(1)	1(1)	1(1)	1(1)
Percopsidae	0(1)	1(1)	1(1)	1(1)	1(1)	1(1)
Gadidae	1(1)	1(1)	1(1)	1(1)	1(1)	1(1)
Cyprinodontidae	0(0)	1(3)	0(0)	1(2)	2(2)	2(4)
Poeciliidae	0(0)	0(1)	0(0)	0(1)	0(0)	0(1)
Atherinidae	0(0)	0(1)	0(1)	1(1)	1(1)	1(1)
Gasterosteidae	0(0)	0(0)	0(0)	0(0)	1(1)	1(1)
Percichthyidae	1(1)	1(2)	1(1)	2(2)	2(2)	2(3)
Centrarchidae	1(6)	7(11)	8(10)	10(12)	8(11)	11(13)
Percidae	3(8)	6(16)	8(13)	13(17)	10(14)	14(22)
Sciaenidae	0(0)	1(1)	1(1)	1(1)	1(1)	1(1)
Cottidae	4(4)	4(4)	4(4)	4(4)	4(4)	4(4)
Totals	44(80)	76(140)	93(115)	101(131)	95(127)	128(177)

The total of six endemic forms for the North American Great Lakes compares poorly with the number of endemics (mostly cichlids) found in the great rift lakes of Africa (e.g. 80% of Lake Tanganyika's 247 species) (Greenwood 1984 and Payne 1986 in Wootton 1990). However, the African Great Lakes have had more time for speciation to occur, since they were formed several million years ago. Due to the relatively recent origin of North American glacial lakes, historical events of colonization have apparently had more effect on their species richness than processes of speciation.

Biological Factors Affecting Species Diversity

Major biological factors affecting species richness and evenness in fish assemblages are productivity, competition, predation, and disturbance. High diversity is possible where there is high production, although an increase in total fish production does not necessarily lead to an increase in fish diversity (Wootton 1990). Rates of primary production and the morphoedaphic index (used by fisheries biologists to predict fishing yields in lakes) are highest in Lake Erie and lowest in Superior (Table 1) in concordance with highest and lowest species richness, giving credence to Wootton's theory. The theory is further supported by a comparison to Africa's Lake Victoria. It is one of the most productive lakes in the world at 1700-3800 mg C m^{-2} day^{-1} (Wetzel 1983) and contains 238 species of fish (Greenwood 1984) — twice as many as Lake Erie.

The effect of interspecific competition on species diversity of fish assemblages is best evaluated in situations where introductions have occurred (Wootton 1990). Invasions of the Great Lakes by planktivorous rainbow smelt (*Osmerus mordax*) and alewife (*Alosa pseuodharengus*) provide two well documented examples.

Smelt were reported first in Lake Michigan in 1923 and soon afterwards in all the other lakes; its initial establishment was attributed to the 1912 stocking of smelt eggs in Crystal Lake, which drains into Lake Michigan (Christie 1973, Emery 1985). Since smelt have the same diet and habitats as lake herring (*Coregonus artedi*) (Anderson 1970 in Smith 1972 and Christie 1974), observed occurrences (in several lakes) of declines in lake herring following increases by smelt were suggestive that competition with smelt was involved in decreases of lake herring (Smith 1972, Christie 1974). Competitive interactions with smelt may have negatively affected other native species, including the coregonid *Coregonus hoyi* in the upper lakes and blue pike in Lake Erie (review in Emery 1985), although factors in addition to competition were involved in declines by these species.

Alewives were first reported in Lake Ontario in 1873, with the Erie Canal as the most probable route of entry (Smith 1970, Emery 1985). They moved into Erie by 1931, and colonized the upper lakes over the next 23 years (Smith 1970, Christie 1974). Alewife populations increased dramatically in Ontario prior to the turn of the century and in Huron and Michigan in the 1950's and 1960's (Smith 1970, Christie 1974). Comparable population increases did not occur in Erie due to its shallowness, or in Superior due to its low temperature (Smith 1972, Bronte et al. 1991).

Competition with alewives was considered one of several factors contributing to population declines of various fishes in the lakes, including lake herring, deep-water ciscoes (*Coregonus* spp.), yellow perch (*Perca flavescens*), emerald shiners (*Notropis atherinoides*), and rainbow smelt (reviews in Smith 1972, Christie 1974 and Emery 1985). However, recent research in Lake Michigan has indicated that predation by alewives on eggs and larvae of the other species may have been more important causes of those species' declines than competition (Crowder 1980, Eck and Wells 1987).

Although smelt and alewife invasions resulted in changes in abundance of native species, there is no evidence that the two species have had a significant effect on species richness of the fish communities (via elimination of native stocks). Documented occurrence of thermal (depth) segregation of coexisting populations of planktivorous fish (including alewife and smelt) in Lake Michigan (Brandt et al. 1980) indicates that resource partitioning permits potentially competing species to coexist, and may explain why competitive interactions do not result in elimination of less favored forms. Structurally complex habitats offer more opportunities for resource partitioning and permit greater species diversity than homogeneous habitats (Wootton 1990). Apparently thermal variations in the waters of Lake Michigan produce a type of habitat heterogeneity that permits the continued but fragile existence there of a diverse assemblage of deep-water planktivores.

Predation may have a negative effect on the diversity of fish communities (Wootton 1990), as exemplified by the previously cited effect of alewife predation on early life stages of other fishes in Lake Michigan (Crowder 1980, Eck and Wells 1987). The effects of the sea lamprey (*Petromyzon marinus*) invasion of the Great Lakes is another example. Lampreys were first recorded in Lake Ontario in the 1830's and were thought to have arrived there via the Erie Canal (Emery 1985). The jawless predator apparently entered Erie through the Welland Canal sometime prior to 1921, and established spawning populations in Huron, Michigan and Superior by 1947 (Lawrie 1970, Smith and Tibbles 1980, Emery 1985).

In the three upper lakes, the establishment of the sea lamprey was followed in the 1940's and 1950's by sharp decreases in populations of lake trout (*Salvelinus namaycush*) (Lawrie 1970, Smith and Tibbles 1980, Emergy 1985). Lake trout populations in Lake Ontario had already been depressed due to the lamprey, and disappeared completely at the same time trout were eliminated in the other lakes (Christie 1974). After the disappearance of lake trout, the lamprey preyed on other deep-water fishes, resulting in declines of burbot (*Lota lota*), lake whitefish (*Coregonus clupeaformis*), and white sucker (*Catostomus commersoni*) (Christie 1974, Smith and Tibbles 1980). With the depletion of burbot stocks in Michigan and Huron, the lamprey subsequently preyed on the two largest cisco species (*C. johannae* and *nigripinnis*), which were reduced nearly to extinction (Christie 1974, Smith and Tibbles 1980). Declines by the major piscivores (lake trout and burbot) in Michigan and Huron subsequently led to in an increase by *C. hoyi,* the smallest of the ciscoes (Smith and Tibbles 1980). A similar increase in populations of the smaller ciscoes occurred in Lake Ontario (Christie 1974).

The lamprey control program initiated in the 1950's (Smith and Tibbles 1980) permitted re-establishment (through stocking) of salmonids (lake trout, chinook and coho salmon, and brown and rainbow trout) in the Great Lakes (Kitchell and Crowder 1986). Salmonid reintroduction helped control alewife populations that had developed in the absence of piscivores in the lakes, and resulted (at least in Michigan) in the recovery of yellow perch and deepwater cisco stocks that had declines due to negative effects of alewives (review in Kitchell and Crowder 1986). This phenomenon was cited by Wootton (1990) as exemplary of predation benefitting fish species richness by preventing a competitively dominant species from reaching levels that might exclude other species.

The negative effects of the lamprey were complicated by human exploitation of their prey (Wootton 1990). Lake trout populations in the upper lakes were already stressed by commercial exploitation in the 1940's (and therefore vulnerable to collapse) when the sea lamprey arrived (Christie 1974, Smith and Tibbles 1980, Kitchell and Crowder 1986). The extirpation of *C. nigripinnis* and *C. johannae* in Huron and *C. kiyi* and *C. reighardi* in Ontario were also considered more the result of overfishing than an effect of the sea lamprey (Christie 1974).

Intermediate levels of disturbance may have a positive effect on fish species diversity by reducing competitive stress and permitting less favored forms to increase (review in Wootton 1990). Certain Wisconsin lakes have more diverse summer mudminnow/cyprinid assemblages because low levels of dissolved oxygen in the winter apparently reduce competitive and/or predatory interactions that might otherwise exclude certain species (Tonn and Magnuson 1982). Diminished oxygen concentrations due to eutrophication in Lake Erie apparently had a similar effect, by favoring pollution tolerant species such as yellow perch, freshwater drum (*Aplodinotus grunniens*), fathead minnow (*Pimephales promelas*), spottail shiner (*Notropis hudsonius*), smelt, and centrarchids (Christie 1974). The previously mentioned effect of reintroducing salmonids on alewife-afflicated communities in the Great Lakes is also exemplary of a disturbance benefitting fish species diversity.

Water quality changes due to eutrophication were not considered important contributors to fishery declines in Huron, Michigan and Superior during the 1950's, but did have negative effects on Erie and Ontario (review in Christie 1974). In Ontario, oxygen depletion associated with nutrient increases contributed to the decline of whitefish and proliferation of the introduced, pollution-tolerant white perch (*Morone americana*) (Christie 1974). In Erie, oxygen depletion was blamed for increased mobility of blue pike stocks (causing greater vulnerability to fishing) and the loss of benthic populations of the burrowing mayfly (*Hexagenia*) (Christie 1974). The blue pike's extinction has been attributed to several factors in addition to changes in water quality, including overfishing (Smith 1972), smelt predation and genetic introgression with the closely related walleye (*Stizostedion vitreum vitreum*).

THE EFFECT OF RECENT INTRODUCTIONS ON LAKE ERIE

Leach and Nepszy (1976) wrote a thorough review of the changes in Lake Erie's fish community through the early 1970's. The lake has a long history of accidental and intentional introductions perceived to have had significant effects on native fish communities. Coho salmon and rainbow trout (steelhead strains) were intentionally introduced to augment a declining sport fishery of native percids and to fill a vacant cold-water niche in the lake's deeper meso-oligotrophic eastern basin. The continued presence of these anadromous fishes is dependent upon recruitment from hatchery sources, because their reproduction in the lake's tributaries is limited and non-sustaining. Addition of these exotics to the native fauna has increased Erie's species richness, but could induce ecological perturbations that may disrupt community structure or contribute to the extirpation of native species. An outcome of this sort has been observed within salmonid/alewife assemblage of Lake Michigan. Depredation by introduced Pacific salmon there resulted in severe depletion of alewife stocks, necessitating the development of approaches to management more cognizant of the sensitivity of this novel predator-prey relationship. Fishery biologists have been confronted by similar problems with introduced salmonids in Lake Ontario (Christie et al. 1987).

Such problems result from fundamental changes in how energy is transferred among trophic levels within the lakes' fish communities, and may have profound future effects on biodiversity of the Great Lakes. Undisturbed, Lake Erie and the other Great Lakes maintained complex, multiple-species food webs linking the various trophic levels within their fish communities. Resource partitioning within the ecosystems permitted the coexistence of diverse assemblages of predator and prey species in relative equilibrium. The disturbance of this equilibrium due to introductions, overfishing, eutrophication and eventual depletion or extirpation of native fishes during the present century (reviewed earlier in this paper) was followed by attempts to restore the communities. Restoration, however, was founded on much simpler predator-prey assemblages. Alewife replaced the once diverse coregonine and sculpin-rich forage base, and fast-growing, voracious Pacific salmon and steelhead were added to lake trout at the terminus of the food web. Such simplified systems, driven by predation and food availability, may be prone to rapid and severe oscillations in predator/prey biomass. Any additional stress imposed upon such systems, due to overstocking predators or overfishing may lead to collapse or severe depletion of either one of the components.

In Lake Erie, rainbow smelt are a major forage species in the eastern basin; native cyprinids and gizzard shad (*Dorosoma cepedianum*), are also important prey species in the fish community of the lake's central and western basins. Introduced coho salmon and rainbow trout, the reintroduced lake trout, percids and burbot all depend upon smelt to maintain their fisheries. The smelt are also heavily exploited by an extended trawl fishery. Thus the decline of native forage species has redirected energy flow through a forage population essentially composed of one species of fish. As indicated earlier, any instability or disturbance within such a simplified system has few buffers against severe oscillations in abundance of fish stocks.

Biologists are presently observing a serious decline of smelt in Lake Erie, evident only during the past few years. The competition for this resource has not yet been resolved, but reduced harvest quotas and stabilized salmonid stocking are being considered. Precise explanations for the change in smelt abundance remain speculative. However, the changes observed in smelt populations seem to be following the same course that has been observed for many introduced and colonizing exotics: a period of niche exploration followed by population explosion and then decline to a lower level at equilibrium with coexisting species and the adopted habitat.

A more recent invader, the zebra mussel (*Dreissena polymorpha*), is now well established in the Great Lakes. First detected in western Lake Erie in 1988, it has now colonized all three basins of the lake, and its initial impacts on the ecosystem are now being documented. The exotic mussel's rapid colonization has been augmented by its occupation of a vacant niche in the lake. There are no native organisms in the productive littoral waters of Erie that utilize firm, exposed substrates (rock shoals, breakwalls, pilings, etc.) as permanent habitat.

Zebra mussel settlement is measured in terms of thousands of individual animals per square meter. By 1991, however, available evidence suggested that the zebra mussel population (in the western end of the lake) had apparently reached its maximum abundance and was leveling off (Ontario Ministry of Natural Resources 1991).

The massive biomass and high rate of colonization of the filter-feeding mussel has enabled it to capture a significant portion of the ecosystem's primary production. Significant declines in standing crop of phytoplankton (indicated by sharp reductions in chlorophyll *a*) have occurred in Erie since the mussel's first appearance in 1988, and seasonal transparencies have increased by more than 75% (Ontario Ministry of Natural Resources 1991). Concern is increasing about the long term effects of the exotic mussel on the lake's food web, since its feeding behavior results in the rapid transfer of organic matter from the pelagic to the benthic zone. In view of the already fragile nature of the fish community previously discussed, the redirection of energy flow through the ecosystem via zebra mussels may have dramatic effects in the future on fish species diversity in the lake.

LITERATURE CITED

Anderson, E.D. 1970. Factors affecting abundance of lake herring *Coregonus artedii* Lesueur, in western Lake Superior. Ph.D. thesis, University of Minnesota: 316 pp.

Bailey, R.M. and G.R. Smith. 1981. Origin and geography of the fish fauna of the Laurentian Great Lakes Basin. Can. J. Fish. Aquat. Sci. 38:1539-1561.

Barbour, C.D. and J.H. Brown. 1974. Fish diversity in lakes. Am. Nat. 108:473-489.

Brandt, S.B., J.J. Magnuson and L.B. Crowder. 1980. Thermal habitat partitioning by fishes in Lake Michigan. Can. J. Fish. Aquat. Sci. 37:1557-1564.

Briggs, J.C. 1986. Introduction to the zoogeography of North American fishes. In: The zoogeography of North American freshwater fishes. C.H. Hocutt and E.O. Wiley (Eds). John Wiley and Sons: 1-16.

Bronte, C.R., J.H. Selgeby and G.L. Curtis. 1991. Distribution, abundance, and biology of the alewife in U.S. waters of Lake Superior. J. Great Lakes Res. 17:304-313.

Christie, W.J. 1974. Changes in the fish species compositions of the Great Lakes. J. Fish. Res. Bd. Can. 31:827-54.

Christie, W.J., K.A. Scott, P.G. Sly and R.H. Strus. 1987. Recent changes in the aquatic food web of eastern Lake Ontario. Can. J. Fish. Aquat. Sci. 44:37-52.

Crowder, L.B. 1980. Alewife, smelt and native fishes in Lake Michigan: competition or predation? Environ. Biol. Fishes 5:225-233.

Eck, G.W. and L. Wells. 1987. Recent changes in Lake Michigan's fish community and their probable causes, with emphasis on the role of alewife (*Alosa pseudoharengus*). Can. J. Fish. Aquat. Sci. 33:53-60.

Emery, L. 1985. Review of fish species introduced into the Great Lakes, 1819-1974. Great Lakes Fish. Comm. Tech. Rept. 45.

Greenwood, P.H. 1984. African cichlids and evolutionary theories. In: Evolution of Fish Species Flocks. A.A. Echelle and I. Kornfield (Eds). University of Maine Press, Orono: 141-154.

Hough, J.H. 1958. Geology of the Great Lakes. University of Illinois Press, Urbana: 313 pp.

Hubbs, C.L. and K.F. Lagler. 1964. Fishes of the Great Lakes region. University of Michigan Press, Ann Arbor: 213 pp.

Hutchinson, G.E. 1957. A Treatise on Limnology Vol. I — Geography, Physics, and Chemistry. John Wiley & Sons, New York: p. 168-169.

Kitchell, J.F. and L.B. Crowder. 1986. Predator-prey interactions in Lake Michigan: model predictions and recent dynamics. Env. Biol. Fish. 16:205-211.

Lawrie, A.H. 1970. The sea lamprey in the Great Lakes. Trans. Amer. Fish. Soc. 99:766-775.

Leach, J.H. and S.J. Nepszy. 1976. The fish community in Lake Erie. J. Fish. Res. Bd. Can. 33:622-638.

Lesht, B.M. and D.J. Brandner. 1992. Functional representation of Great Lakes surface temperatures. J. Great Lakes Res. 18:98-107.

MacArthur, R.H. and E.O. Wilson. 1967. The theory of island biogeography. Monographs in Population Biology 1, Princeton University Press, Princeton, New Jersey: 203 pp.

Ontario Ministry of Natural Resources. 1991. Lake Erie Fisheries Report. mimeo. Lake Erie Committee Meeting, 1992, Erie, Pennsylvania: 96 pp.

Payne, A.I. 1986. The ecology of tropical lakes and rivers. J. Wiley, New York.

Quinn, F.H. 1992. Hydraulic residence times for the Laurentian Great Lakes. J. Great Lakes Res. 18:22-28.

Ryder, R.A. 1972. The limnology and fishes of oligotrophic glacial lakes in North America (about 1800 A.D.). J. Fish. Res. Board Can. 29:617-628.

Scott, W.B. and E.J. Crossman. 1973. Freshwater fishes of Canada. Bull. Fish. Res. Bd. Can. 184:1-966.

Simberloff, D.S. 1974. Equilibrium theory of island biogeography and ecology. Ann. Rev. Ecol. Sys. 5:161-182.

Smith, B.R. and J.J. Tibbles. 1980. Sea lamprey (*Petromyzon marinus*) in Lakes Huron, Michigan and Superior: history of invasion and control, 1936-78. Can J. Fish. Aquat. Sci. 37:1780-1801.

Smith, S.H. 1970. Species interactions of the alewife in the Great Lakes. Trans. Am. Fish. Soc. 99:754-765.

Smith, S.H. 1972. Factors of ecological succession in oligotrophic fish communities of the Laurentian Great Lakes. J. Fish. Res. Bd. Can. 29:717-730.
Tonn, W.M. and J.J. Magnuson. 1982. Patterns of species composition and richness of fish assemblages in northern Wisconsin lakes. Ecology 63:1149-1166.
Underhill, J.C. 1986. The fish fauna of the Laurentian Great Lakes, the St. Lawrence Lowlands, Newfoundland and Labrador. In: The zoogeography of North American freshwater fishes. C.H. Hocutt and E.O. Wiley (Eds). John Wiley and Sons: 105-136.
Wetzel, R.G. 1983. Limnology. Saunders College Publishing, Philadelphia: 767 pp.
Wootton, R.J. 1990. Ecology of teleost fishes. Chapman and Hall, New York: 404 pp.

Biological Diversity: Problems and Challenges. Edited by S.K. Majumdar, F.J. Brenner, J.E. Lovich, J.F. Schalles, and E.W. Miller. © 1994, The Pennsylvania Academy of Science.

Chapter Thirty

BIODIVERSITY AND ZOOGEOGRAPHY OF NON-MARINE TURTLES IN SOUTHEAST ASIA

JEFFREY E. LOVICH
U.S. Department of the Interior
National Biological Survey
Palm Springs-South Coast Resource Area
63-500 Garnet Avenue
P.O. Box 2000
North Palm Springs, California 92258-2000

INTRODUCTION

Analysis of patterns of global and regional biodiversity is an important part of modern conservation programs. Identification of centers of high diversity or endemism provides information for reserve design and location, and targets high priority areas for focused conservation efforts. Additionally, analysis of species richness and distribution provides an important zoogeographic perspective for elucidating the evolutionary history of species groups of interest as well as the geological histories of the land-masses they occupy (Cox et al., 1976).

Several world regions have recently been identified as having exceptional endemism or biodiversity. Using plants as indicators, Myers (1988) identified 10 areas that occupy only 0.2% of the earth's land surface yet contain 34,000 endemic plant species, or about 27% of all tropical plant species. When non-endemics are added to the figure, approximately 35-40% of all plant species are found in these "threatened hotspots." Mittermeier et al. (1992) noted that at least 19% of the world's amphibians and reptiles are also found in the "hotspots" identified by Myers.

Conservation priorities are, by necessity, shaped largely by politics and geo-political boundaries, stressing the need to develop programs on a country-by-country basis. Seven "megadiversity" countries, Mexico, Columbia, Brazil, Zaire, Madagascar, Australia, and Indonesia, contain more than 54% of the world's plant and animal species (Mittermeier, quoted in Anon., 1988; Mittermeier, 1988; McNeely et al., 1990). Several other countries have since been added to the megadiversity list based on high endemism or species richness including Peru, Ecuador, Venezuela, China, India, and Malaysia (Mittermeier et al., 1992). These countries are important targets for focused conservation efforts.

Turtles, as an order of approximately 257 species (Ernst and Barbour, 1989), have received limited attention from the standpoint of regional or global biodiversity conservation, although some groups such as sea turtles (Bjorndal, 1981) and tortoises (Swingland and Klemens, 1989) have been subjects of considerable conservation effort. Iverson and Etchberger (1989) reviewed patterns of distribution of turtles in Florida, USA, and Beaman et al. (1990) discussed conservation issues for selected turtle species. The general lack of interest in turtle biodiversity is surprising for several reasons. First, turtles are an important component of many ecosystems, often contributing significant biomass (Iverson, 1982; Congdon et al., 1986; Congdon and Gibbons, 1989). Second, many turtle species have been exploited as a source of human food for centuries (Clark and Southall, 1920; Carr, 1952; Moll, 1976; Taylor, 1982). Third, turtles form important linkages in ecosystems, providing dispersal mechanisms for plants (Braun and Brooks, 1987; Iverson, 1987; Kaczor and Hartnett, 1990), contributing to environmental heterogeneity (Kaczor and Hartnett, 1990), and fostering important symbiotic associations with a diverse array of organisms (Lago, 1991; Witz et al., 1991).

Two previous studies examined global patterns of turtle distribution and biodiversity. A compendium of species richness maps was generated by Iverson (1992a). His survey identified several world regions as having high turtle diversity, including the southeastern United States (see also Iverson and Etchberger, 1989), and Southeast Asia. Iverson (1992b) examined the relationship between turtle species richness and twelve environmental variables in major river drainage systems on five continents. Annual rainfall was the only highly significant correlate.

The primary objective of this study was to analyze the patterns of biodiversity of non-marine turtles in island-rich Southeast Asia, a megadiversity area for turtles and other organisms. Since modern patterns of species richness are the result of adaptation, dispersal, speciation, human transport, extinction and vicariance events, it is important to place biodiversity studies within the context of zoogeography in an attempt to elucidate the processes shaping today's patterns (Cox et al., 1976). The geological history of Southeast Asia and its effect on various groups of plants and animals have been studied intensively (see reviews by Whitten et al., 1987a, 1987b), thus providing an opportunity to test patterns of one group against that of another in an effort to ascertain congruence. Non-marine turtles provide an interesting group to study due to their potentially high dispersal capability in island systems compared to other non-volant vertebrates.

METHODS AND MATERIALS

For the purpose of this analysis, Southeast Asia was considered to be all of the Southeast Asian land mass and associated islands south of 20 degrees North latitude, east of approximately 92 degrees East longitude, and west of 127 degrees East longitude (Figure 1). Most of the land in this area sits on the Sunda Shelf, a shallow continental shelf approximated by the 120 m bathymetric line. With the exception of the Philippines and the island of Sulawesi, this area is generally west of Wallace's Line (Whitten et al., 1987a) demarcating the separation between the Oriental and Australian zoogeographic regions. The area of interest was divided into seven major land masses based primarily on the insular nature of the Indo-Malay Archipelago including Sumatra, Borneo, Java, Sulawesi, and the Philippines. Due to the depauperate turtle fauna of the Philippines, the island nation was treated as a single unit in subsequent analyses even though it is composed of over 7,000 individual islands. Two other land masses were considered. The first is Peninsular Malaysia south of its connection with mainland Asia at the Isthmus of Kra. Although not an island, its peninsular geometry and faunal similarity to the other large islands on the Sunda Shelf give it an insular quality (Heaney, 1991a). The second continental land mass considered, the Indochinese Peninsula (hereafter Indochina), was defined as mainland Southeast Asia north of the Isthmus of Kra, south of 20 degrees North latitude, and west of approximately 92 degrees East longitude. The great diversity of turtles in the region, coupled with its proximity and faunal similarity to the islands to the south, establish the area as an important mainland reference point. Land area of the various regions was obtained from atlases or measured from an Azimuthal projection map with a compensating polar planimeter.

The list of turtle species occupying the seven geographic regions discussed above (Table 1) was generated from the distribution maps of Iverson (1992c). Questionable records, namely those for *Indotestudo forstenii* from Sulawesi (see Hoogmoed and Crumly, 1984) and *Geoemyda spengleri* from Borneo, were not included. Analyses

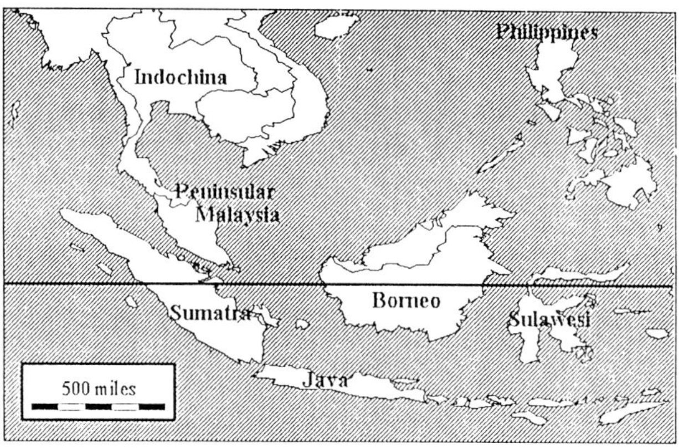

FIGURE 1. Map of the study area defined in the text. The dark horizontal line represents the equator.

were undertaken with full recognition that the ranges of some species are poorly known and that the taxonomy and systematic arrangements of Southeast Asian turtle families, including the Bataguridae (Hirayama, 1984; Sites et al., 1984; Gaffney and Meylan, 1988), Testudinidae (Crumly, 1988), and Trionychidae

TABLE 1
List of Turtle Species Occupying the Geographic Regions Discussed in the Text. Species Presence is Indicated with an "x".

	Region						
Family/Species	Indochina	Malaysia	Sumatra	Java	Borneo	Sulawesi	Philippines
Bataguridae							
Annamemys annamensis	x						
Batagur baska	x	x	x				
Callagur borneoensis		x	x		x		
Cuora amboinensis	x	x	x	x	x	x	x
Cuora galbinifrons	x						
Cyclemys dentata	x	x	x	x	x		x
Cyclemys tcheponensis	x						
Geoemyda spengleri	x						
Heosemys grandis	x	x					
Heosemys leytensis							x
Heosemys spinosa		x	x		x		
Hieremys annandalii	x	x					
Hardella thurji	x						
Kachuga trivittata	x						
Malayemys subtrijuga	x	x	x	x			
Mauremys mutica	x						
Melanochelys trijuga	x						
Morenia ocellata	x	x					
Notochelys platynota	x	x	x	x	x		
Ocadia sinensis	x						
Orlitia borneensis		x	x		x		
Pyxidea mouhotii	x						
Sacalia quadriocellata	x						
Siebenrockiella crassicollis	x	x	x	x	x		
Platysternidae							
Platysternon megacephalum	x	x					
Testudinidae							
Geochelone platynota	x						
Indotestudo elongata	x	x					
Manouria emys	x	x	x	x			
Manouria impressa	x	x					
Trionychidae							
Lissemys scutata	x						
Amyda cartilaginea	x	x	x	x	x		
Chitra indica	x						
Dogania subplana	x	x	x	x	x		
Nilssonia formosa	x						
Palea steindachneri	x						
Pelochelys bibroni	x	x	x	x	x		x
Pelodiscus sinensis	x						

(Meylan, 1987), will change as new species are described and phylogenies are reconstructed. Nevertheless, enough information has accumulated to allow at least a preliminary analysis.

Pairwise comparisons of geographic regions were made using Jaccard Coefficients (Cheetham and Hazel, 1969). Values range from 0 in cases where no species are shared between regions, to 1 in cases where the same suite of species are found in both regions. Faunal similarities among regions were grouped using cluster analysis with the single linkage method (Wilkinson, 1986). Species area relationships were fitted using least squares regression.

RESULTS

A total of 37 non-marine turtle species, representing 31 genera and four families, are found in the area of interest (Table 1). The greatest number of species occur in Indochina (33) followed in decreasing order by Peninsular Malaysia (19), Sumatra (13), Borneo (11), Java (8), the Philippines (4), and Sulawesi (1). Overall, the regions closest to the Indochinese Peninsula show fairly high similarity (Table 2). The most similar faunas occur between Sumatra and Borneo, Sumatra and Java, and Peninsular Malaysia and Borneo. Slightly less than one-half of the species found in Indochina are found below the Isthmus of Kra in Peninsular Malaysia. Faunal similarity declines precipitously as distance between regions increases along an east-west axis.

Only one species, *Cuora amboinensis,* is found in all seven geographic regions. However, recent taxonomic work with the genus *Cuora* suggests that it is more speciose than previously recognized (Ernst and Lovich, 1990; McCord and Iverson, 1991; Pritchard and McCord, 1991). The wide range of *Cuora amboinensis,* coupled with island specific variation, suggests that it may actually consist of more than one species (Carl Ernst, pers. comm.). *Heosemys leytensis* of the Philippines is the only insular endemic in the region, although several species including *Annamemys annamensis, Kachuga trivittata,* and *Geochelone platynota* have relatively restricted geographic distributions.

A dendrogram produced by single linkage clustering of all geographic regions shows that Indochina is distinct from the remaining regions (Figure 2). The next

TABLE 2

Matrix of Jaccard Coefficients for All Pairwise Comparisons of the Geographic Regions Discussed in the Text.

	Indochina	Malaysia	Sumatra	Java	Borneo	Sulawesi
Malaysia	0.44	—	—	—	—	—
Sumatra	0.28	0.68	—	—	—	—
Java	0.24	0.42	0.62	—	—	—
Borneo	0.22	0.58	0.85	0.58	—	—
Sulawesi	0.03	0.05	0.08	0.13	0.09	—
Philippines	0.08	0.15	0.21	0.33	0.25	0.25

branch occurs between the Philippines and Sulawesi combined, and Pensinsular Malaysia, Borneo, Java, and Sumatra. Peninsular Malaysia then branches from the other three, and within the latter group, Borneo and Sumatra form another cluster.

The relationship between land area and number of species (Figure 3) is significantly positive ($r = 0.76$; $P = 0.05$). Land area explains 58% of the variance in number of species when fitted with the following model

$$\text{NUMBER OF SPECIES} = 2.488 + [(2.20401 \times 10^{-5}) \times \text{AREA}].$$

The relationship between the two variables is improved ($r = 0.89$; $P < 0.01$) when Peninsular Malaysia is removed from the analysis as an outlier. Similarly, the

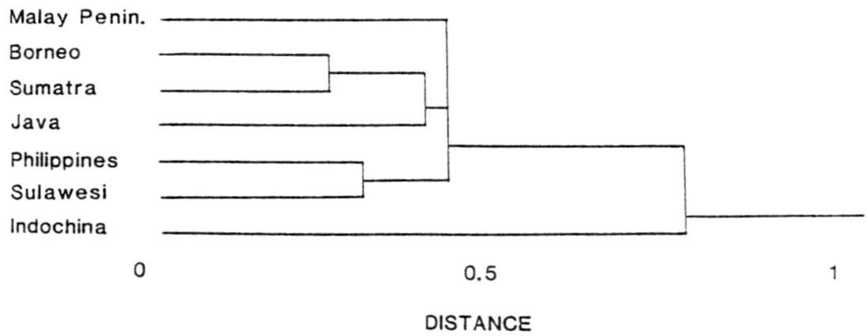

FIGURE 2. Dendrogram of species similarity for the geographic regions discussed in the text. Euclidean distance is shown on the abscissa.

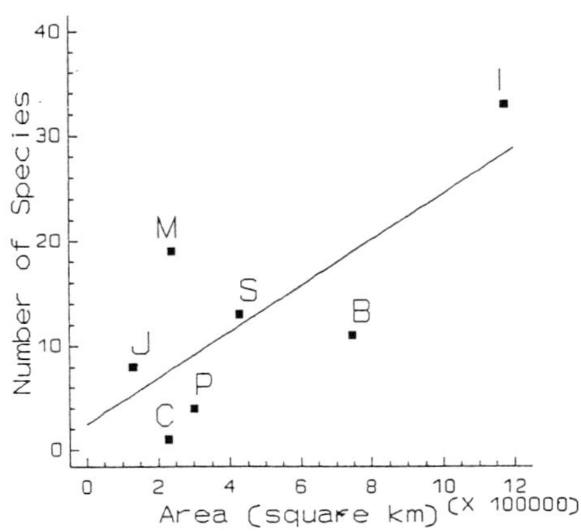

FIGURE 3. Species/area relationship for the geographic regions discussed in the text. Abbreviations are as follows: B = Borneo, C = Sulawesi, I = Indochinese Peninsula, J = Java, M = Peninsular Malaysia, P = Philippines, and S = Sumatra.

relationship between land area and number of genera is significantly positive ($r = 0.77$; $P = 0.04$). Land area explains 60% of the variance in number of genera using the following model (including Peninsular Malaysia)

$$\text{NUMBER OF GENERA} = 2.716 + [(2.00074 \times 10^{-5}) \times \text{AREA}].$$

The relationship between land area and genera is improved ($r = 0.89$; $P < 0.01$) when Peninsular Malaysia is removed from the analysis as an outlier.

DISCUSSION

The modern distribution of the biota of Southeast Asia is shaped largely by the geological history of the area. According to references summarized in Burridge (1992), Peninsular Malaysia, eastern Sumatra, and western Borneo were tectonically stable by the early Tertiary. Some evidence suggests that Sulawesi collided with Borneo in the late Pliocene approximately 3 million years before present (BP), but it is more likely that they have been separated for at least 25 million years (Whitten et al., 1987a), and possibly since the Mesozoic (Inger, 1966). The Philippine Islands originated from the ocean floor during the Miocene and Pliocene (Heaney, 1991a).

During the Pleistocene, most of the islands on the Sunda Shelf were repeatedly connected with and isolated from each other and the mainland due to fluctuating sea levels. In the period of maximum glacial development during the middle Pleistocene, sea levels fell 160 m, and during glacial maxima in the late Pleistocene sea levels fell 120 m (Heaney, 1991b). The most recent exposure of the Sunda Shelf (Sundaland) occurred 11,170-15,000 years BP (Burridge, 1992). The Palawan chain of the Philippines joined Sundaland during the middle Pleistocene, but the other islands of the Philippines remained isolated (Heaney, 1991a). During the last inundation of Sundaland, Sumatra was isolated first from Java, then from Borneo, and finally from Peninsular Malaysia; a sequence mirrored in the degree of similarity between regional biotas (Whitten et al., 1987b).

Most non-marine turtles, or their ancestors, invaded Sundaland from Indochina since none of the modern families of non-marine turtles in Southeast Asia are known from Australia. The Isthmus of Kra was an effective filter to dispersal since less than half of the turtle species found in Indochina today are represented in the fauna south of the Isthmus. The effectiveness of the filter is reflected in the flora of Southeast Asia as 375 plant genera reach the northern limit of distribution at the Isthmus, and 200 genera reach their southern limit at the same point (Whitten et al., 1987b). The Isthmus demarcates relatively dry vegetative communities to the north and relatively wet communities to the south. If other turtle species managed to pass through the filter, they may have become extinct due to rising sea levels in Sundaland and associated changes in the environment.

Non-marine turtle dispersal into Sundaland may have been through large river systems that existed when Sundaland was exposed. For example, the drainage pattern of the North Sunda River once linked the west coast of Sumatra and the east coast of Borneo with west Malaysia, and the East Sunda River linked southern Sumatra with Java and Borneo (Burridge, 1992). As continental glaciers melted and sea levels rose, colonizing turtle species were isolated in the headwaters of these ancient rivers. Several waves of colonization must have occurred due to repeated exposure of the Sunda Shelf. Extinction and speciation undoubtedly occurred following colonization and isolation, as seen in the case of the endemic species *Heosemys leytenis* in the Philippines.

Some turtles are capable, or potentially capable, of dispersal over land and across saltwater barriers. Several Southeast Asian species including *Batagur baska, Callagur borneoensis* (Dunson and Moll, 1980; Davenport and Wong, 1986), and *Pelochelys bibroni* (Ernst and Barbour, 1989) tolerate euryhaline conditions predisposing them for dispersal in marine archipelagos. This capability may explan the wide distribution of *Pelochelys*, but it does not explain the absence of *Batagur* from Java and Borneo, or the absence of *Callagur* from Java. Tortoises (Testundinidae) are well known for their ability to float (Patterson, 1973; Branch, 1991), but this capability does not seem to have fostered wide distribution in Sundaland. The depauperate turtle fauna of Sulawesi, long isolated from Borneo, may be due to the fact that many non-marine turtles, including highly aquatic species, require land connections to facilitate dispersal.

In general, species diversity decreases as distance from mainland Asia increases. The pattern of diversity decreasing with distance from a mainland is often characteristic of peninsulas as demonstrated for turtles on the Florida peninsula by Iverson and Etchberger (1989) (but, see Busack and Hedges, 1984). The connection of modern islands on the Sunda Shelf to mainland Asia would have created a peninsula for colonization during periods of lowered sea level. Apparently, few species of non-marine turtles were able to colonize the eastern end of Sundaland as shown by the depauperate modern fauna. Alternatively, extinction rates may have been higher in eastern Sundaland, but information is currently not available to test this hypothesis.

Patterns of turtle species diversity in Southeast Asia generally parallel those of other biota. For example, Java is more different from Sumatra than are Borneo or Peninsular Malaysia (Whitten et al., 1987b), and the biotas of the Philippines, and particularly Sulawesi, generally share few species with the Asian mainland and the intervening large islands. The pattern of increasing turtle diversity with increasing "island" size is mirrored in numerous other groups including plants, other reptiles, birds, and mammals (Whitten et al., 1987a, 1987b; Heaney, 1991b). Peninsular Malaysia stands out among the geographic regions as having exceptional turtle diversity for its size; a possible consequence of its land connection to species rich Indochina.

The rarity of endemic turtle species on islands in the area contrasts with high insular endemism in other biota. Sixty-one percent of the fruit bats in the Philippines are endemic (Heaney, 1991b) and many species of rats are endemic to Sulawesi (Whitten et al., 1987b). The low degree of island endemism in Southeast Asian turtles may be due to high levels of gene exchange among aquatic turtle populations (Scribner et al., 1984). Alternatively, endemism may be masked by a our current perception of the systematics and taxonomy of Southeast Asian turtles.

Southeast Asia has a "megadiverse" turtle fauna with approximately 14% of the world's species. The factors responsible for such a diverse turtle fauna are unknown, but Iverson and Etchberger (1989) and Iverson (1992b) suggested that high turtle diversity in a geographic area is correlated with abundant rainfall, an environmental characteristic of Southeast Asia. The overall homogeneity of species richness patterns throughout Sundaland implies a commonality of processes shaping the distribution of a diverse biota (Heaney, 1991b). The similarity of patterns for turtles and other groups is of interest because of the diversity of modes of dispersal available to turtles (floating, swimming, overland travel, or a combination thereof).

Humans are exerting an increasingly devastating effect on Southeast Asian ecosystems (Aiken and Moss, 1975; Mittermeier and Konstant, 1982). As a result, some turtle populations are declining (Moll, 1976). The future of a diverse turtle fauna in Southeast Asia depends on implementation of effective conservation strategies and establishment of functional preserves or preserve systems. Given the great diversity of turtles in Indochina and Peninsular Malaysia, immediate attention should be directed to these areas.

ACKNOWLEDGEMENTS

Earlier versions of this manuscript benefitted from comments offered by John Iverson, Earlham College; Ed Moll, Eastern Illinois University; Bill Boarman, Bureau of Land Management, California Desert District; Whit Gibbons, University of Georgia, Savannah River Ecology Laboratory; and Elena Misquez, Bureau of Land Management, Palm Springs-South Coast Resource Area. Research and manuscript preparation were partially supported by contract number DE-AC09-76SROO-819 between the United States Department of Energy and the University of Georgia, Savannah River Ecology Laboratory.

LITERATURE CITED

Aiken, S.R. and M.R. Moss. 1975. Man's impact on the tropical rainforest of Peninsular Malaysia: a review. Biol. Conserv. 8:213-229.

Anon. 1988. The other group of seven. The Economist. 307:73-74.

Beaman, K.R., F. Caporaso, S. McKeown, M. Graff. 1990. Proceedings of the First International Symposium on Turtles and Tortoises: Conservation and Captive Husbandry. Chapman University, CA. 171 pp.

Bjorndal, K.A. (ed.). 1981. Biology and conservation of sea turtles. Smithsonian Institution Press, Washington, D.C. 583 pp.

Branch, W.R. 1991. The herpetofauna of the offshore islands of South Africa and Namibia. Ann. Cape Prov. Mus. (Nat. Hist.). 18(10):205-225.

Braun, J. and G.R. Brooks, Jr. 1987. Box turtles (*Terrapene carolina*) as potential agents for seed dispersal. Am. Midl. Nat. 117:312-318.

Burridge, M.F. 1992. Systematics of the *Acanthophthalmus kuhlii* complex (Teleostei: Cobitidae), with the description of a new species from Sarawak and Brunei. Copeia 1992:172-186.

Busack, S.D. and S.B. Hedges. 1984. Is the peninsular effect a red herring? American Naturalist. 123(2):266-275.

Carr, A. 1952. Handbook of turtles: the turtles of the United States, Canada, and Baja California. Comstock Publ. Assoc., Cornell, New York. 542 pp.

Cheetham, A.H. and J.E. Hazel. 1969. Binary (presence-absence) similarity coefficients. Journal of Paleontology. 43:1130-1136.

Clark, H.W. and J.B. Southall. 1920. Fresh-water turtles: a source of meat supply. U.S. Bureau Fish. Doc. 889:3-20.

Congdon, J.D. and J.W. Gibbons. 1989. Biomass productivity of turtles in freshwater wetlands: a geographic comparison. *In,* pp. 583-592. Freshwater wetlands and wildlife. R.R. Sharitz and J.W. Gibbons (eds.). U.S. Dept. of Energy Symp. Ser. No. 61. USDOE Office of Scientific and Technical Information, Oak Ridge, Tennessee.

Congdon, J.D., J.L. Greene and J.W. Gibbons. 1986. Biomass of freshwater turtles: a geographic comparison. Am. Midl. Nat. 115:165-173.

Cox, C.B., I.N. Healy and P.D. Moore. 1976. Biogeography: an ecological and evolutionary approach. John Wiley and Sons, New York. 194 pp.

Crumly, C.R. 1988. A nomenclatural history of tortoises (Family Testudinidae). Smithsonian Herpetological Information Service No. 75:1-17.

Davenport, J. and T.M. Wong. 1986. Observations on the water economy of the estuarine turtles *Batagur baska* (Gray) and *Callagur borneoensis* (Schlegel and Muller). Comp. Biochem. Physiol. 84A(4):703-707.

Dunson, W.A. and E.O. Moll. 1980. Osmoregulation in sea water of hatchling emydid turtles, *Callagur borneoensis,* from a Malaysian sea beach. J. Herpetol. 14:31-36.

Ernst, C.H. and R.W. Barbour. 1989. Turtles of the world. Smithsonian Institution Press, Washington, D.C. 313 p.

Ernst, C.H. and J.E. Lovich. 1990. A new species of *Cuora* (Reptilia: Testudines: Emydidae) from the Ryukyu Islands. Proc. Biol. Soc. Wash. 103:26-34.

Gaffney, E.S. and P.A. Meylan. 1988. A phylogeny of turtles. *In,* pp. 157-219, The phylogeny and classification of the tetrapods, Vol. 1. M.J. Benton (ed.). Clarendon Press, Oxford.

Heaney, L.R. 1991a. An analysis of patterns of distribution and species richness among Philippine fruit bats (Pteropodidae). Bulletin of the American Museum of Natural History. (206):145-167.

Heaney, L.R. 1991b. A synopsis of climatic and vegetational change in Southeast Asia. Climatic Change. 19:53-61.

Hirayama, R. 1984. Cladistic analysis of batagurine turtles. Studia Geologica Salmanticensia, Vol. Especial 1. Studia Palaeocheloniologica. 1:141-157 [Published in 1985].

Hoogmoed, M.S. and Crumly, C.R. 1984. Land tortoise types in the Rijksmuseum van Naturrlijke Historie with comments on nomenclature and systematics (Reptilia: Testudines: Testudinidae). Zool. Med. Leiden 58:241-259.

Inger, R.F. 1966. The systematics and zoogeography of the amphibia of Borneo. Fieldiana: Zoology. 52:1-402.

Iverson, J.B. 1982. Biomass in turtle populations: a neglected subject. Oecologia 55:69-76.

Iverson, J.B. 1987. Tortoises not dodos, and the tambalocoque tree. J. Herpetol. 21:229-230.

Iverson, J.B. 1992a. Species richness maps of the freshwater and terrestrial turtles of the world. Smithsonian Herpetological Information Service. (88):1-18.

Iverson, J.B. 1992b. Global correlates of species richness in turtles. Herpetological Journal. 2:77-81.

Iverson, J.B. 1992c. A revised checklist with distribution maps of the turtles of the world. Privately Published, Richmond, Indiana. 363 p.

Iverson, J.B. and C.R. Etchberger. 1989. The distribution of the turtles of Florida. Florida Scientist 52:119-144.

Kaczor, S.A. and D.C. Hartnett. 1990. Gopher tortoise (*Gopherus polyphemus*) effects on soils and vegetation in a Florida Sandhill community. Am. Midl. Nat. 123:100-111.

Lago, P.K. 1991. A survey of arthropods associated with gopher tortoise burrows in Mississippi. Entomol. News 102:1-13.

McCord, W.P. and J.B. Iverson. 1991. A new box turtle of the genus *Cuora* (Testudines: Emydidae) with taxonomic notes and a key to the species. Herpetologica 47:407-420.

McNeely, J.A., K.R. Miller, W.V. Reid, R.A. Mittermeier and T. B. Werner. 1990. Conserving the world's biological diversity. International Union for the Conservation of Nature and Natural Resources, World Conservation Union, Gland, Switzerland. 193 pp.

Meylan, P.A. 1987. The phylogenetic relationships of soft-shelled turtles (Family Trionychidae). Bull. Amer. Mus. Nat. Hist. 186:1-101.

Mittermeier, R.A. 1988. Primate diversity and the tropical forest: case studies from Brazil and Madagascar and the importance of the megadiversity countries. *In,* pp. 145-154. Biodiversity. E.O. Wilson, (ed.). Natl. Acad. Press, Washington, D.C.

Mittermeier, R.A. and W.P. Konstant (eds.). 1982. Species conservation priorities in the tropical forests of Southeast Asia. Occ. Pap. International Union for the Conservation of Nature, Species Survival Commission. Gland, Switzerland. 58 pp.

Mittermeier, R.A., J.L. Carr, I.R. Swingland, T.B. Werner and R.B. Mast. 1992. Conservation of amphibians and reptiles. *In,* pp. 59-80. Herpetology: current research on the biology of amphibians and reptiles. K. Adler, (ed.). Proceedings of the First World Congress of Herpetology. Soc. Stud. Amphib. Rept., Oxford, Ohio

Myers, N. 1988. Threatened biotas: "hotspots" in tropical forests. Environmentalist 8:187-208.

Moll, E.O. 1976. West Malaysian turtles: utilization and conservation. Herpetol. Rev. 7:163-166.

Patterson, R. 1973. Why tortoises float. J. Herpetol. 7:373-375.

Pritchard, P.C.H. and W.P. McCord. 1991. A new emydid turtle from China. Herpetologica 47:139-147.

Scribner, K.T., M.H. Smith and J.W. Gibbons. 1984. Genetic differentiation among local populations of the yellow-bellied slider turtle (*Pseudemys scripta*). Herpetologica 40:382-387.

Sites, J.W., Jr., J.W. Bickham, B.A. Pytel, I.F. Greenbaum and B.B. Bates. 1984. Biochemical characters and reconstruction of turtle phylogenies: relationships among batagurine genera. Syst. Zool. 33:137-158.

Swingland, I.R. and M.W. Klemens. 1989. The conservation biology of tortoises. Occ. Pap. Inter. Union Cons. Nat., Species Survival Comm. No. 5. 204 pp.

Taylor, R.W., Jr. 1982. Human predation on the gopher tortoise (*Gopherus polyphemus*) in north-central Florida. Bull. Florida State Mus., Biol. Sci. 28:79-102.

Whitten, A.J., M. Mustafa and G.S. Henderson. 1987a. The ecology of Sulawesi. Gadjah Mada University Press, Yogyakarta, Indonesia. 777 p.

Whitten, A.J., S.J. Damanik, J. Anwar and N. Hisyam. 1987b. The ecology of Sumatra. Gadjah Mada University Press, Yogyakarta, Indonesia. 583 p.

Wilkinson, L. 1986. SYSTAT: The system for statistics. SYSTAT, Inc. Evanston, Illinois.

Witz, B.W., D.S. Wilson and M.D. Palmer. 1991. Distribution of *Gopherus polyphemus* and its vertebrate symbionts in three burrow categories. Am. Midl. Nat. 126:152-158.

Biological Diversity: Problems and Challenges. Edited by S.K. Majumdar, F.J. Brenner, J.E. Lovich, J.F. Schalles, and E.W. Miller.

Chapter Thirty-One

VERTEBRATES IN THE PACIFIC NORTHWEST: Species Richness, Endemism and Dependency on Old-Growth Forests

R. BRUCE BURY
U.S. Fish and Wildlife Service
National Biological Survey
4512 McMurry Avenue
Fort Collins, Colorado 80525

INTRODUCTION

The Pacific Northwest region is generally considered to encompass Washington and Oregon in the United States, but I also include northern California (north of San Francisco Bay) because its biota has many northern elements. Southern parts of British Columbia in southwest Canada are also similar in climate and biota, but are not specifically addressed here.

There is national concern about the loss of remaining old-growth forests in the Pacific Northwest (Wilcove 1987, Maser 1988, Norse 1990). These forests are productive, diverse ecosystems that provide essential habitat for many species of plants and animals (Carey 1989, Ruggiero et al. 1991). Controversy continues over protection of the Northern spotted owl, which requires large tracts of old-growth forest to survive (Gutierrez and Carey 1985, Thomas et al. 1990). Although the spotted owl is the "flagship" for wildlife losses from timber harvest in the region, there are many other vertebrate species and environmental issues in the Pacific Northwest that merit attention for both intrinsic and conservation reasons.

This paper addresses biological diversity in the Pacific Northwest and is based on my diverse background as: (1) A native Oregonian, who grew up in a timber family and its boom-bust economy; (2) An avid outdoorsman with a deep concern for wilderness and wildlife; and (3) A research biologist whose career is based on objectivity and data. These concerns or involvement with a timber economy, the environment and science may seem to be mutually exclusive, but I hope these contrasts will provide a perspective to interpret the interrelated issues facing people, wildlife, and biodiversity in Pacific Northwest forests. If not, I admitted my biases.

My objectives are to: (1) describe the prominent features of the region; (2) provide a synopsis of the species richness (total number of species) of vertebrates; (3) identify the endemic species of animals; (4) determine the association or dependency of vertebrates with old-growth stands; and (5) evaluate the effects of timber harvest on the biological diversity and offer suggestions for improved management of wildland resources.

TOPOGRAPHIC AND VEGETATIVE FEATURES

Old-growth temperate forests of the Pacific Northwest are among the most productive in the world. They are characterized by a wide range of tree sizes and ages with deep multilayered canopies, many large-sized trees and snags, and deep accumulations of coarse woody debris (Maser and Trappe 1984, Harmon et al. 1986, Franklin and Spies 1991). The forest floor becomes laden with fallen trees and decaying wood, and both redwood and Douglas fir forests accumulate more plant biomass than tropical forests (Franklin 1988). Pacific Northwest forests contain the world's tallest trees (some that can reach a height of over 100 m) and once were among the world's most extensive stands (Harlow and Harrar 1969).

There are many headwaters, streams and rivers except in xeric areas along the eastern slopes of mountain ranges and southern latitudes. Large amounts of coarse woody debris also occur in most streams and rivers (Sedell and Swanson 1984, Bryant 1985). The forests also provide protection to the soils (particularly on steeper slopes), and maintain high water quality with low sediment loads.

The physical diversity of the region is exceptionally high for many reasons. The mountain ranges are oriented along a north-south axis and are dissected by rivers than generally flow east to west. The Olympic Mountains dominate the landscape in northwestern Washington; to the south, the Coast Ranges extend from southwestern Washington to southwestern Oregon. The Cascade Mountains are an interior north-south chain in each state. The Klamath Mountains in southern Oregon and northern California form another massive range. The Coast Range in northern California has coastal and interior ridge lines that are roughly parallel to each other. The great topographic relief led to development of varied soils, slopes and aspects, resulting in a mosaic of many biotic communities in close proximity.

Most of the region has a temperate climate but ranges from wet and cool near the ocean to subalpine/alpine areas (frequent freezing temperatures) in the Cascade Range and dry, hot valleys in the interior. Plant communities also vary regionally along several gradients: (1) continentally, from the coast (wetter) to inland (drier);

(2) elevationally (usually wetter and cooler with increasing elevation), and (3) latitudinally from north (cooler) to south (warmer).

The marked variation in physical and biological conditions in the Pacific Northwest led to one of the most diverse floras in North America (Barbour and Billings 1988). Diversity is enhanced because the region is at a major biotic interface between northern (conniferous forests) and southern (chaparral, oak-woodland) elements (Barbour 1988, Franklin 1988). For example, in a 100-km transect across northernmost California, one passes through coastal scrub, redwood forest, mixed conifers-hardwoods (oak, madrone), chaparral, subalpine forests and meadows on mountain peaks, ponderosa pine forest, oak-woodland and grassland. Plant endemism in the Klamath-Siskiyou Mountains in southern Oregon and northern California is among the highest known in the United States (Whittaker 1960).

VERTEBRATE SPECIES

There are more than 640 vertebrate species (excluding marine forms) in the Pacific Northwest, including about 180 species of freshwater and anadromous fishes and about 460 amphibians, mammals, birds, and reptiles (Figure 1). The number of

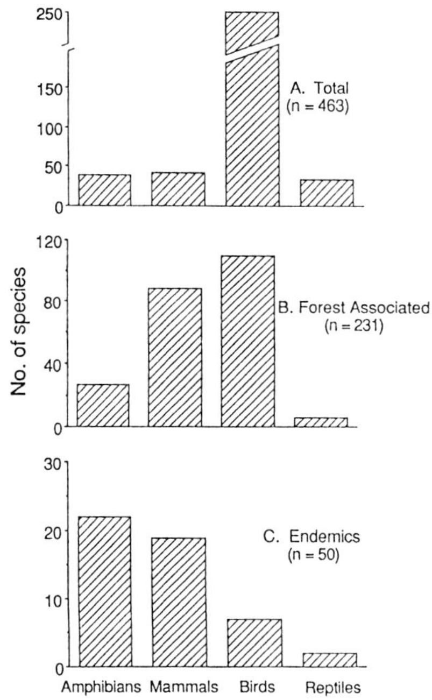

FIGURE 1. Number of species of vertebrates in the Pacific Northwest (Washington, Oregon and northern California). A. Species richness (total number of species). B. Species associated with forests to some extent. C. Endemic species.

species in the Pacific Northwest is comparable to other rich faunal areas of about the same geographic size in the southwestern and southeastern United States (Bury et al. 1980). Many of the species have forest association, i.e., occuring almost exclusively in forests or numerically more common in forests than other habitats (following Ruggiero et al. 1991).

Among ecosystems in the United States and Canada, the Pacific Northwest has the highest number of birds (about 250 species) and the most bird families (Harris 1984). However, Ralph et al. (1991) pointed out that this estimate was for all breeding birds, not just forest-dwelling species, and that most species of forest birds occur in the northeastern United States (140 species), followed by the Rockies and northwest (each with 110 species) and then the southeast (100 species). Ralph et al. (1991) also found that the Pacific Northwest has the greatest number of endemic birds (7 species) of any North American forest region (Table 1), followed by the southeast (6 species), northeast (1 species) and Rocky Mountains (none). There are few endemic birds in regional forests of North America because most species of birds have large or continental geographic ranges.

Mammals are an important group regionally. Raphael (1990) reported that mammalian species richness in the Pacific Northwest is exceptional, second only to the Sierra Nevada of California within the United States and Canada. He reported that 110 species occur in western Washington and Oregon, and 97 in northern California. About 75% of the mammals are associated with forests to some extent. Calculations from range maps (Hall 1981) indicate that there are at least 19 species of mammals endemic to Washington and Oregon west of the Cascade crest, and northwestern California (Table 1).

The herpetofauna has relatively few species (Table 2) but most amphibians are endemics (Bury 1988, Welsh 1990, Bury et al. 1991). Some of those previously known as one species now consist of four recognized species in the Giant salamanders,

TABLE 1

Endemic Amphibians, Mammals and Birds of the Pacific Northwest

Amphibians (22 species)	Mammals (19 species)	Birds (7 species)
Olympic salamander (4 species)	Townsend's mole	Hermit warbler
Giant salamanders (3 species)	Shrew-mole	Chestnut-backed chickadee
Northwestern salamander	Shrews (2 species)	Red-breasted sapsucker
Oregon slender salamander	White-footed vole	White-headed woodpecker
Woodland salamanders (6 species)	Red tree mouse	Rufous hummingbird
Shasta salamander	Red-backed mouse	Vaux's swift
California newt	Voles (3 species)	Northern spotted owl
Tailed frog	Mountain beaver	
Cascade frog	Pacific jumping mouse	
Undescribed (3 species)	Forest deer mouse	
	Chipmunks (3 species)	
	Olympic marmot	
	Mazama pocket gopher	
	Cascade golden-mantled squirrel	

genus *Dicamptodon* (Good 1989), and 3-4 species of Olympic salamanders, genus *Rhyacotriton* (Good et al. 1987, Good and Wake *in press*). These salamanders are all endemic species associated with forested habitats. Amphibians are important because in forested ecosystems they often exceed both mammals and birds in density and biomass (Burton and Likens 1975, Bury 1988). There are few species of reptiles. Most reptiles in the Pacific Northwest are marginal in occurrence because they tend to be most abundant in arid areas to the east (Great Basin) and south (Central Valley of California).

To compare taxa regionally, I first examined species over a broad geographic area (Washington, Oregon and northern California from San Francisco north and east to Nevada), following Ralph et al. (1991), who compared birds over this region. I compiled a list of the amphibians, mammals and reptiles of this area to compare with the bird data (Figure 1A). The regional fauna (463 species) is predominantly comprised of birds (54%) and mammals (31%), followed by amphibians (8%) and reptiles (7%). Half of the fauna (231 species) is associated with forests to some degree. The frequency of forest species (Figure 1B) is reduced because vertebrates in this broad area include a number of arid land species occurring in the Great Basin and the Central Valley of California. For endemics (50 species), the order of frequency (Figure 1C) is reversed for all taxa except mammals: amphibians (58% are endemic), mammals (13%), reptiles (6%) and birds (3%).

If the geographic coverage is restricted to northwestern California and west of the Cascade Mountains crest in western Oregon and Washington (hereafter "westside"), 66-70% of the fauna (excluding birds) are associated with forests (Table 2). In westside areas, 70-80% of the amphibians and mammals are associated with forests to some extent. Still, some species are not forest-associated because they are wide-ranging or occur in semiarid and arid enclaves (e.g., grassland, chaparral) in interior valleys and on southern or eastern slopes of the mountain ranges. There is high endemism for amphibians, but few reptiles (Table 2). Mammals have high endemism. Although there are a moderate number of total vertebrate endemics (50 species) in the Pacific Northwest, the occurrence and ecological roles played by most of these species are barely known.

TABLE 2

Resident Species of Forest Vertebrates in Two Regions of the Pacific Northwest, Excluding Birds

Region	Amphibians	Mammals	Reptiles
Western Oregon and Washington			
Total species	25	110	19
Forest associated (and %)	20 (80)	83 (75)	5 (25)
Endemics (and %)	16 (64)	17 (16)	2 (11)
Northern California			
Total species	20	97	20
Forest associated (and %)	14 (70)	71 (73)	5 (20)
Endemics (and %)	9 (45)	12 (17)	1 (5)

TIMBER HARVEST

"In the United States, the management of old-growth forests in the Pacific Northwest, and political events that surround this issue, could emerge as the most significant natural resources issue of this century" –Thomas (1991).

Managed forests are not as structurally complex or biologically diverse as the forests they replace because logging practices do not replicate natural processes of wild forests. Clearcutting, the dominant forestry practice of recent decades, removes all standing trees and is usually followed by burning of slash (limbs, snags and cull logs). Slash burning reduces ground cover by 50% or more (Bartles et al. 1985, Maser 1988). Predicted losses of snag and log abundance are 70% after the first harvest period and 94% by the end of the second harvest (Spies and Cline 1988). Thus, managed stands have little or no large coarse woody debris.

Most managed forests consist of even-aged stands with only 1-2 dominant tree species and, increasingly, logged areas are being replanted with cloned saplings (i.e., monocultures) thereby reducing genetic variability. Managed forests have younger trees (smaller size), less ground litter, fewer or no snags, and a more simplified ecological system than old-growth forests (Hansen et al. 1991).

The landscape is also influenced by past logging practices. Logging 30 or more years ago left cull logs and snags behind, and there was little or no slash burning following logging. The second-growth stands resulting from earlier logging practices retained more plant biomass in the forest than managed stands from recent forestry practices. Earlier practices were not driven by any inherent conservation ethic about the environment, but generally were based on ecomonic decisions to harvest the larger, more valuable trees, and often ignored smaller or cull trees.

Tree plantations and harvested acreage now predominate on private lands of the Pacific Northwest, and are increasing on public lands. The average age of harvested trees from Forest Service lands has been more than 150 years, but within the next 50 years the average age of trees harvested may drop to 90 years or less (Sessions et al. 1990). Timber harvest occurs on shorter rotations on private lands.

DEPENDENCY OF WILDLIFE ON OLD-GROWTH FORESTS

Old-growth forests provide an environment for a diverse assemblage of plants and animals that includes many endemic and unique species of birds, mammals, amphibians, invertebrates, fungi and higher plants (Harmon et al. 1986, Carey 1989, Norse 1990, Ruggiero et al. 1991). The river systems also contain an abundance of salmonid fishes, nongame fishes, amphibians and other biota. Water is usually both abundant and permanent, but is ephemeral or limited in southern and inland locales. Clearcut logging usually increases sedimentation in streams, which has negative impacts on aquatic biota (Bisson and Sedell 1984, Hartman et al. 1984, Schrivener and Anderson 1984, Murphy et al. 1986, Corn and Bury 1989).

Timber harvesting has fragmented remaining old-growth forests into habitat "islands" in a sea of managed forests (clearcut or young stands). Many animals, particularly smaller-sized species, usually do not move across clearcuts, perhaps leading to isolated populations with potential for local extinction by stochastic events (e.g., drought). Fragmentation of forests threatens extinction of several species and subspecies of animals (Thomas et al. 1990).

The loss of habitat and fragmentation of forests by timber harvest have led to reductions in populations of species dependent on older forests. Many studies now clearly indicate that clearcut and managed stands have fewer species and lower abundance of wildlife than old-growth forests (Bury 1983; Raphael 1984, 1988; Corn and Bury 1989; Ruggiero et al. 1991). The use and protection of remaining ancient forests is an urgent concern because up to 80-87% of the original forest has been logged (Norse 1990), and many of the remaining stands are not secure. Thus, species that depend on old-growth forests have already lost the majority of their preferred habitat.

Although seldom discussed, managed forests are now the predominant forest type in the Pacific Northwest. The fact is, when properly managed, these logged stands have great potential to maintain populations of vertebrates. Harvested lands are vital to the welfare of wildlife and biodiversity in general. Many species may recover faster and closer to former levels in logged forests when there is retention of essential habitat features (e.g., snags, buffer strips on streams). Old-growth parcels near harvested stands are probably important as reservoirs for recolonization of wildlife in logged areas. Clearly, protection and enhancement of wildlife will be more effective when we integrate protection of old-growth reserves (habitat islands) with management programs on harvested forest (the majority of the landscape).

VIEWPOINT

There has been a major loss of forest habitat in the Pacific Northwest because most of the old-growth forests of Douglas-fir (up to 87%) and redwood (over 90%) have been harvested for timber in the last 150 years. Almost all of these lands are open to logging; thus the forests are a treasure house for both natural resources values and the economic livelihood of the region. However, the consequences of large-scale timber harvest have not been adequately addressed from a long-term perspective.

Recovery of forests and wildlife depends on carryover of biomass (mostly large downed woody debris) into harvested stands, protection of buffer zones, and retention of live trees. Such silvicultural practices can be implemented or developed in harvest plans. Although some successional species (e.g., deer) may benefit from the opening of forest canopies, clearcutting in Pacific Northwest forests appears to be detrimental to most wildlife resources, especially endemic species. For example, clearcutting appears to have reduced greatly the occurrence and abundance of the spotted owl (Thomas et al. 1990), red tree vole (Carey 1989), and Olympic salamander (Bury and Corn 1988).

Streams and riparian zones are particularly sensitive to clearcut logging and should have the highest priority for protection (Bury and Corn 1988, Raedeke 1988, Bury et al. 1991). Often streams become choked by sedimentation from excessive logging in one watershed; this effect may be mitigated by reducing the size of harvested areas and staggering the times of logging in individual watersheds. The value of small waterways (those without salmon spawning areas) are often ignored in timber plans and, thus, timber harvest appears to have clogged the headwaters (the arteries of the system). However, there is increased attention being paid to basin-scale design considerations for streamside management (Swanson and Franklin 1992), which may improve watershed integrity from headwaters to rivers.

The future of the few remaining stands of old-growth forests in the Pacific Northwest are increasingly debated by two divergent camps: total protection versus immediate harvest. We need to break out of this deadlock and realize that there are other options and considerations. The majority (85% or more) of the Pacific Northwest has been logged and will continue to be logged. Thus, the regional economic vitality should be based primarily on sustained use of this large landscape, and not on the harvest of remaining patches of old-growth forest.

On the other hand—and seldom recognized—many forest lands thought to be "old-growth" (e.g., most stands along the Oregon coast) actually are second-growth forests that were logged around the turn of the century. Thus, old-growth characteristics—much downed wood, many snags, and large-sized trees—have persisted because of "sloppy" logging practices and harvest about 100 years ago. Forests regenerate and grow fast in the Pacific Northwest. Thus, forests can recover from negative effects of logging, and today we need to focus on ways to accelerate the recovery process in harvested stands to favor wildlife, fisheries and other societal values.

Logging is a process that can be changed. Thus, I challenge foresters, land managers and biologists to develop improved techniques to harvest forest products while retaining wildlife and other values of forests. I offer three examples of actions that may address both economic and environmental concerns.

First, employ selective logging (removal of only the most valuable trees) to maintain coarse woody debris, protect buffer strips and retain natural features (e.g., seeps, creekside foliage) in stands during and after timber harvest. This technique may be more labor intensive than clearcutting, but it may fulfill the cry to retain more jobs regionally.

Second, it would be prudent to leave large pieces of coarse woody debris intact during logging, rather than follow current policies that require loggers to haul the material into piles for burning. It seems logical to burn highly flammable slash (limbs, small pieces) resulting from tree harvest, but it is ecologically unsound to pile up and burn the residual coarse woody debris (especially material already on the forest floor before logging). Loggers can save time-consuming jobs by leaving the large pieces of downed wood in place and thus have less to burn. At the same time, the large debris retains valuable plant biomass for recovery of the ecosystem.

Third, effective protection of streams and streamside corridors is obtainable because riparian zones require retention of only 5-10% of the forest landscape (Bury and Corn 1988). This minor portion of the landscape is crucial to the survival of sensitive aquatic species and most of the forest species, which concentrate in riparian zones.

Important steps are being taken to improve timber harvest practices. The size of individual clearcuts are being reduced in size and buffer strips are increasingly provided along larger streams with salmonids. Another change is improved forestry practices on private lands, which is important because animals do not read boundary markers.

"New Perspectives" is a U.S. Forest Service initiative to balance timber harvest with ecological and other values in forests (Kessler et al. 1992, Swanson and Franklin 1992). Although some quarters question if results have fully lived up to the program's promise (Foss 1992), the program provides guidelines for sustained yields of timber and greater attention to wildlife, watersheds and other forest values.

To aid all of these efforts, we need innovative research and management programs to replicate old-growth conditions in harvested stands, to improve protection of waterways, and to determine the lower limits of habitat area required by both vertebrate species and wildlife communities. As Hall et al. (1985) stated, "Creative skill and cooperation is required by both the forester and biologist to describe optimum habitat conditions, to prescribe management objectives, and . . . to accomplish management on the ground." Key elements are to experiment with new technologies and to sell their worth as cost-effective to resource managers and of protective value to conservationists.

Although many relict and endemic species are on state lists of concern or candidate species for federal listing, we are woefully ignorant of where these species now occur, their status, and habitat requirements. However, the task to conserve them is feasible because there are only about 50 endemic species of vertebrates in the Pacific Northwest, many of which share the same patches of remaining habitat. To guard against loss of the most important biological diversity in the Pacific Northwest, I suggest that a regional strategic plan is needed to: (1) identify areas with high species richness (Scott et al. 1987).; (2) document occurrence of endemics (often not in areas of highest species richness); (3) determine habitat requirements of endemic species; (4) protect unique or specialized habitats, and (5) coordinate interagency efforts.

Some endemic species may persist in areas with appropriate logging practices (e.g., employ selective logging and avoid sensitive habitats). If an endemic species is threatened by resource exploitation, however, I think that full habitat protection is the prudent decision. With a few notable exceptions (e.g., spotted owls need large areas of old-growth), the total land area involved may be of little or minor consequence. Most endemic species occupy small areas or specialized habitats.

Soon, we may reach our consumptive limit of harvesting old-growth trees in the Pacific Northwest because only a small fraction (10-15%) of the original forest will be left when planned timber sales are completed by the year 2000. Some of the remaining wildland resources in the Pacific Northwest are irreplaceable and, to me, there is no urgency to exploit them. For all the products yielded and profit earned from the forests of the Pacific Northwest, we collectively could leave the remaining ancient groves as a tithe.

ACKNOWLEDGEMENTS

I thank Jeffrey Lovich, Gordon Rodda, and Jaclyn Wolfheim for their reviews and comments. This paper does not represent the opinion nor policy of the Fish and Wildlife Service or any other federal agency. All conclusions, suggestions and interpretations are my own.

LITERATURE CITED

Barbour, M.G. 1988. California upland forests and woodlands. Pp. 131-164 *in* M.G. Barbour, and W.D. Billings, eds. 1988. North American Terrestrial Vegetation. Cambridge Univ. Press, Cambridge.

Barbour, M.G., and W.D. Billings, eds. 1988. North American Terrestrial Vegetation. Cambridge Univ. Press, Cambridge. 434 pp.

Bartles, R., J.D. Dell, R.L. Knight and G. Schaefer. 1985. Dead and down woody material. Pp. 171-186 *in* E.R. Brown, ed. Management of Wildlife and Fish Habitats in Forests of Western Oregon and Washington. U.S. Forest Service, Pacific Northwest, No. R6-FW&L-192.

Bisson, P.A. and J.R. Sedell. 1984. Salmonid populations in streams in clearcut vs. old-growth forests of western Washington. Pp. 121-129 *in* W.R. Meehan, T.R. Merrell, Jr., and T.A. Hanley, eds., Fish and wildlife relationships in old-growth forests. Am. Inst. Fish. Res. Biol., Juneau, AK.

Bryant, M.D. 1985. Changes 30 years after logging in large woody debris, and its use by salmonids. Pp. 329-334 *in* R.R. Johnson, C.D. Ziebell, D.R. Patton, P.F. Ffolliot, and R.H. Hamre, tech. coords. Riparian ecosystems and their management: reconciling conflicting uses. First North American Riparian Conference. U.S. Forest Serv. Gen. Tech. Rep. RM-120.

Burton, T.M. and G.E. Likens. 1975. Salamander populations and biomass in the Hubbard Brook Experimental Forest, New Hampshire. Copeia 1975:541-546.

Bury, R.B. 1983. Differences in amphibian populations in logged and old growth redwood forest. Northwest Sci. 57:167-178.

Bury, R.B. 1988. Habitat relationships and ecological importance of amphibians and reptiles. Pp. 61-76 *in* K. Raedeke, ed. Streamside management: riparian wildlife and forestry interactions. Univ. Wash. Inst. For. Resour., Contrib. 59.

Bury, R.B., H.W. Campbell and N.J. Scott. 1980. Role and importance of nongame wildlife. Proc. N. Am. Wildlife Nat. Resour. Conf. 1980:197-207.

Bury, R.B. and P.S. Corn. 1988. Responses of aquatic and streamside amphibians to timber harvest: a review. Pp. 165-184 *in* K. Raedeke, ed. Streamside management: riparian wildlife and forestry interactions. Univ. Washington Inst. For. Resour., Contrib. 59.

Bury, R.B., P.S. Corn, K.B. Aubry, F.F. Gilbert and L.L.C. Jones. 1991. Regional comparisons of aquatic amphibian communities. Pg. 353-362 *in* L.F. Ruggierio, K.B. Aubry, A.B. Carey, and M.H. Huff, tech. coords. Wildlife and vegetation of unmanaged Douglas-fir forests. U.S. For. Serv. Gen. Tech. Rep., PNW-GTR-285.

Carey, A.B. 1989. Wildlife associated with old-growth forests in the Pacific Northwest. Nat. Areas J. 9:151-161.

Corn, P.S. and R.B. Bury. 1989. Logging in western Oregon: responses of headwater habitats and stream amphibians. For. Ecology Manage. 29:1-9.

Foss, T. 1992. Whither new perspectives? Inner Voice 4(4):1,7.

Franklin, J.F. 1988. Pacific northwest forests. Pp. 103-130 *in* M.G. Barbour, and W.D. Billings, eds. North American Terrestrial Vegetation. Cambridge University Press, Cambridge.

Franklin, J.F. and T.A. Spies. 1991. Composition, function, and structure of old-growth Douglas-fir forests. Pp. 71-80 *in* L.F. Ruggierio, K.B. Aubry, A.B. Carey, and M.H. Huff, tech. coords. Wildlife and vegetation of unmanaged Douglas-fir forests. U.S. For. Serv. Gen. Tech. Rep., PNW-GTR-285.

Good, D.A. 1989. Hybridization and cryptic species in *Dicamptodon* (Caudata: Dicamptodontidae). Evolution 43:728-744.

Good, D.A., G.Z. Wurst and D.B. Wake. 1987. Patterns of geographic variation in allozymes of the Olympic salamander, *Rhyacotriton olympicus*. Fieldiana, Zoology, new ser. 32:1-15.

Gutierrez, R.J. and A.B. Carey, eds. 1985. Ecology and management of the spotted owl in the Pacific Northwest. U.S. For. Serv. Gen. Tech. Rep. PNW-185, 65 pp.

Hall, E.R. 1981. The Mammals of North America. John Wiley & Sons, New York. 1181 pp.

Hall, F.C., C. McComb and W. Ruediger. 1985. Silvicultural options. Pp. 291306 *in* E.R. Brown, ed. Management of Wildlife and Fish Habitats in Forests of Western Oregon and Washington. U.S. For. Serv. Pac. Northwest, No. R6-FW&L-192.

Hansen, A.J., T.A. Spies, F.J. Swanson and J.L. Ohmann. 1991. Conserving biodiversity in managed forests. BioScience 41:382-392.

Harlow, W.M. and E.S. Harrar. 1969. Textbook of dendrology. 5th ed. McGraw-Hill, New York. 512 pp.

Harmon, M.E., J.F. Franklin, F.J. Swanson, P. Sollins, S.V. Gregory, J.D. Lattin, N.H. Anderson, S.P. Cline, N.G. Aumen, J.R. Sedell, G.W. Lienkaemper, K. Cromack, Jr. and K.W. Cummins. 1986. Ecology of coarse woody debris in temperate ecosystems. Adv. Ecol. Res. 15:133-302.

Harris, L.D. 1984. The Fragmented Forest: Island Biogeography Theory and the Preservation of Biotic Diversity. Univ. Chicago Press, Ill.

Hartman, G.F., L.B. Holtby and J.C. Scrivener. 1984. Some effects of natural and logging-related winter stream temperature changes on the early life history of Coho salmon (*Oncorhynchus kisutch*) in Carnation Creek, British Columbia. Pp. 141-149 *in* W.R. Meehan, T.R. Merrell, Jr., and T.A. Hanley, eds. Fish and wildlife relationships in old-growth forests. Am. Inst. Fish. Res. Biol., Juneau, AK.

Kessler, W.B., H. Salwasser, C.W. Cartwright, Jr. and J.A. Caplan. 1992. New perspectives for sustainable natural resources management. Ecol. Appl. 2:221-225.

Maser, C. 1988. The Redesigned Forest. R&E Miles, San Pedro, CA.

Maser, C. and J.M. Trappe. 1984. The seen and unseen world of the fallen tree. U.S. For. Serv. Gen. Tech. Rep. PNW-164.

Murphy, M.L., J. Heifetz, S.W. Johnson, K.V. Koshi and J.K. Thedinga. 1986. Effects of clear-cut logging with and without buffer strips on juvenile salmonids in Alaskan streams. Can. J. Fish. Aquat. Sci. 43:1521-1533.

Norse, E.A. 1990. Ancient Forests of the Pacific Northwest. Island Press. The Wilderness Society, Washington, DC.

Raedeke, K.J., ed. 1988. Streamside management: riparian wildlife and forestry interactions. Univ. Wash. Inst. For. Resour., Contrib. 59. 277 pp.

Ralph, C.J., P.W.C. Paton and C.A. Taylor. 1991. Habitat association patterns of breeding birds and small mammals in Douglas-fir/hardwood stands in northwestern California and southwest Oregon. Pp. 379-393 *in* L.F. Ruggierio, K.B. Aubry, A.B. Carey, and M.H. Huff, tech. coords. Wildlife and vegetation of unmanaged Douglas-fir forests. U.S. For. Serv. Gen. Tech. Rep., PNW-GTR-285.

Raphael, M.G. 1984. Wildlife populations in relation to stand age and area in Douglas-Fir forests of northwestern California. Pp. 259-274 *in* W.R. Meehan, T.R. Merrell, Jr., and T.A. Hanley, tech. eds. Fish and Wildlife Relationships in Old-Growth Forests. Am. Inst. Fish. Resour. Biology, Juneau, AK.

Raphael, M.G. 1988. Long-term trends in abundance of amphibians, reptiles, and mammals in Douglas-fir forests of northwestern California. Pp. 23-31 *in* R. Szaro, D.P. Patton, and K.E. Severson, eds. Management of Amphibians, Reptiles, and Small Mammals of North America. U.S. For. Serv. Gen. Tech. Rep. RM-166.

Raphael, M.G. 1990. Mammals of the ancient forests. Pp. 95-100 *in* E.A. Norse, Ancient Forests of the Pacific Northwest. Island Press. The Wilderness Soc., Washington, DC.

Ruggiero, L.F., K.B. Aubry, A.B. Carey and M.H. Huff, tech coords. 1991. Wildlife and vegetation of unmanaged Douglas-fir forests. U.S. For. Serv. Gen. Tech. Rep. PNW-GTR-285. 533 pp.

Scott, J.M., B. Csuti, J.D. Jacobi and J.E. Estes. 1987. Species richness - a geographic approach to protecting future biological diversity. BioScience 37:782-788.

Scrivener, J.C. and B.C. Anderson. 1984. Logging impacts and some mechanisms that determine the size of spring and summer populations of Coho salmon fry (*Oncorhynchus kisutch*) in Carnation Creek, British Columbia. Can. J. Fish. Aquat. Sci. 41:1097-1105.

Sedell, J.R. and F.J. Swanson. 1984. Ecological characteristics of streams in old-growth forests of the Pacific Northwest. Pp. 9-16 *in* W.R. Meehan, T.R. Merrell, Jr., and T.A. Hanley, eds. Fish and Wildlife Relationships in Old-growth Forests. Am. Inst. Fish. Res. Biol., Juneau, AK.

Sessions, J., K.N. Johnson, J. Beuter and G. Lettman. 1990. Timber for Oregon's Tomorrow: The 1989 Update. Forest Research Lab., Oregon State Univ., Corvallis, OR.

Spies, T.A. and S.P. Cline. 1988. Coarse woody debris in forests and plantations of coastal Oregon. Pp. 5-24 *in* C. Maser, R.F. Tarrant, J.M. Trappe, and J.F. Franklin, tech. eds. From the forest to the sea: a story of fallen trees. U.S. For. Serv. Gen. Tech. Rep. PNW-GTR-229.

Swanson, F.J. and J.F. Franklin. 1992. New forestry principles from ecosystem analysis of Pacific Northwest forests. Ecol. Appl. 2:262-274.

Thomas, J.W. 1991. Research on wildlife in old-growth forests: an attempt at perspective. Pp. 471-473 *in* Ruggierio, L.F., K.B. Aubry, A.B. Carey, and M.H. Huff, tech. coords. Wildlife and vegetation of unmanaged Douglas-fir forests. U.S. For. Serv. Gen. Tech. Rep. PNW-GTR-285.

Thomas, J.W., E.D. Forsman, J.B. Lint, E.C. Meslow, B.R. Noon and J. Verner. 1990. A Conservation Strategy for the Northern Spotted Owl: Report of the Interagency Committee to Address the Conservation of the Northern Spotted Owl. Portland, Oregon, U.S. Govt. Printing Office. Document #1990-791-171/20026.

Welsh, H.H., Jr. 1990. Relictual amphibians and old-growth forests. Conservation Biology 4:309-319.

Whittaker, R.H. 1960. Vegetation of the Siskiyou Mountains, Oregon and California. Ecol. Monogr. 30:279-338.

Wilcove, D.S. 1987. Public lands management and the fate of the spotted owl. Am. Birds 41:361-367.

Biological Diversity: Problems and Challenges. Edited by S.K. Majumdar, F.J. Brenner, J.E. Lovich, J.F. Schalles, and E.W. Miller.

Chapter Thirty-Two

BIODIVERSITY OF PRESQUE ISLE

JOHN M. CAMPBELL[1], JAMES K. BISSELL[2], HARRY N. CUNNINGHAM, JR.,[3] EDWIN C. MASTELLER[3] and JEAN STULL[4]

[1]Department of Biology
Mercyhurst College
Erie, PA 16546
[2]Department of Botany
Cleveland Museum of Natural History
University Circle
Cleveland, OH 44106
[3]Division of Science, Engineering and Technology
The Pennsylvania State University at Erie
The Behrend College
Erie, PA 16563-1200
and
[4]661 Benson Road
Waterford, PA 16441

INTRODUCTION

Presque Isle (PI) is a 1300 ha recurved flying sandspit on the southern shore of Lake Erie in Erie County, PA (Figure 1). It lies at the boundary of two major vegetational provinces in Eastern North America (Bailey 1978): the Laurentian mixed (Northern hardwood) forest and Eastern deciduous (Beech-maple) forest provinces. A second major geographical coincidence ocurring a few miles from the peninsula is the divide between the drainage basins of the St. Lawrence and Mississippi Rivers.

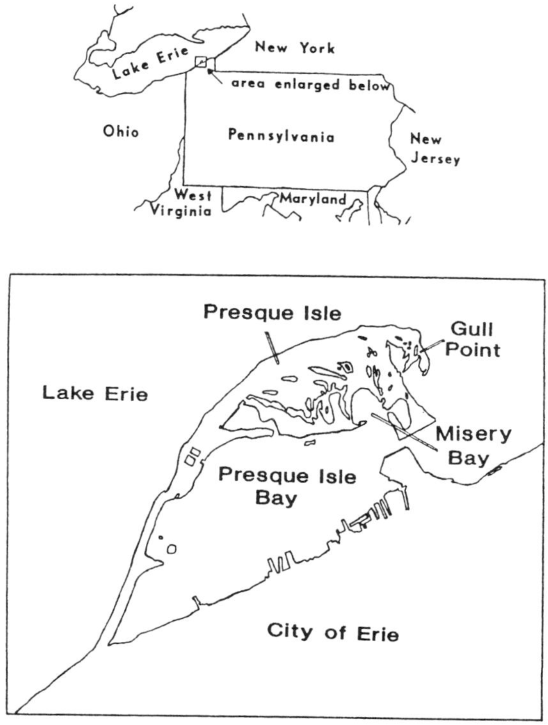

FIGURE 1. Map of Presque Isle and its location in Pennsylvania.

During its recent geological history, Erie County was covered by at least three different glaciers. The last glaciation (the Late Wisconsin - 15 thousand years ago) disrupted drainage patterns and deposited glacial till that has affected soils and vegetation in the region (Taylor 1960). The biotic communities inhabiting the area of PI today represent populations established during only 14,000 years since the glacier receded.

PI was designated a State Park in 1921 and in 1967 was declared a National Natural Landmark. The park contains a mixture of beach, sand dune, wetland and forest habitats that have been disturbed to varying degrees. Jennings (1909) published the first description of the natural processes which formed PI, and Kormondy (1969) reviewed more modern knowledge regarding its geomorphology and ecology. Loeb and Ortiz (1991) summarized the history of water level changes and human activities on the peninsula.

The peninsula encloses a 1287 ha embayment named Presque Isle Bay (PIB), which contains a major Great Lakes port. The bay has been degraded by municipal and industrial pollution and was recently designated as the 43rd Area of Concern on the Great Lakes by the U.S. Department of State. Since the water of PIB is confluent with some of the peninsula's wetlands, the bay will be considered as part of PI.

A long history of biological research begun more than 100 years ago, coupled with numerous investigations in the last 20 years suggests that PI may be a valuable case study in biodiversity. The main purpose of this chapter is to review available information on the biodiversity of PI and to evaluate the evidence for its change during the current century. Wherever possible the effects of human activities on PI's biodiversity will be discussed.

BIODIVERSITY OF PRESQUE ISLE AND EVIDENCE OF CHANGE

Prokaryotes, Algae and Fungi

Twenty-five species of aquatic pathogenic bacteria have been identified at PI beaches, PIB, and MillCreek, which empties into the bay (Zagorski and Galus 1972, Zagorski et al. 1979, Zagorski et al. 1981) (Table 1). Studies of the algal flora of PI ponds and PIB (Morrissey 1929, Gottschall 1933, Gottschall and Jennings 1933, Van Dersal and Cartledge 1933, Kormondy 1969, Zagorski and O'Toole 1970, Zagorski and Gance 1972, Walter 1972, Bardarik et al. 1973) identified 22 species of bluegreen bacteria and 161 species of eukaryotic algae, with green algae (Division Chlorophyta) and diatoms (Division Chrysophyta) having the most representatives (Table 1). Bardarik et al. (1973) compared PIB algae between 1972 and Gottschall's (1933) data and found that Chlorophyta and bluegreen bacteria populations had increased, while Chrysophyta populations decreased during the 40-year span. The compositional shift in the phytoplankton community was attributed to increased eutrophy (nutrient enrichment) of the bay.

Jennings (1909) listed only 17 species of fungi (including one ascomycete and 16 basidiomycetes) in his dissertation on the plants of PI and apparently collected other fungi not named in his paper. No other published records of fungi from PI were found. The lack of published information on PI fungi does not signify a poorly developed fungal flora, as a mycologist from Erie has identified over 230 species at PI in casual collecting from 1988 through 1991 (L.P. Gauriloff, pers. comm.). Gauriloff's list (not documented with voucher specimens) (Table 1) include 27 species of Ascomycetes and over 170 species of Basidiomycetes. The data for prokaryotes, algae and fungi are not sufficient to judge whether biodiversity of these groups has changed.

Plants

A compilation of the published work on bryophytes at PI (H. Webster, pers. comm.) indicates that 76 species of mosses (Jennings 1913, 1951, Boardman 1977) and 7 species of hepatics (Lanfear 1933a, 1933b) have been reported from the peninsula (Table 2). Most of the PI bryophyte taxa have widespread distribution and are weedy to cosmopolitan in nature (H. Webster, pers. comm.). Although good

TABLE 1

Biodiversity of prokaryotes, algae and fungi of Presque Isle. Totals for each major group are shown in parentheses.

Group	*Number of Genera*	*Number of Species*
Prokaryotes		
True Bacteria	12	25
Bluegreen Bacteria	17	22
Total	(29)	(47)
Eukaryotic Algae		
Chlorophyta	58	80
Chrysophyta	36	74
Euglenophyta	3	3
Pyrrophyta	3	3
Cryptophyta	1	1
Total	(101)	(161)
Fungi		
Eumycotina		
Eumycota		
Ascomycetes	20	27
Deuteromycetes	3	5 +
Basidiomycetes	83	171 +
Zygomycota	4	8 +
Mastigomycotina		
Oomycota	4	8 +
Chytridiomycota	5	8 +
Mycetazoans		
Myxamycota		
Protostelida	1	1
Myxomycetes	2	2 +
Acrasiomycetes	2	2
Total	(124)	(232)

TABLE 2

Biodiversity of plants of Presque Isle. Totals for major groups are shown in parentheses.

Group	Number of Families	Number of Species
Bryophytes		
Mosses	23	76
Hepatics	5	7
Total	(28)	(83)
Vascular Plants		
1860-1910 collections	83	407
Carnegie Museum post 1910 additions	15	173
Cleveland Museum 1985-1991 additions	10	219
Total	(108)	(799)

historical records of the bryophytes of the peninsula exist, recent collections and verificiation of previous reports are not complete enough to judge whether changes have occurred.

The diverse vascular plant communities of PI were originally described by Jennings (1909) as exemplary of the ecological theory of succession. The plant communities have recently been studied because of the large number of rare plants (61 species) they contain (Bissell 1992). Early vascular plant surveys at PI were conducted in the 1860's by A.P. Garber and T.C. Porter, and from 1878-1910 by G. Guttenberg, J.A. Shafer, O.E. Jennings and J. Miller. A compilation of the records of these collections from Jennings (1909) and Miller (1923) indicates that 407 species of plants in 83 families were identified at PI prior to 1910 (Table 2). Herbarium specimens in the Carnegie Museum of Natural History (CMNH) of plants collected from PI after 1910 by various botanists document 173 additional species and 15 families that were not reported in Jennings (1909) and Miller (1923). A recent (1985-1991) survey of the vascular plants of PI by J.K. Bissell of the Cleveland Museum of Natural History found 219 more species and 10 families not previously reported for PI. These records increase the cumulative number of vascular plants known to have inhabited the park to 799 species.

The total number of plants found at PI by Bissell during his 7-year survey (605), exceeds the pre-1910 total (407) compiled from Jennings (1909) and Miller (1923). These totals may not be comparable, however, since Bissell's collecting effort (over 100 field days) was probably greater than the early surveys. Bissell's modern survey did not find 80 species of plants collected in pre-1910 studies and 106 of the species that had been deposited in the CMNH after 1910. The 186 species and 12 families of vascular plants presumably lost from the historical flora of PI are approximately balanced by the 219 species and 10 families apparently "gained" (Tables 2 and 3), although the loss of all club mosses (Lycopodiaceae) and violets (Violaceae) from the PI flora is both puzzling and disturbing.

TABLE 3.

Families of vascular plants lost and gained at Presque Isle, based upon historical records and the 1985-91 survey of J.K. Bissell. Numbers in parentheses are the number of species in families lost or gained.

Families Lost	Families Gained
Lycopodiaceae (5)	Zannichelliaceae (1)
Isoetaceae (1)	Commelinaceae (1)
Ophioglossaceae (3)	Aisoaceae (2)
Taxaceae (1)	Oxalidaceae (1)
Juncaginaceae (1)	Cannabinaceae (1)
Aristolochiaeceae (1)	Simaroubaceae (1)
Fumariaceae (2)	Rhamnaceae (1)
Hamamelidaceae (1)	Moraceae (1)
Violaceae (10)	Bignoniaceae (1)
Droseraceae (1)	Verbenaceae (2)
Monotropaceae (1)	
Cucurbitaceae (1)	

The number of alien plant species on PI has increased significantly during the current century. Alien plants composed 24% of the 605 species found in Bissell's recent survey, compared to 10% of the 407 species listed by Jennings (1909) and Miller (1923). One exotic species, *Lonicera morrowi* (Japanese bush-honeysuckle) has extensively invaded sand plains of the peninsula, while another, *Typha* X *glauca* (hybrid cattail) has replaced native plants in marshes of PI. The modern flora of the sand dunes and beaches of PI is very similar to Jennings' (1909) accounts, since these habitats have been little invaded by exotics.

The white-tailed deer population has negatively affected vascular plants in nearly every terrestrial plant community on PI. Overbrowsing has eliminated tree sapling populations in the forests. *Salix* spp. in beach and dune areas where deer frequently graze seldom exceed 30 cm. in height, and *Vaccinium corymbosum* (high bush blueberry) and *Ilex verticillata* (winterberry) have been heavily browsed along marsh shores. *Lithospermum caroliniense* (hispid gromwell) grazed on dry sand plains may soon be extirpated from the peninsula and PA. Deer overpopulation is the suspected cause of extirpation of *Arctostaphylos uvi-ursi* (bearberry - also extirpated in PA), *Prunus pumila* (beach plum), and *Cypripedium acaule* (pink lady's slipper).

Invertebrates

Of approximately 1200 species of invertebrates estimated by E.C. Masteller to occur at PI (Table 4), 850 species can be verified from records detailed below. Masteller estimates that at least 300 species of protozoa may occur at PI, although few voucher records exist due to difficulties in preserving these organisms. Published studies of the plankton of PIB list several species of protozoans (Zagorski and O'Toole 1970, Zagorski and Welz 1973). Encrusting sponges (Porifera) and hydra (Cnidaria) have been collected by E.C. Masteller and Zagorski and Wilcox (1973), respectively. Free-living flatworms, flukes and tapeworms (Platyhelminthes) have also been collected, but no verified records exist. Zagorski and O'Toole (1970) provided a list of rotifers (Rotifera) from zooplankton samples and Bardarik et al. (1973), list the oligochaetes and leeches found in bottom samples from PIB.

Extensive studies have been conducted at PI on many groups within the Phylum Arthropoda, including spiders (Arachnida: Araneae) (Truman 1941), Crustacea (Kormondy 1969, Zagorski and O'Toole 1970, Bardarik et al. 1973, Zagorski and Wilcox 1973) and Insecta. The best known groups are the Odonata (Emery 1933, Kormondy 1969, unpublished data of Masteller and Curtis), Trichoptera (Masteller and Flint 1991), Coleoptera (Gamble 1931, Shively 1938, Kormondy 1969, Gaumer et al. 1970), digger wasps (Pompilidae) (Kurcewski and Kurcewski 1963, 1968) and cuckoo wasps (Chrysididae) (Kurcewski and Kurcewski 1970) in the order Hymenoptera, Diptera (Kormondy 1969, Masteller 1977), and Lepidoptera (butterflies) (McWilliams 1982, Prescott 1984). Most of the Lepidoptera are thought to be temporary residents of PI, although some species may breed there, including the rare eyed-brown (*Satyrodes eurydice*) (G.M. McWilliams, pers. comm.), the wave moth (*Idaea violaccaria*) (C. Bier, pers. comm.), and the PA Threatened regal fritillary (*Speyeria idalia*) (J. Prescott, pers. comm. in Genoways and Brenner 1985).

TABLE 4

Biodiversity of invertebrates of Presque Isle.

TAXON				Number of Species			
				Worldwide	United States	Pennsylvania	Presque Isle
Ciliophora				8000			∫100
Sarcomastigophora				18000			
	Sarcodina			13500			∫ 50
	Mastigophora			8500			
		Phytomastigophora					∫ 50
		Zoomastigophora					∫100
Apicomplexa				4600			∫ 10
Porifera			Spongillidae	9000-150			∫ 1
Cnidaria	Hydrozoa			9000	16		∫ 1
Platyhelminthes				14500-20000			
	Turbellaria			3000-4500			∫ 10
	Trematoda			8000-9000			∫ 5
	Cestoda			3500-5000			∫ 5
Nematoda				12000-16000			∫100
Nematomorpha				230-250			∫ 2
Acanthocephala				700-1200			∫ 2
Rotifera				1500-1800			17
Gastrotrichia			Chaetonotoidae	200 fresh H_2O			∫ 5
Annelida				8700-15000			
	Oligochaeta			3100-3500			26
	Hirudinea			400-500			6
	Polychaeta						1
Mollusca				100000			
	Gastropoda			10000-35000			14
	Bivalvia		Unionidae	25000	40		26
			Sphaeriidae				5
Arthropoda				65000-74000			
Chelicerata	Arachnida			74000			

TABLE 4 *continued*

			Araneae	35000			93
		Acari		30000			3
		Opilione		5000			5
		Pseudoscorpions		2000			1
Uniramia				960,000			
	Chilopoda			2500-3000			5
	Diplopoda			7500-10000			5
	Insecta			860,000			
		Odonata		5500		171	37
		Ephemeroptera		2000		175	3
		Plecoptera		1600		97	
		Trichoptera		7000		311	34
		Coleoptera		339000			100
			Cicindelidae				6
		Diptera	Culicidae	1800-3000	120	41	13
			Chironomidae				25
			Ceratopogonidae				3
			Ephydridae				8
			Chaboridae				1
		Neuroptera	Myrmeleontidae				1
		Lepidoptera		140000			64
		Hymenoptera		130000			250
		Siphonoptera					2
		Heteroptera					15
Crustacea	Branchiopoda	Cladocera		4000			32
	Ostracoda			2000-6650			5
	Copepoda			7500-9000			10
	Peracarida	Isopoda		10000		100	4
		Amphipoda		6000		100	5
	Eucarida	Decapoda		10000			3
Tardigrada				400		90	1
Bryozoa				4000			
	Phylactolaemata				50	20	2

The insect order Neuroptera is represented at PI by the ant lion, an inhabitant of sand dunes. One representative of the order Hemiptera, *Protenor belfragi,* is rare in PA but in present at PI due to the abundance of its host plant, *Calamagrostis canadensis.* The PA extirpated American burying beetle (*Nicrophorus americanus*) was once reported in Erie County (A. Wheeler, pers. comm.) and may have been collected from PI. Although only 10 of the more than 30 orders of insects are listed in Table 4, Masteller believes it is probable that representatives of at least 18 insect orders occur, based upon the habitats available.

Adequate data exists to evaluate changes in the biodiversity of the insect order Odonota, whose adults are conspicuous summer inhabitants of PI ponds. Emery (1933) collected 17 species of dragonflies and 9 species of damselflies in the early 1930's. Kormondy (1969) identified 6 and 9 species of dragonflies and damselflies, respectively in 1960-61. Masteller and Curtis in 1990-91 (unpublished data) found 13 species of dragonflies and 8 species damselflies, producing a total of 37 species of Odonata for PI (23 dragonflies and 14 damselflies). A summary of odonates lost and gained at PI since the early 1930's (Table 5) indicates that the family Gomphidae (Anisoptera) has been most affected, as none have been collected since Emery's study in 1932-33. These changes may reflect human influences or changes due to the natural processes of pond succession.

Data on snails and clams (Phylum Mollusca) inhabiting PI and its surrounding waters (Ortman 1919, Kormondy 1969, Bardaric et al. 1973, Zagorski and Wilcox 1973, unpublished report of Masteller 1990) permit analysis of changes in the biodiversity of unionid clams (Pelcypoda: Unionidae). Of the 25 species of unionids known to occur in Lake Erie waters adjacent to PI, 6 species found in 1919 were not represented among the specimens (over 1000) collected by Masteller during 1990-91 (Table 6), and 4 species identified in the most recent study were not found in 1919.

The invasion (via accidental introduction) of the zebra mussel (*Dreissena polymorpha*) in Lake Erie has negatively affected unionid populations of PI. Zebra mussels, which colonize exposed surfaces of shells of bottom-dwelling unionids, were observed covering the siphons of native unionids in bay waters at the eastern end of PI during the summer of 1990. After a storm in September 1990, nearly 1000 unionids heavily encrusted with zebra mussels were found stranded on the beach at Gull Point. Further decline in unionids is anticipated at PI, similar to what has occurred in western Lake Erie (D.W. Schloesser, pers. comm., Nalepa et al. 1991).

Vertebrates

Sixty species of fish are known from PI and PIB (Table 7), based upon unpublished surveys conducted from 1951 to 1991 by the Pennsylvania Fish Commission (PFC), Bardarik et al. (1973), and personal observations by H.N. Cunningham. This represents more than one half of the 100 species known for Lake Erie and about one-third of the 177 species known to occur in the Great Lakes (Underhill 1986). Twelve of the 60 PI species are considered "Species of Special

TABLE 5

Species of Odonota (Arthropoda: Insecta) lost and gained at Presque Isle since 1932 based upon Emery (1933), Kormondy (1969) and unpublished data of Masteller and Curtis (1991).

Species Lost	Species Gained
Lestes disjunctus australis Southern spread-wing	*Enallagma geminatum* Twin-spot bluet
L. forcipatus Forcipate spread-wing	*E. vesperum* Vesper bluet
Enallagma hageni Hagen's bluet	*Epiaeschna heros* Hero darner
E. signatum Brown bodied bluet	*Erythemis simplicicollis* Green-jacket skimmer
Celithemis eponina Eponia spotted skimmer	*Libellula semifasciata* Half-banded skimmer
Pantala flavescens Common globe skimmer	*Perithemis tenera* Common amber wing
P. hymenaea Brown-spotted globe skimmer	*Sympetrum internum* Northern red skimmer
Sympetrum rubicundulum Common red skimmer	
Tramea carolina Carolina raggedy skimmer	
Gomphus exilis Spiked clubtail	
G. spicatus Spiked clubtail	
Hagenius brevistylus Black dragon	

TABLE 6

Species of Unionidae (Mollusca: Pelecypoda) lost and gained at Presque Isle since 1919 based upon Ortman (1919) and unpublished data of Masteller (1991).

Species Lost	Species Gained
Anodontoides ferussacianus (Lea) Cylindrical papershell	*Lampsilis ovata* (Say) Pocketbook
Epioblasma triquetra (Raf.) Snuffbox	*Toxoplasma parvus* (Barnes) Lilliput
Lasmigona compressa (Lea) Brook lasmigona	*Truncilla truncata* (Raf.) Deer-toe
Pleurobema coccineum (Conrad) False pig-toe	*Truncilla donasiformes* Fawnsfoot (I. Lea)
Ptychobranchus fasciolaris (Raf.) Kidneyshell	
Villosa iris iris (Lea) Rainbow shell	

Concern" by the PFC, including: lake sturgeon and eastern sand darter (endangered); burbot (threatened); and spotted gar, longnose gar, bowfin, mooneye, hornyhead chub, warmouth, black bullhead, tadpole madtom, and Iowa darter (candidates for listing as endangered).

An 1896 report by the State Commission of Fisheries indicated that "black bass" (probably largemouth bass) were numerous, and that a large number of fish used the ponds on the peninsula for spawning, especially "grass pike" (probably northern pike). The usual catch in the bay in 1896 included "bass, perch, catfish, pike,

TABLE 7

Biodiversity of classes and orders of vertebrates of Presque Isle. Names and number for classes are shown in parentheses.

Class and Order	Number of Families	Number of Genera	Number of Species
(Osteichthyes)	(20)	(39)	(60)
(Amphibia)	(7)	(7)	(13)
Urodela	4	4	4
Salienta	3	3	9
(Reptilia)	(5)	(14)	(17)
Testudines	4	8	9
Squamata	1	6	8
(Aves)	(46)	(173)	(318)
Gaviiformes	1	1	2
Podicipediformes	1	3	5
Pelecaniformes	3	3	3
Ciconiiformes	2	9	13
Anseriformes	1	15	35
Falconiformes	3	9	18
Galliformes	1	4	4
Gruiformes	2	6	7
Charadriiformes	5	25	62
Columbiformes	1	2	2
Cuculiformes	1	1	2
Strigiformes	2	7	8
Caprimulgiformes	1	2	3
Apodiformes	2	2	2
Coraciiformes	1	1	1
Piciformes	1	5	8
Passeriformes	18	78	143
(Mammalia)	(14)	(23)	(26)
Marsupialia	1	1	1
Insectivora	2	3	3
Chiroptera	1	3	4
Carnivora	3	4	5
Artidactyla	1	1	1
Rodentia	5	10	11
Lagomorpha	1	1	1

and sturgeon." Known changes in the PI fishery include introductions by the PFC in the early 1970's of coho salmon, steelhead trout and chinook salmon. Significant numbers of several species native to Lake Erie have also been stocked into the bay, which provides a high quality sport fishery (PFC 1983, 1988).

Bardarik et al. (1973) found greater diversity of fish on the peninsular side of the bay (37 species) than on the industrialized south shore (30 species), and 31 species were found in Misery Bay (Figure 1), an area rich in aquatic vegetation and apparently unaffected by the municipal and industrial wastes that affect other areas of PIB. Little is known about fish communities in landlocked ponds of PI's interior, although banded killifish have been reported in some of the newly formed ponds on Gull Point (Figure 1) (R. Wellington, pers. comm.).

Unpublished reports of the CMNH and R. McPherson indicate that 27 species of amphibians and reptiles were identified at PI from 1900-1933 (McKinstry et al. 1991). Studies during the last 20 years increased the cumulative total to 30 species (Table 7). The most recent survey in 1988-1990 by McKinstry et al. (1991) found only 20 species, despite more intensive sampling than earlier studies (Hudson 1930, Williams 1931, McKinstry 1975, McKinstry and Felege 1974, McKinstry et al. 1987).

The Blanding's turtle found in McKinstry et al.'s (1991) study (Table 8) had been considered extirpated from PA (Genoways and Brenner 1985), and the specimen was thought to be a waif from across Lake Erie and not a member of an established population. The first records of bullfrog and gray treefrog in the early 1980's coincided with last records of the red-spotted newt and wood frog. McKinstry et al. (1991) suggested that red-spotted newts, wood frogs and queen snakes might be found with more intensive sampling of lagoons, ponds and less accessible terrestrial areas. They thought that nest depredation by skunks, opossum and raccoons might have caused the loss of spotted and eastern box turtles, and that queen snakes may be absent or scarce due to the decline of their preferred food (crayfish). The absence of eastern hognose snakes puzzles McKinstry et al. (1991) because the sandy soil and abundance of toads at PI make ideal habitat for this species. The apparent absence of wood frogs, pickerel frogs and red-spotted newts at PI may be a local manifestation of the global decline of many amphibian populations for unknown reasons (Phillips 1990).

TABLE 8

Amphibian and reptile species lost and gained since 1900 based upon summary in McKinstry 1991.

Species Lost	Species Gained
Red-spotted newt	Gray treefrog
Northern spring salamander (*)	Bullfrog
Pickerel frog	Blandings turtle
Wood frog	
Spotted turtle	
Eastern box turtle	
Queen snake	
Eastern milk snake	
Eastern hognose snake	

(*) Early Carnegie Museum record: It is questionable whether this was from an established population since it was not in its typical habitat.

In 1900, PI was known to be an excellent site for "water birds," and the first formal compilation of birds seen on the peninsula and vicinity listed 237 species (Todd 1904). A modern compilation (Table 7) lists 318 species on PI (Stull et al. 1985), including 36 "Species of Special Concern" (PA Game Commission 1991). Half of the orders and more than a third of PI bird species prefer aquatic habitats, indicating that the proximity of Lake Erie and presence of wetlands accounts for much of the park's avian diversity.

The difference in the number of bird species listed by Todd (1904) and Stull et al. (1985) does not indicate that bird diversity has increased at PI during this century, since the totals resulted from different numbers of observers (fewer than ten and hundreds respectively). In fact, long-term records of birds captured by mist-netting at PI indicate decreasing counts of 63% of the neotropical migrants (G. Czarnecki, pers. comm.) even though the number of kinds of migrating birds has remained high. The numbers of long-distant migrant shorebirds stopping at Gull Point have declined in recent years, along with the duration of their stay (unpublished records of the International Shorebird Survey 1976-1991, compiled by the Manomet Bird Observatory, Manomet, MA). Observers at Gull Point frequently report seeing birds circle and continue their flights without stopping. A possible reason for declining shorebird numbers at PI is degradation of feeding areas by accretion of silty upland sands and gravels used to replenish beaches.

Records of bird nesting activities indicate that as many as 31 breeding species may have been lost since the turn of the century (Table 9), and 18 appear to have been gained (Todd 1904, 1940, Stull et al. 1985, Brauning, 1992). Four of the species gained are southern species whose ranges have extended northward, and ten are species associated with urban and fragmented landscapes. Most birds listed in Table 9 as "breeding species lost" may still be observed at PI, but they no longer nest there. Construction of the park marina and a beach replenishment operation in 1956 eliminated the wetland nesting sites of the Bald Eagle, Sedge Wren (Threatened - PGC 1991), and Black Tern (Endangered). Human activities (boating, picnicing and swimming, running dogs) at Gull Point that interfered with nesting activities may have contributed to the loss of resident Piping Plovers (now Federal Great Lakes Endangered - US Fish and Wildlife Service 1985) and Common Terns (PA Extirpated - Genoways and Brenner 1985).

The earliest inventory of mammals of PI (1934 field notes of J.K. Doutt from CMNH) listed 14 species. Studies and confirmed observations since that time (Czarnecki 1988, Wesdock et al. 1988, Andersen 1991) have increased the number of mammal species known to have inhabited the park to 26 (Table 7). This total compares favorably to numbers of mammals found elsewhere in Erie County (Richmond and Rosland 1949, Ropski and Andersen 1984, Cunningham 1989, 1990). The actual total number of mammal species on the park may be higher than 26, in light of unverified reports of gray fox and coyote, and the likely but undocumented presence of brown rat and house mouse.

One mammal species well represented in Doutt's 1934 study that has probably been lost from the park is the prairie deer mouse, which was last observed in 1959. The mouse was rarely found far from the open beach, suggesting that its absence or scarcity may be due to habitat disturbance caused by the increased recreational use of the park's beaches.

Among the information regarding furbearing mammals of PI found in minutes of meetings of the Park and Harbor Commission from 1927-1963 is a notation that in 1938 over 1500 muskrat houses existed in the park. Apparently the muskrat were so abundant that they interfered with a bridle path that then encircled the peninsula (Fred Ralph, Jr. pers. comm.). The beaver population of PI recently gained attention when a pair felled a row of cottonwood trees adjacent to the main road through the park. Because of the damage, the beavers were live-trapped and relocated in Erie County (Erie Times-News, December 30, 1991). A recent investigation of PI's red fox population by G. Czarnecki (1988) provided details regarding the effects of park visitors on fox behavior.

The most important historical change in the mammal community of PI has been the growth of the white-tailed deer population, following its intentional introduction in 1928 and supplemental additions by the PA Game Commission through 1938 (Minutes of the Park and Harbor Commussion 1927-1963). By 1940 is was suggested that periodic removal of a portion of the deer herd would be necessary to prevent overpopulation (Erie Dispatch-Herald, January 31, 1940). A severe winter die-off of deer in 1974 prompted an annual effort to determine the size of the herd, producing estimates ranging from 48 (1986) to 200 (1975). In 1975 and six times since, removals (6-127 deer) have been made, justified by evidence that the vegetation of PI is severely overbrowsed, and the fact that the deer is the major host of the adult deer tick, *Ixodes dammini* (the vector of the etiological agent of Lyme disease).

TABLE 9

Breeding bird species believed to have been lost and gained at Presque Isle since 1900.

Breeding Species Lost	Breeding Species Gained
American Bittern	Canada Goose
American Black Duck	Common Moorhen
Bald Eagle	Mourning Dove
Northern Harrier	Red-headed Woodpecker
Ruffed Grouse	Red-bellied Woodpecker
Piping Plover	Eastern Phoebe
Common Tern	Tree Swallow
Black Tern	Barn Swallow
Barred Owl	Tufted Titmouse
Belted Kingfisher	Carolina Wren
Acadian Flycatcher	European Starling
Sedge Wren	Warbling Vireo
Veery	Northern Cardinal
Chestnut-sided Warbler	Chipping Sparrow
Rufous-sided Towhee	Common Grackle
Field Sparrow	Brown-headed Cowbird
	House Finch
15 additional species believed to have been lost since 1900 are not listed since their nesting frequency was unclear in Todd (1904).	House Sparrow

SUMMARY AND CONCLUSIONS

To date, 3007 species have been documented to have occurred at PI, including 882 plants, 1251 invertebrates, 434 vertebrates and 440 prokaryotes, algae and fungi. Plants and vertebrates were the most thoroughly investigated major groups, although detailed inventories were available for algae, some insects and molluscs. The number of species lost during the past century exceeded the number gained from odonate insects, unionid clams, reptiles and amphibians, and breeding bird populations. Most of the losses were unexplained, except for some breeding birds lost due to human disturbance of nesting areas. Introductions by humans (both intentional and unintentional) have had important effects on the biodiversity of the park, especially white-tailed deer and exotic plants that have flourished to the detriment of native species. Although the total number of plant species on the park has not changed much during the past century, the percentage of aliens in the flora has increased, and several families (e.g. club mosses) and formerly conspicuous species (e.g. bearberry) have been lost. The fact that nearly 120 "species of special concern" inhabit Presque Isle underscores the need for park management programs that will insure the preservation of the peninsula's diverse natural communities.

ACKNOWLEDGEMENTS

The authors thank Evelyn Anderson for ideas and comments during preparation of the manuscript and valuable contributions toward developing a base of historical and scientific information about Presque Isle.

LITERATURE CITED

Andersen, K. 1991. The seasonal activity of bats on Presque Isle. First Presque Isle Research Symposium (Abstract).

Bailey, R.W. 1978. Description of the ecoregions of the United States, U.S.D.A. Forest Service Intermountain Region, Ogden, Utah.

Bardarik, D.G., J.C. Alden and R.L. Shema. 1973. A study of the operation of a steam electric generating station on the aquatic ecology of Presque Isle Bay, Erie, Pennsylvania. Aquatic Ecology Associates, Pittsburgh, PA:588p.

Bissell, J.K. 1992 (in press). Rare plants and rare natural communities of Presque Isle. Bartonia.

Boardman, C.M. 1977. Mosses not previously reported from Western Pennsylvania. The Bryologist 80:351-352.

Brauning, D.W. (Ed.). (1992). Atlas of Breeding Birds in Pennsylvania. University of Pittsburgh Press, Pittsburgh, PA.

Cunningham, H.N., Jr. 1989. The mammals of a forest-agricultural site in Erie County, PA. J. Pa. Acad. Sci. 63:81-87.

Cunningham, H.N., Jr. 1990. Small mammals on various successional stages of disturbed lands in northeastern Ohio and northwestern Pennsylvania. J. Pa. Acad. Sci. 64:11-18.

Czarnecki, G. 1988. Denning habits of red fox (*Vulpes vulpes*) on Presque Isle, Erie County, Pennsylvania. J. Pa. Acad. Sci. 62:42-43. (Abstract).

Emery, R.G., Jr. 1933. A systematic survey of the dragonflies and damselflies of Presque Isle, with notes on their ecology and distribution. M.S. Thesis, University of Pittsburgh, Pittsburgh, PA: 44pp.

Gamble, J.T. 1931. A systematic survey of the aquatic coleoptera of Presque Isle, Erie, Pennsylvania, with notes on their ecology and distribution. Ph.D. Dissertation, University of Pittsburgh, Pittsburgh, PA: 74pp.

Gaumer, G.C., E.J. Kurczewski and F.E. Kurczewski. 1970. The tiger beetles of Presque Isle State Park, Pennsylvania, Cicindela 2:4-7.

Genoways, H.H. and F.J. Brenner. 1985. Species of Special Concern in Pennsylvania. Special Publication of Carnegie Museum of Natural History 11:1-430.

Gottschall, R.Y. 1933. Limnological studies at Erie, Pa. Ph.D. Thesis, University of Pittsburgh, Pittsburgh, PA.

Gottschall, R.Y. and O.E. Jennings. 1933. Limnological studies at Erie, Pa. Trans. Amer. Microsc. Soc. 52:181-191.

Hudson, H.B. 1930. The distribution and habitat preference of the urodele amphibian *Triturus viridescens* at Presque Isle, Erie, Pennsylvania. Proc. Pa. Acad. Sci. 4:55-58.

Jennings, O.E. 1909. A botanical survey of Presque Isle, Erie County, PA. Annals of the Carnegie Museum 5:145-159.

Jennings, O.E. 1913. A manual of the mosses of Western Pennsylvania. City Mission Publishing Company, Pittsburgh, PA: 429pp.

Jennings, O.E. 1951. A manual of the mosses of Western Pennsylvania and adjacent regions. American Midland Naturalist Monograph No. 6.

Kormondy, E.J. 1969. Comparative ecology of sandspit ponds. The American Midland Naturalist 82(1):28-60.

Kurcewski, F.E. and E.J. Kurcewski. 1963. An annotated list of digger wasps from Presque Isle State Park, Pennsylvania. Entomol. Soc. Wash. 65:141-149.

Kurcewski, F.E. and E.J. Kurcewski. 1968. Host records for some North American Pompilidae (Hymenoptera) with a discussion of factors in prey selection. Kansas Entomol. Soc. 41:1-33.

Kurcewski, F.E. and E.J. Kurcewski. 1970. An annotated list of the cuckoo-wasps from Erie County, Pennsylvania (hymenoptera: Chrysididae). Proc. Entomol. Soc. Washington 72:190-201.

Lanfear, L.H. 1933. Preliminary list of Hepaticae of Western Pennsylvania. Proc. Pa. Acad. Sci. 7:1-5.

Lanfear, L.H. 1933. A manual of the Hepaticae of Western Pennsylvania. Bulletin of the University of Pittsburgh 30:1-10.

Loeb, R.E. and J.A. Ortiz. 1991. Development of vegetation on Presque Isle: Was Otto Jennings seduced by succession? The Journal of Erie Studies 20:20-35.

Masteller, E.C. 1977. Mosquitoes collected with CDC traps in Erie County, Pennsylvania. Proc. Pa. Acad. Sci. 51:117-121.

Masteller, E.C. and O.S. Flint, Jr. 1991. Trichoptera biodiversity of Pennsylvania. Report to Wild Resource Conservation Fund and PA Fish Commission Vol. I and II: 500pp.

McKinstry, D.M. 1975. Notes on the herpetology of Presque Isle State Park, Erie, PA. Bull. Md. Herp. Soc. 11:21-26.

McKinstry, D.M. and S. Felege. 1974. Snakes of northwestern Pennsylvania. Bull. Md. Herp. Soc. 10:29-31.

McKinstry, D.M., R. McPherson, and N. McAllister. 1987. Herpetology of Presque Isle State Park, Erie, Pennsylvania. Bull. Md. Herp. Soc. 23:58-64.

McKinstry, D.M., M.A. Lethaby, and H.N. Cunningham, Jr. 1991. Amphibians and reptiles of Presque Isle State Park, Erie County, Pennsylvania. J. Pa. Acad. Sci. 65:17-23.

McWilliams, G.M. 1982. Checklist of butterflies recorded in Erie County, PA including Preqsue Isle State Park. 5pp.

Miller, J. 1923. Flora of Erie County, Pennsylvania. Published by Miller. 47pp.

Morrissey, R.V. 1929. Survey of algae in ponds on Presque Isle, Erie, PA — Summer, 1928. Proc. Pa. Acad. Sci. 3:47-51.

Nalepa, T.F., B.A. Manny, J.C. Roth, S.C. Mozley and D.W. Schloesser. 1991. Long-term decline in freshwater mussels (Bivalvia: Unionidae) of the Western Basin of Lake Erie. J. Great Lakes Res. 17:214-219.

Ortman, A.E. 1919. A monograph of the naiades of Pennsylvania, Part 3: Systematic account of the genera and species. Memoirs of the Carnegie Museum 8(1):1-384.

Pennsylvania Fish Commission. 1983. Presque Isle Bay (215A) Management Report. Pennsylvania Fish Commission, Division of Fisheries, Fisheries Management Section, January 1983.

Pennsylvania Fish Commission. 1988. Presque Isle Bay Fisheries Assessment. Pennsylvania Fish Commission, Division of Fisheries, Fisheries Management Section, Spring 1988.

Pennsylvania Game Commission. 1991. Pennsylvania's wild birds and mammals: Special Concern Species, Extirpated Species, Established Exotics. PA Game Commission, Harrisburg, PA.

Phillips, K. 1990. Where have all the frogs and toads gone? Bioscience 40:422-424.

Prescott, J.M. 1984. The butterflies of Presque Isle State Park, Erie County, PA. Melsheimer Entomol. Ser. 34:19-23.

Richmond, N.P. and H.R. Rosland. 1949. Mammal survey of northwestern Pennsylvania. PA Game Commission: 67pp.

Ropski, S.J. and K.W. Andersen. 1984. A survey of mammals of the Wattsburg Fen Natural Area. Proc. Pa. Acad. Sci. 58:53-54.

Shiveley, W.F. 1938. Survey of the carrion beetles on Presque Isle, with notes on their ecology. M.S. Thesis, University of Pittsburgh, Pittsburgh, PA: 33pp.

Stull, J., J.A. Stull and G.M. McWilliams. 1985. Birds of Erie County Pennsylvania including Presque Isle. Allegheny Press, Elgin, Pennsylvania: 173 pp.

Taylor, D.C. 1960. Soil Survey: Erie County Pennsylvania. U.S.D.A. Soil Conservation Service (Series 1957, No. 9): U.S. Government Printing Office, Washington, D.C.

Todd, W.E.C. 1904. The birds of Erie and Presque Isle, Erie County, PA. Annals of the Carnegie Museum 2:481-596.

Todd, W.E.C. 1940. Birds of Western Pennsylvania. University of Pittsburgh Press, Pittsburgh, PA.

Truman, L.C. 1941. A taxonomic and ecological study of the spider fauna of Presque Isle. Ph.D. Dissertation, University of Pittsburgh, Pittsburgh, PA: 91 pp.

Underhill, J.C. 1986. The fish fauna of the Laurentian Great Lakes, the St. Lawrence Lowlands, Newfoundland and Labrador. In: The zoogeography of North American freshwater fishes. C.J. Hocutt and E.O. Wiley (Eds). John Wiley and Sons: pp 105-136.

U.S. Fish and Wildlife Service. 1985. Determination of endangered and threatened status for the Piping Plover. Federal Register 50(238):50720-34.

Van Dersal, W.R. and J.L. Cartledge. 1933. The blue-green algae of Pennsylvania. Proc. Pa. Acad. Sci. 7:82-88.

Walter, R.A. 1972. Pond succession on Presque Isle, Erie, Pennsylvania as determined by sediment core analyses for diatoms and various minerals. Senior Thesis, Department of Geology, Allegheny College, Meadville, PA.

Wesdock, P.S., T.E. Zahniser and S.J. Ropski. 1988. A study of the small mammals of Presque Isle State Park, Erie County, Pennsylvania. J. Pa. Acad. Sci. 62:62 (Abstract).

Williams, S.H. 1931. Preliminary report on the annual ecology of Presque Isle, Lake Erie, PA. Proc. Pa. Acad. Sci. 5:88-97.

Zagorski, S.J. and A.J. O'Toole. 1970. A study of phytoplankton and zooplankton of the Bay of Lake Erie at Erie, Pennsylvania, Summer 1969. Proc. Pa. Acad. Sci. 44:127-140.

Zagorski, S.J. and C.B. Galus. 1972. A bacteriological analysis of Presque Isle Bay at Erie, Pennsylvania, 1971. Proc. 15th Conf. Great Lakes Res. 1972:214-220.

Zagorski, S.J. and R.W. Gance. 1972. A comparative study of algae in sandspit ponds on Presque Isle State Park at Erie, Pennsylvania Fall, 1971. Proc. Pa. Acad. Sci. 46:37-39.

Zagorski, S.J. and N.A. Welz. 1973. A comparison of the net plankton method and the plankton centrifuge method in determining plankton populations at Erie, Pennsylvania, Fall, 1972. Proc. PA. Acad. Sci. 47:93-95.

Zagorski, S.J. and G. Wilcox. 1973. Quantitative and qualitative analysis of the macroinvertebrates of Lake Erie, at Erie, Pennslyvania: September-December, 1972. Proc. Pa. Acad. Sci. 47:99-101.

Zagorski, S.J., C. McCutcheon, F. Morosky and R. Raubenstrauch. 1979. A bacterial analysis of Millcreek Stream in Erie, Pennsylvania. Proc. Pa. Acad. Sci. 53:61-64.

Zagorski, S.J., R.J. Gammon, and G. Kraus. 1981. The effect of bacterial population on the quality of two fresh water beaches. Proc. Pa. Acad. Sci. 55:57-61.

Biological Diversity: Problems and Challenges. Edited by S.K. Majumdar, F.J. Brenner, J.E. Lovich, J.F. Schalles, and E.W. Miller.

Chapter Thirty-Three

THE PLATTE RIVER: A Conflict Between a Managed Water Resource and Regional Biodiversity

JOHN F. SCHALLES
Associate Professor of Biology
Creighton University
Omaha, NE 68178-0103

INTRODUCTION

The Platte River Valley in Nebraska has played an important role during the past 160 years in the settlement and development of the western United States. In the nineteenth century, the valley was the gateway to passes in the Rocky Mountains for the Oregon and Mormon trails, the Pony Express, and the nation's first transcontinental railway, overland mail, and telegraph routes. Currently, Interstate Highway 80 and a main line of the Union Pacific Railroad traverse the Platte corridor. Passage of the Homestead Act in 1862 encouraged an influx of settlers and promoted rapid conversion of the central Great Plain's prairies and steppes to agriculture and cattle grazing lands. Demand for irrigation water in the Platte Basin increased steadily and a reservoir storage system holding about 8800 hm^3 (= 7.2 million acre feet) of water was built between 1909 and 1957. This storage is equivalent to about 7 years of recent average annual flow in the Platte in central Nebraska. By 1975, 2.5 million ha were irrigated by surface diversions from the Platte (U.S. and Wildlife Service 1981; hereafter USFWS). The success of this human ingenuity and technological prowess was accompanied, however, with heavy costs to the natural environment. Today, society is only beginning to appreciate the magnitude and continued growth of such costs. In reshaping and controlling the physical landscape, humans have reshaped the biological landscape and dramatically escalated the assault on our planet's biodiversity. The major focus of this chapter is to relate physical landscape disturbance in the Platte River system to species requirements and community structures.

This chapter examines biodiversity issues in the "Big Bend" reach of the Platte River (Figure 1) in central Nebraska. The Big Bend area is the spring staging site for the largest migratory gathering of cranes in the world. An estimated 550,000 sandhill cranes (*Grus canadensis*) spend four to six weeks roosting in the shallow, braided channels of the Platte and feeding in nearby native grasslands and croplands in preparation for migration to high latitude nesting sites in North America and northern Asia. No other location along the crane's migratory route can sustain these vital spring activities. Whooping cranes (*Grus americana*) also use the Big Bend during migration, but more sporadically and for short durations. In addition, at least seven million ducks and geese, including almost all mid continental White-fronted geese (*Anser albifrons*), and many shore and wading bird species use the Big Bend in their migrations. Ninety-eight migratory bird species using the Platte are protected by international treaties and conventions (Platte River Whooping Crane Habitat Maintenance Trust 1990; hereafter Platte River Trust). Two endangered birds, the Interior Least Tern (*Sterna antillarum*) and Piping Plover (*Charadrius melodus*), use the Platte's exposed sand bars and bare islands for nesting and adjacent aquatic habitats for foraging. These facts, coupled with large scale changes to the river and adjacent riparian habitats, have resulted in an intense examination of the Platte system by diverse and often confrontational interests. At

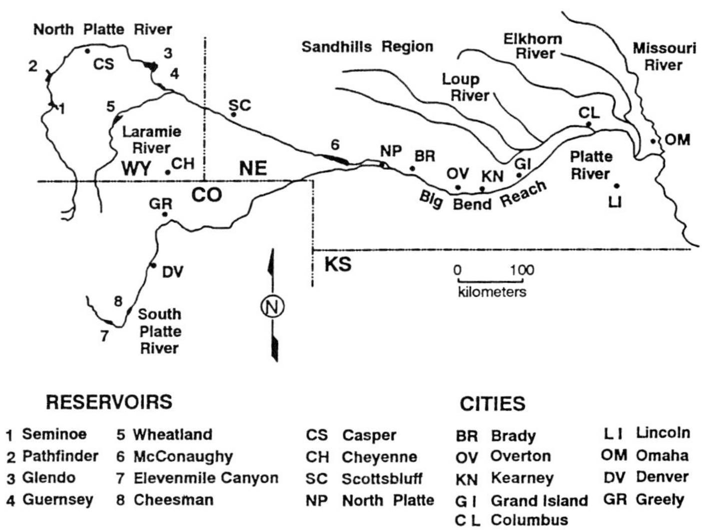

FIGURE 1. Location of the Platte River and its major tributaries (the South Platte, North Platte, Laramie, Loup, and Elkhorn Rivers) and the locations of major storage reservoirs and cities in the Platte drainage. The Big Bend reach of the Platte River Valley is the large arc extending from about Brady to Grand Island in southcentral Nebraska.

stake in the controversies are water resources needed for Denver and other rapidly growing Front Range cities, irrigation supplies in three states, the municipal water supply for 1/3 of Nebraska's population, proposed water projects in the Platte basin, important habitat for five endangered bird species, and, most importantly, one of the largest remaining prairie stream ecosystems in the North American grassland biome.

GEOGRAPHIC SETTING

The entire Platte river system is about 1460 km long and drains 223,000 km^2 before joining the Missouri River 23 km south of Omaha, Nebraska (Figure 1). Most of its water comes from a few sources: snowmelt in the Colorado and Wyoming Front Range, groundwater fed streams draining the porous Nebraska Sandhills, and the Elkhorn River and Salt Creek in eastern Nebraska. The upper and central reaches of the Platte are highly controlled and provide water for irrigation and other uses. Irrigation leads to substantial net water losses (often 70-75%) through cropland evapotranspiration and seepage and evaporation losses from reservoirs and unlined distribution and drainage canals. In the Nebraska portion of the Platte Basin, irrigation accounts for 87% of all water withdrawals, compared to 9% for power plants and 4% for munipical and industrial uses (Gilliland et al. 1985). More importantly, irrigation accounted for about 96% of all water consumption in 1980. The headwater areas of the North Platte River in Wyoming and the South Platte River in Colorado supply most of the water available to the important wildlife habitats in the Big Bend reach in Nebraska.

The North Platte tributary is about 1080 km long, and originates in the Rocky Mountains in northern Colorado and southcentral Wyoming. The river flows north through the semiarid Wyoming Basin to Casper and then arcs southeast across high plains grasslands before joining the South Platte near its namesake city of North Platte, Nebraska (Figure 1). The North Platte and its tributaries are impounded by many reservoirs, with summer water diversions for irrigation. Seminoe and Pathfinder Reservoirs in Wyoming and Lake McConaughy in Nebraska have the largest storages (Figure 1). In addition to flow reduction from irrigation, the North Platte is affected by changes in the timing of flows, with water released to serve the requirements of irrigators and for hydroelectric power generation. Another problem is reservoir sediment entrapment, which contributes to downsteam channel degradation. For example, Lake McConaughy traps about 1.28 million metric tons of North Platte sediment per year (Lyons and Randall 1988). Downstream on the Platte at Overton, Lyons and Randall found a 2/3 reduction in the amount of bed material transported since Lake McConaughy was created.

The South Platte is 730 km long and originates in the Colorado Rockies at Hoosier Pass near Alma, flows southeast for about 95 km, then veers northeast along the eastern flank of the Front Range. A number of mountain tributaries, including the Big and Little Thompson and Cache La Poudre Rivers, join the South Platte in this northward passage. This area has the largest concentration of people in the Platte drainage and includes the Denver metropolitan area, Boulder, Loveland,

and Fort Collins. To increase water supplies in central and eastern Colorado, about 414 hm^3 per year are diverted from the upper Colorado River using a system of four storage reservoirs, tunnels under the Continental Divide, and a storage and distribution system on the east side of the Front Range serving agricultural and municipal users. The unconsumed portion of this interbasin diversion represents a significant supplement (at least 100 hm^3/year) to South Platte flows. Near Greeley, the river arcs eastward towards Nebraska. In eastern Colorado several off-channel reservoirs, fed by diversion canals, store water for irrigation. Although Colorado irrigation diversions substantially reduce net flow, the seasonality of the South Platte's flow regime is less controlled and sediment capture less severe than on the North Platte because of minimal main channel reservoir storage.

The Platte River flows east across Nebraska. In the Big Bend reach, the valley arcs southeast to about Kearney and then northeast past Grand Island (Figure 1). Here, the valley's alluvial bottomlands are broad (up to 24 km in width) and flat, with gently sloping valley sides grading to upland tablelands. Tallgrass prairie formerly dominated the bottomlands. In the Big Bend, 72% of native valley grasslands are farmed and much of the remainder damaged by overgrazing and fire suppression (Currier et al. 1985). Active river channel in this reach has declined 67% (10,500 ha) due to reduced water flows and sediment transport and subsequent woody vegetation encroachment (Currier et al. 1985). Flows in the Platte are much greater and less erratic and channel losses become less severe downstream of the Loup River confluence near Columbus and the Elkhorn River confluence near Ashland. Largely due to these two rivers, average flows increase about 450% in the reach between Columbus and Ashland (Bentall 1982). From Ashland east to the Platte's confluence with the Missouri, the valley is narrower and bordered by heavily wooded, steeper bluffs. In this picturesque setting, many private and public retreats, including Nebraska's two newest state parks, have been established.

Substantial sand and gravel mining occurs in the Platte Valley. Active and abandoned quarries cover 5 ha/km of former river channel in the Big Bend, compared to 28 ha/km of active channel surface in the same reach (Peake et al. 1985). Groundwater fed lakes occupy nearly 50% of the quarry site areas. These lakes are a significant and novel lentic habitat. The lakes are widely used for recreation and bring people closer to important wildlife habitats. A large number of permanent and weekend homes have been built on these lakes. The lakes vary in depth, age, and trophic state (unpublished data and personal observations of author).

In addition to irrigation supply, the Platte is the primary water source for 1/3 of Nebraska's residents. The capital city of Lincoln has highly saline groundwater and streams (i.e. the Salt Creek drainage) and imports water from wellfields adjacent to the Platte near Ashland. Omaha draws some water from wellfields near Plattsmouth and a new field, between Valley and Ashland, is under development for the city's rapidly growing western suburbs.

In its natural state, the Platte River channel is braided and transports substantial bed sediment. Braiding refers to a wide, shallow condition with multiple shifting channels and sand bars. Braiding in the Platte is promoted by a moderate and relatively uniform bed slope (about 1.3 m per km in central Nebraska), large amounts of imported sediments, and scouring from ice flows and high water events. Early descriptions of the Platte indicate the entire channel approached or exceeded 1.5 km in width, with annual floods from mountain snowmelt covering the channel to a depth of about 2 m (Eschner et al. 1983). Since the mid 1800's, the channel has greatly narrowed and sand bars have stabilized and been transformed into forested island and riparian habitats. Horizontal and vertical island growth has occurred due to sediment accretion, organic soil enrichment, increased erosion resistance from plant roots, and channel excision. Willows (*Salix* sp.) and other shoreline plants stabilize island margins and promote lateral island expansion. Island sediment accretion and channel excision from bed scour cause an increase in relative height of islands and lower water tables on islands and the adjacent valley floor (Platte River Trust 1990). Nagel and Dart (1980) calculated an increase of 0.8 cm per year (40 year period from 1938-1978) in island height above active channel elevation in the central Big Bend reach. Below Lake McConaughy at Sutherland, channel elevation decreased an average of 1.8 cm per year between 1960 and 1990 (USFWS data cited in Platte River Trust 1990). Also, median size of residual sediment particles has increased substantially (Platte River Trust 1990), leading to increased resistance (termed armoring) to channel shifting.

REGIONAL CLIMATE AND EDAPHIC PATTERNS

In general, the Platte basin is affected by three climatic regimes: (1) the colder, alpine zones of the Rocky Mountains which intercept moisture through adiabatic expansion of the atmosphere and water vapor condensation, (2) the semiarid mountain rainshadow of the high plains in eastern Wyoming and Colorado and western Nebraska, and (3) a mid-continental regime in the eastern basin with moderate moisture. Borchert (1950) refers to these distinctive physiographic zones as western highland, short grass steppe, and tall grass prairie. Average annual precipitation varies from 50-100 cm in the Front Range, to 30-38 cm in eastern Colorado, to 60 - 75 cm in east central Nebraska. Interannual variation in precipitation can be quite high and droughts are often persistent and regional in scale. Drought is the most significant climate element in the Great Plains and largely determines the carrying capacity of the region, from archeological and historic as well as ecological perspectives (Barry 1983). Summer temperature and moisture conditions have strong negative correlation in the Great Plains, especially in Nebraska.

On average, annual snowfall in the foothills and mountains ranges from 300-600 cm and much of the winter snowpack melts to form runoff in late spring and early summer. In the summer, thunderstorms are common in the Front Range. Snowfall is much less in the plains (60-120 cm) and about 65% of annual precipitation occurs in the growing season. Runoff in the basin is highly variable and depends on amount and timing of precipitation, slope and permeability of the land, and E-T rates. Watersheds in the Rockies may yield 50-75 cm of runoff per year, compared to 12 cm in the Nebraska Sandhills, and only 3-5 cm in the high plains. As expected, growing seasons are inversely correlated with both latitude and elevation: about

160 days in east central Nebraska, 130 days in western Nebraska, 100 days in the Wyoming basin, and 60 days in higher alpine zones (data from U.S. Geological Survey 1970).

The geochemistry and soils of the basin are quite varied, and differences are reflected in the chemistry of tributaries and mainstem reaches. For example, the South Platte at the Nebraska state line has an average specific conductance [measure of total dissolved salts] of 1900 μmhos/cm and a chloride concentration of 71 mg/L (Enberg 1983). By contrast, the North Platte at the Nebraska state line is much more dilute, averaging 800 μmhos/cm and 18 mg/L. Solutes are low in groundwater fed tributaries of the Nebraska Sandhills. For example, the Loup River has conductance and chloride values of about 250 μmhos/cm and 2.5 mg/L. The large Loup River inflow near Columbus (Figure 1) dilutes total salt levels in the Platte by about 50% and chloride values by nearly 75%. In contrast, Salt Creek near Lincoln has chloride values of 1700-3000 mg/L and Platte chloride values increase by 700% downstream of the Salt Creek confluence (Enberg 1983). Nutrient levels of Platte River water are relatively high, reflecting substantial inputs from agriculture and municipal waste discharges. Nitrate levels often exceed 0.5 mg/L and total phosphorous averages 0.2-0.4 mg/L in the Big Bend reach. The effect of agriculture is evident in data from irrigation drains in the North Platte drainage, where average nitrate is typically 2.5-7.5 mg/L. The Elkhorn River (Figure 1) flows though agriculture land in eastern Nebraska. Near its confluence with the Platte, the Elkhorn averages about 1.6 mg/L nitrate and 1.1 mg/L total phosphorous.

HISTORICAL CHANGES IN HYDROLOGY AND CHANNEL MORPHOLOGY

The reasons for the present biodiversity crisis in the Platte Valley are rooted in the aggressive development of the basin's land and water resources during the westward expansion of the United States in the mid to late 1800's and the subsequent appearance of large scale irrigation projects, mechanized agriculture, and transportation systems to link farm producers with urban consumers. In the mid nineteenth century, successive waves of gold miners, farmers, ranchers, and other emigrants crossed the plains via the Platte Valley's famous Oregon and Mormon trails and the Union Pacific line, the nation's first transcontinental railroad. In this period, native plains tribes were decimated by introduced diseases and the elimination of the large, nomadic bison herds they hunted, and to a lesser extent by hostile encounters with settlers and their military guardians and the federally imposed, apartheid system of tribal reservations. The declines in native tribes and bison greatly altered two important regulators of the prairie ecosystems: fire and grazing.

Changes in the Platte's hydrology (total water supply and the timing of its delivery) have caused the greatest changes in ecosystem structure and function. However, documenting the linkage of flow reductions to habitat disturbance is difficult. Flow measurements began in the late 1800's on the North Platte and early 1900's for areas of the central Platte. However, direct channel diversions for irrigation began as early as the 1830's on tributaries of the South Platte. The numbers and cumulative impact of irrigation diversions in the basin increased dramatically between the 1860's and 1890's, a period coincident with rapid population growth. Summer flows in the South Platte were overappropriated by 1885, in the North Platte in Wyoming by 1901, and the North Platte in Nebraska by 1917 (Eschner et al. 1983). Population growth in Nebraska illustrates the rapid settlement of the Platte basin. The state had 28,000 residents in 1860, but just over one million were counted in the 1890 census. By contrast, Nebraska did not reach 1.5 million residents until the 1980 census.

Investigators such as Eschner et al. (1983) and Currier et al. (1985) carefully examined expedition accounts, immigrant journals, and early maps and photographs to document and infer characteristics of the river and its drainage basin before and during the nineteenth century colonization. Early written accounts and turn-of-the-century photographs reveal that trees were rare in the Big Bend reach. In central Nebraska, trees were restricted to a few larger islands and nearby valley swales and bluff ravines of the Platte and its tributaries. Such locations were more protected from prairie fires and grazing and had higher moisture levels. By the 1860's, much of this scarce timber may have been cut by settlers and railroad workers. Ironically, a century later woody plant colonization was a serious problem for native Platte Valley species and habitat conservation.

Most hydrologic and morphometric information from the last century is anecdotal. However, comprehensive government maps from the 1860's (see Peake et al. 1985) and aerial photography since 1938 provide useful data to assess channel width and habitat changes. In the 120 year period from the 1860's to the 1980's, active channel width decreased 82% in the section from North Platte to Lexington, 76% between Lexington and Chapman, and 43% between Chapman and Duncan (Sidle et al. 1989). Channel narrowing was most rapid between 1938 and 1957 (Platte River Management Joint Study 1989, Sidle et al. 1989), coincident with the greatest increases in upstream reservoir storage and irrigation diversion and water consumption. Peake et al. (1985) applied digital mapping techniques to aerial photographs of the Big Bend reach to compute two dimensional changes in former active channel. In 1860, active channel and unvegetated island and bar habitat covered about 26,000 ha in a 220 km length of channel between the present day cities of Brady and Chapman. By 1938, inactive channel containing other cover types (Figure 2) occupied 45% of the historic channel. Inactive channel increased to 70% by 1957, and 77% by 1983. Woody vegetation (tree and shrub classes) invaded former channel and covered 19,000 ha by 1981. Open-water, lentic habitats (primarily flooded quarries) increased from 0.24% of former channel in 1938 to 2.3% in 1983. Herbaceous vegetation rapidly colonized former channels, but in turn was largely replaced by woodland and shrub communities (Figure 2).

Platte River discharges are highly variable due to seasonal and interannual precipitation and temperature dynamics and human designed storage and diversion systems. Sorting out the relative contributions of these factors is difficult. Upper basin precipitation and runoff have varied substantially in this century. Runoff from the Rockies was generally quite high between 1910 and 1930, elevated in the early

1970's, and sharply higher again in 1983 and 1984. In contrast, upper basin runoffs were quite low in the 1930's and moderately low in the mid 1950's. During this period, new water projects affected discharge rates and contributed to total variation. Flows at Overton, Nebraska (a station just downstream of return flows from Tri-County system) averaged 62.1 m^3/sec (= 1.59 million acre feet, or maf) per year) between 1926-1940, but only 39.1 m^3/sec between 1941 and 1981. Comparable data for the North Platte at Bridgeport showed a decline from 44.6 to 34.0 m^3/sec whereas the South Platte at Julesburg, Colorado increased from 8.6 to 14.5m^3/sec. The increase in South Platte flows was largely related to the advent of interbasin transfer of western slope Colorado River water. The steeper declines in Platte flows were coincident with the completion of Lake McConaughey and diversions from the South Platte for the Tri-County Supply Canal. This canal diverts, on average, 43 m^3/sec (= 1.1 maf/year). Irrigation and canal losses from this project consume at least 18.5 m^3/sec (Platte Trust 1990). A small portion of unused water from the Tri-County system is returned by the Jeffrey Power Station Canal near Brady and a much larger fraction by the J-2 Canal just upstream of Overton.

Although lack of data from pre-irrigation flows and changing precipitation patterns make quantitative estimates of total historical flow reductions in the Platte River difficult, channel narrowing estimates and limited flow data from the late nineteenth century led Williams (1978) to conclude that flows declined at least 70%

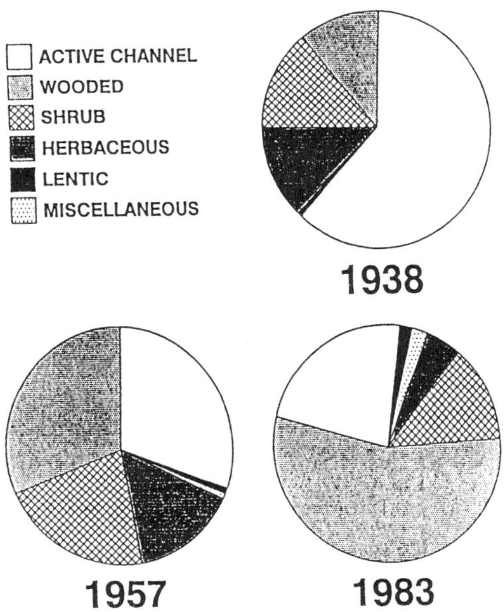

FIGURE 2. Changes in the relative proportions of habitats found in the former active channel of the Big Bend reach of the Platte River, as referenced to the area classified as active channel in survey maps from the 1860's.

in the Big Bend reach downstream of the J-2 Canal return flows. Platte River flows in the past 4 decades appear to have stabilized (Eschner et al. 1983). Changes in flow regimes have generally decreased peak flows and perhaps low flows as well (Kircher and Karlinger 1983), although this latter conclusion has been challenged (see Currier et al. 1985; Platte River Trust 1990).

ECOLOGICAL COMMUNITIES

(A) Vegetation

The Platte's wide, shallow, braided channel habitat with sparsely vegetated sand bars and adjacent moist and dry soil tall grass prairie is the condition that supports many native migratory and non-migratory species. However, woody plants dominate the vegetation cover in exposed parts of the former channel, while half of the Big Bend native prairie is now tilled and remnant prairie has declined in productivity and species richness and is often reduced to patches too small for wildlife use (Currier et al. 1985). The grasslands are important foraging sites for sandhill cranes and support many breeding bird populations.

The woody plant communities are initially established through seed germination of Eastern cottonwood (*Populus deltoides*) and several riparian willow species (*Salix* sp.). Most successful seedling establishment appears to involve seeds transported by the river. These seeds become stranded on moist substrates during a falling river stage (Currier 1982). Dry microsites don't support seedling establishment and established seedlings and saplings are effectively removed by bed-scouring high water flows and the break up and transport of ice flows. The issue of why woody vegetation was so rare before this century has been debated. One hypothesis invokes the former condition of annual late spring floods with bed scouring flows capable of undercutting and burying rooted plants (Currier 1982). The second hypothesis proposes that low and zero summer flow conditions were more common prior to water storage and irrigation releases and that woody species were primarily suppressed by inadequate moisture for germination and rooting (Ecological Analysts 1983). Both mechanisms are plausible and not contradictory. Adult mortality from episodic droughts (Borchert 1950), prairie fires, and native herbivores also probably limited trees. Tree ring dating studies clearly demonstrate that massive woody vegetation encroachment in upper reaches of the Platte River historic channel was coincident with the construction of storage reservoirs in this century (USFWS 1981, Currier 1982), with a large majority of trees established between 1930 and 1960. A few old trees (established between 1870 and 1900) encountered in demographic studies were found on uplands or former bank shorelines (U.S. Fish and Wildlife Service 1981). Tree diversity and seed sources were apparently increased through plantings by settlers, an activity promoted by the Timber Culture Act of 1873 (USFWS 1981). Channel encroachment is a rapid process, with herbaceous annuals and perrenials overgrown by shrubs and trees within a few years. Since the 1930's, vegetation encroachment has increased and progressed downstream through the Big Bend reach, in spite of a more stabilized (albeit greatly reduced) mass balance of sediment in the channel system in recent years (Platte River Trust 1990). Rapid vegetation growth in exposed channel areas may be linked to favorable moisture availability, low incidence rate of soil flooding, and the high nutrient levels of the river water. Incremental analysis of tree rings (Currier 1982) showed that cottonwood growth was relatively insensitive to elevation in the floodplain and interannual changes in river hydrology whereas willow growth increased and red cedar decreased in years of high river flows.

Tree and shrub diversity are relatively low in Platte River Valley forest stands. Currier (1982) found 16 arboreal and 15 shrub species in a comprehensive survey of 320 km of Platte and North Platte floodplains. Community composition varied as a function of soil texture, elevation, and stand age with cottonwood, diamond willow (*Salix rigida*) and Russian olive (*Elaeagnus angustifolium*) the dominants on lower, moist sites and cottonwood and red cedar (*Juniperus virginiana*) on higher, drier sites. Green ash (*Fraxinus pennsylvanica*), red mulberry (*Morus rubra*), slippery elm (*Ulmus rubra*) and American elm (*Ulmus american*) were also common, especially in older stands. These latter species become more important as stands age (Nagel et al. 1980). Five shrubs were common: rough-leafed dogwood (*Cornus drumondii*), coyote willow (*Salix exigua*), indigo bush (*Amorpha fruticosa*), wild rose (*Rosa woodsii*), and prickly ash (*Zanthoxylum americanum*). Currier (1982) observed that present Platte valley communities resemble valley floras in eastern portions of the central Mississippi drainage, except for the absence of certain prominent species such as American sycamore (*Platanus occidentalis*) and honey locust (*Gleditisia triacanthos*).

Comprehensive surveys of Platte Valley herbaceous vegetation were made by Bunde et al. (1975), Currier (1982), and Nagel and Kolstad (1987). No rare endemic species were identified. A plant listed on the federal threatened list, the western prairie fringed orchid (*Plantanthera praeclara*), was found in the Platte Valley at one large, protected wet meadow at the Mormon Island Crane Meadows near Grand Island. Non agricultural herbaceous vegetation occurs as understory in tree stands, as the principal cover of newly exposed or disturbed river channel substrates, as ruderals in agricultural land and other disturbance sites, and in native grassland stands. Marsh vegetation was probably never common and today is relatively rare except in certain disturbed areas such as remnant backwaters and channel areas impeded by canal diversion and dam structures. Platte Valley grassland communities are maintained by high groundwater levels, fire, and grazing. Vegetation abundance and community structure display interannual behaviors linked to groundwater behavior (Currier 1989). Channel excision and a consequent decline in groundwater levels (Platte River Trust 1990) may cause further declines in wet meadows, a productive habitat important to cranes and other wildlife.

(B) Aquatic Animals

Fish and aquatic invertebrates, although not well studied in the highly impacted channels of the Big Bend reach, are clearly important to Platte Valley food chains. Although Great Plains prairie streams are highly variable and challenging environments even in their natural state, native species appear well adapted to contend with stressors such as temperature, turbidity, salt content, and low and high flow conditions (Matthews 1988). Many midwestern stream communities have experienced long term declines in fish populations (Karr et al. 1985, Cross and Moss 1987), largely attributable to physical habitat alteration, flow reductions, and agricultural water pollution. The Big Bend reach of the Platte appears to have undergone a decline (Platte River Trust 1990), although a recent survey (Chadwick & Associates 1992) challenges the position that fish species richness has decreased significantly.

Several comprehensive surveys of fish in the central Platte were made in the last 50 years. Numerically, cyprinids dominate the total number of species and total numbers of fish in the river. Sand shiners (*Notropis stramineus*) and red shiners (*Cyprinella lutrensis*) are the most common minnows. Important non-cyprinid species include channel catfish (*Ictalurus punctatus*), river carpsucker (*Carpiodes carpio*), and the plains killifish (*Fundulus zebrinus*). Functionally, most fish in the central Platte River can be placed in one of three guilds: slow/shallow habitat, fast/shallow habitat, and a slow/deep habitat (Platte River Joint Study 1990). Johnson (1942) found 64 species in the entire Platte, including several species not encountered in subsequent collections. Morris (1960) found 50 species in the central Platte, including 2 species not noted by Johnson. Cochnar and Jensen (1981), sampling only one location near Grand Island, found 26 species in the Platte channels and 13 species in a sandpit lake connected to the river. Seven of the 13 lake species were not found in the river. The Chadwick and Associates (1992) survey found 41 species in the river and connected back channels. The Chadwick and Associates study failed to find 25 species reported by Johnson and 13 species reported by Morris. In contrast, the 1992 survey documented 6 species not found by Johnson and 7 not found by Morris. Most new species found since 1942 are attributable to fish stocking programs. Although differences in survey outcomes probably reflect differences in methodology, intensity of sampling effort, and locations and microhabitats sampled, shifts in community composition and a decline in the number of native species are apparent.

Any overall perspective of fish biodiversity must take into account the long history of human manipulations of fish stocks. The common carp (*Cyprinus carpio*) is well established, but a lack of information prior to introduction prevents an assessment of its impact. A number of game fish either uncommon or not native to the central Platte have been stocked by fishery biologists and amateurs in the Platte and its tributataries, upstream reservoirs, farm ponds, and quarry lakes in the valley. Some of these species (for example: bluegill - *Lepomis macrochirus,* freshwater drum - *Aplodinotus grunniens,* and yellow perch - *Perca flavescens*) are common in impoundments, floodplain lakes, and backwaters are found in low numbers in the central Platte. Others, such as white bass (*Morone chrysops*), several salmonids, and smallmouth bass (*Micropterus dolomieu*) are seen irregularly and represent fugitives from upstream reservoirs (True 1974). Non-game species may be transplanted to serve as forage fish, released from bait buckets, or introduced for other reasons. Occasionally these species survive, and more rarely become so abundant that they threaten others. Species introductions are the second leading cause of fish extinctions in North America (Miller et al. 1989).

Recently, Lynch (1989) documented the introduction and spectacular dispersal of western mosquitofish (*Gambusia affinis*) in the central Platte. Inadvertent releases to the river in the Brady and North Platte areas probably began about 1973 from the stocking of a cold tolerant race in lakes, ponds, and ditches for mosquito control. In 1987, Lynch found mosquito fish, often in great abundance, at virtually all sites inspected between North Platte and Central City, Nebraska, an area bracketing the entire Big Bend reach. Lynch estimated downstream dispersal rates of 15-20 km per year and upstream rates of 5-10 km per year. In recent work, Lynch (personal communication) found serious declines in the plains topminnow (*Fundulus sciadicus*) at sites with abundant mosquitofish and concluded that mosquitofish predation on eggs and young threatens the topminnow and the rarer brook stickleback (*Culaea inconstans*) with extirpation from the Platte and other area streams. In the Chadwick and Associates (1992) survey, only sand shiner were more abundant than mosquito fish.

Fish species declines in the central Platte are less dramatic than those reported in the western and central reaches of the Arkansas and Kansas Rivers in Kansas (Cross and Moss 1987). These rivers experience more persistent low and no flow conditions. A significant additional factor may be abundant refugia for fishes of the central Platte, including constant flow streams from the Nebraska Sandhills, the lower Platte and Missouri Rivers, and quarry lakes, reservoirs, and irrigation canals. Fish kills are common in the Big Bend (Platte Trust 1990). Recolonization must come from refugia. This mechanism may also explain the relatively large proportion of species with low densities and patchy distributions. Rare fish populations are usually peripheral members of larger populations in other watershed locations (Sheldon 1987). Rarity is the norm for many species in many locations. In streams, Sheldon concluded that rare species are at much greater risk of localized extirpation and even global extinction because of widespread habitat damage and fragmentation.

The large reductions in flows and channel dimensions in the Big Bend reach must have caused reductions in fish trophic support and total fish production, although historical data are not available. Fish kills must also sharply reduce average fish abundance. Eight kills, some severe, were noted near Grand Island between 1985 and 1990 (Platte River Trust 1990). In one episode in late June and early July of 1990, river flows were consistently less than 8.5 m^3/sec. Maximum water temperatures above 36°C occurred for eight consecutive days beginning June 25 and reached 38.3° on one date. A fish kill in this reach began June 29. These water temperatures equalled or exceeded the known Critical Thermal Maximum for Great Plains stream fishes, as determined by Matthews (1987).

Aquatic invertebrates of the Platte River are poorly characterized. Nagel (1974), in a comprehensive review, concluded that no systematic survey of aquatic invertebrates had ever been attempted in Nebraska. Recently, the Nebraska Department of Environmental Control (1991) conducted a fish and invertebrate survey in all major streams in the state. The Platte was poorly represented in this study (one station each in the Big Bend and North Bend reaches), but all major Platte tributaries were sampled. The Big Bend station in Polk County, had 28 macroinvertebrate species, including 4 diptera and 6 total mayfly, stonefly, and caddisfly species. Similar values were obtained at the North Bend station, as well as 6 other stations on the North and South Platte Rivers. Macroinvertebrate densities and biomass are much greater in bank and backwater habitats of the Platte, compared to the unstable inorganic substrates of the main channel (Nagel 1974). Beer et al. (1972) found very low invertebrate abundance at five stations in the Big Bend: average density = 3.1 organisms/m^2 and average dry weight biomass = 2.85 mg/m^2. Water column drift was also low: density = 3.3 organisms/m^3 and dry weight biomass = 0.09 mg/m^3. Numerically, midges, mayflies, and oligochaetes dominated benthos and cladocerans, midges, nematodes, and mayflies dominated the drift. Nagel (1974) cited a study by Chapman where annual macroinvertebrate biomass averaged 10 mg/m^2 in the Platte channel compared to 580 mg/m^2 in a backwater.

Recently, Freeman and Perkins (1992) documented eleven species of unionid mussels in the Platte system. Ten native species were exclusive to or most common in the Big Bend reach. These ten native species have parasitic glochidia stages important for both larval development and dispersal. Five species were relatively abundant and widespread. These mussels may be especially sensitive indicators of habitat conditions because of the sessile benthic adult stage and need for a proper fish host for larval development. Most of the Platte unionids seem to be peripheral western populations of widespread eastern species (Freeman and Perkins 1992). Only one mussel, the pond horn shell (*Uniomerous tetralasmus*) is considered a true prairie species. The south channel of the Platte and several of its offshoots in Hall County contained the highest diversity (eight species). Mussels were uncommon or absent in the main channel, preferring more stable substrates and cooler waters in shaded or groundwater seep areas of shallows and backwaters (Rodel 1990). Such areas, as well as valley lakes and ponds, may be important refugia. Mussels may withstand brief exposures through burrowing to stay in moist substrate. Some species were also common in regional lakes and reservoirs. Mortality during low water periods appears high (Rodel 1990), perhaps due to a combination of desiccation, tissue anoxia, overheating, and increased predation losses. One species found by Freeman and Perkins, the Asiatic clam *Corbicula fluminea,* is a problem exotic. This species was collected in the Cozad canal near the Platte and in two lakes near Lincoln. Although very adaptable and dispersed as a free living veliger stage, it may be limited in Nebraska by cold temperatures.

(C) Avifauna

Much of the recent research on Platte River biota examined the biology and habitat requirements of migratory and endangered bird species (see USFWS 1981, Currier et al. 1985, Platte River Joint Management Study 1990). For many, the birds are the "main show" and the specific attraction is the mass staging of sandhill cranes in late winter and early spring and the possibility of sighting small groups of migrating whooping cranes. Sandhill cranes are large birds with wingspans of 2 m.

They are gregarious and communicate with a ringing, trumpet call produced by a unique intrasternal coiling of their trachea (Johnsgard 1985). Sandhill cranes often roost in groups of thousands to tens of thousands on the Platte. Increasing numbers of people visit the Platte Valley when sandhill cranes are present in March and early April, bringing at least $15 million in new revenue and $40 million in total impact per year to local economies (Gary Lingle, personal communication). Unfortunately, "crane mania" causes traffic congestion and safety hazards near viewing sites (especially at bridges on the Platte) and brings added stress to the shy birds, who may leave their nocturnal river roosts if disturbed.

Sandhill cranes funnel into the Big Bend reach from wintering grounds in Mexico, Texas, and New Mexico. The USFWS (1981) estimated 540,000 cranes used the Platte Valley in early April of 1980, with 74% using the eastern Big Bend reach between Kearney and Grand Island. Three subspecies of sandhill use the Platte: the lesser sandhill (*Grus. c. canadensis*) are the most common, followed by the Canadian sandhill (*Grus. c. rowani* - but may not be a distinct race; see Johnsgard 1985), and mid continental populations of the greater sandhill crane (*Grus. c. tabida*). Sandhill cranes stage on the Platte primarily in the spring, with far fewer present during fall migration. To prepare for spring migration to the far north and reproduction, cranes in the Platte Valley gain about 25% in body weight (Reinecke and Krapu 1986) and increase fat from less than 8% to about 23% of body weight (USFWS 1981) over a 4-6 week period.

During daylight, sandhill cranes in the Platte Valley spend about 53% of their time in corn fields gleaning waste grain, 28% in native grasslands (especially wet meadows near the river) and 19% in alfalfa/tamegrass hayfields (Krapu et al. 1984). Waste grain (principally corn) accounts for 95% or more of total food intake (Reinecke and Krapu 1986). However, the cranes spend about 45% of their active period in grasslands and haylands (USFWS 1981), where they probe for food in the ground litter and soil and feed almost exclusively on invertebrates (Reinecke and Krapu 1986, Davis 1991). Earthworms were the dominant invertebrate food in Reinecke and Krapu's study, whereas insects (principally scarab beetle larvae) were the most common item in Davis's study. Davis attributed the difference to dry soil conditions in his study, which kept earthworms deeper and less available. Reinecke and Krapu reasoned that the low intake of invertebrates per unit of feeding time must be offset by securing vital nutrients — most likely essential amino acids and calcium. The calcium comes mainly from snails and empty shells in native meadows. Thus, croplands and native grasslands both play key roles in present day sandhill crane nutrition. Local farmers are quite tolerant of crane feeding.

Sandhill cranes roost between sunset and sunrise in shallow water (commonly 4-24 cm) sections of the Platte, and require wide channels and bars unobstructed by tall vegetation. Estimates of minimum preferred channel widths vary from 50 m in the North Platte (Folk and Tacha 1990) to 150 m in the central Platte (USFWS 1981). Loss of acceptable roosting habitat for cranes is one major consequence of channel changes in the Platte. Currier et al. (1985) estimated that the loss of acceptable roosting habitat in the Big Bend reach has been proportionately much greater than the direct loss of open channel. For example, between 1938 and 1982,

48% and 52% of active channel was lost at Odessa and Denman, but both areas experienced a 97% reduction in suitable crane roosting habitat. Historically, cranes used all of the Big Bend reach, but they progressively abandoned degraded habitats in the western portion and are now heavily concentrated in several open channel areas between Kearney and Grand Island (Currier et al. 1985). In recent years, sandhill crane numbers have held steady or increased slightly, but concern is growing for the crowding of cranes into shrinking roost habitats. Overcrowding may lead to disease epidemics, vulnerability to changes in land use and loss of food resources, and natural disasters such as severe hailstorms (USFWS 1981). Avian cholera (*Pasteurella multocida*) is the main disease concern. The bacterium has caused sporadic, serious mortality in ducks and other waterfowl in the Rainwater Basin wetlands just south and east of the Big Bend, and occasional outbreaks in cranes are reported (USFWS 1981).

Whooping cranes also funnel through the Big Bend in their migrations (Currier et al. 1985), but recent use of the Platte Valley is low. This species, perhaps the most famous of all endangered North American wildlife, narrowly avoided extinction in the early 1940's to mid 1950's when the population size of the last viable wild flock dropped to 15-20 birds. By the early 1990's, numbers exceeded 130 on their winter reserve at Aransas, Texas. During most of this century, confirmed sightings of whoopers in the Platte Valley have declined, even in the period since 1950 when the population was growing (USFWS 1981). Historic data suggest that whooping cranes were far more common in the Platte Valley in the 1800's and early 1900's. Indeed, the Platte may have been their principal way station (Currier et al. 1985). Prior to federal protection, more whooping cranes were reported shot in Nebraska than any other state. Based on historical use patterns, the Big Bend reach is federally classified as critical habitat for whooping cranes. The birds are mostly carnivorous and feed in wet areas during migration. Based on sighting records, whoopers prefer larger open channels and unobstructed viewing areas than sandhill cranes, even though the two species will roost together. The destruction of backwaters and riparian wetlands and the loss of open channel are the probable reasons for the recent low use of Platte habitats by whooping cranes (USFWS 1981). Their use of Rainwater Basin wetlands has increased, and officials are concerned about exposure to avian cholera.

Two other endangered bird species, the interior least tern and the piping plover, require the open channel conditions provided by the Platte and some other streams in the Great Plains. These birds nest on exposed sand bars and sparsely vegetated islands (Faanes 1983). The terns feed on small fish and the plovers on aquatic invertebrates in shallow channels and backwaters. Channel degradation and woody vegetation encroachment have greatly diminished the habitat for these birds. Reproductive success has also decreased in plovers and terns because of the loss of predictable late spring floods and more erratic river flows and nest flooding (Platte River Trust 1990). Another endangered species, the Eskimo Curlew (*Numenius borealis*), was once common on the Platte, particularly in native grasslands. This wide-ranging bird is close to extinction and was thought to be extirpated from Nebraska. However, a single eskimo curlew was spotted on an island in the

Platte near Grand Island in April, 1987 (Platte River Trust 1990).

The Big Bend reach and adjacent Rainwater Basins attract 7-9 million ducks, geese, and other waterfowl each spring and fall. Mallard ducks (*Anas platyrhynchos*) and Canadian geese (*Brantus canadensis*) are the most common. Thousands of mallards and Canadian geese often overwinter, with numbers positively correlated to the amount of open water. Waste grain is an important food source for these birds (USFWS 1981).

Several hundred endangered bald eagles (*Haliaeetus leucocephalus*) use the Platte and North Platte Rivers, primarily in the winter. Eagles perch and roost in woody riparian vegetation, which is now abundant. Fish are their most important food and they need patches of ice free water to gather fish. Bald eagles switch to waterfowl and other birds, as well as mammals, when the river freezes or water levels are high (Lingle and Krapu 1986). Eagle populations show fewer signs of damage from habitat changes in the Platte Valley than the other endangered species. A number of other bird species utilize forest, shrub, and shelterbelt habitats in the Platte Valley. A USFWS (1981) survey in 1979 and 1980 found 142 breeding species of birds in Platte Valley habitats, with many favoring wooded habitats. Paradoxically, total bird diversity, and possibly total numbers, may have increased in the Platte Valley due to the spectacular increase in forest and shrub growth, with the Platte Valley now acting as a functional extension of eastern deciduous forest. However, riparian grasslands provide some of the best nesting and feeding habitat for native Great Plains birds, and many endemic populations have been adversely affected by the loss and fragmentation of these sites (Currier et al. 1985).

CONCLUSIONS AND PROSPECTS FOR PLATTE RIVER BIODIVERSITY

Although research efforts on the Platte have increased dramatically in the past two decades, much of the work focused on the status and habitat requirements of endangered avian species (for example — USFWS 1981, Platte River Management Joint Study 1989 and 1990). Many other strands of the system's community structure are poorly known. We are largely ignorant of the magnitudes, regulatory mechanisms, and interconnections of many vital ecological processes in the Platte system, including: basic nutrient cycles, primary and secondary productivity patterns, feeding behaviors of non-endangered organisms, microbial components and detrital processing, successional patterns and long term responses to natural and human disturbances, and landscape scale couplings of upland, riparian, and stream channel systems. Surface and subsurface hydrology and channel geomorphology are reasonably well understood in the Platte system, but direct and indirect interactions of the physical realm with the biological components demand far more attention. Attempts to model these interactions have helped frame research issues and provide guidance for water management decisions and biological conservation (Gilliland et al. 1985, Platte River Joint Management Study 1990).

Humans exert powerful controls over the behavior and fate of water in the Platte system and any consideration of future biological diversity and ecological health begins with the premise that water management decisions affect all other components of this system. Gilliland (1984), Farrer (1985), and Schalles (1991) presented overviews of Platte River water policies, development projects, legal disputes, and habitat conservation efforts. Struggles betweeen conservation groups and water development interests to control the remaining flows and re-examine the system's current status quo have become increasingly impassioned, litigious, and expensive. The controversies are exemplified by two recent cases: the current Federal Energy Regulatory Commission's relicensing process for Lake Mcconaughy and the related Tri-County irrigation and power system in Nebraska and the recently rejected Denver Water Board proposal for the U.S. Army Corps of Engineers to construct the Two Forks reservoir on the upper South Platte River to store and deliver more water for the Denver metropolitan area.

The issue of how to best protect biodiversity is exemplified by the problems of the Big Bend reach of the Platte. Currently, the most effective legal instrument to protect the Platte system and endemic biota is the Endangered Species Act. Unfortunately this highly visible and controversial federal legislation dictates a myopic approach to research and conservation efforts. We can't save nature one species at a time. Our seeming preoccupation with rare and seriously jeopardized species allows chronic, subversive damage to less rare species, whole community structure, and ecosystem processes to proceed under less scrutiny and with few legal impediments. This is not to negate the symbolic value of endangered species for conservation causes, nor their intrinsic right to exist. Protection and restoration of native Big Bend habitats (Currier et al. 1985, Farrar 1985) and well publicized protection efforts by the National Audubon Society, The Nature Conservancy, and other groups were catalyzed by the importance of this ecosystem for the popular sandhill cranes and endangered birds — especially the whooping crane. Removal of woody vegetation and restoration of dynamic braided channels and riparian grasslands have been very successful at Audubon's Row Sanctuary near Gibbon and the Platte Trust's Mormon Island Crane Meadows and adjacent reaches near Grand Island. These efforts were undertaken primarily to protect birdlife. Would efforts to protect the Platte as valuable remnant component of prairie landscapes be as forceful if the endangered species came from less charismatic taxonomic groups such as reptiles, insects, or lower plants; or, if ecological damage was slower to unfold, and damage to species less obvious?

The dangers to Platte River biodiversity seem obvious (Table 1), but remedies are complex and difficult. Increasing instream flows by reducing irrigated acreage and improving water use efficiency could help reverse damage to the Platte system, but would not address the reservoir effects of altered flow regimes and sediment trapping. Moreover, other serious threats to the system's biodiversity exist (Table 1) and may become more forceful as human population and non-agricultural development proceeds in the basin and as regional and global scale disturbances increase. If regional temperatures increase and drought episodes become more common and severe as a result of human induced climate change, past and current efforts to save the Platte River ecosystem and its biodiversity could be negated. Indeed, if climate change should reduce the available water supplies, results may be catastrophic for both societal and natural systems in the Platte Valley. The environmental challenges to the Platte basin vividly illustrate the need for comprehensive analysis and solutions to the biodiversity crisis.

TABLE 1

Key threats to biological diversity in ecosystems of the Platte River Valley.

•*Biodiversity Threat* / ®Reasons For / ©Consequences
•Loss of braided channel/shifting sand bar habitat, channel excision ®Reductions in flow, change in flow regime (esp. loss of bank full flows), reduced bed load ©Loss or reduction of waterfowl, fish, and aquatic invertebrates species
•Destruction, degradation, and spatial fragmentation of riparian moist and dry grasslands ®Tillage, lower water table, fire suppression, livestock overgrazing, agricultural chemicals ©Loss or reduction of plant, herbivore, bird, and soil invertebrate species
•Exotic species introductions or range extensions of non-native species ®Accidental or planned releases, altered habitats which can support new species ©Increased competition, predation, or disease effects on native populations
•Toxic chemical exposure and biomagnification in food chains ®Municipal and agricultural water pollution, non-point regional atmospheric drift ©Impaired competitive fitness and reproduction, reduced disease resistance, acute toxicity
•Increased frequency and severity of droughts ®Regional scale changes in climate from atmospheric damage and planet warming ©Increased reduction in soil moisture and river flows — stress or eliminate species

ACKNOWLEDGEMENTS

I am indebted to the staff of the Platte River Whooping Crane Habitat Maintenance Trust, particularly Paul Currier, Beth Goldowitz, and Gary Lingle, for help in obtaining references and generously sharing perspectives gained through their research. Mike Duever of the National Audubon Society Ecosystem Research Center provided valuable advice and ideas. Dave Bowman and Justin King of the U.S. Fish and Wildlife Service in Grand Island encouraged my participation in annual Platte River workshops and greatly facilitated exchanges of ideas and data between workers. I also wish to thank Craig Fischenich, Barbara Hayes, Rich Holland, John Lynch, Jack Meeder, Jeff Peake, Ed Peters, Ken Strom and Duane Woodward for explaining their work and helping me to understand various aspects of the Platte system. John Sidle of the U.S. Fish and Wildlife Service provided a thorough and thoughtful review of this paper.

LITERATURE CITED

Barry, R.G. 1983. Climatic environment of the Great Plains, past and present. pp. 45-55. In: W.W. Caldwell, C.B. Schultz, and T.M. Stout (Eds.) *Man and the Changing Environments in the Great Plains.* Trans. Nebr. Acad. Sci. 11 — Special Symposium Issue. Lincoln, NE. 144 p.

Beer, R., R. Cox, L. Flathman, R. Jasnoch, R. Kupper, and R. Lemmerman. 1972. *Impact of Proposed Water Resource Developments on Microflora, Fauna, and*

Water Quality of the Platte River in Central Nebraska. Kearney St. College, Kearney, NE. 34 p.

Bentall, R. 1982. *Nebraska's Platte River: A Graphic Analysis of Flows.* Nebraska Water Survey Paper 53. Inst. of Agr. and Natural Resources, Univ. of Nebraska, Lincoln, NE. 47 p.

Borchert, J.R. 1950. The climate of the central North American grassland. *Ann. Assoc. Amer. Geog.* 40:1-39.

Bunde, M., G. Deyle, J. Gray, T. Imel, M. Stines, T. Sweeney and L. Weiss. 1975. *The Effects of the Selective Removal of Cottonwood Trees Along the Platte River.* Biology Department, Kearney State College, Kearney, Nebraska. 110 p.

Chadwick and Associates. 1992. *Forage Fish Monitoring Study, Central Platte River, Nebraska, 1990-1991.* Report prepared for the Nebraska Public Power District and the Central Nebraska Public Power and Irrigation District. Chadwick and Associates, Littleton, CO. 66 p.

Cochnar, J., and D. Jensen. 1981. *1980 Mormon Island Crane Meadows Fish Survey.* Cross, F.B. and R.E. Moss. 1987. Historic changes in communities and aquatic habitats of plains streams in Kansas. pp. 155-165. In: W.J. Matthews and D.C. Heins (Ed.) *Community and Evolutionary Ecology of North American Stream Fishes.* Univ. of Oklahoma Press, Norman, OK. 269 p.

Crowley, K.D. 1981. *Large Scale Bedforms in the Platte River Structure Process, and Relation to Channel Narrowing.* U.S. Geological Survey Open File Report 81-1059. 33 p.

Currier, P.J. 1982. *The Flooplain Vegetation of the Platte River: Phytosociology, Forest Development, and Seedling Establishment.* Ph.D. Dissertation. Iowa State Univ., Ames, IA. 178 p.

Currier, P.J. 1989. Plant species composition and groundwater levels in a Platte River wet meadow. pp. 19-24. In: *Proceedings of the Eleventh North American Prairie Conference.* Lincoln, NE. August 7-11, 1988.

Currier, P.J., G.R. Lingle, and J.G. VanDerwalker. 1985. *Migratory Bird Habitat on the Platte and North Platte Rivers in Nebraska.* Platte River Whooping Crane Habitat Maintenance Trust. Grand Island, NE. 177 p.

Davis, C.A. 1991. *The Ecology of Macroinvertebrates Inhabiting Native Grasslands and their Role in the Feeding Ecology of Sandhill Cranes.* M.S. Thesis. Iowa State Univ., Ames, IA. 88 p.

Ecological Analysts. 1983. *An Evaluation of Historical Flow Conditions in the Platte River as Related to Vegetatin Growth and Habitat Use by Endangered Whooping Crane and Bald Eagle and the Threatened Least Tern.* Report Submitted to the Central Platte Natural Resource District. Lincoln, NE. 96 p.

Enberg, R.A. 1983. *A Statistical Analysis of the Quality of the Surface Water in Nebraska.* U.S. Geol. Surv. Professional Paper 2179. 252 p.

Eschner, T.R., R.F. Hadley, and K.D. Crowley. 1983. *Hydrologic and Morphologic Changes in Channels of the Platte River Basin in Colorado, Wyoming, and Nebraska: A Historical Perspective.* U.S. Geol. Surv. Professional Paper 1277-A, GPO, Washington, D.C. 39 p.

Faanes, C.A. 1983. Aspects of the nesting ecology of least terns and piping plovers in central Nebraska. *Prairie Nat.* 15(4): 145-154.

Farrar, J. 1985. Partners on the Platte. *Nature Conserv. News* 35(1): 13-18.

Folk, M.J., and T.C. Tacha. 1990. Sandhill crane roost site characteristics in the North Platte River Valley. *J. Wildl. Manage.* 54(3): 480-486.

Freeman, P.W., and K. Perkins. 1992. *Survey of Molluscs of the Platte River.* Report to the U.S. Fish and Wildlife Service. Univ. of Nebraska State Museum, Lincoln, NE. 37 p.

Gilliland, M.W. 1984. Trouble downstream: migrating cranes force a showdown on Platte River water projects. *Sierra* 69(2): 12-18.

Gilliland, M.W., L. Becker, R. Cady, J. Gabig, J. Gilley, R.A. Kern, A. Larsen, Q.M. Nguyen, and R. Supalla. 1985. Simulation and decisionmaking: The Platte River Basin in Nebraska. *Water Res. Bull.* 21(2): 281-290.

Johnsgard, P.A. 1983. *Cranes of the World.* Indiana Univ. Press, Bloomington, IN. 257 p.

Johnson, R. 19742. *The Distribution of Nebraska Fishes.* Ph.D. Thesis. Univ. of Michigan, Ann Arbor, MI. 145 p.

Karr, J.R., L.A. Toth, and D.R. Dudley. 1985. Fish communities of midwestern rivers: a history of degradation. *Bioscience* 35:90-95.

Kircher, J.E., and M.R. Karlinger. 1983. *Effects of Water Development on Surface-Water Hydrology, Platte River Basin in Colorado, Wyoming, and Nebraska Upstream from Duncan, Nebraska.* U.S. Geol. Surv. Professional Paper 1277-B, GPO, Washington, D.C. 77 p.

Krapu, G.L., D.E. Facey, E.K. Fritzell, and D.H. Johnson. 1984. Habitat use by migrant Sandhill Cranes in Nebraska. *J. Wildl. Manage.* 48(2): 407-417.

Lingle, G.R., and G.L. Krapu. 1986. Winter ecology of Bald Eagles in southcentral Nebraska. *Prairie Nat.* 18(2): 65-78.

Lynch, J.D. 1989. Introduction, establishment, and dispersal of western mosquitofish in Nebraska (Actinopterygii: Poeciliidae). *Prair. Nat.* 20(4):203-216.

Lyons, J.K., and T.J. Randall. 1988. *Platte River Channel Characteristics in the Big Bend Reach.* U.S. Bureau of Reclammation, Denver, CO. 28 p.

Matthews, W.J. 1988. North American prairie streams as systems for ecological study. *J. of N. Am. Benthol. Soc.* 7(4): 387-400.

Matthews, W.J. 1987. Physiological tolerance and selectivity of stream fishes as related to their geographic ranges and local distributions. pp. 111-120. In: W.J. Matthews and D.C. Heins (Eds.) *Community and Evolutionary Ecology of North American Stream Fishes.* Univ. of Oklahoma Press, Norman, OK. 269 p.

Miller, R.R., J.D. Williams, and J.E. Williams. 1989. Extinctions of North American fishes during the past century. *Fisheries* 14(6): 22-38.

Morris, L.A. 1960. *The Distribution of Fish in the Platte River, Nebraska.* M.A. Thesis. Univ. of Missouri, Columbia, MO. 73 p.

Nagel, H.W. 1974. Aquatic Invertebrates of the Nebraska Mid-State Division. pp. I-1 to I-60. In (O.A. Kolstad, project director): *Identification and Evaluation of Present Zoologic Resources, Nebraska Mid-State Division, Pick-Sloan Missouri Basin Program and Associated Areas.* Report to U.S. Department of the Interior, Bureau of Reclammation. Kearney St. College, Kearney, NE.

Nagel, H.W., and M.S. Dart. 1980. Platte river evapotranspiration: a historical perspective in central Nebraska. *Trans. Nebr. Acad. Sci.* 8: 55-76.

Nagel, H.W., and O.A. Kolstad. 1987. Comparison of plant species composition of Mormon Island Crane Meadows and Lillian Annette Rowe Sanctuary in central Nebraska. *Trans. Nebr. Acad. Sci.* 15: 37-48.

Nagel, H.W., K. Geisler, J. Cochran, J. Fallasen, B. Hadenfeldt, J. Matthews, J. Nickle, S. Stec, and A. Walters. 1980. Platte River island succession. *Trans. Nebr. Acad. Sci.* 8: 77-90.

Nebraska Department of Environmental Control. 1991. *Nebraska Stream Classification Study.* Water Quality Division, Dept. of Environmental Control. Lincoln, NE. 342 p.

Peake, J., M. Peterson, and M. Laustrup. 1985. *Interpretation of Vegetation Encroachment and Flow Relationships in the Platte River by Use of Remote Sensing Techniques.* Remote Sensing Applications Laboratory, Univ. of Nebraska-Omaha, Omaha, NE. 182 p.

Platte River Management Joint Study. 1989. *Evaluating Management Alternatives: Sediment, Flow, and Channel Geometry Considerations.* Hydrology Workgroup. U.S. Bureau of Reclamation. Grand Island, NE. 84 p.

Platte River Management Joint Study. 1990. *Biology Workgroup Report.* Denver, CO. 131 p.

Platte River Whooping Crane Habitat Maintenance Trust. 1990. *Preliminary Comments of the Platte River Trust on the Deficiency Response of the CNPPID and the NDPPD For Projects Numbers 1417 and 1835.* Brand and Lowell. Washington, DC. 140 p.

Reinecke, K.J., and G.L. Krapu. 1986. Feeding ecology of sandhill cranes during spring migration in Nebraska. *J. Wildl. Manage.* 50(1): 71-79.

Rodel, M.D. 1990. Unionid molluscs in the Big Bend reach of the Platte River, Nebraska. *Prairie Nat.* 22(1): 27-32.

Schalles, J.F. 1991. The Platte: river at risk. *Creighton Univ. Window Magazine.* 7(3): 4-13.

Sheldon, A.L. 1987. Rarity: patterns and consequences for stream fishes. pp. 203-209. In: W.J. Matthews and D.C. Heins (Eds.) *Community and Evolutionary Ecology of North American Stream Fishes.* Univ. of Oklahoma Press, Norman, OK. 269 p.

Sidle, J.G., E.D. Miller, and P.J. Currier. 1989. Changing habitats in the Platte River Valley of Nebraska. *Prairie Nat.* 21: 91-104.

True, C. 1974. Fishes of the Nebraska Mid-state Division. pp. II-1 to II-62. In (O.A. Kolstad, project director): *Identification and Evaluation of Present Zoologic Resources, Nebraska Mid-State Division, Pick-Sloan Missouri Basin Program and Associated Areas.* Report to U.S. Department of the Interior, Bureau of Reclamation. Kearney State College, Kearney, NE.

U.S. Fish and Wildlife Service. 1981. *The Platte River Ecology Study: Special Research Report.* Northern Prairie Research Center, Jamestown, N.D. 187 p.

U.S. Geological Survey. 1970. *The National Atlas of the United States of America.* GPO, Washington, D.C. 417 p.

Williams, G.P. 1978. *The Case of the Shrinking Channels - the North Platte and Platte Rivers in Nebraska.* U.S. Geol. Surv. Circular 781, GPO, Washington, D.C. 48 p.

Biological Diversity: Problems and Challenges. Edited by S.K. Majumdar, F.J. Brenner, J.E. Lovich, J.F. Schalles, and E.W. Miller.

Chapter Thirty-Four

BIODIVERSITY OF THE MISSISSIPPI DELTA: Past, Present, and Future

LEONARD A. BRENNAN*

Department of Wildlife and Fisheries
Drawer LW
Mississippi State University, MS 39762

INTRODUCTION

The Mississippi River Delta (MRD) extends north nearly 1,000 km from the estuarine interface of the main river channel and the Gulf of Mexico to southern Illinois (Figure 1). Reinecke et al.[1] classified the upper 800 km of this system that is beyond the influence of saltwater intrusion as the Mississippi Alluvial Valley (MAV). Chabreck et al.[2] considered the lower, estuarine region of this sytem a central, key part of the southern Gulf Coast estuaries. Here, I consider this coastal region the Mississippi River Estuary (MRE). When taken together, the MRE and MAV constitute the MRD.

The main channel of the Mississippi River is the only tenth-order stream in the United States.[3] Since the Pleistocene, the main channel of the Mississippi River has been continuously depositing and eroding sediments from approximately one half of the land mass of the conterminous 48 states.[4] These rich sedimentary soils form the basis of the vegetation, landscape, and hence biological diversity of the MRD. The high fertility of the rich sedimentary soils of the MRD, and long, humid growing seasons of the region, combined to create a highly productive and diverse ecosystem that in many ways was a temperate, and in some places, sub-tropical analog to Amazonia.

Harris and Gosselink[4] claimed that principles of tropical ecology are directly

*Present address: Tall Timbers Research Station, Route 1, Box 678, Tallahassee, FL 32312

applicable to systems such as the MRD. They argued that since areas such as the MRD lie below 35 degrees north latitude and have seasonal cycles dominated by precipitation and flooding, rather than temperature, there are many parallels to tropical systems. From the standpoint of terrestrial vertebrate diversity, the MRD supports relatively small percentages of species of North American reptiles, amphibians, and mammals, and a large percentage of its avifauna (Table 1).

The productive ecological properties of the MRD attracted settlement, agriculture, and industry to this region of south-central North America.[5] Widespread settlement by Europeans, and the resultant clearing of vast tracts of land for agriculture, wrought profound changes on this region. This region was also quickly exploited as a convenient, central location for shipping and industry. Therefore, any assessment of biodiversity in this region requires that it be assessed in terms of land use and settlement patterns. My goal in this chapter is to provide a brief, historial sketch of how humans have influenced the biodiversity of the MRD. Since I am a wildlife ecologist, my primary focus will center on issues pertaining to terrestrial vertebrates, and how changes in vegetation and alteration of ecological processes has influenced their distribution and abundance in the MRD.

There are three primary parts to this chapter. First, I will begin with some educated speculation about the landscape and ecological processes of the MRD before invasion by Europeans. Second, I will assess, in the broadest of terms, the current status of biodiversity in the contemporary MRD landscape. Third, I will look to the future and assess potential opportunities for maintaining and restoring components of biodiversity to this spectacular, yet highly modified, region of North America.

THE DELTA PRIMEVAL

A comprehensive survey and inventory of the plants and animals of the MRD

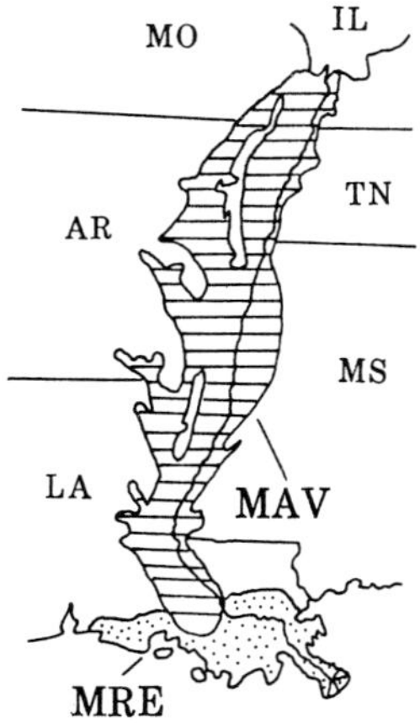

FIGURE 1. Geographic locations of the Mississippi Alluvial Valley (MAV) and Mississippi River Estuary (MRE). Together, these two regions constitute the Mississippi River Delta. State abbreviations are as follows: AR-Arkansas, IL-Illinois, KY-Kentucky, LA-Louisiana, MO-Missouri, MS-Mississippi, TN-Tennessee.

was not undertaken prior to the clearing of the vast bottomland forests and open coastal marshes for agriculture. In this section, my purpose is to reconstruct an approximate "mind's eye view" of what the MRD must have looked like, and hence how it functioned ecologically, prior to invasion and settlement by Europeans.

Historically, the MAV was dominated by approximately 5 million hectares of bottomland hardwood forests that were seasonally flooded during late winter and early spring.[1] This vast, moist, humid forest must have presented a formidable barrier to the first Europeans who encountered it. Natchez, the earliest permanent settlement in the MRD region, and the Natchez Trace, which provided overland access to it, were located on generally well-drained, upland areas adjacent to the MRD.[6] The early settlers gravitated toward the high ridges and bluffs on the perimeter of the MRD to avoid the physical and pathological hazards present in the seasonally-flooded lowlands. In addition to flooding, the incidence of disease (primarily yellow fever and malaria) was much higher in those people who lived year-round in the lowland, than those who settled on the higher surrounding uplands. Thus, it is clear that the early settlers had numerous, motivating reasons to control the seasonal flooding, and remove the largest amounts of bottomland hardwood forests possible so that agricultural opportunities could be maximized.

The journal of John James Audubon provides some of the earliest written records of the MRD landscape.[7] Although land-clearing for agricultural development had been underway for several decades, Audubon was one of the first naturalists to travel systematically throughout the MRD. His observations of the Mississippi River during a spring flood provide a glimpse of the MRD landscape during the early part of the nineteenth century (7:162-163).

> *"I have travelled on the Mississippi. . . in a canoe [during] flood and travelled across immense portions of the country when it was under water. Spread over neighboring swamps, the Mississippi looks like an ocean overgrown with stupendous trees. On the Mississippi bottomlands, little or no current is met with, except on a bayou. All is silent, melancholy, except for the mournful whinny of the hemmed-in deer, the dismal scream of an eagle . . . Bears, cougars, . . . and other quadrupeds that can ascent trees may be seen crouched among the top branches. Fatigued by the effort of reaching dry land, some will stand near the hunter's fire . . . they fall by the hundreds."*
>
> *"The gradual subsiding of the flood effects mighty changes. Everywhere,*

TABLE 1

Taxonomic summary of numbers of terrestrial vertebrate species found in the Mississippi River Delta (MRD) in relation to numbers of species of terrestrial vertebrates in the conterminous United States and Canada.

Taxonomic Group	Number species in MRD	Approximate percentage of species in North America
Amphibians	36	19
Reptiles	49	18
Mammals	49	13
Birds	233	38

the earth is covered by deep deposits of muddy loam. New, large streams are found to exist where, before, none had been seen. Large sand banks have been removed and deposited elsewhere. Trees bend over the stream like arms of giants."

Clearly, the above passages from Audubon's journal create an image of the MRD before widespread settlement and modification. The seasonal deposition of sediments, dynamic nature of the landscape, and relative (compared to today) abundance of large predators, were all present in what once was the largest system of undisturbed North American riparian forest.

THE DELTA TODAY

The current condition of the MRD landscape reflects the cumulative impacts from hundreds of thousands of management and development actions. The actions responsible for the current condition of the MRD fall broadly into three categories; 1) forest clearing for agriculture, 2) river channelling and diking for flood control, and 3) industrialization for manufacturing and shipping. The MRD of today has been transformed from a large, intermittently-flooded bottomland forest that interfaced a huge complex of brackish, coastal marshes to a landscape dominated by vast expanses of open cropland and small fragments of remnant hardwood forests.

Seasonal flooding has been minimized or eliminated from over 80 percent of the MAV.[1] This, in turn, has caused an increase in sediment deposition downstream in the MRE. The cumulative impacts of settlement, land-clearing, industrialization, and flood control have had a profound effect on the biodiversity of this region. From the standpoint of terrestrial vertebrates, the impact of these actions has been most detrimental to the large, upper trophic level carnivores, such as the red wolf (*Canis niger*), mountain lion (*Felis concolor*), and black bear (*Ursus americanus*) and other species that have large home ranges, or require habitats with old-growth forest characteristics such as the ivory-billed woodpecker (*Campephilus principalis*). Surprisingly, there have been a relatively small number of actual extinctions of vertebrates in the MRD, especially when one considers the extent of forest removal and fragmentation that has occurred.

Vertebrates in the MRD that have been driven to extinction include the ivory-billed woodpecker, passenger pigeon (*Ectopistes migratorius*), Carolina parakeet (*Cornuropis carolinensis*), and bison *(Bison bison)*.[8,9]

Although remnant, highly dispersed populations of large carnivores such as the red wolf, mountain lion, and black bear persist in or near the MRD, their populations are so small that ecologically, they no longer play a significant role as upper trophic level predators. Presently, these vertebrates can, for all practical purposes, be considered functionally extinct as ecological components of the MRD landscape. Potential opportunities for restoration of upper trophic level carnivores exist, but may be limited by negative cultural attitudes towards these animals by human residents of the MRD.

"New" Wildlife Habitats in the MRD

European settlers have had a vast impact on the landscape of the MRD. Their patterns of settlement have replaced huge areas of seasonally-flooded forest with intensively managed agricultural fields. Additionally, borrow pits for material used in levee construction, and catfish ponds from this newly-thriving industry are now found across much of the MRD landscape.

Settlement and forest clearing for agriculture has created and maintained millions of hectares of open field environments in the MRD. This has undoubtly created a favorable environment for open-country birds such as meadow larks (*Sturnella magna*), and mourning doves (*Zenaida macroura*), and reduced the populations of birds (both resident and neotropical migrants) that live in woodland habitats of the MAV. The relatively new, open-field environments of the MRD created a habitat that is extremely favorable for introduced, exotic birds from Europe such as the European starling (*Sturnus vulgarus*), and house sparrow (*Passer domesticus*). Conversion of bottomland forest to agricultural fields undoubtly eliminated large numbers of small mammals and herpetiles that require forest habitats.

Landin[10] observed that borrow pits along the lower Mississippi River provided habitat used by 23 species of mammals, and 186 species of birds. These borrow pits have become a permanent feature of the MRD landscape, encompassing approximately 20,000 hectares in 1985. Borrow pits have been a permanent feature of the MRD landscape since the late 17th century, and will continue to be developed and maintained as long as levee-building and maintenance efforts in the MRD continue.

Dubovsky[11] conducted the first comprehensive survey of waterfowl associated with the complexes of catfish ponds in the MAV. He found that over 150,000 individuals from 8 species of duck and 1 species of coot used these ponds, and habitats surrounding them, for a variety of life history requisites during winter. Dubvosky (11:27) was clear in providing a caveat that stated ". . . catfish ponds cannot be considered ecological substitutes for natural wetlands." This is because many species of waterfowl, especialy dabbling ducks . . . more commonly use flooded agricultural fields and wetlands."

Green-tree Reservoirs (GTR) are a relatively new component of the MAV landscape. Their purpose is to create, in a controlled situation, seasonally-flooded landscapes for waterfowl during winter.[1,12,13] GTR management represents an example of the extraordinary efforts being made by waterfowl research ecologists and managers to create favorable wintering habitat for migrant ducks and wildfowl.

Regeneration of bottomland hardwood forests in the MAV creates early successional mosaics on a scale that is probably larger than what one would encounter in old-growth bottomland hardwood forests. This is because natural regeneration in these forests mostly occurs when individual trees fall and open small (1-5 ha) gaps in the canopy.[14] High winds are usually the responsible agent in this process. The exposed soil from torn-up roots, and newly-abundant sunlight on the forest floor stimulates the growth of several hundred seedlings that will eventually succumb to one or a few individuals that grow to maturity.

Regeneration of hardwood forests after logging typically results in large (20 + ha) patches that are dominated by seedlings and saplings. Bourland[15] found that, in regeneration areas, the highest density and diversity of breeding bird species occurred in stands that were four-years post harvest. However, bird species diversity in bottomland hardwood forests was consistently highest in the mature, and old-growth (100 + year) age classes.[16,17]

Chemical Contamination of the MRD Landscape

The widespread agricultural operations, heavy use of pesticides on non-food commodities such as cotton, and huge refinery/industrial operations in the MRE, have combined to allow portions of the MRD to join ranks with some of the most polluted ecosystems on earth. The initial chemical contamination of the MRD environment began with the widespread application of arsenic to cotton for control of boll weevil. The organochlorine compounds such as DDT and its derivatives were widely applied throughout the MRD, and continue to persist in mammals from this region.[18]

Today, most agrochemicals that are currently used in the MRD do not bioaccumulate like the organochlorine compounds. Nevertheless, many of the pesticides used in the MRD are highly toxic for short periods of time. Their effects on non-target organisms in the wild remain virtually unknown, even though their potential to inflict mortality on non-target vertebrates is high.[19,20] Virtually nothing is known about the effects of agrochemicals on non-target populations of herpetiles, small mammals, and passerine birds in the MRD.

THE DELTA OF THE FUTURE

In their review of broad-scale patterns of vertebrate biodiversity. Pimm and Gittleman (21:940) stated that "We clearly know too little about where the diversity is, why it is there, and what it will become." The foregoing discussion on patterns of diversity indicates that the MRD is clearly no exception to this problem. Nevertheless, I believe that it is both possible and worthwhile to speculate on the most likely directions and prospects for issues related to biodiversity in the MRD as we enter the 21st century.

In my opinion, the primary issues relating to the biodiversity of ecosystems in the MRD fall into the following categories; 1) restoration of upper trophic level vertebrate populations as functional ecosystem components, 2) monitoring populations and providing sustained habitats for nongame species such as herpetiles, small mammals, passerine birds, and wintering waterfowl; 3) improving water quality, and minimizing non-point source pollution for agro- and industrial chemical contamination. Three excellent mechanisms for addressing the above three topics are 1) restoration of bottomland hardwood forests in the MAV, 2) restoration of estuarine coastal marsh in the MRE, and 3) systematic, enlightened management of agrochemicals that is integrated with sustainable agricultural systems.

It is hard to imagine a region in North America that has more problems and opportunities than the MRD with respect to maintaining and restoring biological diversity. The problems faced by residents of the MRD, and those charged with managing its resources are legion. From a sociological perspective, the MAV is economically the poorest region of the United States. Thus, private capital for restoration and enhancement of projects related to biodiversity is extremely limited. Furthermore, with the widespread, abject poverty faced by many residents in the MAV, comes a spate of other limitating factors such as low-quality public schools, and limited opportunities for higher education. It is difficult for many residents of the MAV to become concerned about issues relating to biodiversity when they are in the midst of a struggle for survival. Support for projects related to maintaining and enhancing biodiversity in the MRD will most likely have to come from the state and federal sectors. Any project that is successful in this arena will no doubt require inclusion of components that will increase the standard of living of the people in the MAV, as well as the quality of the environment in the MRD.

The recent Lake George Forest Restoration project is a prime example of the kind of project that needs to be undertaken to promote maintenance and enhancement of biodiversity in the MRD. This project, which is coordinated by the U.S. Army Corps of Engineers, the U.S. Fish and Wildlife Service and Mississippi Department of Wildlife, Fisheries, and Parks, calls for 9,000 acres of bottomland to be replanted to hardwood forest during the next 5 years. Perhaps the most significant aspect of this study was that it was designed as a corridor to provide a linkage of bottomland hardwood habitat between portions of the Delta National Forest and Panther Burn National Wildlife Refuge. Other, smaller scale bottomland hardwood forest restoration projects are also underway in the MRD.

Overall, I perceive several trends that will hopefully result in maintaining or restoring biodiversity in the MRD, while at the same time enhancing the quality of life for MRD residents. Recent concerns about environmental quality, non-point source pollution, and chemical pesticides in general are forcing agricultural producers in the MRD to practice more stewardship in the ways they handle agrochemicals. The initiation of both large and small scale bottomland hardwood forest restoration projects should serve as a springboard for further work in this area if these initial projects meet with success. Additionally, recreational demand for activities associated with wildlife (primarily those that are hunting-related) will most likely continue to increase. The high productivity of the forest systems within the MAV results in white-tailed deer (*Odocoileus virginianus*) densities that are nearly twice as great as those in many non-delta areas. In Mississippi, the MAV is the only region of the state that supports wild turkey (*Meleagris gallopavo*) populations which are sufficiently abundant to support both spring gobbler and fall either-sex hunting.

The Conservation Reserve Program (CRP) holds a tremendous amount of opportunity for conserving and enhancing biodiversity in the MRD. By taking land out of agricultural production for ten-year blocks of time, pesticide applications and soil erosion problems are dramatically reduced. Recent changes in CRP guidelines that allow hardwood planting (formerly pines were only allowed) also bode well for forest restoration opportunities in the MRD. Although the MRD will clearly never return to a contiguous, ten million hectare block of seasonally-flooded bottomland hardwood forest, the aforementioned opportunities point to at least some hope for conserving and enhancing the remaining vertebrate resources of the MRD.

Perhaps most importantly, the formation of private interest groups that are concerned with wildlife and associated resources and values, such as the Delta Wildlife Foundation, indicates that the private sector may ultimately play a significant role in preservation and enhancement of biodiversity in the MRD.

ACKNOWLEDGEMENTS

Teresa Pruden helped with editing and provided comments that greatly improved an early draft of this chapter. Susan Hall, of the Mitchell Memorial Library, Mississippi State University, performed the initial computer search for literature on biodiversity in the MRD. The Department of Wildlife and Fisheries at Mississippi State University provided support and a productive working environment that was essential for writing this chapter.

LITERATURE CITED

1. Reinecke, K.J., R.M. Kaminski, D.J. Moorehead, J.D. Hodges and J.R. Nasser. 1989. Mississippi alluvial valley, pp. 203-248. In: L.M. Smith, R.L. Pederson and R.M. Kaminiski (Eds.) *Habitat Management for Migrating and Wintering Waterfowl in North America.* Texas Tech University Press, Lubbock, TX, pp. 560.
2. Chabreck, R.H., T. Joanen and S.L. Paulus. 1989. Southern coastal marshes and lakes. pp. 249-280. In: L.M. Smith, R.L. Pederson and R.M. Kaminski (Eds.) *Habitat Management for Migrating and Wintering Waterfowl in North America.* Texas Tech University Press, Lubbock, TX, pp. 560.
3. Leopold, L.B., M.G. Wollman and J.P. Miller. 1964. *Fluvial Processes in Geomorphology.* Freeman, San Francisco, CA. pp. 522.
4. Harris, L.D. and J.G. Gosselink. 1990. Cumulative impacts of bottomland hardwood forest conversion on hydrology, water quality, and terrestrial wildlife. pp. 259-322. In: J.G. Gosselink, L.C. Lee and T.A. Muir. *Ecological Processes and Cumulative Impacts: Illustrated by Bottomland Wetland Ecosystems.* Lewis Publishing, Chelsea, MI pp. 708.
5. Fisher, J.L. 1966. Natural resources and economic development: the web of events, policies, and policy objectives. pp. 261-276. In: F.F. Darling and J.P. Milton. (Eds.) *Future Environments of North America.* Natural History Press, Garden City, NY. pp. 767.
6. Kane, H.T. 1947. *Natchez on the Mississippi.* William Morrow and Company, NY. pp. 373.
7. Ford, A. (Ed.). 1969. *Audubon, By Himself: A Profile of John James Audubon, from Writings Selected, Arranged, and Edited by Alice Ford.* Natural History Press, Garden City, NY. pp. 276

8. American Ornithologists' Union. 1983. *Check-list of North American Birds.* 6th Edition. American Ornithologists' Union, Washington, D.C. pp. 877.
9. Leopold, A. 1929. *Report on a Game Survey of Mississippi.* Sporting Arms and Ammunition Manufacturers Institute. Report on File, Mississippi Department of Wildlife, Fisheries, and Parks, Jackson, MS pp. 65.
10. Landin, M.C. 1985. *Bird and Mammal Use of Lower Mississippi River Borrow Pits.* Dissertation. Mississippi State University, MS. pp. 105 + Appendices.
11. Dubovsky, J.A. 1987. *Wintering Waterfowl Abundance and Habitat Associations with Catfish Ponds in the Alluvial Valley Region of Mississippi.* Thesis. Mississippi State University, MS. pp. 101.
12. Mitchell, W.A. and C.J. Newling. 1986. *Greentree Reservoirs.* Sect 5.5.3, U.S. Army Corps of Engineers Wildlife Resource Management Manual. Technical Report EL-86-9. Vicksburg, MS. pp. 22.
13. Wigley, T.B. Jr. and T.H. Filer, Jr. Characteristics of greentree reservoirs: a survey of managers. *Wildlife Society Bulletin.* 17:136-142.
14. Shugart, H.H., Jr. 1984. *A Theory of Forest Dynamics.* Springer-Verlag. NY. pp. 278.
15. Bourland, T.R. 1978. *An Evaluation of Three Site Preparation Treatments Applied to Naturally Regenerated Areas of Bottomland Hardwood Forests in the Mississippi River Delta in Terms of Seedling Species Composition and Density, Deer Food Abundance, and Passerine Bird Usage.* Thesis. Mississippi State University, MS. pp. 50.
16. Johnson, R.L., D.E. Wesley, C.J. Perkins and A.D. Sullivan. 1978. *Effects of Even-aged Cottonwood Culture on Certain Non-game Species.* Final Report U.S. Forest Service Cooperative Agreement 16 USC 581. New Orleans, LA. pp. 91.
17. Hunter, M.L., Jr. 1990. *Wildlife, Forests, and Forestry: Principles of Managing Forests for Biological Diversity.* Prentice Hall, Englewood Cliffs, NJ. pp. 370.
18. Ford, W.M. and E.P. Hill. 1990. Organochlorine Residues in Mississippi Raccoons. *Journal of Wildlife Management* 54:591-594.
19. Ludke, J.L., E.F. Hill and M.P. Deiter. 1975. Cholinesterase (ChE) responses and related mortality among birds fed ChE inhibitors. *Archives of Environmental Contamination and Toxicology.* 3:1-21.
20. Hudson, R.H., R.K. Tucker and M.A. Haegele. 1984. *Handbook of Toxicity of Pesticides to Wildlife.* U.S. Fish and Wildlife Service Resource Publication 153. pp. 90.
21. Pimm, S.I. and J.L. Gittleman. 1992. Biological diversity: where is it? *Science* 255:940.

Addendum

(Chapter 12, Miguel F. Acevedo)

References continued from page 150.

Murphy, D.D. and S.B. Weiss. 1992. Effects of climate change on biological diversity in Western North America: species losses and mechanisms, pp. 355-368. In: R.L. Peters and T.E. Lovejoy (Eds.) *Global Warming and Biological Diversity.* Yale University Press, pp. 386.

Myers, N. 1992. Synergisms: joint effects of climate change and other forms of habitat destruction, pp. 344-354. In: R.L. Peters and T.E. Lovejoy (Eds.) *Global Warming and Biological Diversity.* Yale University Press, pp. 386.

Nelson, B.W., C.A.C. Ferreira, M.F. da Silva and M.L. Kawasaki. 1990. Endemism centres, refugia and botanical collection density in Brazilian Amazonia. *Nature.* 345:714-716.

Neilson, R.P. 1991. Climatic constraints and issues of scale controlling regional biomes, pp. 31-51. In: M.M. Holland, P.G. Risser and R.J. Neiman (Eds.). *Ecotones.* Chapman & Hall.

Norby, R.J., C.A. Gunderson, S.D. Wullschleger, E.G. O'Neill and M.K. McCracken. 1992. Productivity and compensatory responses of yellow-poplar trees in evlevated CO_2. *Nature.* 357:322-324.

Parsons, P.A. 1991. Evolutionary rates: stress and species boundaries. *Ann. Rev. Ecol. Syst.* 22:1-18.

Peters, R.L. 1988. The effect of global climatic change on natural communities, pp. 450-459. In: E.O. Wilson (Ed.). *Biodiversity.* National Academy Press. pp. 521.

Peters, R.L. 1992. Conservation of biological diversity in the face of climate change, pp. 15-30. In: R.L. Peters and T.E. Lovejoy (Eds.) *Global Warming and Biological Diversity.* Yale University Press, pp. 386.

Ray, G.C., B.P. Hayden, A.J. Bulger, Jr., M.G. McCormick-Ray. 1992. Effects of global warming on the biodiversity of coastal-marine zones, pp. 91-104. In: R.L. Peters and T.E. Lovejoy (Eds.) *Global Warming and Biological Diversity.* Yale University Press, pp. 386.

Richerson, P.J. and K. Lum. 1980. Patterns of plant species diversity in California: relation to weather and topography. *Am. Nat.* 116(4):504-536.

Riclefs, R.E. and R.E. Latham. 1992. Intercontinental correlation of geographic ranges suggest stasis in ecological traits of relict genera of temperate perennial herbs. *Am. Nat.* 139(6):1305-1321.

Rind, D., R. Goldberg, J. Hansen, C. Rosenzweig and R. Ruedy. 1990. Potential evapotranspiration and the likelihood of future drought. *J. Geophys. Res.* 95(D7):9983-10,004.

Robinson, P.J. 1991. Scenarios of future climate for impact assessment. *Env. Profess.* 13(1):8-15.

Robock, A., R.P. Turco, M.A. Harwell, T.P. Ackerman, R. Andressen, H. Chang and M.V.K. Sivakumar. 1992a. Use of general circulation model output in the creation of climate change scenarios for impact analysis. Manuscript.

Robock, A., R. Andressen, M. Acevado and R. Jaimez. 1992b. General circulation model-based scenarios of future climate for Venezuela. Manuscript.

Rubenstein, D.I. 1992. The greenhouse effect and changes in animal behavior: effects on social structure and life-history strategies, pp. 180-192. In: R.L. Peters and T.E. Lovejoy (Eds.) *Global Warming and Biological Diversity.* Yale University Press, pp. 386.

Sheldon, P.R. 1990. Shaking up evolutionary patterns. *Nature* 345-772.

Shugart, H.H. and T.M. Smith. 1992. Using computer models to project ecosystem response, habitat change, and wildlife diversity, pp. 147-157. In: R.L. Peters and T.E. Lovejoy (Eds.) *Global Warming and Biological Diversity.* Yale University Press, pp. 386.

Schneider, S.H. 1992. The climatic response to greenhouse gases. pp. 1-32. In: F.I. Woodward (Ed.). *The Ecological Consequences of Global Climate Change.* Advances in Ecological Research Vol. 22 Academic Press, pp. 337.

Schneider, S.H., L. Mearns and P.H. Gleick. 1992. Climate-change scenarios for impact assessment, pp. 38-58. In: R.L. Peters and T.E. Lovejoy (Eds.) *Global Warming and Biological Diversity.* Yale University Press, pp. 386.

Schubert C., H.O. Briceño and P. Fritz. 1986. Paleoenvironmental aspects of the Caroni-Paragua river basin (Southeastern Venezuela). *Interciencia* 11(6):278-289.

Seagle, S.W. and H.H. Shugart. 1985. Faunal richness and turnover on dynamic landscapes: a simulation study. *J. Biogeogr.* 12:499-508.

Shmida A. and M.V. Wilson. 1985. Biological determinants of species diversity. *J. Biogeogr.* 12:1-20.

Smith, J.B. and D. Tirpak. 1989. *The Potential Effects of Global Climate Change on the United States.* Chapter 8 pp. 149-163. Report to Congress. EPA-230-05-89-050. pp. 412.

Smith, T.M., H.H. Shugart, G.B. Bonan and J.B. Smith. 1992. Modeling the potential response of vegetation to global climate change, pp. 93-116. In: F.I. Woodward (Ed.) *The Ecological Consequences of Global Climate Change.* Advances in Ecological Research Vol. 22 Academic Press. pp. 337.

Solbrig, O.T. 1992. The IUBS-SCOPE-UNESCO program of research in biodiversity. *Ecol. Appl.* 2(2):131-138.

Stevens, G.C. 1989. The latitudinal gradients in geographic range: how so many species coexist in the tropics. *Am. Nat.* 133(2):240-256.

Steyermark, J.A. 1979. Plant refuge and dispersal centers in Venezuela: their relict and endemic element, pp. 185-221. In: K. Larsen and L.B. Holm-Nielsen (Eds.), *Tropical Botany,* Academic Press.

Terborgh, J. 1992. Maintenance of diversity in tropical forests. *Biotropica* 24(2b): 283-292.

Tracy, C.R. 1992. Ecological responses of animals to climate, pp. 171-179. In: R.L.

Peters and T.E. Lovejoy (Eds.) *Global Warming and Biological Diversity.* Yale University Press, pp. 386.

Turner, J.R., C.M. Gatehouse and C.A. Corey. 1987. Does Solar Energy Control Organic Diversity? Butterflies, moths and the British climate. *Oikos* 48:195-205.

Uchijima, Z. and H. Seino. 1988. An agroclimatic method of estimating net primary productivity of natural vegetation. *JARQ* 21(4):244-250.

Vitousek, P.M. and P.A. Matson. 1992. Tropical forests and trace gases: potential interactions between tropical biology and the atmospheric sciences. *Biotropica.* 24(2b):233-239.

Webb III, T. 1992. Past changes in vegetation and climate: lessons for the future, pp. 59-75. In: R.L. Peters and T.E. Lovejoy (Eds.) *Global Warming and Biological Diversity.* Yale University Press, pp. 386.

Westman, W.E. and G.P. Malanson. 1992. Effects of climate change on Mediterranean-type ecosystems in California and Baja California, pp. 258-276. In: R.L. Peters and T.E. Lovejoy (Eds.) *Global Warming and Biological Diversity.* Yale University Press, pp. 386.

Wigley, T.M.L. and S.C.B. Raper. 1992. Implications for climate and sea level of revised IPCC emisisons scenarios. *Nature* 357:293-300.

Wilson, E.O. 1988. The current state of biological diversity. In: E.O. Wilson (Ed.) *Biodiversity.* pp. 3-18. National Academy Press.

Whittaker, R.H. 1977. Evolution of species diversity in land communities. *Evol. Biol.* 10:1-67.

Woodward, F.I. 1992. A review of the effects of climate on vegetation: ranges, competition and composition, pp. 105-123. In: R.L. Peters and T.E. Lovejoy (Eds.) *Global Warming and Biological Diversity.* Yale University Press, pp. 386.

Wright, D.H. 1983. Species-energy theory: an extension of species-area theory. *Oikos* 41:496-506.

Subject Index

A

Accidental species introductions, 94
Acid precipitation, 77
Actual evapotranspiration (AET), 141-143, 146
Adaption and migration response, 143, 144, 146
Alpha and beta diversity, 214
Alpha, beta, and gamma diversity, 19, 22
Altering hydrology, 102
Amphibian biodiversity, 354-356, 359, 363-365
Amphibian censusing, 357
Analytical methods, 124, 125
ANC (acid neutralizing capacity) 78, 79
Antifoulants, 126, 127
Approaches to diversity, 52, 53
Aquatic diversity, 30
Aquatic life, 290-292
Article XIV, 165

B

Background on Platte River and Big Bend, 423-425
Background of Presque Isle (PI) and Presque Isle Bay (PIB), 405, 406
Background research, 7, 8
Biodiversity, 65, 66, 69-71, 163, 164, 168, 169, 200, 202, 209-217, 265-268, 272, 302-307, 311
Biodiversity and the tropics, 13-16
Biodiversity monitoring, 1-4, 6-9
Biodiversity of Presque Isle and evidence of change
- invertebrates, 410-414
- plants, 407-410
- prokaryotes, algae, and fungi, 407, 408
- vertebrates, 413, 415-418
Biodiversity status of aquatic insects, 344, 347
Biological diversity, 12, 13, 344, 349
Biological diversity of Pacific Northwest, 392, 394, 397, 398, 400
Biological effects of acidification, 79, 80
Biological factors affecting species diversity, 373
Biological invasion, 92, 93, 97, 99
Biosphere preserves, 271, 273
Biosphere reserves, 42
Bottomland hardwood forests, 445
Broken stick model, 61
Buffer zones and upland habitats, 294, 295

C

Calculations of diversity, 65, 66
Captive breeding programs, 254-256
Captive propagation, 213, 270
Carbon and nutrient cycling, 114, 116
Causes of insect diversity, 327
Chemical contamination, 448
Chemical effects of acidification, 78, 79
Chemical prospecting, 339
Chromomere gradient, 227, 228
Chromosomal recombination, 230
Chromosome evolution, 227
Chromosome gradient hypothesis, 223
Clean Water Act, 29-31, 35
Co-dependency, 203
Community ecosystem diversity, 305
Community similarity, 64
Comparative diversity of species, 316-318
Competitive superiority of introduced plant species, 98
Components of biological diversity, 51
Comprehensive biodiversity strategies, 37
Concept of diversity, 50, 51
Conditions for storage, 45
- collection, 46
- drying and moisture content, 45, 46
- herbarium facilities, 46
- longterm storage, 45
- medium storage, 45
Conflicting purposes of management, 186, 187
Conservation, 163, 164, 166
Conservation biology, 190, 191
Conservation community, 152
Conservation of genetic resources, 41-44
Conserving biodiversity, 32, 33
- proposal of the NIE, 33
- proposal of the SBA, 33
- research, 33
Construction of wetlands, 293, 294

Contaminants in the environment, 126, 127
Contaminant pressures on land, 153
Cool water fish, 370
Cooperative breeding programs, 254, 258
Critical habitat designation, 269
Cumulative effects, 194

D

Decrease in biodiversity, 79, 80
Degraded habitat, 213
Delta of the Future, 448-450
Delta Primeval, 444-446
Delta Today, 446-448
Demographic and environmental stochasticity, 99
Dependency of wildlife on old-growth forest, 397
Derivatization, 125
Detectors, 125
Dichotomy, 52, 57
Discussion, 386-388
- geological history of area, 386
- non-marine turtle dispersal into Sundalana, 386, 387

Disturbance, 65, 69
Disturbance hypothesis, 97
Diversity, 64-71, 221, 222, 226, 233
Diversity and evenness estimation, 357
Diversity below the species level, 19
Diversity measures of Patil and Tallie, 57
Diversity measures of Rao, 59
DNase hypersensitive sites, 226

E

Ecological fundamentals of diversity, 18, 19
Ecological recovery (recovery of endangered species), 36
Ecological Society of America (ESA), 165
Ecologists Union, 165, 166
Ecology of aquatic insects, 347
Economic consequences of conservation, 156-158
Economic significance of local use on limited products, 155
Ecosystem biotic components, 80-85
consumers, 81-85
- amphibians, birds, and mammals, 84, 85
- benthic macroinvertebrates, 82, 83
- fish, 84
- microorganisms, 81
- zooplankton, 82

producers
- macrophytes, 81
- phytoplankton, 80, 81

Ecosystem fragmentation, 192, 193
Ecosystem management, 268
Ecosystem stability, 193, 194
Ecosystem stability and resiliency, 355
Ecosystems and communities, 1-9
Ectones and edges, 20, 21
Education programs, 258, 259
Effect of climate on biological diversity, 136, 137, 139, 140, 143-147
Effects of acidification, 116
Effects of heavy metals, 117
Effects of petroleum hydrocarbons, 118, 119
Effects on humans, 128, 129
Emerging and existing programs, 243-247
Endangered Species Act, 27, 28, 166-168
Endemic species
- of amphibians, 396, 397
- of mammals, 395-397
- of reptiles, 396, 397
- of vertebrates, 393, 400

Endemism, 380, 381
Endomitotic replication, 224
Entropy, 67
Entropy measure of Havrda and Charvat, 59, 61
Environmental application, 71
Environmental assessment, 268, 270
Environmental education, 36, 37
Environmental ethics, 200, 201
Environmental policy, 26, 27
Environmental pressure, 224
European settlers, 447
Eutrophication and extirpation of native fishes, 376
Evenness, 66
Ex situ conservation, 268, 274
Extent of insect diversity, 327
External threats, 187, 189
Extinctions, 13, 144, 168, 152, 153
Extirpated amphibian species, 318
Extirpated bird species, 318, 319
Extirpated mammalia species, 318-320
Extraction, 124

F

Factors controlling extent and rate of acidification, 77

Faunal and floristic changes, 2, 3
Feeding guild, 63
Field surveys, 2, 9
Fish habitat improvement, 291, 292
Flagship species, 212
Flaws of the ESA, 28
Florida Game and Fresh Water Commission, 5
Fluid chromosome model, 223
Food web diversity, 21, 22
Forest clearing, 446, 447
Forest fragmentation, 308
Forest regeneration, 355, 364
Fragmentation of forests by timber harvest, 398
- loss of habitat by timber harvest, 398
Fragmentation of the biosphere, 15
Functional diversity, 23

G

Gene amplification, 228, 229
Gene banks, 41, 43, 47, 48
- selection of plants, 48
General Circulation Model (GCM), 137-139, 143, 146, 147
Genetic diversity, 18, 19, 95, 98, 303, 304
Genetic variability, 222
Geographic setting, 425-427
Geographical factors affecting species richness, 369
Global biodiversity decline, 354, 355
Global dispersion of biota, 93
Global interdependence, 344
Global patterns of species diversity, 13
Global perspectives, 156-158, 163, 164
Goals and tools of conserving biodiversity, 266
Goals and tools of endangered species recovery, 269
Goals of the ESA, 28
Great Lake fishes, 365, 369
Greenhouse effect, 136, 143, 145, 147
Gringard reaction, 125

H

Habitat and colonization enhancement, 283
Habitat Conservation Plans (HCPs), 271, 272
Habitat degradation, 320
Habitat destruction and fragmentation, 137
Habitat modification, 323
Habitat protection, 8, 9
Habitat requirements, 4
Habitat structural diversity, 363, 364
Height class, 63
Heterochromatin, 222, 223
Heterogeneity, 66, 67, 69
Heterogeneity and biodiversity, 24
Historical changes in diversity, 318-320
Historical changes in hydrology and channel morphology, 428
Historical factors affecting species richness, 372
Human alteration of ecosystems, 344
Hydraulic forces, 283, 285
Hydrology, 283

I

Impacts of aquatic insects and man, 345
Imposing monocultures, 102
Inclusive fitness, 204
Increasing biodiversity through stream restoration, 282, 283
Index Seminum, 247
Indicator species, 3, 4, 8
Industrialization, 446
Insect biodiversity, 326, 327, 331-333, 336-338, 340
In situ conservation programs, 259-261, 267, 273, 274, 277
Interactions of microorganisms with tin compounds, 129, 130
Internal threats, 185, 189
International coordination, 247
International trade and botanical gardens, 246, 247
Introduction of alien species, 93-95, 98, 103, 334
In vitro technique, 48
Island biogeographic theory, 211, 371
Isolation, removal, transfer, and dilution, 287, 288

J

John James Audubon, 445, 446

K

Key common species, 3, 9
Key threats to biodiversity in ecosystem of the Platte River Valley, 440
Keystone groups, 23
Keystone species, 3, 9

Keystone taxa, 365
Kin selection, 204
Knob formation, 228
Knowledge lost, 178, 179

L

Lack of interest in turtle biodiversity, 381
Lake George Forest restoration project, 449
Land management plans, 2
Landscape and local scales, 142, 147
Landscape diversity, 19-21
Landscape regional diversity, 305
Latitude location, 369
Leaf litter, 116, 117
Limestone treatment, 116
Limiting similarity hypothesis, 330
Linkage profession, 176-178
Log-normal distribution, 65
Long term intensive sampling, 5-7
Long term research, 6-9
Loss and degradation of habitat, 340
Loss of biodiversity, 184, 185

M

MacArthur's broken stick law, 61
Macroinvertebrates, 290
Maintaining biological diversity, 306
Managed forests, 397, 398
Management communication, 173, 174
Management technology, 211, 212
Marine Mammal Protection Act, 29
Marketing and strategic planning, 242
Materials and methods, 382-384
- area of analysis: Southwest Asia, 382
- comparisons, 384
- turtle species occupying area, 382, 383

Measurement of biodiversity, 315, 316
Mendelian principles, 226
Methyl and butyltins, 126
Microbial diversity, 22, 114-117, 119
Mine drainage effects on microbial processes, 116, 117
Mississippi Alluvial Valley (MAV), 443, 445-449
Mississippi River Delta (MRD), 443, 450
Mississippi River Estuary, 443, 446
Mitochondria genome, 232, 233
Monitoring changes and impacts, 2-9
Multiple control agents, 105
Multiple use areas, 35, 36
- agricultural landscape, 35

N

Natchez, 445
National Bureau of Plant Genetic Resources (NBPGR), 47, 48
National Environmental Policy Act (NEPA), 29, 166
National Facility for Plant Tissue Culture Repository (NFPTCR), 48
National Forest Management Act, 31
National Park Organic Act, 185-187
Nature conservancy, 166
Net Primary Productivity (NPP), 141, 146
Neutral mutation, 223
New economic incentives and policies, 152
No reduction scenario, 138
Non-equilibrium conditions, 66, 69
North American botanical gardens, 241-243
Nuclear genome, 221, 223-226

O

Objective of study, 381
Open field environments, 447
Organotins, 123-125, 127-129, 130, 131
Other laws, 32

P

Partial policies, piece meal practices, 26
Persistence, 68
Phenotypic plasticity, 221
Physics envy, 67
Physiological evolution of diapause, 332
Phytochemical prospecting, 202
Phytophagy by insects, 329, 330
Phytophthora cinnamoni - an introduced pathogen, 102
Plant genetic resources, 43, 44
Plastid genome, 232
Platte River aquatic animals, 432-435
Platte River avifauna, 435-437
Platte River channel braiding, 426, 431
Platte River discharges, 429
Platte River irrigation, 425, 428, 434, 439
Platte River vegetation, 431, 432
Pollution abadment, 283
Pollution by human waste water and agricultural run off, 115, 116
Pollution ecology, 348
Polynemy, 225
Population growth, 153

Population viability analysis, 270, 272
Potential evapotranspiration (PET), 141-143, 146, 147
Preservation, 163, 164, 166
Preservation and restoration of habitat, 340
Preservation of biodiversity, 18, 19, 21, 25
Prior to invasion and settlement, 445
Priority areas for conservation, 153
Properties of diversity indices, 53
Proposed solution: the value of change, 176
Protected areas, 34, 35
Protection and enhancement of wildlife, 398

Q

Quantification of diversity, 53-61
- Brillouin's index, 55
- diversity measures of Hill, 56, 57
- entropy measures, 55, 56
- Shannon-Wiener index, 55
- Simpson index, 54, 55
Quantitive ecology, 64

R

Ranking, 52
Rapoport rule, 142
Reciprocal altruism, 204
Reclamation of disturbed land, 292, 293
Recolonization of streams, 290
Recording organizational memory of the manager, 179, 180
Regional climate and endaphic patterns, 427
Regional diversity, 320-323
Regional scale, 141, 142
Reiterated sequences, 225, 226
Relative abundance, 51
Remnant habitat, 212
Repetitive DNA, 223
Research programs, 256-258
Reservoir sediment entrapment, 425
Resilience, 68
Resistance, 68
Resources for conservation, 153
Restoration ecology, 155, 156, 282
Restoration of water quality, 286-288
Restoration or rehabilitation of ecosystems, 281
Results, 384-386
Reverse transcriptase, 229, 231
Richness and eveness, 61
Riparian vegetation, 288-291, 294
River channelling and diking, 446
Role of science, 188
Roles of botanic gardens, 248-250

S

Sampling issues, 65, 67
Sampling techniques, 5-9
Satellite DNA's, 223
Scenario-consequence analysis, 137
Science communication, 172, 173
Science and management, 170-172
Selective logging, 399
Sensitive life history stages, 4-7, 9
Sexual crosses, 229, 230
Shannon's index, 67
Simpson's index, 67
Site specificity and longevity, 242
Slash burning, 397
Slow information transfer, 172
Smithsonian Institution, 5
Soft Waters, 77
Somatic mutation, 222
Soules postulates, 190
Sound grazing management, 107
Spatial scale, 140
Spatial-temporal changes, 66
Species diversity, 304
Species richness, 66, 369, 370, 372, 373, 376
- of amphibians, 395
- of birds, 395
- of plants, 394
- of vertebrates, 393
Species selection, 295
Stability, 65, 68
"*State of the Parks*" report, 188, 189
Statistical population, 51
Steps to preserve insect biodiversity, 336, 339
Strategies for addressing biological diversity concerns, 307
- endangered species approach, 310
- game species approach, 307
Strategy of conservation, 43, 44
Stream bank and bed stabilization, 283
Struggle for survival, 449
Subsistence hunting, 214, 215

T

Taxon Advisory Groups (TAGs), 255-258
Taxonomic analysis of aquatic insects, 346

Taxonomic and geographic focus, 4
TBT (tributylin compounds), 123, 124, 128-131
Theoretical and empirical impacts, 66, 68, 69
Threats of introduced species to native plant populations, 98
Threats to insect diversity, 333, 334
Threats to national parks, 188-190
Toxicity, 127, 128
Transposable elements, 231, 232
Transposition, 224
Trophic level, 63
Tropical deforestation, 151-154
Tropical forest, 145, 147
Types of germplasm, 45
Types of seeds, 44, 45
- orthodox species, 44
- recalcitrant, 44, 45

U

Unifying theory, 60

V

Validation of the analytical procedure, 125
Vegetational structural diversity, 363, 364
Vicariant diversification and radiation, 93
Vigilance of the environment, 7
Volunteering, 242

W

Warm water fish, 369
Wilderness Act, 166, 167
Wildlife use, 296
Wise use versus preservation, 154-156

Z

Zoos and Aquariums, 253-261